The subject of mineralogy is moving away from the traditional systematic treatment of mineral groups toward the study of the behaviour of minerals in relation to geological processes. A knowledge of how minerals respond to a changing geological environment is fundamental to our understanding of many dynamic earth processes.

By adopting a *materials science* approach, *Introduction to Mineral Sciences* explains the principles underlying the modern study of minerals, discussing the behaviour of crystalline materials with changes in temperature, pressure and chemical environment. The concepts required to understand mineral behaviour are often complex, but are presented here in simple, non-mathematical terms for undergraduate mineralogy students.

After introductory chapters describing the principles of diffraction, imaging and the spectroscopic methods used to study minerals, the structure and behaviour of the main groups of rock-forming minerals are covered, and the role of defects in the deformation and transformation of a mineral are explained. The energy changes and the rate of transformation processes are introduced using a descriptive approach rather than attempting a complete and rigorous treatment of the thermodynamics and kinetics.

Examples and case histories from a range of mineral groups are set in an earth science context, such that the emphasis of this book is to allow the student to develop an intuitive understanding of the structural principles controlling the behaviour of minerals.

Cover from the top:
Crystals of Vanadinite, $Pb_5(VO_4)_3Cl$
The structure of garnet
Optical micrograph (crossed polars) of olivine crystals in an igneous rock
High resolution transmission electron micrograph of cordierite

Introduction to mineral sciences

Introduction to mineral sciences

ANDREW PUTNIS

University of Cambridge

CAMBRIDGE
UNIVERSITY PRESS

Published by the Press Syndicate of the University of Cambridge
The Pitt Building, Trumpington Street, Cambridge CB2 1RP
40 West 20th Street, New York, NY 10011–4211, USA
10 Stamford Road, Oakleigh, Melbourne 3166, Australia

First published 1992
Reprinted 1993

Printed in Great Britain by BAS Printers Ltd, Over Wallop, Hampshire

A catalogue record for this book is available from the British Library

Library of Congress cataloguing in publication data

Putnis, A.
    Introduction to mineral sciences / Andrew Putnis.
        p.   cm.
    Includes bibliographical references and index.
    IBN 0-521-41922-0. – ISBN 0-521-42947-1 (pbk.)
    1. Mineralogy.   2. Minerals.   3. Crystallography.   I. Title.
    QE363.2.P88   1992
    549–dc20   92-8420   CIP

ISBN 0 521 41922 0 hardback
ISBN 0 521 42947 1 paperback

GE

For Jan
    Soni
    Anna
    Sven
    Thor
    Nina
      and Christine

Contents

Preface

The area of Earth Sciences traditionally referred to as Mineralogy has changed dramatically over the last two decades. Although the foundations of the subject still lie in the systematic description of the structures and compositions of the mineral groups, this taxonomic approach no longer fulfills the needs of Earth Scientists. Merely recognising minerals in rocks is equivalent to being able to put a name to a fossil or a plant, without appreciating its function or significance. The changes that have taken place in Mineralogy can be broadly summarized as a shift in emphasis towards understanding the *behaviour* of minerals within the Earth – how they respond to changes in the physical and chemical environment during geological processes. This response generally involves structural and chemical changes within individual minerals in a rock as well as reactions between minerals.

To understand these processes the student of Mineralogy should ideally be familiar with aspects of crystallography, materials science, solid state chemistry and solid state physics. This is reflected in the new titles given to Mineralogy courses, which often include the terms Mineral Physics or Mineral Chemistry. The complexity of mineral structures as well as their unique thermal history provides a range of phenomena which cover the whole spectrum of physical and chemical behaviour, and mineral behaviour is increasingly attracting the attention of physicists and chemists. This interdisciplinary approach is proceeding at a rapid pace and needs to be reflected in the way Mineral Science courses are taught.

This need has been widely recognized. The Mineralogical Society of America runs a regular series of Short Courses published as *Reviews in Mineralogy*. The aim of these multi-author volumes is to summarize the progress that has been made in the study of a particular mineral group, or the application of current theories and modern techniques to the subject. Other Mineralogical Societies are pursuing similar projects, on a smaller scale. These volumes are an invaluable resource and have done a great deal to promote the subject. However they are generally too detailed to answer a student's needs.

The aim of this book is to provide an introduction to the main concepts required to understand minerals and their behaviour. This is a rather ambitious claim and I should therefore restate these aims in more specific terms. First, I restrict the topics to those involving solid minerals, and exclude mineral–fluid interactions. Furthermore, most of the discussion relates to the processes which take place within minerals, rather than reactions between minerals. While some aspects of the latter are discussed, including the general concepts of the thermodynamics and kinetics of intermineral reactions, a more detailed treatment lies within the field of metamorphic petrology.

This book deals mainly with solid state transformations in minerals which take place in response to changes in temperature and pressure, as well as introducing the basic ideas of crystallography, physics and chemistry needed to understand these processes. The approach is designed for an undergraduate studying Mineral Sciences within the context of Earth Sciences and does not assume a knowledge of physics, chemistry or mathematics beyond that which would normally be expected from a student who had taken science subjects at school. The book would also be useful in materials science courses.

As far as possible I have tried to begin each chapter at the simplest level, introducing all of the concepts which will be needed in a more advanced discussion. Some of these concepts, are often regarded as 'difficult' because they are usually dealt with in more advanced physics and chemistry books. However, I try to deal with them in a simple (sometimes perhaps over-simplified) way in the belief that when a student has a descriptive understanding of the basic ideas, he or she is more likely to understand more mathematical accounts in books listed in the bibliography of each chapter.

The level of the discussion in the book is therefore aimed at undergraduates, even in their first year, and extends to a level which includes current research, and hence will be relevant to final year undergraduates and post-graduates. The book is written in a 'linear' way, each chapter building on the previous one. Some topics are therefore discussed many times from various points of view. For example, a process such as cation ordering may be introduced as part of the early description of the mineral structure and composition (Chapter 6), then again in the context of the planar defects which it generates (Chapter 7), then from the point of view of the thermodynamics (Chapter 9) of the process, and its kinetics (Chapter 10) before being discussed again in terms of some aspect of the mechanism of the process (Chapter 12).

There are two separate chapters systematically outlining the compositions and structures of the important mineral groups. The emphasis however is on the way in which the mineral structure is likely to respond to changes in temperature and pressure i.e. on those structural features which explain why a mineral has a particular range of chemical composition and why its structure might distort when it is cooled or compressed. In most of the book the role of the minerals is to provide examples of processes which take place and the geological context within which these processes occur. I believe that there is a distinction to be made between the traditional Mineralogy book which systematically lists minerals and their properties (essentially a manual or a reference to be consulted) and a text book which attempts to teach the essentials of the subject. I have aimed for the latter approach.

My views on the subject have been influenced by many colleagues and co-workers over the years, especially by Desmond McConnell, to whom I owe my enthusiasm for minerals as dynamic materials with a behaviour which transcends their inanimate chemistry. Without Desmond's initial inspiration this book would never have been contemplated.

More recently I have benefited greatly from my association with Michael Carpenter, Tim Holland and Ekhard Salje with whom discussions over morning coffee are a regular source of information and enlightenment. I am also grateful to my colleagues all over the world who have so generously sent original electron micrographs which are included in the book, especially to André Authier, Alain Bourret, Bill Brown, Michael Carpenter, Pam Champness, Michael Czank, Martyn Drury, Jean-Claude Doukhan, Peter Heaney, Masao Kitamura, David Kohlstedt, Ken Livi, Alex McLaren, Hiroshi Mori, Wolfgang Müller, Gordon Nord, Trevor Page, David Price, Peter Robinson, Werner Skrotzki, Gustav Van Tendeloo, Richard Tilley, David Veblen, Mary Wegner and Christian Willaime.

Mike Bown unstintingly read the entire manuscript and made many suggestions for its improvement, in matters of substance as well as in style. I am grateful for this generous effort on his part, but any remaining errors are of course mine.

Finally I thank my family who have put up with this project for almost two years, especially Christine without whose support the book may never have been completed.

Andrew Putnis
Cambridge
May 1992

Figure 1.1. This image of the crystal structure of the mineral cordierite, $(Mg_2Al_4Si_5O_{18})$ has been taken with a high resolution transmission electron microscope. It is a projection, through a very thin ($\sim$200 Å) slice, of the atomic distribution, the black spots representing hollow channels through the structure while the white spots can be equated with the regions of high atomic density, arranged around the channels in 6-fold rings. It is shown here to illustrate some aspects of the periodicity and symmetry of crystalline materials. (Scale: The distance between the black spots is $\sim$ 9.7Å or 0.97nm.)

1   Periodicity and symmetry

The photograph on the opposite page (Figure 1.1) is a high resolution electron micrograph of a crystal of the mineral cordierite. The image represents a two-dimensional projection of the atomic structure, and its most obvious feature illustrates a fundamental characteristic of all crystalline materials. The distribution of atoms is *periodic*. Most minerals are crystalline and an ideally perfect single crystal is made up of a repetition of identical building blocks, or *unit cells*, filling space, each unit cell in an identical orientation and related to the next by translation along one of three *crystallographic axes*.

As well as this *translational symmetry*, the pattern of atoms may itself have symmetry due to the relationship between neighbouring atoms and their bonds. In Figure 1.1 the black spots represent empty channels in the structure, normal to the page, and around each channel the six white spots represent projections of rows of atoms. The symmetry of such atomic arrangements is a consequence of the sym-

metry of electron orbitals and the role they play in bonding the atoms together in the structure. In Figure 1.1 we see a hexagonal arrangement in the distribution of the 'white atoms', i.e. a 6-fold rotation axis through the centre of the black channels. You may also be able to find other *symmetry elements* in this pattern, such as 3-fold and 2-fold rotation axes and mirror lines.

In well-formed crystals, the internal symmetry of the atomic distribution results in the symmetrical relationship between the orientation of external faces (Figure 1.2). The fascination of crystals began here, and the scientific study of their external morphology, from the seventeenth century onwards, led to the complete mathematical description of symmetry many years before it was possible to demonstrate their internal symmetry experimentally.

More importantly however, the symmetry of a mineral strongly influences its physical properties (e.g. mechanical, thermal, optical, magnetic and

(a) |_ 0.5 cms _| (b)

Figure 1.2. These two photographs illustrate the symmetry of the external shape, or morphology, of a well-developed crystal of the tetragonal mineral vesuvianite, $Ca(Mg,Fe)_2Al_4\{SiO_4\}_5\{Si_2O_7\}_2(OH,F)_4$. (a) A general view of the crystal showing the 4-fold axis and mirror planes. (b) A view of the crystal looking down the 4-fold axis. The orientations of the faces are related by the same symmetry elements which describe the atomic distribution.

electronic properties), as well as the way the mineral can respond to temperature and pressure changes in the Earth. This is the basis for the technological importance of minerals and their synthetic equivalents, and for the central role played by minerals in the physical processes taking place in the Earth's crust and mantle. In this chapter therefore, we will cover the most important aspects of symmetry which will be used later in the book.

1.1 The lattice, the unit cell and the motif

Symmetry operations are much easier to see in two dimensions, especially on a two-dimensional page, so we will develop the general principles in this way, and then extend the same ideas to three dimensions. If we look again at Figure 1.1 we can choose *any* point in this pattern (e.g. the centre of the dark spot) and then proceed to mark all other points which have an identical environment (i.e. all the other dark spots). This geometrical array of points defines a *lattice*, drawn out in Figure 1.3.

This translational symmetry is described by choosing a unit cell, which is a parallelogram with lattice points at its corners. The size and shape of the unit cell is specified by its *lattice parameters*. The *crystallographic axes*, x and y, are parallel to the sides of the unit cell, and hence describe the directions of the translations of the unit cells to fill space. The choice of unit cell is somewhat arbitrary, but it is usual to choose the smallest unit cell which describes the periodicity. A unit cell with lattice points only at its corners is called *primitive*. Figure 1.3 shows various

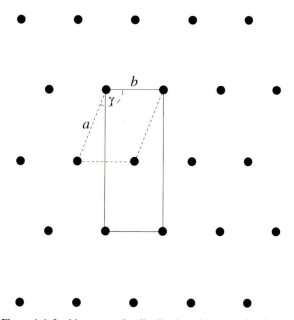

Figure 1.4. In this pattern the distribution of lattice points is such that the smallest primitive unit cell (shown dashed: $a \neq b$, $\gamma \neq 120°$) does not describe the symmetry, and hence the rectangular unit cell (solid line) is conventionally chosen to describe the lattice. This rectangular cell, with a lattice point at the centre as well as at the corners is termed non-primitive, and is obviously larger than the smallest repeat unit.

possible choices of unit cell, as well as the conventional one with lattice parameters $a = b$, and an interaxial angle γ of 120°. This unit cell is termed *hexagonal*.

For some patterns *non-primitive* unit cells are more appropriate. Figure 1.4 shows a pattern with a non-primitive rectangular lattice. While a primitive unit cell is obviously smaller, the fact that the interaxial angle $\gamma \neq 120°$ in this case means that the primitive cell has no special symmetry properties. On the other hand, the non-primitive *centred* rectangular cell shows the symmetry of the lattice, and the convenience of orthogonal crystallographic axes makes it easier to use. The same convention will apply when we look at lattices in three dimensions.

To describe a crystal structure, we begin with the lattice and then we specify the arrangement of atoms, their spacings and bond angles i.e. their atomic coordinates, relative to each lattice point, taking the origin at a lattice point. The atomic arrangement or pattern associated with each lattice point is called the *motif*. Figure 1.5 illustrates this using a two-dimensional pattern similar to the atomic distribution in Figure 1.1. Thus the structure can be reproduced by drawing the lattice, which defines

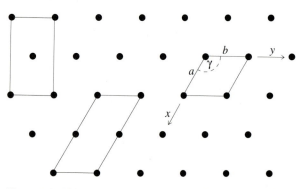

Figure 1.3. This two-dimensional array of lattice points shows some of the ways in which a unit cell could be defined. The conventionally chosen hexagonal unit cell has lattice parameters a and b equal and an interaxial angle $\gamma = 120°$. This is the smallest cell which describes the symmetry in the distribution of the lattice points.

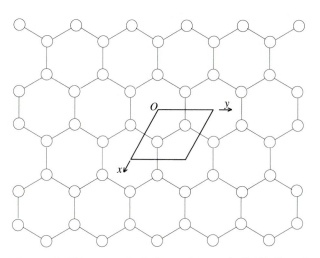

Figure 1.5. This pattern is similar to the atomic distribution of 'white atoms' shown in Figure 1.1. If point O is chosen as a lattice point, the array of identical points defines a hexagonal unit cell as shown in bold. The motif consists of 2 atoms with coordinates (2/3, 1/3) and (1/3, 2/3). The whole pattern can be generated by specifying the hexagonal lattice, and applying this motif to each lattice point. Note that only 2 atoms are needed for the motif, but that their coordinates depend on the choice of origin for the unit cell.

the translational periodicity, and then adding the motif at each lattice point.

i.e.   Lattice + Motif = Structure

1.2 Two-dimensional lattice symmetries

One of the important consequences of looking at symmetry in patterns is the realisation that the number of combinations of symmetry elements is limited, and so therefore is the number of possible structures that could be devised for a given chemical compound. The most efficient way of handling combinations of lattice symmetry operations is to use a branch of mathematics known as group theory, but it would be inappropriate to develop the algebraic notation here, as our aim is merely to become sufficiently comfortable with the basic ideas of crystallography to be able to appreciate their applications.

Linear arrays of points can only be combined in five distinguishable ways to create a two-dimensional

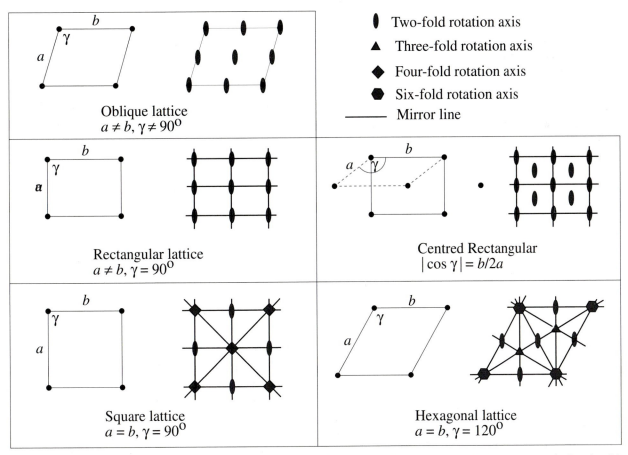

Figure 1.6. The five two-dimensional lattices, and the symmetry of the distribution of lattice points in each. In each case a single unit cell is shown on the left, and the symmetry elements within the cell are shown in the figure on the right.

Table 1.1. *Two-dimensional crystal systems, lattices and space groups*

Crystal system	Point groups compatible with crystal system	Lattices in system	Space groups compatible with lattice
Oblique $a \neq b, \gamma \neq 90°$	1, 2	p (primitive)	p1, p2
Rectangular $a \neq b, \gamma = 90°$	1m, 2mm	p (primitive)	pm, p2mm, pg, p2mg p2gg
		c (centred)	cm, c2mm
Square $a = b, \gamma = 90°$	4, 4mm	p (primitive)	p4, p4mm, p4gm
Hexagonal $a = b, \gamma = 120°$	3, 3m, 6, 6mm	p (primitive)	p3, p31m, p3m1 p6, p6mm

lattice. These are shown in Figure 1.6. The most general lattice is *oblique* with no special relationship between the lattice parameters *a*, *b* and *γ*. When *a* and *b* are equal or when *γ* = 90° or 120°, the other lattices are generated. The centred rectangular lattice is the special case when cos *γ* = *b/2a*, resulting in orthogonal axes when the centred unit cell is chosen. All two-dimensional periodic patterns belong to one of these five lattice types, which can be assigned to 4

crystal systems defined by the shape of the unit cells (Table 1.1).

In addition to their translational symmetry, lattices also have other symmetry elements at particular points. When a symmetry element operates on a point in a lattice, it restores the lattice to its original position (i.e. leaves it unaltered). The symmetry elements which we need to consider, and the way they operate, are illustrated in Figure 1.7. Note that only 1-, 2-, 3-, 4- and 6-fold rotations (360°, 180°, 120°, 90° and 60° rotation) are consistent with the translational symmetry of a lattice. A lattice cannot possess 5-fold symmetry, for example, because it is not possible to fill a plane completely with a network of regular pentagons. A simple geometric proof of this is outlined in Appendix A at the end of this chapter.

In Figure 1.6, the point symmetry is shown for each of the five lattice types. Rotation axes and/or mirror planes occur at the lattice points, and there are other locations in the unit cell with comparable or lower degrees of symmetry with respect to rotation and reflection.

1.3 Two-dimensional point groups and space groups

The complete symmetry of a structure is not defined by the lattice alone, since the arrangement of the atoms around each lattice point may also be symmetrical, as in Figure 1.1. If we now consider the

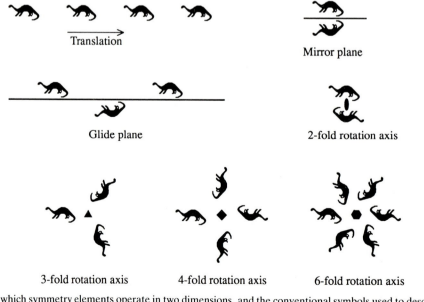

Figure 1.7. The way in which symmetry elements operate in two dimensions, and the conventional symbols used to describe them.

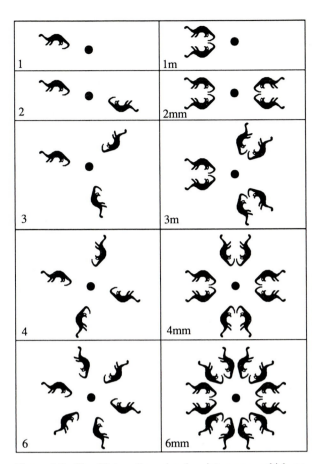

Figure 1.8. The 10 two-dimensional point groups which are consistent with the symmetries of a plane lattice. These represent the total number of combinations of mirror lines with rotation axes about a point. The symmetry elements themselves are not shown, but are indicated by the point group symbol.

possible combinations of symmetry elements of such groups of atoms, which are still consistent with the symmetry elements of the lattice, we again need only to derive the number of ways of combining 1-, 2-, 3-, 4-, and 6-fold rotations and mirror lines. Each different combination of symmetry elements about a point, without translational repetition, defines a *point group*. Inspection of Figure 1.8 shows that there are ten such two-dimensional point groups.

To complete the analysis of two-dimensional symmetry we now combine the ten point groups with the five lattice types. Here we also must introduce the glide operation (a compound operation involving reflection plus a translation of one half of a lattice vector) illustrated in Figure 1.7, to generate a total of seventeen two-dimensional *space groups*. Any reader attempting to derive these will notice that some combinations of symmetry elements produce

identical space groups (e.g. in a centred lattice two orthogonal mirror lines generate the same space group as two orthogonal glide lines, i.e. cmm = cgg). The 17 two-dimensional space groups are illustrated in Figure 1.9, and it is a worthwhile exercise to spend some time on this to appreciate how point symmetry and translational symmetry can be combined. All repetitive two-dimensional patterns must be based on one of these space groups.

1.4 Two-dimensional space groups – an example

Because of the importance of space groups in any discussion of crystal structure and atomic positions it is worthwhile to illustrate how the number of *equivalent points* is determined from the complete symmetry. Space groups are conventionally represented by diagrams such as Figure 1.10. The diagram on the right shows one unit cell of the lattice, with symmetry elements displayed, while that on the left shows the distribution of equivalent points, produced by the operation of these symmetry elements on a general position. The space group in this example is p2mm.

A general position is chosen with coordinates (x,y) (labelled in Figure 1.10 as a circle). Operation of a mirror line involves an inversion (see Figure 1.7) and this is denoted by a comma in the circle. Notice that it is not necessary to operate every symmetry element in the cell to generate the four *general equivalent positions* in this space group. The diads are automatically present at the intersections of mirror lines. The operation of the lattice translations reproduces the same distribution of equivalent positions in adjacent cells, and these are often added to the figure. The four general equivalent positions are listed by their fractional coordinates, in this case $\pm (x,y); \pm (x,\bar{y})$.

If the original point was at a special position with respect to some symmetry element, e.g. on a mirror line, or on a diad axis, the number of equivalent positions would be reduced. These are referred to as *special equivalent positions*, and are also listed as fractional coordinates. For example, a position on a mirror line at $(x, 1/2)$ is equivalent to $(x, \bar{1/2})$ i.e. there are only two special equivalent positions of this type. A position on a diad axis is not reproduced in any other part of the cell in this space group (i.e. a one-fold special position).

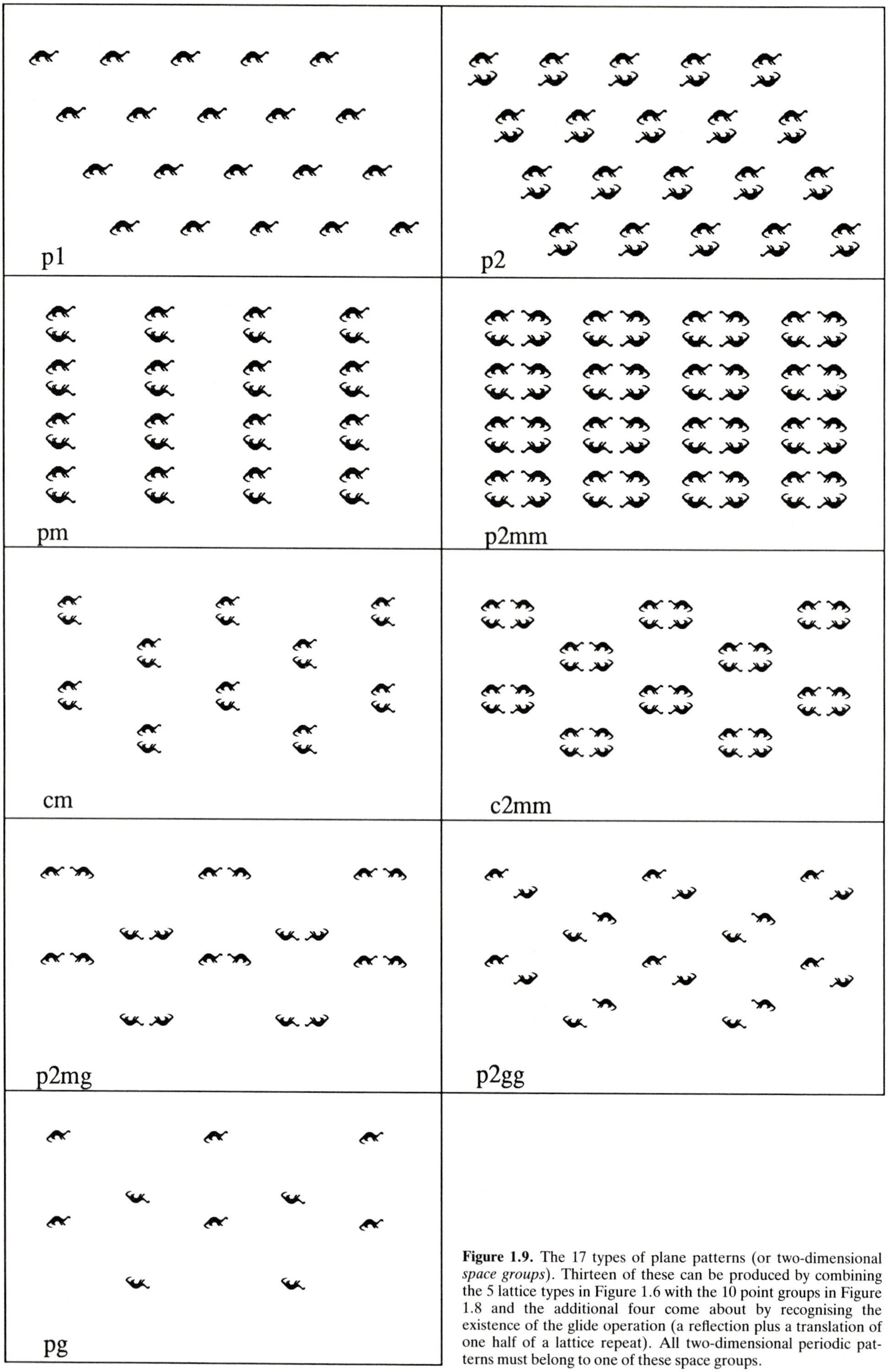

Figure 1.9. The 17 types of plane patterns (or two-dimensional *space groups*). Thirteen of these can be produced by combining the 5 lattice types in Figure 1.6 with the 10 point groups in Figure 1.8 and the additional four come about by recognising the existence of the glide operation (a reflection plus a translation of one half of a lattice repeat). All two-dimensional periodic patterns must belong to one of these space groups.

p6mm

p6

p31m

p3

p3m1

Figure 1.9 (cont'd)

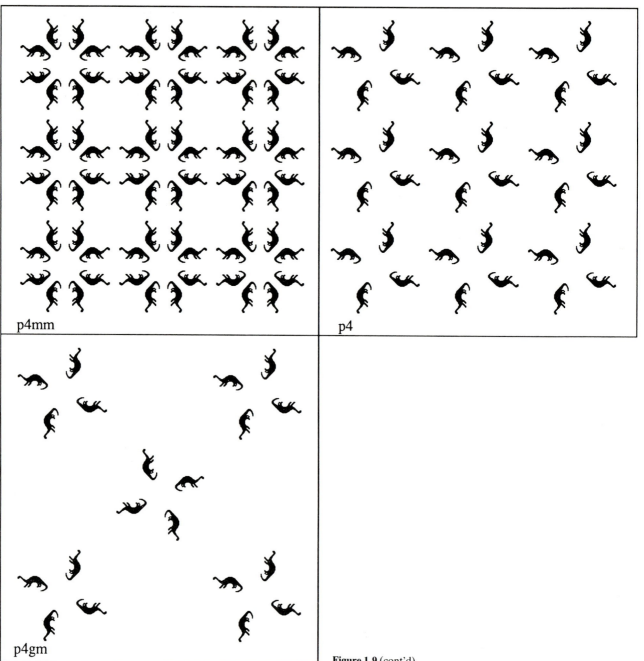

p4mm

p4

p4gm

Figure 1.9 (cont'd)

1.5 Three-dimensional lattices and crystal systems

The development of the concepts of lattice types, point groups and space groups in three dimensions is very similar to our two-dimensional discussion above, except that the extra dimension introduces extra complexity by increasing the number of symmetry element combinations. This makes three-dimensional symmetry more difficult to visualize, although the principles are the same. Many crystallography texts are largely devoted to this development and its various subtleties; here we will take the briefest possible route merely to demonstrate the general principles involved.

In two dimensions we found that there were five different lattices possible, within four crystal systems. The three-dimensional lattices can be derived

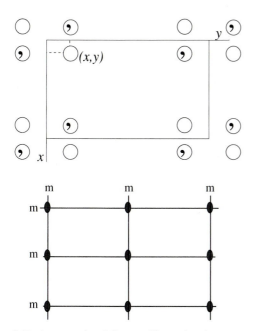

Figure 1.10. A conventional diagram illustrating the space group p2mm. The symmetry elements in one unit cell are shown on the right, while the diagram on the left shows the operation of these symmetry elements on a general position, with coordinates (x,y), in the unit cell.

Table 1.2. *The 7 crystal systems and 14 Bravais lattices in 3 dimensions*

Crystal system	Unit cell dimensions	Essential symmetry	Bravais lattices
Triclinic	$a \neq b \neq c$; $\alpha \neq \beta \neq \gamma$	None	P
Monoclinic	$a \neq b \neq c$ $\alpha = \gamma = 90°$ $\neq \beta$	A diad (2-fold) axis	P, C
Orthorhombic	$a \neq b \neq c$ $\alpha = \beta = \gamma$ $= 90°$	Three mutually perpendicular diad axes	P, C, I, F
Tetragonal	$a = b \neq c$ $\alpha = \beta = \gamma$ $= 90°$	A tetrad (4-fold) axis	P, I
Cubic	$a = b = c$ $\alpha = \beta = \gamma$ $= 90°$	Four triad (3-fold) axes	P, I, F
Trigonal*	$a = b = c$ $120° > \alpha$ $= \beta = \gamma \neq$ $90°$	A triad axis	R (rhombo-hedral)
Hexagonal	$a = b \neq c$ $\alpha = \beta$ $= 90°,$ $\gamma = 120°$	A hexad (6-fold) axis	P

* Crystals in the trigonal system may be described by an hexagonal unit cell, even though they do not have a hexad rotation axis.

by considering the various ways of regularly stacking each of these two-dimensional lattices, such that their superposition preserves the symmetry elements present. For any particular stacking sequence the choice of the most appropriate unit cell must be made, and this again raises the issue of primitive and non-primitive cells. As well as unit cells with an extra lattice point in the centre of one face (centred A, B or C), three-dimensional lattices may be body-centred with an extra lattice point at the centre of the cell (denoted I, German *Innenzentrierte*), and face-centred, with extra lattice points at the centre of each face (denoted F).

The three-dimensional lattice types were derived by Bravais in 1848, who found that there were 14 distinguishable space lattices within seven crystal systems. Other apparent possibilities can be shown to be equivalent to one of these 14 lattices (body-centred monoclinic for example, can be shown to be equivalent to the C monoclinic cell). The choice of cell is again a matter of convention. These 14 Bravais lattices are illustrated in Figure 1.11, and their grouping into seven crystal systems summarized in Table 1.2.

Returning briefly to non-primitive lattices, the decision to define a face-centred cubic unit cell, for

example, is based on the fact that if we chose instead the appropriate primitive unit cell to describe the lattice, its symmetry would not be obvious and the overwhelming convenience of using a cubic unit cell would be lost. We must note however, that whenever we describe a lattice by a non-primitive cell we define a repeat unit which is larger than the fundamental lattice repeat. We will need to take this into account when we describe diffraction from a non-primitive lattice.

1.6 Three-dimensional point groups and their representation

The combination of symmetry elements which relate one atom to another in a crystal structure, considered separately from the translational symmetry

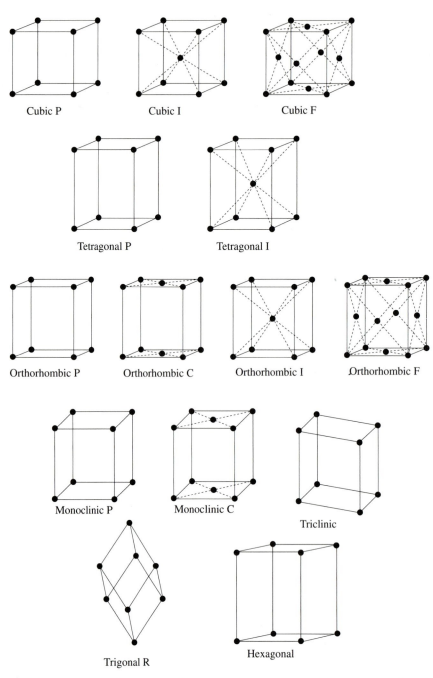

Figure 1.11. The 14 Bravais lattices. All crystalline solids can be described by unit cells which belong to one of these 14 types.

of the lattice, is described by the point group. Extending the ten two-dimensional point groups (outlined in Section 1.3) to three dimensions involves comparable symmetry operations; mirror lines are replaced by mirror planes, and rotation about a point is replaced by rotation axes about a line. Additionally, a new symmetry element, termed an inversion axis is introduced. The operation of an inversion axis involves a rotation followed by an inversion through a point lying in the axis. Inversion axes are symbolically written with a 'bar' above the axis, i.e. $\bar{1}, \bar{2}, \bar{3}, \bar{4}, \bar{6}$. Thus $\bar{1}$ (bar one) means 'rotate through 360° and invert'; $\bar{3}$ means 'rotate through 120° and invert' etc. The operation of these symmetry elements may be clearer after the discussion in Appendix B.

The derivation of the three-dimensional point groups involves the combination of mirror planes with 1-, 2-, 3-, 4- and 6-fold axes and $\bar{1}$, $\bar{2}$, $\bar{3}$, $\bar{4}$, and $\bar{6}$ inversion axes. This generates 32 point groups compatible with the symmetry of the 7 crystal systems.

Point groups are not easy to visualize in three dimensions, but being able to appreciate the angular relationships between symmetry elements in a point group is important for understanding the effect of symmetry on physical properties. The representation of the symmetry elements in three-dimensional point groups on a two-dimensional page is traditionally done by means of a *stereographic projection*. A brief outline of the principle of the stereographic projection is given in Appendix B, although this may be best read after Section 1.8 on planes and directions in a crystal.

The 32 point groups and their stereographic representations are shown in Figure 1.12. The point group symbol defines the distinctive symmetry elements it contains and their mutual orientations within the crystal system to which the point group belongs. We are not concerned here about the conventions for naming the point groups (which are different in each crystal system) but it is worthwhile to understand how any general direction is operated on by the point group symmetry elements. For this purpose we choose an example: point group mm2 belongs to the orthorhombic system and contains two mirror planes at right angles to each other with a twofold axis passing along their line of intersection (see Figure 1.12). A general direction pointing above the plane of the page is denoted by a solid black point and belongs to the set of four equivalent directions, generated by these symmetry elements. For the point group mmm the total number of equivalent directions would be eight, as the horizontal mirror plane operates to generate directions pointing below the plane of the page, shown as circles in the projection. Special directions parallel to symmetry elements will obviously have fewer equivalents.

Of particular interest in the context of physical properties is whether a point group has a *centre of symmetry* (is every direction equivalent to its opposite?), whether it is *enantiomorphic* (can right-handed and left-handed motifs be distinguished?) or whether it is *polar* (are there any directions for which the opposite is not equivalent?). Enantiomorphic point groups have no centre of symmetry,

Table 1.3. *Centrosymmetric, enantiomorphic and polar point groups*

Crystal system	Centro-symmetric	Enantio-morphic	Polar
Triclinic	$\bar{1}$	1	1
Monoclinic	2/m	2	2, m
Orthorhombic	mmm	222	mm2
Tetragonal	4/m, 4/mmm	4, 422	4, 4mm
Trigonal	$\bar{3}$, $\bar{3}$m	3, 32	3, 3m
Hexagonal	6/m, 6/mmm	6, 622	6, 6mm
Cubic	m3, m3m	23, 432	—

mirror plane or inversion axis. A polar axis in a point group can exist if there are no symmetry elements that generate its opposite. The centrosymmetric, enantiomorphic and polar point groups are listed in Table 1.3.

1.7 Three-dimensional space groups

A combination of the 32 point groups with the 14 Bravais lattices generates the complete set of three-dimensional space groups, analogous to the similar procedure for two dimensions in Section 1.3. For two-dimensional space groups we introduced the the glide operation; in three dimensions we introduce another compound symmetry element termed a *screw axis*, which as the name implies, involves a translation imposed on a rotation (of 60°, 90°, 120°, or 180°) to produce an invariance of the structure. The result of all possible permutations of lattice type with permissable symmetry operations gives rise to 230 space groups. Their derivation at the end of the nineteenth century was the independent achievement of Schoenflies, Barlow and Federov, as a theoretical exercise, before their application to crystal structures was anticipated. The space group is thus the ultimate sub-division in the classification of symmetry elements. Each space group is associated with a point group, which in turn belongs to one of the seven crystal systems. All crystalline materials have a structure which belongs to one of these space groups.

A complete tabulation of the 230 space groups, and their general and special equivalent positions is given in the *International Tables for X-ray Crystallography* (Vol.1, Symmetry Groups). The generation of equivalent positions in a three-dimensional space group is exactly analogous to the two-dimensional

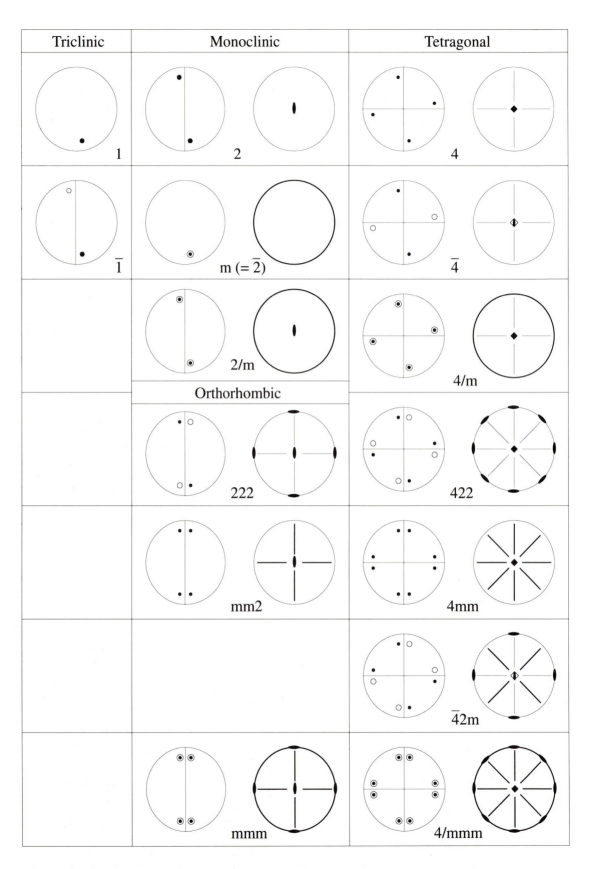

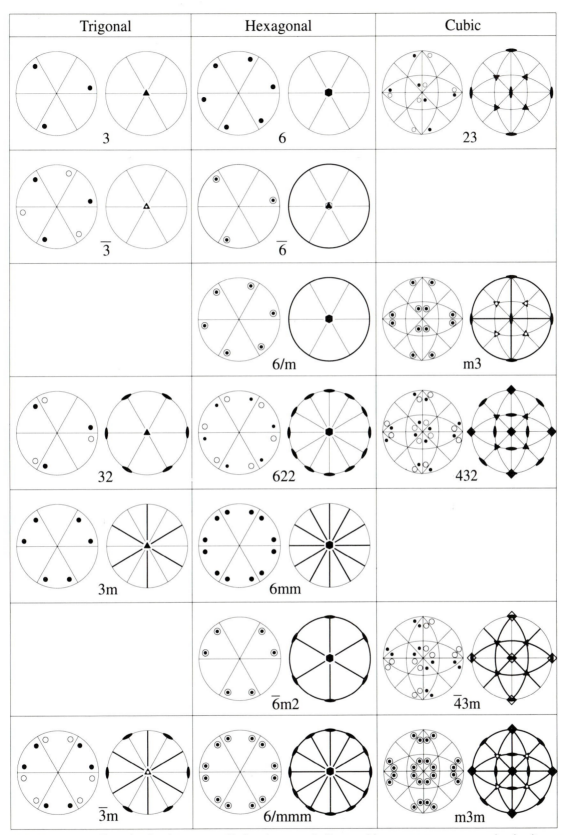

Figure 1.12. The 32 three-dimensional point groups . Each point group is illustrated by two stereograms, one showing its symmetry elements, the other showing how this symmetry operates on a general direction (or pole) denoted by a spot. Appendix B describes the principle behind representing the point groups in this way.

example given in Section 1.4, apart from the extra complexity associated with the visualisation of symmetry operations in three dimensions. The written symbol for a space group is a list of characters beginning with the lattice type denoted by a capital letter, P, I, F etc. followed by some of the symmetry elements present. Space group diagrams in three dimensions are much the same as in two dimensions (Figure 1.10): usually drawn as parallelograms corresponding to the *xy* plane of a unit cell, one showing the symmetry elements, the other showing the general equivalent positions.

The space group is fundamental in determining the positions of atoms in a crystal structure. If the unit cell constants and space group have been determined, a knowledge of the chemical composition and density defines the number of atoms of each type in the unit cell. The space group limits the positions of each type of atom to a set of general or special equivalent positions, and combined with an understanding of cation – anion coordination a possible trial structure for simpler compounds can often be proposed.

1.8 Planes and directions in a crystal

Having followed the path from lattice to point group to space group, we now return to some of the more elementary geometrical concepts in the description of crystals. It is often necessary to refer to a particular crystallographic plane or direction in a crystal

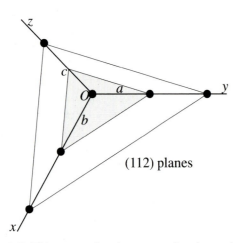

Figure 1.13. This construction shows a set of equi-spaced planes (the first of which is shaded) relative to a general lattice with lattice repeats of *a,b* and *c* along the crystallographic axes *x,y,* and *z* respectively. The small black circles are lattice points along the three crystallographic axes. The Miller indices of this set of planes is (122).

and in this section we describe the necessary symbolism.

Crystal planes are known by their *Miller indices* which are determined as follows. Consider the general three-dimensional lattice in Figure 1.13, for which the lattice constants are *a*, *b*, *c*. If a set of planes intersects these axes and divides the *a* lattice translation into *h* parts, the *b* lattice translation into *k* parts and the *c* lattice translation into *l* parts, the plane is represented by the Miller indices (*hkl*). Another way of expressing this is that the plane (*hkl*) has intercepts of *a/h*, *b/k*, and *c/l* on the three crystallographic axes, i.e. the Miller indices are the reciprocals of the fractional intercepts.

The origin is defined at a lattice point, but otherwise its choice is arbitrary. Once the positive sense of the axes has been defined, planes may intersect the negative directions of the axes leading to Miller indices such as (*h̄kl*), [pronounced 'bar' *h*,...], indicating that the plane cuts the *a* axis on the negative side and the *b* and *c* axes on the positive side of the origin. A plane (*hk0*) does not intersect the *c* axis i.e. is parallel to it. Some further examples are given in Figure 1.14.

We often refer to a set of equivalent planes related by symmetry, in which case curly brackets are used. Thus {100} represents all of the planes equivalent to (100) by symmetry. In the cubic system {100} is the set of planes (100), (1̄00), (010), (01̄0), (001) and (001̄). In the tetragonal system {100} would mean the set of four planes (100), (1̄00), (010) and (01̄0). Inspection of the point group diagram in Figure 1.12 will show that the point group symmetry defines the number of planes in the set {*hkl*}.

If we are only concerned about the orientation of a set of planes we choose the smallest indivisible set of integers *h,k,l* for the Miller indices. Thus since planes (200) are parallel to (100), the latter indices are used to describe their orientation. The orientation relations of the external faces of a crystal are described in this way. However, the set of (200) planes within the crystal has an interplanar spacing (perpendicular distance between the planes) which is half that of a set of (100) planes (see Figure 1.15) and this will be an important distinction to remember when discussing diffraction in Chapter 3.

Directions in a lattice have similar symbols but are described somewhat differently. Directions are given by three integers in square brackets [UVW] but it is worth emphasising from the outset that the

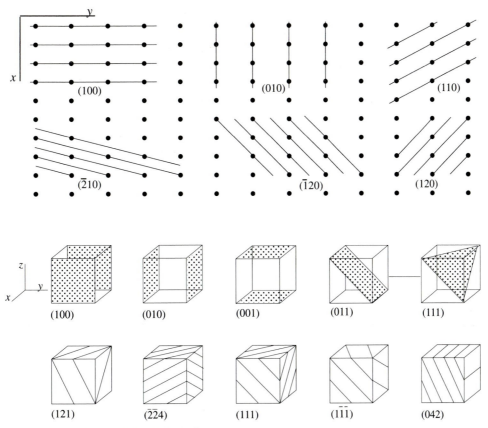

Figure 1.14. Examples of lattice planes and their Miller indices.

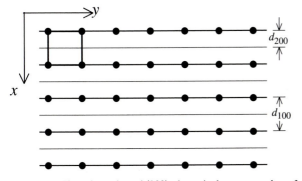

Figure 1.15. The orientation of (200) planes is the same as that of (100) planes, but the interplanar spacing of (200) planes is clearly one-half of the (100) interplanar spacing. The set of (200) planes includes all of the planes in this diagram whereas the set of (100) planes is shown by the darker lines. The bold square is the unit cell.

direction [UVW] is **not** in general the direction normal to the set of planes indexed (*hkl*) with the same integers, except in the cubic system, i.e. [123] is not, in general, the direction normal to (123).

The direction [UVW] in a lattice with unit cell base vectors **a**, **b**, **c** is a direction parallel to the vector U**a** + V**b** + W**c**, where U,V,W are integers.

(Figure 1.16). To illustrate this further, Figure 1.17 shows a two-dimensional lattice in the *xy* plane with various directions [UVW] labelled.

The *set* of symmetry related directions [UVW] is designated by different brackets i.e. <UVW> Thus in the orthorhombic system point group mmm (Figure 1.12) the symbol <111> means a set of 8 equivalent directions. Note however that in point group 222 this would be reduced to 4.

As well as being able to describe the orientation of planes and directions in a crystal, we also often need to deal with the **relationship between planes and directions**, and of most importance is the concept of a *zone* and a *zone axis*. A set of planes lie in the same zone if they are all parallel to a common direction [UVW]. The direction [UVW] is called the zone axis. We can visualize this easily in relation to the six faces of a cube. The planes (100), (010), ($\bar{1}$00) and (0$\bar{1}$0) are all mutually parallel to the direction [001], the *c* axis direction, and are said to lie in the zone [001]. All planes (*hk*0), as they do not intersect the *c* axis, lie in the [001] zone.

For any crystal system, the condition that a plane

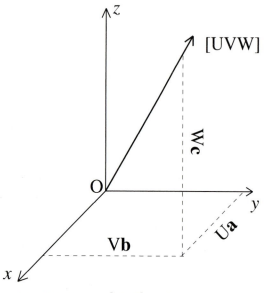

Figure 1.16. The direction [UVW] in relation to its components along the crystallographic axes.

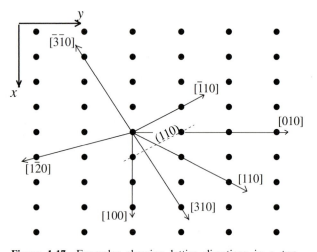

Figure 1.17. Examples showing lattice directions in a two-dimensional lattice. Note that in general, a lattice direction is not parallel to the normal to a plane indexed with the same integers. This is illustrated by the relationship between the plane (110), shown as a dashed line, and the direction [110].

(hkl) is parallel to the direction [UVW] is given by the *Weiss zone law*:

$$h\mathrm{U} + k\mathrm{V} + l\mathrm{W} = 0$$

If two planes (hkl) and ($h'k'l'$) lie in the same zone [UVW] then both the above equation and

$$h'\mathrm{U} + k'\mathrm{V} + l'\mathrm{W} = 0$$

must be satisfied. We can solve these equations for U,V,W using the following device. First set up the matrix

$$\begin{bmatrix} \mathrm{U} & \mathrm{V} & \mathrm{W} \\ h & k & l \\ h' & k' & l' \end{bmatrix}$$

from which we can write that

$$\begin{aligned} \mathrm{U} &= kl' - k'l \\ -\mathrm{V} &= hl' - h'l \\ \mathrm{W} &= hk' - h'k \end{aligned}$$

For example, the planes (111) and (212) lie in the zone [$10\bar{1}$]
since

$$\begin{bmatrix} \mathrm{U} & \mathrm{V} & \mathrm{W} \\ 1 & 1 & 1 \\ 2 & 1 & 2 \end{bmatrix}$$

hence
$$\begin{aligned} \mathrm{U} &= 1\times2 - 1\times1 = 1 \\ -\mathrm{V} &= 1\times2 - 2\times1 = 0 \\ \mathrm{W} &= 1\times1 - 2\times1 = -1 \end{aligned}$$

which is the same as [$\bar{1}01$], the result obtained if (hkl) and ($h'k'l'$) had been reversed in the matrix.

The determination of the plane (hkl) common to two directions [UVW] and [U'V'W'] is found in a similar way from the matrix

$$\begin{bmatrix} h & k & l \\ \mathrm{U} & \mathrm{V} & \mathrm{W} \\ \mathrm{U'} & \mathrm{V'} & \mathrm{W'} \end{bmatrix}$$

so that

$$\begin{aligned} h &= \mathrm{VW'} - \mathrm{V'W} \\ -k &= \mathrm{UW'} - \mathrm{U'W} \\ l &= \mathrm{UV'} - \mathrm{U'V} \end{aligned}$$

The concepts of zones and zone axes are also discussed in Appendix B.

Appendix A
Allowed rotations in a periodic lattice

A crystal has an *n*-fold axis of rotation if rotation through an angle 360°/*n* about this axis produces an atomic array identical to the original. The allowed rotation operations in a crystal with a periodic lattice are *n* = 1, 2, 3, 4, and 6. The case of *n* = 1 is the trivial one of rotation through 360°. The other allowed cases (*n* = 2,3,4,6) are equivalent to considering the filling of space by rectangles, triangles, squares and hexagons respectively. Attempts to fill an area with other polygonal shapes lead to wasted space or overlapping polygons (e.g Figure 1.18).

A simple geometric verification is presented below.

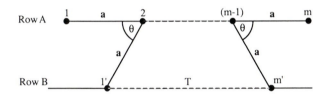

Consider a row of lattice points with spacing **a** (Row A). Along this row lattice points 1 and m are separated by a distance (m−1)**a**. Suppose there is an

n-fold rotation axis through each lattice point in this 'crystal'. An anticlockwise rotation of angle θ = 360°/*n* about point 2 would move point 1 to a new position 1'. Similarly a clockwise rotation of angle θ about point (m−1) would move point m to a new position m'. Lattice points 1' and m' are in row B and their separation T must be an integral multiple of the lattice repeat **a**, i.e. p**a** where p is an integer.

From simple trigonometry applied to the figure,

$$T = pa = (m-3)\,a + 2a\cos\theta$$
$$\cos\theta = (3+p-m)/2 \quad \text{and} \quad \theta = 360°/n$$

Since m and p are integers this equation has only 5 solutions, tabulated below.

(p−m)	−1	−2	−3	−4	−5
cos θ	1.0	0.5	0	−0.5	−1.0
θ°	360	60	90	120	180
n	1	6	4	3	2

Other rotation axes, with *n* = 5 or *n* >6 are not consistent with lattice translations and so cannot occur in crystals although they are found elsewhere in nature, for example in isolated molecules, 5-pointed starfish, flowers etc.

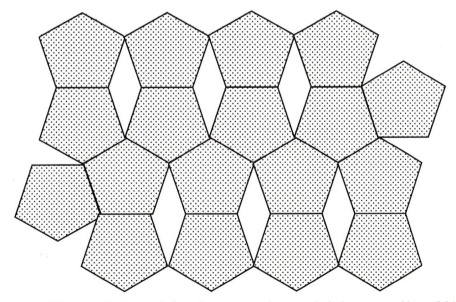

Figure 1.18. Any attempt to fill space with close-packed regular pentagons always results in free space and hence 5-fold symmetry is not allowed in a crystal lattice. The same is true of *n*-fold axes with *n*>6.

Appendix B
The stereographic projection

Directions and planes in three dimensions are not easy to visualize, and the stereographic projection provides a convenient way of representing three-dimensional angular relationships on a two-dimensional plane. When accurately constructed, a stereogram allows quantitative measurements of angles between directions in any crystal system. Here we will limit ourselves to the general principles of stereographic projection, to illustrate the relationships between symmetry elements in the three-dimensional point groups shown in Figure 1.12.

In the stereographic projection, all directions to be plotted are considered as originating from the centre of a sphere. The point at which each of these radii touches the sphere is then projected back through the equatorial plane, which will be the plane of the projection, to the projection point in the lower hemisphere, as shown in Figure 1.19. The point where this line intersects the equatorial plane is termed the *pole* of that particular direction, and is represented by a filled circle. When a direction projects into the lower hemisphere, its pole is represented by an open circle (in which case the projection pole becomes the one in the upper hemisphere.)

Planes in a crystal are often represented by the pole of the direction normal to the plane. Figure 1.20 shows the stereographic projection of the six faces of a cube. It is easiest to visualize the cube at the centre of the sphere such that the normal to any plane becomes a radius of the sphere. The cube is oriented such that the xy plane lies in the horizontal plane as shown. Planes whose normals project into the upper hemisphere are represented by filled circles in the projection plane, and those whose normals project into the lower hemisphere are represented by open circles.

In the simple case here the normals to the four vertical faces of the cube [(100), (010), ($\bar{1}$00) and (0$\bar{1}$0)] are horizontal and hence each touches the sphere at the equator. This point becomes the pole. All vertical planes will obviously have horizontal normals which will have poles on this equatorial circle, usually called the *primitive circle*. The normal to the top face (001) is vertical and hence its projection will lie at the centre of the stereogram.

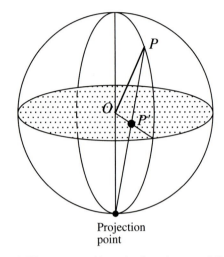

Figure 1.19. The stereographic projection of a general direction in a lattice, considered as originating from the centre of the sphere, touching its surface at P and then projected to the 'south' pole. The point P' represents the pole of the direction OP, in the shaded equatorial plane, which is the plane of the projection.

The bottom face, (00$\bar{1}$), is represented by an open circle, also at the centre. Poles to planes are usually labelled as the indices with the brackets removed.

If we now extend this same process to consider a vertical face with index (110) (Fig.1.20b), its projection will lie on the primitive circle at 45° to the x and y axes, i.e. between the poles 100 and 010. The fact that the normals to (100), (010) and (110) are co-planar means that their poles will plot on the same great circle (a great circle is one whose plane passes through the centre of the sphere), in this case the primitive great circle. In the same way, a plane (011) will project to a point between the poles 001 and 010, such that it lies at 45° to both. Note that in this type of projection, angular relationships are preserved although they are not linearly related to distances in the projection plane (i.e. the pole 011 is not half-way between 001 and 010 although it is equally inclined to each). The other planes symmetrically equivalent to (110) i.e. formed by cutting off all eight edges of the cube at 45° can be added to the stereogram in the same way.

We can now add the great circles which pass through these poles. A set of faces whose normals

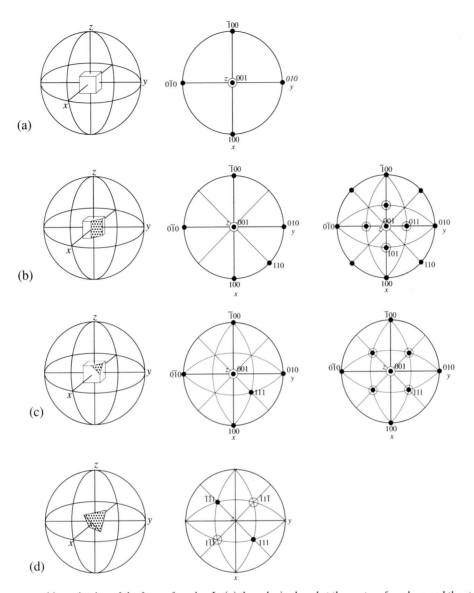

Figure 1.20. The sterographic projection of the faces of a cube. In (a) the cube is placed at the centre of a sphere and the stereogram shows the poles to the six cube faces. In (b) the shaded face has Miller indices (110) formed by cutting off one edge of the cube. The pole of this face is added to the stereographic projection of the cube, followed by the rest of the faces of the faces symmetrically equivalent to (110), i.e. the set of faces {110}. In (c) the face (111) has been added to the cube, followed by its stereographic projection and the rest of the faces of the form {111}, totalling eight. (d) shows the stereogram of the four {111} faces of a tetrahedron with the crystallographic axes in the same orientation as in (c).

are co-planar are said to lie in the same *zone*, and their normals will all plot on the same great circle.

Finally we consider the projection of the plane formed by cutting off a corner of the cube at equal inclination to the *x*, *y* and *z* axes. The pole to this (111) plane will plot between 001 and 110 along the same great circle. Similarly, as the normal to (111) is also coplanar with normals to 011 and 100 it must also lie along the great circle between them. The same applies to its relationship to 101 and 010. Thus the pole to (111) can be located at the intersection of

the great circles as shown in Figure 1.20c. In this way the pole to any plane can be plotted.

Implicit in the above discussion is the fact that if we consider any two poles *hkl* and *h'k'l'*, then the pole to *h+h'*, *k+k'*, *l+l'* must lie between them along the same great circle. This is often called the *addition rule*.

As well as representing a plane by a pole to its normal, a plane may also be represented as a great circle. The plane of this great circle passes through the origin of the original sphere and is parallel to the

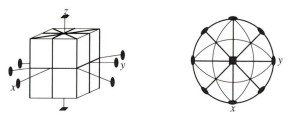

Figure 1.21. The symmetry elements of a tetragonal prism and the stereographic projection illustrating them. The great circles drawn in heavy line indicate mirror planes.

plane in the crystal. When we use the stereographic projection to illustrate the symmetry elements of a tetragonal prism, we depict the mirror planes as great circles in the projection and rotation axes as poles (Figure 1.21).

The stereographic projection in other crystal systems operates in the same way except that the change in the shape of the unit cell means that angular relationships between directions will also change, e.g. in the tetragonal system the direction [011] is no longer at 45° to [010], nor is it the same direction as the normal to the plane (011).

One important application of stereograms is to illustrate the symmetry elements in the 32 point groups, as in Figure 1.12. The way in which individual symmetry elements operate on a general direction can be best appreciated by examining these stereograms in some detail.

Bibliography

BLOSS, F.D. (1971). *Crystallography and crystal chemistry*. Holt, Reinhart & Winston Inc.

BUERGER, M.J. (1970). *Contemporary crystallography*. McGraw-Hill Book Co.

DENT-GLASSER, L.S. (1977). *Crystallography and its applications*. Van Nostrand Reinhold.

KELLY, A. AND GROVES, G.W. (1970). *Crystallography and crystal defects*. Longman.

McKIE, D. AND McKIE, C. (1986). *Essentials of crystallography*. Blackwell Scientific Publications.

STEADMAN, R. (1982). *Crystallography*. Van Nostrand Reinhold.

2 Anisotropy and physical properties

Throughout the previous chapter we emphasised the idea of the symmetry of crystalline materials. Another way of referring to the translational symmetry of crystals is to say that a perfect crystal has *long range order* in the arrangement of atoms. A solid with no long range order is said to be *amorphous*. Ordinary glass is a good example. The randomness of the atomic arrangement and the consequent lack of symmetry means that, on average, every direction in an amorphous structure is equivalent to every other. If we were to measure some physical property which depends on direction, it would not vary with orientation in glass, i.e. an amorphous solid is *isotropic*. Crystalline materials, on the other hand, are generally *anisotropic*, which means that the magnitude of many physical properties will depend on direction in the crystal.

Some physical properties are clearly non-directional, i.e. they are *scalar* properties. For example the density of a mineral, or its heat capacity can be measured without reference to direction. Other properties such as surface energy and hence chemical reactivity do depend on the crystallographic orientation of the surface plane, but are not in themselves direction-dependent quantities. However, the thermal conductivity of a mineral is defined as the ratio of the heat flow to the temperature gradient, both of which need to be specified by direction as well as magnitude (i.e. they are vector quantities). The thermal conductivity therefore must also be defined in relation to direction in a crystal, which may be anisotropic with respect to this property. However, in crystals belonging to the cubic system, thermal conductivity is in fact equal in all directions – cubic crystals are isotropic for thermal conductivity, while only crystals with lower symmetry are anisotropic. Cubic crystals *are* anisotropic for elastic properties, which describe the elastic deformation of a crystal due to an applied force, and there exists a formal ranking of the physical

properties of crystals in relation to the mathematical quantities defining them.

In general, these physical properties are described or represented by mathematical quantities called *tensors*. In this formalism, a property which is non-directional, i.e. a scalar property such as density, can be specified by a single number and is called a *zero rank tensor*. Vector properties, for which both magnitude and direction must be specified, for example the temperature gradient at some point in the crystal, are termed *first rank tensors*. Properties such as thermal conductivity, which relate two vector quantities are termed *second rank tensors*, and must be specified in a different way. We will see in this chapter that by choosing a suitable axial system we can specify the property by three principal values. This classification or ranking of physical properties includes third and fourth rank tensors whose description is beyond the scope of this book. Elastic properties belong to fourth rank tensors and their description is correspondingly more complex.

The aim of introducing physical properties in this way is to illustrate the phenomenon of anisotropy with more than merely a passing reference, and to provide some background to an understanding of optical properties of minerals. A full mathematical treatment would not be appropriate and we will also restrict ourselves to second rank tensors. Elastic properties will only be mentioned briefly.

2.1 Anisotropy – a mechanical analogy

Before discussing physical properties of minerals, a simple two-dimensional mechanical analogy may help to describe some general features of anisotropy and how to specify a property when it varies with direction.

Consider a mechanical system where a central ring is held into a frame by 2 pairs of springs at right angles (Figure 2.1). The springs on opposite sides are

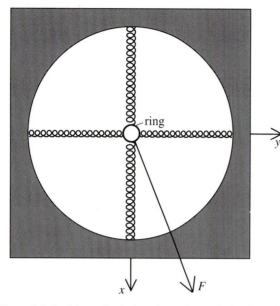

Figure 2.1. In this mechanical analogy of an anisotropic system four springs hold the central ring at the centre of the circular hole. If the springs along the x and y directions have different spring constants, a force F (the *cause* vector) will not result in a displacement (the *effect* vector) of the central ring in the same direction as F, except when F is parallel to either x or y. In a general direction the *cause* and the *effect* are not parallel.

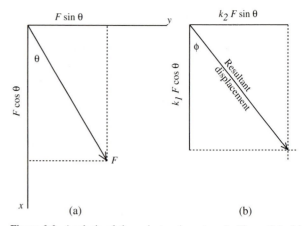

Figure 2.2. Analysis of the anisotropic system in Figure 2.1. (a) The components of the applied force F along the principal axes of the system, which produce (b) the two different displacements $k_1 F \cos\theta$ and $k_2 F \cos\theta$ along the principal axes. k_1 and k_2 are the relevant spring constants. The resultant displacement will not in general be parallel to the applied force F.

identical but have a different spring constant to the perpendicular pair. When a force F is applied to the central ring in this plane, the ring will be displaced from its central position. We will refer to the force F as the *cause* vector and the resultant displacement of the ring as the *effect* vector. The relationship between the cause and effect defines the spring stiffness.

If a force is applied in a general direction in this plane, it may be intuitively obvious that the displacement of the ring will depend in some way on the relative stiffness of both pairs of springs, and that the ratio of the cause and effect i.e. applied force/ displacement will depend on the orientation of the force. The next step is to show that, in general, the displacement will not be in the same direction as the applied force.

The way such a problem would be treated is first to find the components of the force F in the direction of each of the two springs, next to work out the displacement which each force component would produce parallel to each spring, and finally to combine these two orthogonal displacements to find the resultant displacement.

The components of the applied force F in the directions x and y are $F \cos\theta$ and $F \sin\theta$ respectively

(Figure 2.2(a)). If the spring stiffness along x is k_1 (defined as the displacement per unit applied force), then

Displacement along $x = k_1 F \cos\theta$

Similarly, if the spring stiffness along y is k_2,

Displacement along $y = k_2 F \sin\theta$

The resultant displacement is shown in Figure 2.2(b), and the angle between the resultant displacement and the x axis (ϕ) is given by

$$\tan\phi = \frac{k_2}{k_1} \tan\phi$$

Thus in general, the direction of the displacement will not be the same as the direction of the applied force. Only when $\theta = 0°$ or $90°$ will the displacement (the effect vector) be parallel to the applied force (the cause vector).

On the basis of this very simple analogy we can make a number of observations about anisotropy:

(i)   In an anisotropic system the effect vector is not, in general, parallel to the applied cause vector.

(ii) In our two-dimensional example there are two orthogonal directions along which the effect is parallel to the cause. By simple mathematics it can be shown that in a three-dimensional example there would be three such orthogonal directions.

(iii) An anisotropic system can be analysed in terms of components along these orthogonal principal directions, termed principal axes.

Along these principal axes the values of the physical property are termed the principal values. In the case above these are k_1 and k_2. In three dimensions the principal values along the three orthogonal principal directions would be k_1, k_2, and k_3.

We next need to describe how the physical property, which in the above example is the spring stiffness, varies with direction in the system. We will stay with our mechanical analogy but extend it to three dimensions.

In three dimensions a general direction can be defined by its direction cosines, explained in Figure 2.3(a). Consider a system in which a force F is applied in a general direction, resulting in a displacement D at some angle φ to F (Fig.2.3(b)). The component of this displacement in the direction of F is related to the applied force F by a constant K (which in our mechanical analogy depends on the three spring stiffnesses).

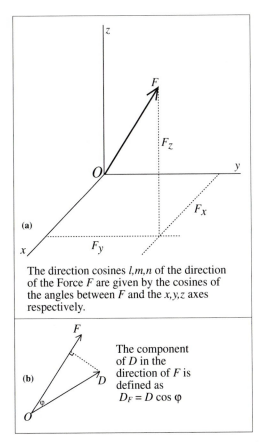

The direction cosines l,m,n of the direction of the Force F are given by the cosines of the angles between F and the x,y,z axes respectively.

(b) The component of D in the direction of F is defined as $D_F = D \cos \varphi$

Figure 2.3. The effect of a force F in a general direction in an anisotropic system can be analysed in terms of the separate effects of its 3 components along the 3 principal directions. The resultant displacement D, produced by the force F, lies at an angle ϕ to the direction of F.

Since the component of D in the direction of F is $D_F = D \cos \varphi$

$$K = \frac{D \cos \varphi}{F} = \frac{D_F}{F}$$

We can calculate the value of K in terms of the principal values k_1, k_2 and k_3:
The components of F along the principal axes are

$$F_x = Fl$$
$$F_y = Fm$$
$$F_z = Fn$$

The components of the displacement along the principal axes are therefore

$$D_x = k_1 Fl$$
$$D_y = k_2 Fm$$
$$D_z = k_3 Fn$$

We now need to find the resultant displacement from these three orthogonal components.

The magnitude of D in the direction of F (D_F) can be written in terms of its components as:

$$D_F = D_x.l + D_y.m + D_z.n$$

Therefore $D_F = k_1 F.l^2 + k_2 F.m^2 + k_3 F.n^2$
and since $K = D_F/F$
then $\mathbf{K = k_1.l^2 + k_2.m^2 + k_3.n^2}$ (2.1)

This equation defines the value of the property K in any direction in terms of the three principal values k_1, k_2, k_3 and the direction cosines l, m, n.

2.2 Second rank tensor properties and their variation with direction

The analogy above can be applied to describe the variation of any physical property which relates two vectors. Such properties are described by second rank tensors and are treated in the same way as the variation of K above. Thermal conductivity, relating the heat flow to the temperature gradient has already been mentioned. Other similar properties are electrical conductivity (relating electric current density to electric field) and the diffusivity (relating the atomic flux to the concentration gradient). In all such cases, three principal values of the property must be specified to describe it, and the value in any general direction is given by equation (2.1) above.

It is easier to visualize the variation of a property K, with direction by representing the magnitude of K as a distance from a point origin. The result will be

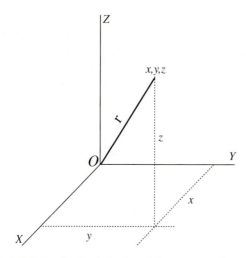

Figure 2.4. Figure for the derivation of the representation surface which describes the variation of a second rank tensor property with direction in a crystal.

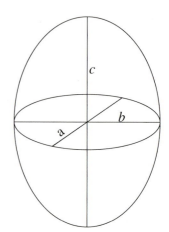

Figure 2.5. The general form of an ellipsoid with semiaxes a, b and c. This is the most general form of the representation surface for a second rank tensor. The distance from the centre to the surface is $1/\sqrt{K}$, where K is the property being represented.

a three-dimensional surface (the representation surface) whose shape gives an immediate indication of the anisotropy of the property. Thus if we had an isotropic material the property would be equal in all directions and the resulting representation surface would be a sphere.

We derive the shape of the representation surface by considering a point with coordinates x,y,z at a distance r from the origin in the direction of F (Figure 2.4). The direction cosines l,m,n may be written in terms of x,y,z as

$$l = x/r, \ m = y/r \ \text{and} \ n = z/r.$$

Substituting into the general equation

$$K = k_1 . l^2 + k_2 . m^2 + k_3 . n^2$$
$$= k_1 . x^2/r^2 + k_2 . y^2/r^2 + k_3 . z^2/r^2$$

If we write that $\qquad r^2 K = 1,$

i.e. that $r = 1/\sqrt{K}$, then

$$\mathbf{k_1 x^2 + k_2 y^2 + k_3 z^2 = 1} \qquad (2.2)$$

If the three principal values k_1, k_2, k_3 are all positive, then equation (2.2) describes an ellipsoid. The normal form of the equation to an ellipsoid is

$$\frac{x^2}{a^2} + \frac{y^2}{b^2} + \frac{z^2}{c^2} = 1$$

where $a, b,$ and c are the semiaxes (Figure 2.5).
Thus the semiaxes a,b,c of the representation surface are

$$\frac{1}{\sqrt{k_1}}, \ \frac{1}{\sqrt{k_2}}, \ \frac{1}{\sqrt{k_3}} .$$

In any general direction the radius is equal to the value of $1/\sqrt{K}$ in that direction.

2.3 Symmetry control of physical properties

We have now established that variation of a physical property with direction in a crystal can be described in terms of a representation surface which, in the most general case, is an ellipsoid with semiaxes $a \neq b \neq c$. The principal axes of this ellipsoid are always orthogonal and we will denote them X, Y, Z to distinguish them from the crystallographic axes of a crystal. In describing how the symmetry of the crystal controls physical properties we need to describe the way in which the principal axes of the representation ellipsoid are related to the crystallographic axes.

The fundamental concept here is that the variation in the value of a physical property must be consistent with the point group symmetry of the crystal. Therefore, if for example a crystal has a 4-fold axis of symmetry, then the variation in the physical property must also obey the 4-fold axis. Expressed more formally, this is known as *Neumann's Principle*: the symmetry elements of any physical property of a crystal must include the symmetry elements of the point group of the crystal. We have already seen that we can describe second rank tensor properties by a representation ellipsoid even in the most general case, and as the ellipsoid has three mirror planes it therefore follows that the physical properties may have more symmetry than the crystal.

We will consider each crystal system in turn.

(i) In a *triclinic* crystal for example, there are no symmetry restrictions in the way in which the physical property varies with direction in the crystal, i.e. the ellipsoid may have any orientation with respect to the crystallographic axes.

(ii) In the *monoclinic* system, there is a diad axis of symmetry along one of the crystallographic axes (usually taken to be the y axis) and therefore one of the principal axes of the ellipsoid must also be parallel to this axis. There is no restriction however on the orientation of the other two axes, which will lie in the xz plane.

(iii) In the *orthorhombic* system the three principal axes of the ellipsoid must be parallel to the three crystallographic axes in order to satisfy the symmetry requirement.

(iv) When we consider the *tetragonal*, *hexagonal* and *trigonal* systems it is apparent that the general ellipsoid cannot satisfy the symmetry requirement of a 4-, 6- or 3-fold symmetry axis. However if the general ellipsoid had $k_1 = k_2$, it would have a circular section in the XY plane i.e. it would be an ellipsoid of revolution. The XY plane of the ellipsoid must therefore be normal to the high symmetry axis of the crystal.

(v) In the *cubic* system the representation surface can only be a sphere, i.e. $k_1 = k_2 = k_3$ and the crystal is isotropic for second rank tensor properties.

The relationship between physical properties and crystal system are most apparent when describing optical properties of crystals, which we describe in the next section. In this context the general ellipsoid is referred to as *biaxial*, while the ellipsoid of revolution is termed *uniaxial*.

2.4 Examples of symmetry control of physical properties

(a) Heat flow in a crystal

This example will serve to review some of the points discussed above, as well as provide an easily visualized demonstration of a representation surface.

Consider a point source of heat on the face of a crystal of a tetragonal mineral. The thermal gradient will result in a heat flow out from the point source, but since the thermal resistivity varies with direction this heat flow may not be equal in all radial directions. If the surface of this crystal was coated in wax, and the point source of heat was a hot wire, the wax would melt as the heat was conducted through the crystal, and the shape of this melted outline would indicate the anisotropy of the heat flow (Figure 2.6(a)).

The tetragonal symmetry constrains the thermal resistivity representation surface to be a unixial ellipsoid with a circular section normal to the c axis. When we consider heat flow on any two-dimensional surface of the crystal, as is the case here, the variation of resistivity with direction is given by the equivalent two-dimensional section through this ellipsoid.

Therefore, if the face of the crystal was the (001) plane, the variation of resistivity with direction in this plane is described by a circular section through the ellipse, i.e. in this plane the resistivity must be equal in all directions. The melting wax would form a circle around the point source of heat. On a (100) face of a tetragonal crystal the thermal resistivity will be different along the z axis than normal to it, and the melt figure would be an ellipse, the tetragonal symmetry restricting the semimajor and semiminor axes to be parallel to the crystallographic axes. In Figure 2.6(a), the resistivity along the c axis is less than that normal to it.

The heat flow direction is always radial from the hot point, P, although the magnitude varies with direction. An important observation is that the melt figure is an isothermal surface, and therefore the temperature gradient is always normal to it. It follows that, in general, the heat flow is at an angle to the thermal gradient. Only along the two principal axes of the ellipse are they parallel, just as in the mechanical analogy in Section 2.1.

Throughout this section the property of thermal resistivity has been referred to, rather than thermal conductivity. In an isotropic medium, resistivity and conductivity are reciprocal, and different experimental arrangements are used to determine each. In an experiment designed to measure the resistivity, the heat is constrained to flow along a long rod and the temperature gradient is measured. In the determination of conductivity a thin plate of material is

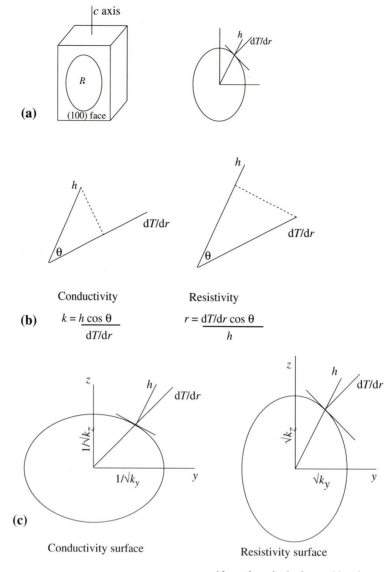

Figure 2.6. (a) A point source of heat P on the (100) face of a tetragonal crystal. Conduction in this plane results in an elliptical isothermal surface. The heat flow is always radially away from P, while the thermal gradient is normal to the isothermal surface. Thus in general, the cause vector (the thermal gradient) is not parallel to the effect vector (the heat flow). (b) Definitions of conductivity and resistivity in an anisotropic system. (c) For the derivation of the conductivity surface the heat flow is resolved in the direction of the thermal gradient, while for the resistivity surface the thermal gradient is resolved in the direction of the heat flow. Notice that in both cases the directions of the heat flow and thermal gradient are the same. In an isotropic material the resistivity and conductivity are always reciprocal, but in an anisotropic material this is only the case along the principal axes.

used so that the temperature gradient is normal to the plate, but the heat flow is unconstrained. In an isotropic medium this distinction may seem irrelevant as the heat flow is always parallel to the temperature gradient. However this is not the case in an anisotropic crystal and the distinction between resistivity and conductivity becomes important.

In the anisotropic case (see Figure 2.6(b)) we define the resistivity r as the ratio of the resolved temperature gradient onto the heat flow, i.e.

$$r = \frac{\mathrm{d}T/\mathrm{d}r \cos\theta}{h}$$

The conductivity, k is defined as the ratio of the resolved heat flow onto the temperature gradient, i.e.

$$k = \frac{h \cos\theta}{dT/dr}$$

In the long rod experiment the temperature gradient which is measured is that resolved along the direction of the heat flow, and hence it is resistivity which is being determined. In the flat plate, the heat flow is unconstrained but the temperature gradient is normal to the plate. The heat flow measured is that resolved in the direction of the temperature gradient, and the experiment determines the conductivity.

The form of the conductivity surface and resistivity surface is shown in Figure 2.6(c). Note that the conductivity and resistivity are only simple reciprocals of each other along the principal axes. The resistivity is the inverse property to conductivity.

If we now return to consider the situation of the flow of heat from a point source, (Figure 2.6(a)), the heat flow is radial from the centre while the temperature gradient is tangential to the isothermal surface, defined by the melt figure. The melt figure therefore represents the resistivity surface.

(b) The anisotropic diffusion of Ni in olivine

Diffusion of atoms through a crystal plays a fundamental role in the chemical and structural changes induced by changes in the external environment. The rate of diffusion is measured in terms of the diffusion coefficient D, which relates the flow of atoms to the concentration gradient:

$$J = -D \, dC/dx$$

where J is the flux of an atomic species across unit area of a plane normal to the concentration gradient, dC/dx (C is the concentration and x is distance). The minus sign indicates that the flow is from high to low concentrations (see Section 11.2.1). The diffusion coefficient thus relates two vectors and is a second rank tensor.

A polycrystalline solid, in which the crystals are randomly oriented, is isotropic and only one value of the diffusion coefficient needs to be specified. In a single crystal with symmetry lower than cubic, diffusion is anisotropic and in general, three principal values of D must be quoted to fully describe it. We can illustrate this with data on the diffusion of Ni atoms at 1150°C, through a single crystal of olivine, $(Mg,Fe)_2SiO_4$, which has an orthorhombic structure. The diffusion coefficients are:

$$D_x = 4.40 \times 10^{-14} \text{ cm}^2/\text{s}$$
$$D_y = 3.35 \times 10^{-14} \text{ cm}^2/\text{s}$$
$$D_z = 124.0 \times 10^{-14} \text{ cm}^2/\text{s}$$

where the values refer to the diffusion coefficient along the x, y, and z crystallographic axes respectively. The representation surface for the diffusion coefficient is therefore an ellipsoid with semiaxes

$$a{:}b{:}c = \frac{1}{\sqrt{D_x}} : \frac{1}{\sqrt{D_y}} : \frac{1}{\sqrt{D_z}} = 0.48 : 0.55 : 0.09.$$

This is shown in Figure 2.7.

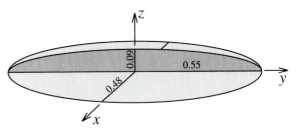

Figure 2.7. The two principal sections of the representation surface which describes the variation of the diffusion coefficient in the orthorhombic mineral olivine.

2.5 Optical properties of minerals

Optical properties of minerals and the consequences of their optical anisotropy form the basis for the identification of minerals in thin sections of rocks, and hence their study has played a central role in Mineralogy. The correlation of optical properties with chemical composition and structural state, as well as the obvious role of optical microscopy in the study of mineral intergrowths has made this simple technique the starting point in most mineralogical investigations. Many textbooks have been devoted to the practical aspects of optical microscopy, and we will not attempt to cover the same ground here. The aim of this section is to further illustrate mineral anisotropy using this important example, to point out the relationship between optical properties and the tensor properties discussed above, and finally to explain in outline some of the unique consequences of the transmission of light waves through crystals.

2.5.1 The refractive index variation in a crystal – the optical indicatrix

The refractive index of an isotropic medium is defined as the ratio of the velocity of light *in vacuo* to that in the medium,
i.e. $n = c/v$.

In an isotropic medium such as glass, there is a single value of the refractive index and the velocity of light through the glass is independent of the orientation of the incident beam. In an anisotropic crystal this is not the case, and the velocity and hence the refractive index may vary with direction.

In order to be able to use the theory of the tensor properties of crystals described above, we must express the optical properties in terms which relate two vectors, one the cause vector and the other the effect vector. Before we go any further, certain assumptions regarding the nature of light and its interaction with matter will have to be very briefly outlined.

1. Light is an electromagnetic wave motion consisting of an electric vector **E** and a magnetic vector **B** at right angles to it. Both are normal to the direction of propagation of the light wave. Only the electric vector **E** is important in the transmission of light through a transparent medium.

2. In plane polarised light the electric vector **E** is fixed to vibrate in a single direction. Plane polarised light is produced by passing ordinary light through a polaroid filter, or other polarising material.

3. The electric vector of the incident light beam induces a response in the crystal which involves a distortion of the internal charge distribution within the crystal. This generates electric dipoles (the separation of positive and negative charges). The generation of these electric dipoles produces a net dipole moment per unit volume, termed the *electric polarisation* **P**. The electric dipoles are forced into oscillation by the applied field **E** and act as sources of secondary wavelets. In this way the light is propagated through the medium. The relationship between **E** and **P** is via the permittivity ε:

$$(\varepsilon - \varepsilon_o)\, \mathbf{E} = \mathbf{P}$$

where ε is the permittivity of the medium and ε_o is the permittivity of a vacuum.

4. The existence of the electric dipoles contributes an additional electric field component within the medium, and this results in an electric flux density **D** inside the medium given by

$$\mathbf{D} = \varepsilon\, \mathbf{E}$$

5. The relative permittivity of the medium $\varepsilon/\varepsilon_o$ is usually termed the *dielectric constant K* of the medium. Hence

$$\mathbf{D} = \varepsilon_o K \mathbf{E}$$

In terms of our previous general discussion, **E** is the cause vector and **D** is the effect vector. In an isotropic material, *K* is a constant value i.e. an isotropic material has a single value of the dielectric constant, and **D** is always parallel to **E**. Since the dielectric constant of a crystal is a second order tensor property, in an anisotropic material it must be specified by three principal values which can be represented by a representation surface just as we have already described for the other properties above. However, the dielectric constant is not a familiar property, and it is much more useful to express the optical properties in terms of the *refractive index* variation with direction.

6. The refractive index of a non-magnetic material is equal to $\sqrt{K}$.

The representation surface for the dielectric constant would in the most general case be an ellipsoid with semiaxes, $1/\sqrt{K}$ for each principal axis. The equation to the ellipsoid has the general form

$$\frac{x^2}{a^2} + \frac{y^2}{b^2} + \frac{z^2}{c^2} = 1$$

where a, b, c are the semiaxes. Therefore, $a = 1/\sqrt{K_x}$, $b = 1/\sqrt{K_y}$ and $c = 1/\sqrt{K_z}$.

However, we wish to construct an ellipsoid which describes the variation of refractive index such that the semiaxes are the three principal values of the refractive index α,β,γ i.e. an ellipsoid whose equation is

$$\frac{x^2}{\alpha^2} + \frac{y^2}{\beta^2} + \frac{z^2}{\gamma^2} = 1$$

Since $\alpha = \sqrt{K_x}$, $\beta = \sqrt{K_y}$ and $\gamma = \sqrt{K_z}$ such an ellipsoid would be the representation surface for the property which is the inverse to the dielectric constant, i.e. $1/K$, which is termed the relative dielectric impermeability.

The ellipsoid describing the variation of refractive index is a very useful way of visualizing the optical anisotropy in a mineral, although it is not, strictly speaking, a representation surface for the refractive index, since refractive index is not the second rank tensor property. This ellipsoid is generally known as the *optical indicatrix*.

2.5.2 Symmetry control on the variation of refractive index in a crystal

The situation is exactly the same as the general case described in Section 2.3.

1. A cubic crystal is optically isotropic and hence the variation in refractive index with direction is described by a sphere. Only one value of refractive index is needed.

2. In tetragonal, hexagonal and trigonal crystals the indicatrix is *uniaxial* which means that it is an ellipsoid of revolution with the circular section of the ellipsoid normal to the 4-fold, 6-fold or 3-fold axis of the crystal (Section 2.3). Two principal values of refractive index must be specified. It is traditional, for reasons which will be mentioned below, to denote these two principal values in a uniaxial crystal as *o* (for *ordinary*) and *e* (for *extraordinary*). The *o* refractive index is that in the *xy* crystallographic plane of the crystal, the *e* refractive index is along the *z* axis (Figure 2.8(a)).

3. For crystals of lower symmetry three principal values of refractive index must be specified. In this case the indicatrix is termed *biaxial* (Section 2.3). These principal values are generally referred to as α, β and γ where $\alpha < \beta < \gamma$ (Figure 2.8(b)). In orthorhombic crystals the three principal values of refractive index must coincide with the three crystallographic axes *x,y,z* but not necessarily in that order.

4. In monoclinic crystals one of the principal axes of the indicatrix must be parallel to the *y* crystallographic axis, due to the diad axis which must be satisfied by the indicatrix orientation. The other two principal values must therefore lie in the *xz* plane but are not constrained by symmetry requirements.

5. In triclinic crystals there is no constraint on the orientation of the indicatrix relative to the crystallographic axes.

It is often helpful to imagine the indicatrix as a surface within the crystal and it has often been depicted in this way as an aid in visualizing the orientational relationships between the crystallographic axes and the indicatrix. Figure 2.9 shows two such illustrations, both for uniaxial crystals. In quartz the indicatrix is elongated along the *c* axis because *e > o*. Such a crystal is referred to as *optically positive*. In calcite, the opposite is the case, and hence calcite is *optically negative*.

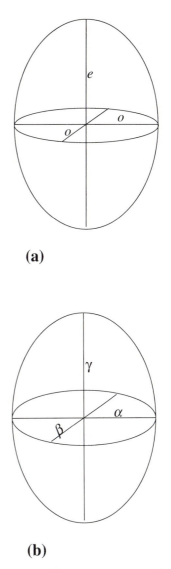

(a)

(b)

Figure 2.8. The variation of the refractive index with direction in a crystal is described by an ellipsoid termed the optical indicatrix. (a) The uniaxial indicatrix which is an ellipsoid of revolution with two principal values *o* and *e*. When *e > o* the crystal is termed optically positive, as shown here. The uniaxial indicatrix describes the variation of refractive index in the tetragonal, trigonal and hexagonal crystal systems. (b) The general case, appropriate for orthorhombic, monoclinic and triclinic systems, is the biaxial indicatrix in which the principal values are $\gamma > \beta > \alpha$.

The cubic system is optically isotropic and the representation surface is a sphere.

The distinction between optically positive and negative biaxial crystals is also based on the shape of the indicatrix but with three principal values is defined somewhat differently. This will become apparent after the following few paragraphs and the introduction of the concept of an *optic axis*.

In a uniaxial crystal, light travelling along the *z* crystallographic axis must have the electric vector **E**

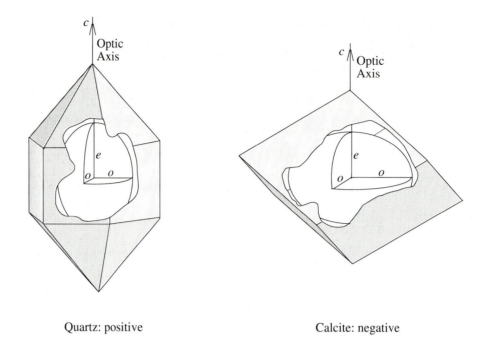

Quartz: positive   Calcite: negative

Figure 2.9. Visualisation of the relationship between the optic orientation of a uniaxial indicatrix and the crystallographic orientation : (a) in quartz, which is optically positive, and (b) calcite, which is optically negative. The principal value *e* is always parallel to the *c* crystallographic axis.

vibrating in the *xy* plane, where the refractive index is always the same. Hence for this light beam the crystal will appear isotropic. The axis normal to an isotropic section is termed the *optic axis*. In a biaxial crystal there will be two incident beam orientations which will encounter isotropic sections, as Figure 2.10 shows. Due to the fact that $\alpha < \beta < \gamma$, a section of the indicatrix containing α and γ must have values of refractive index equal to β somewhere between its two principal values. There must therefore be two circular sections of the ellipsoid with radius β, and hence two optic axes. The angle between these two optic axes is written as $2V\gamma$ where the γ refers to the fact that the γ axis of the indicatrix bisects $2V$. When $2V\gamma < 90°$ the biaxial crystal is optically positive; for $2V\gamma > 90°$ the crystal is optically negative.

The origin of the terms uniaxial and biaxial refer to the existence of one and two optic axes respectively.

2.5.3 Birefringence

The optical behaviour of crystals is a consequence of a remarkable property of anisotropic crystalline materials. Stated simply, *an incident light ray passing into an anisotropic crystal gives rise to two refracted rays, rather than one. These two refracted rays are*

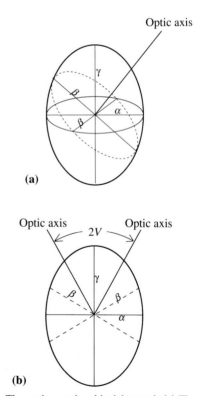

Figure 2.10. The optic axes in a biaxial crystal. (a) The optic axis is defined as that normal to a circular section of the indicatrix. Since $\gamma > \beta > \alpha$, the optic axis lies in the plane defined by γ and α, normal to a circular section with radius β. (b) A section through the indicatrix containing the principal values γ and α and the traces (dashed lines) of the two circular sections. The angle between the two optic axes is defined as the *optic axial angle 2V*.

Figure 2.11. Double refraction in calcite. Light from a single spot beneath the calcite crystal is split into two refracted beams on passing through the crystal. The two beams are plane polarised at right angles to each other and travel with different velocities due to the difference in the refractive indices of these two permitted vibration directions.

always plane polarized at right angles to one another, and hence each encounters a different refractive index. If the two refractive indices are different enough, and the crystal is thick enough, the two beams can be readily separated due to the difference in the degree of refraction. This phenomenon of *double refraction* is best illustrated by the mineral calcite, as shown in Figure 2.11. All anisotropic crystals have this property to some extent, but in calcite the difference in the two refractive indices is particularly large. The numerical difference between the two refractive indices encountered by an incident beam as it passes through a crystal is termed the *birefringence*.

Although double refraction is easy to demonstrate experimentally, it is neither intuitively obvious nor is it easy to justify theoretically without an understanding of Maxwell's theory of electromagnetic radiation and then its application to anisotropic media. We will accept the bold statement above as axiomatic; see the bibliography section for more details.

It is important to be able to interpret birefringence in terms of the relationship between the orientation of the incident light ray and the orientation of the indicatrix, and hence to see how the birefringence depends on orientation. Note that the electric light vector **E** which interacts with the crystal is always

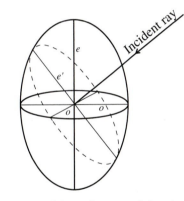

Figure 2.12. When a light ray in a general direction is incident on a uniaxial mineral the two operative refractive indices are given by the semimajor and semiminor axes of the elliptical section normal to the light ray. This elliptical section of the uniaxial ellipsoid is shown by the dashed line, and the two relevant refractive indices are given by o and e'.

normal to the ray direction, and that the two refracted beams vibrate at right angles to one another. The plane in which the two refracted beams vibrate is that plane normal to the incident beam direction.

Figure 2.12 illustrates this with an incident beam relative to a uniaxial indicatrix – we choose uniaxial in the first instance for geometrical convenience. The plane normal to the incident beam is drawn through the centre of the indicatrix. This plane defines a two-dimensional section elliptical through the indicatrix, shown as a dashed line in Figure 2.12. The **E** vibrations of the two refracted rays must lie at right angles within this elliptical section. *The two principal axes of this elliptical section define the two vibration directions, and the lengths of the semimajor and semiminor axes of the ellipse define the two refractive indices.*

It follows then that if the incident beam direction was at right angles to the optic axis in Figure 2.12, the relevant elliptical section would contain the optic axis and the two refractive indices would be *e* and *o*. This section would have the maximum birefringence. For an incident beam parallel to the optic axis the section through the indicatrix is circular and this section is isotropic (i.e. zero birefringence). For any general direction in a uniaxial indicatrix, one of the refractive indices must always be *o*, while the other will have a value between *e* and *o*, and this is usually denoted *e'*. A method for calculating the value of *e'*, given the principal refractive indices, is outlined in Appendix A to this chapter.

For biaxial crystals, the principles are the same and the elliptical section normal to the incident ray contains the vibration directions and refractive indices for the two refracted beams. The geometry is more complex because in a general section neither of the two refractive indices will be one of the principal values, and their determination becomes a little more cumbersome. References are given in the

Bibliography for methods of determining the refractive indices in a general biaxial section.

Finally, we need to consider one further aspect of this general treatment of the double refraction phenomenon. So far we have not imposed any restriction on the polarization state of the incident beam. However, if we wish to describe the phase relationships between the two refracted beams, the incident beam must be plane polarized. A plane polarised incident beam is produced by passing ordinary light through a polaroid filter (the polariser) before it enters the crystal (Figure 2.13).

When this polarised incident beam enters the mineral section it is split into two refracted beams, each initially with the same phase. As the two beams pass through the mineral, each experiencing a different refractive index, the difference in their velocity will cause one beam to lag behind the other and a phase difference will develop between them. We will discuss the consequences of this phase difference in the next section, but here we need to point out how this modifies some aspects of the discussion above.

Let us consider Figure 2.12 again. If the polarization direction of the **E** vector of the incident beam was in some general direction relative to the indicatrix, the situation would be the same as that described above, i.e. the incident beam would be split into two components, vibrating along the principal axes of the elliptical section. If however the incident electric vector was parallel to either of these

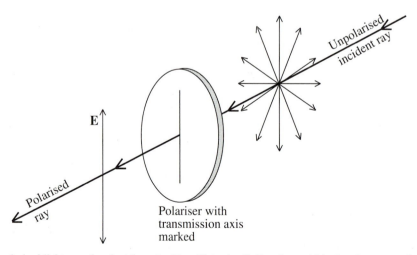

Figure 2.13. In an unpolarised light ray, the electric vector **E** oscillates in all directions within the plane normal to the beam direction. When an unpolarised light ray passes through a polariser only the components of **E** along the transmission axis of the polaroid are transmitted, resulting in a polarised ray of light.

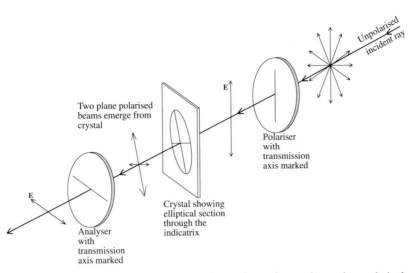

Figure 2.14. When polarised light passes, in a general direction, through an anisotropic crystal, two plane polarised rays are produced, their **E** vectors along the directions of the axes of the elliptical section through the indicatrix. When these rays reach a second polaroid (the analyser) only the components along the transmission axis of the analyser are transmitted. This combines the two rays, which have a phase difference imposed on them after passing through the crystal.

two principal axes only this component would be transmitted, and double refraction would not occur. By using a plane polarized incident beam and changing the orientation of the crystal, the two orthogonal components can be observed separately, as well as together.

To analyse the polarisation direction of light passing through a crystal, another polaroid filter (the analyser), is placed in the light path after the crystal, as shown in Figure 2.14. The polarisation direction of the analyser is set at right angles to the polariser (the situation is termed *crossed polars*). With this arrangement we can now distinguish between a number of possible situations.

1. If the crystal is isotropic, the incident **E** vector will pass through the crystal irrespective of its orientation, but will be blocked by the analyser. No light will be transmitted beyond the analyser. The crystal is said to be in *extinction*. An isotropic crystal is always in extinction, in this sense.

2. With an anisotropic crystal, the general case is that the incident **E** vector will be split into two orthogonal components. However if the incident **E** vector is parallel to one of these directions only one component will be transmitted, the polarisation state will remain unchanged, and the analyser will block the beam. This is an *extinction position*.

3. There will also be another extinction position when the crystal is rotated through 90° so that the incident **E** vector is parallel to the other component.

All anisotropic crystals have two orthogonal extinction positions when the polars are crossed.

Thus the extinction positions indicate that only one component is being transmitted. This is the method used to determine the orientation of each component relative to the crystallographic axes, and to isolate each component when measurements of refractive index are made.

The general case, when the polarisation direction of the incident beam is at some angle to the polarisation directions of the two refracted beams is described in Appendix B to this chapter, and its consequences are discussed in the following section.

2.5.4 The birefringence of minerals in thin section – interference colours

Minerals in rocks are usually studied in thin sections (polished to 30 μm thickness) using a polarizing microscope which produces a plane polarized incident beam. Unlike calcite, the principal refractive indices of most minerals are similar to one another, and in thin sections the double refraction effect will not be noticeable in the same way as in the calcite demonstration in Figure 2.11. Instead, the two refracted beams interfere with one another, and the difference in the refractive index and hence velocity of each results in a phase difference between them as they emerge from the thin section. The extent of this phase difference depends on the difference in the two velocities and hence refractive indices (i.e. on the birefringence), and on the thickness of the crystal section.

The two emerging beams, with **E** vectors orthogonal and with a phase difference imposed on them by the mineral, are recombined in the analyser which is set at right angles to the polarization direction of the incident beam (see Figure 2.14). When using white light, the phase difference between the two beams will result in the destructive interference of some wavelengths and lead to a characteristic *interference colour*. When minerals are studied in thin sections of standard thickness, the interference colours can be related directly to their birefringence.

2.6 A note on elastic properties of minerals

At the beginning of this chapter we mentioned that elastic properties were in a different category to the second rank properties described above, and although we will not discuss elastic properties in detail, it is worthwhile to make some brief reference to them. Elastic properties describe the relationship between an external force applied to a crystal and the consequent change in its shape. Elasticity describes the regime in which the removal of the force results in a recovery of the original shape. Permanent deformation i.e. plasticity occurs when a material exceeds its elastic limit, or yield strength.

The external force acting on a crystal is defined in terms of the *stress*, or force per unit area acting on it. The applied force is transmitted through the bulk of the crystal and to describe its effect on the surfaces of some small volume element within the crystal requires a description of the stress as a tensor property. A more complete analysis of this situation can be found in books cited in the Bibliography, and here we merely state that by choosing a suitable coordinate system, the stress tensor can be described in terms of thee principal stresses, σ_1, σ_2 and σ_3.

The effect of the applied stress is to cause a displacement of the atomic positions within the crystal. If this displacement was the same for every atom, the whole crystal would be merely translated. If however the displacement within the crystal is not uniform, the distance between any two points changes and consequently there is a change in the dimensions and shape of the crystal. This change in shape is defined by the *strain*, which is also a tensor property since the extent of the deformation depends on direction within the crystal, i.e. on the extent of anisotropy. If the distortion is the same everywhere within the crystal, the strain is said to be homogeneous and a spherical volume element within the crystal is deformed into an ellipsoid. A sphere of unit radius deforms into an ellipsoid termed the strain ellipsoid, whose principal axes define the principal values ε_1, ε_2, ε_3, of the strain tensor.

Within the elastic limit the stress and strain are linearly related by the *elastic constants* of the crystal. Since the elasticity relates two tensor properties, it is itself defined as a fourth rank tensor, and the number of values which must be specified depends on the symmetry of the structure. In the most general triclinic case there are 21 independent constants but these reduce to three for cubic crystals. For isotropic solids such as glasses or polycrystalline aggregates such as rocks, there are only two elastic constants, Young's modulus (σ/ε) and Poisson's ratio ($-\varepsilon'/\varepsilon$, where ε is the strain in the direction of the applied force, and ε' is the strain at right angles). Further discussion of elastic properties can be found in the more specialized texts referred to in the Bibliography.

2.7 The relationship between anisotropy and crystal structure

Throughout this chapter we have taken a macroscopic approach and have correlated anisotropy with the symmetry of the crystal system rather than with the atomic arrangement itself. Obviously the origin of anisotropic properties must ultimately be able to be explained in terms of the atomic structure.

In some cases the relationship between crystal structure and anisotropy is fairly obvious. Graphite for example has a layer structure (Figure 2.15(a)) and the properties within the layer are very different from those normal to the layers. Within the layers, carbon atoms are strongly bonded, with a C–C interatomic distance of 1.42Å. The layers are held together by weak bonds, and the nearest C–C distance between the layers is 3.35Å. These layers are easily disrupted mechanically, giving graphite its softness. This extreme anisotropy is responsible for an electrical resistivity of 5×10^{-3} Ωm perpendicular to the layers and 5×10^{-6} Ωm parallel to the layers, i.e. graphite is a semiconductor perpendicular to the layers and a metallic conductor parallel to the layers.

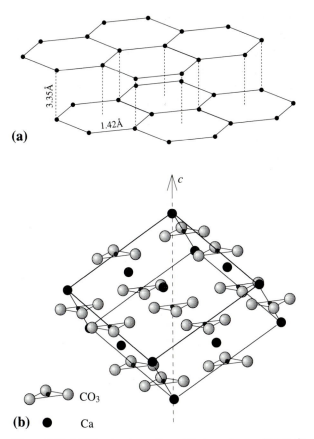

(a)

(b)

Figure 2.15. (a) The structure of graphite is made up of layers in which the carbon atoms lie at the corners of a hexagonal mesh. Within the layers the carbon atoms are strongly bonded with a C-C distance of 1.42Å. The layers are weakly bonded to each other, and the shortest C-C distance between atoms in adjacent planes is 3.35Å. This anisotropy in the structure results in strongly anisotropic properties. (b) The structure of calcite, $CaCO_3$, can be described in terms of a rhombohedron (essentially a cube which has been shortened along one of the triad axes). The Ca atoms have a face-centred distribution on this rhombohedron, and the triangular CO_3 groups lie at the centres of each edge. This results in layers of CO_3 groups lying normal to the *c* axis of calcite, with layers of Ca atoms lying between them. This layering of the CO_3 groups is responsible for the marked optical anisotropy of calcite.

Similarly the thermal expansion coefficient at 600K is 28.3×10^{-6} K^{-1} perpendicular to the layers and virtually zero parallel to the layers.

In Section 2.6 we illustrated optical anisotropy with the example of calcite. The structure of calcite (trigonal, therefore uniaxial) consists of planar CO_3 groups arranged in layers perpendicular to the *c* axis (Figure 2.15(b)). The Ca atoms lie between the layers of CO_3 groups. Within the planes containing the CO_3 groups the distortion of the internal electric field by an applied light vector **E** is markedly different from that normal to the planes, due mainly to the effect of the interaction of neighbouring oxygen

atoms. The refractive index for an **E** vibration within the planes (1.658) is much larger than for light vibrating at right angles to the planes (1.486) resulting in the large birefringence.

These two rather extreme examples serve to illustrate the point that in a qualitative way at least, it is possible to relate the relative magnitudes of properties to crystal structure. As the degree of anisotropy in the structure becomes less the problem becomes more subtle. However, using the computational methods discussed in Section 8.5.4 it is possible to calculate absolute values of the dielectric constants (and hence the refractive indices) for most minerals, from a knowledge of their structures and the polarizability of the constituent atoms.

As our final topic on anisotropy in minerals we return to a point made at the beginning of this chapter regarding surface related properties, which are non-directional but yet do depend on orientation relative to the crystal structure. The most obvious of these is the external shape of crystals i.e. their morphology. When a crystal grows freely into a fluid medium the orientational relationship between external faces reflects the symmetry of the internal structure (Figure 1.2). The study of the morphology of minerals was once a central theme in Mineralogy, especially when the external form was the only clue to the internal symmetry. The advent of diffraction techniques, and the fact that minerals with well developed faces are rare and reside in mineral collections, not rocks, has diminished interest in morphological studies. However, the key question of why a mineral grows with a particular morphology is not well understood, and the increasing emphasis on the structure of mineral surfaces and their reactivity makes some reference to this problem relevant.

2.8 Anisotropy and the external shape of minerals

The scientific interpretation of morphology began with Bravais in around 1850, who stated that the most prominent faces of a crystal should be those with the highest density of lattice points. In a cubic crystal with a primitive lattice the list of planes with decreasing lattice point densities is {100}, {110}, {200}, {210},... This is the same as the list of planes with decreasing interplanar spacing d_{hkl}. Such a crystal would be expected to show these faces in decreasing order of prominence. In a non-primitive lattice, planes with the highest number of lattice points do not have the largest d_{hkl} spacings, merely because the Miller indices hkl are not defined in relation to the smallest possible unit cell (see Figure 1.4 for example). In a body-centred cubic crystal the {110} faces have the highest lattice point density, followed by {200}, {211},... In a face-centred cubic crystal the sequence is {111}, {200}, {220}, ...

Although there are many instances where this sequence fits the observed occurrence of crystal faces, the discrepancies led Donnay and Harker, in 1937 to extend the theory to include not just the translational symmetry of the lattice but also the space group symmetry, by taking into account the density of points made equivalent by translational symmetry elements such as glide planes and screw axes. These theories can be regarded as only approximate because they take no direct account of atomic arrangement or bonding, but more seriously, they are not applicable to the prediction of surface structure.

The first attempt to relate morphology to the actual crystal structure was proposed by Hartman and Perdok (1955) and has since been refined further so that it is now the basis on which future studies will develop. In Hartman's approach crystal morphology is analysed in terms of structurally important directions in a crystal rather than the density of planes. A structurally important direction is one which contains an uninterrupted chain of strong bonds which form on crystallisation.Such a chain is called a periodic bond chain (PBC). The strong bonds are thermodynamically preferred i.e. more stable, and might therefore be expected to form faster than weaker bonds. In its simplest form this can be demonstrated by the observation that the minerals which tend to grow as long fibres (e.g. asbestos) do have chains of strongly bonded atoms running parallel to the fibre axis. At the other extreme, platy crystals such as mica have networks of strong bonds lying in sheets, with weaker bonds between the sheets. In more complex structures the recognition of PBC's is less obvious, and other formulations of similar ideas have been proposed (see Bibliography).

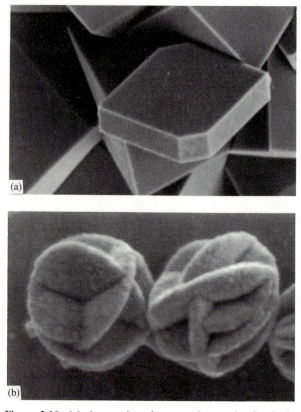

Figure 2.16. (a) A scanning electron micrograph of a barite ($BaSO_4$) crystal grown by the controlled diffusive mixing of Ba^{2+} and SO_4^{2-}. (b) A multiply nucleated barite crystal with curved surfaces, grown under the same conditions as above but with the addition of 100 parts per million of an organic polymer. The size of the crystal in each case is $\sim 2\mu m$.

Growth morphology can be altered by the presence of trace amounts (~ 10 parts per million) of certain impurity atoms or molecules which are adsorbed onto the surface of growing crystals. In terms of the PBC theory it is likely that these molecules are selectively adsorbed and disrupt the periodic bond chain, reducing the rate of growth and possibly favouring the development of other growth directions. A practical application of this effect is the use of chemical inhibitors which modify the morphology (Figure 2.16) or entirely prevent the growth of unwanted mineral scale deposits from supersaturated solutions in industrial processes (e.g $CaCO_3$ deposits from hard water, $BaSO_4$ deposits in paper production etc.).

Appendix A
How to determine the birefringence of a uniaxial mineral for any general orientation of the incident light ray

In a uniaxial mineral the refractive indices are defined by two principal values, o and e, normal and parallel to the optic axis respectively. The variation of refractive index with direction is described by the uniaxial indicatrix shown in Figure 2.17(a). An incident beam in any general direction will be split into two refracted beams, plane polarized at right angles to each other and in the plane normal to the incident beam. This plane is drawn as a section through the origin of the indicatrix and defines an elliptical section through the indicatrix, shown in Figure 2.17(a) by the dashed line. The two refractive

indices we are trying to determine are given by the semimajor and semiminor axes of this dashed ellipse.

In a uniaxial mineral, one of these values must be the o vibration, whatever the orientation of the incident beam. The other value must lie somewhere between o and e, denoted here as e'.

To determine e', given the values of o and e, we draw a section of the indicatrix containing e, e' and o. This is shown in Figure 2.17(b). The equation of an ellipse with semiaxes o and e is:

$$\frac{x^2}{o^2} + \frac{z^2}{e^2} = 1$$

If the angle between the incident beam direction and the optic axis is θ, the coordinates of the point P on the surface of the ellipse are $(-e'\cos\theta, e'\sin\theta)$. Since P lies on the ellipse we substitute these values into the above equation giving

$$\frac{(e'\cos\theta)^2}{o^2} + \frac{(e'\sin\theta)^2}{e^2} = 1$$

Since $\cos^2\theta = 1 - \sin^2\theta$

$$\left(\frac{1}{e^2} - \frac{1}{o^2}\right)\sin^2\theta = \frac{1}{(e')^2} - \frac{1}{o^2}$$

from which e' can be calculated.

Example. In an optically positive tetragonal crystal with $a = 4$Å, $c = 8$Å, a thin section is cut parallel to (011) planes. When viewed normally in plane polarised light, the two values of refractive index associated with the two extinction positions (i.e. the values associated with the two orthogonal vibration directions) are 1.65 and 1.55. What are the principal values of the refractive index for this material?

Solution. One of the measured values of the refractive index must be the o value, and since the crystal is optically positive it must be the smaller value.

Therefore $o = 1.55$.

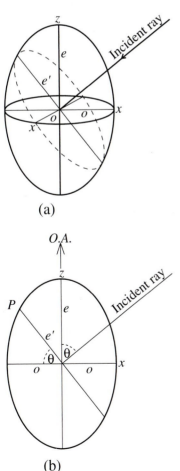

Figure 2.17   (a)   (b)

The angle θ must be calculated geometrically from the lattice parameters:

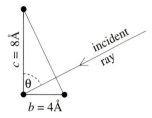

The angle θ is 63.4°, and the coordinates of point P in Figure 2.17 above are therefore $(-1.65\cos\theta, 1.65\sin\theta)$.

Substituting this into the equation for an ellipse gives $e = 1.68$

Therefore the principal refractive indices are $o = 1.55$ and $e = 1.68$.

Appendix B
Analysing polarisation directions as plane polarised light passes through a crystal section

Consider the general case shown in the figure below, in which a plane polarised ray of light, travelling up out of the page has a polarisation direction shown by the vector **E**, and passes through a crystal slice. The two allowed vibration directions in the crystal (defined by the semimajor and semiminor axes of the elliptical section in the plane of the crystal slice) are shown as OX and OY, and the angle ϕ is the angle between the **E** vibration and OX.

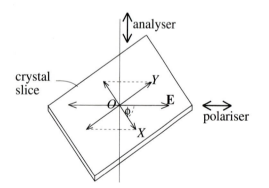

The polarised light wave incident normally on the lower face of the crystal slice can be represented by
$y = A \sin \omega t$
where A is the amplitude of the light, ω is the frequency and t is the time measured with respect to some arbitrary zero.

The components of the incident beam along the permitted vibration directions OX and OY are
$$y_1 = A \cos\phi \sin \omega t$$

and
$$y_2 = A \sin\phi \sin \omega t$$
respectively.

Since the refractive indices along OX and OY are different, the velocities of the two waves will be different and the waves will emerge from the top of the crystal with a phase difference δ.

At the top of the crystal
$$y_1 = A \cos\phi \sin \omega t$$
$$y_2 = A \sin\phi \sin (\omega t + \delta)$$

The waves will travel with constant phase difference until they enter the analyser. Here only the component of each wave which vibrates along the transmission axis of the analyser will be transmitted, and the resultant transmitted wave will be the sum of these two components.

If the analyser is crossed with respect to the polariser, the components of the two transmitted waves in the direction of the analyser are
$$y_1' = A \cos\phi \sin \omega t \cos (90 + \phi)$$
$$y_2' = A \sin\phi \sin (\omega t + \delta) \cos \phi$$

so that the resultant wave polarised in the direction of the analyser is

$$y_R = A \cos\phi \sin \omega t \cos (90 + \phi) + A \sin\phi \sin (\omega t + \delta) \cos \phi$$
$$= A \cos\phi \sin\phi [\sin(\omega t + \delta) - \sin\omega t]$$
$$= A \sin2\phi \cos (\omega t + \frac{\delta}{2}) . \sin \frac{\delta}{2}$$

The intensity (I) of the transmitted wave (which is proportional to the square of the amplitude) is therefore

$$I = A^2 \sin^2 2\phi \sin^2 \frac{\delta}{2}.$$

When $\phi = 90°$ i.e. when the incident vibration direction is parallel to one of the allowed vibration directions in the crystal, the intensity of the transmitted wave is zero, i.e. the crystal is in an *extinction position*.

The phase difference δ

If the refractive indices corresponding to the two vibration directions OX and OY are μ_1 and μ_2 respectively, and the crystal thickness is d, then the optical path difference introduced between the two waves will be $(\mu_2 - \mu_1)d$.

The phase difference is therefore given by

$$\delta = \frac{2\pi}{\lambda}(\mu_2 - \mu_1)d$$

Hence the intensity of light transmitted when a crystal is viewed between crossed polars in monochromatic light is

$$I = A^2 \sin^2 2\phi \sin^2 \frac{\pi d}{\lambda}(\mu_2 - \mu_1)$$

The maximum intensity is observed when $\sin^2 2\phi = 1$, i.e. when $\phi = 45°$.

In white light each component wavelength has a different value of δ. The intensity of each component wavelength in the transmitted beam is modified on passing through the crystal so that the resultant beam has an interference colour. The interference colour depends on the thickness of the crystal, d, and the value of $(\mu_2 - \mu_1)$.

Bibliography

DOWTY, E. (1976). Crystal structure and crystal growth: I. The influence of internal structure on morphology. *Amer. Mineral.* **61**, 448–459.

HARTMAN, P. (1987). Modern PBC theory. In: *Morphology of crystals* (I. Sunagawa ed.) 269–319. Terra Scientific Pub. Co.

HARTMAN, P. AND PERDOK, W.G. (1955). On the relations between the structure and morphology of crystals I: *Acta Cryst.* **8**, 49–52 and **8**, 521–524.

HECHT, E. (1987). *Optics*. Addison-Wesley.

KELLY, A. AND GROVES, G.W. (1970). *Crystallography and crystal defects*. Longman.

LOVETT, D.R. (1989). *The tensor properties of crystals*. Adam Hilger.

McKIE, D. AND McKIE, C. (1974). *Crystalline solids*. Nelson.

NYE, J.F. (1985). *Physical properties of crystals*. Oxford University Press.

POIRIER, J-P. (1985). *Creep of crystals*. Cambridge University Press.

SUNAGAWA, I. (1987). Morphology of minerals. In: *Morphology of crystals* (I. Sunagawa ed.) Terra Scientific Pub. Co.

WOOSTER, W.A. (1973). *Tensors and group Theory for the physical properties of crystals*. Oxford University Press.

3 Diffraction and imaging

This chapter is concerned with the way in which we can determine the positions of the atoms in a mineral structure. In the process we will also determine the size of the unit cell and the symmetry elements present. Structure determination is essentially an 'imaging' process – our aim is to produce a picture showing the relative distribution of the atoms in a unit cell. An obvious question is to what extent could we do this with a very powerful microscope? Figure 1.1 is in fact such an image of the distribution of atoms in a structure. It is a high resolution electron microscope image of the projection of the structure of hexagonal cordierite, $Mg_2Si_4Al_5O_{18}$, looking down the c axis. Although the unit cell is easily imaged, as well as some aspects of the atomic distribution within the unit cell, there is nowhere near the detail required to 'see' the 29 atoms in the formula unit. This is a fundamental limitation of electron microscopy, and is not merely due to the microscope not being 'powerful enough' to have the necessary resolution. To understand the reasons for this limitation we need to discuss the imaging process in some detail, and this will lead us through the principles involved in the determination of crystal structure using X-ray methods.

3.1 The ideal imaging system – Abbe theory

Throughout this chapter we will illustrate the principles involved in structure determination by using an optical analogy of a simple imaging system consisting of an object, a lens and an image (Figure 3.1). The early history of optical microscopy was dominated by the belief that the resolving power of a microscope was limited only by the quality of the lens, a view brought to an end by Ernst Abbe in 1873 who proposed a theory of image formation which will be the basis of our treatment of the topic.

In the Abbe theory, image formation is described as a two-stage process.

(i) The object scatters the incident light by diffraction, and the objective lens collects some of these diffracted rays and focuses them on the focal plane (Figure 3.1). This is the diffraction pattern of the object. *This diffraction pattern contains all the information about the object which will appear in the image.*

(ii) The diffracted beams propagate beyond the focal plane and arrive at the image plane. There they overlap and interfere to form an inverted image of the object.

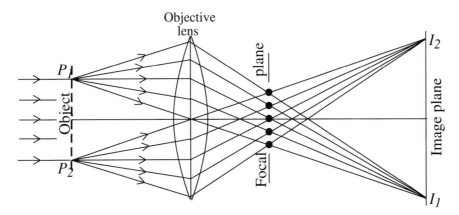

Figure 3.1. Simple imaging system illustrating the two stage process in image formation: first, the formation of the diffraction pattern in the focal plane of the objective lens, and second, the recombination of the diffracted beams to form the image.

In Figure 3.1 the object is periodic, e.g. a grating, and for simplicity, diffracted beams are shown from only two points P_1 and P_2 which are imaged at I_1 and I_2, respectively. A non-periodic object can be treated in the same way – the only difference being that in a periodic object the scattered light has intensity maxima in specific directions, while in a non-periodic object the scattered light is continuous. We will return to this below. The role of the objective lens is to collect the diffracted rays from the object, and hence the objective lens must have an aperture large enough to collect as much of this information as possible. Any rays diffracted at an angle larger than can be collected by the objective lens represent information which will not appear in the image. Thus the imaging process is concerned firstly with diffraction of the incident light within the object and secondly its collection. We need therefore to turn our attention to the diffraction process.

3.2 Optical diffraction patterns

We will consider diffraction as a process where the incident beam interacts with the object, and the scattering process transfers spatial information about the object to the beam. The simplest way to see this is initially to consider the object to have a periodic structure, and hence behave as a diffraction grating. The conclusions we draw however, will be applicable to all objects.

Consider a one-dimensional grating, such as a set of parallel slits all with the same width and regularly spaced, illuminated by light from a laser beam (Figure 3.2). This is similar to the well-known Young's double slit experiment, in which each slit acts as a source of secondary wavelets which overlap to produce an interference pattern. The pattern is produced because there are only certain directions for which constructive interference between the wavelets can take place. As shown in Figure 3.2(a), the condition for constructive interference, i.e. for a diffraction maximum to occur, is that the diffracted beams from each slit are in phase. The path difference between the beams from each slit must therefore be equal to an integral number of wavelengths:

$$a \sin \phi = h \lambda \qquad (3.1)$$

where a is the spacing between slits, ϕ is the angle

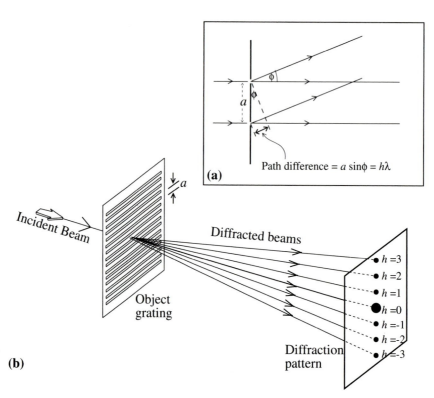

Figure 3.2. The formation of an optical diffraction pattern from a one-dimensional grating illuminated by a laser beam. (a) The condition for constructive interference producing a diffraction maximum, (b) Schematic of the diffraction experiment and the formation of a row of diffraction spots.

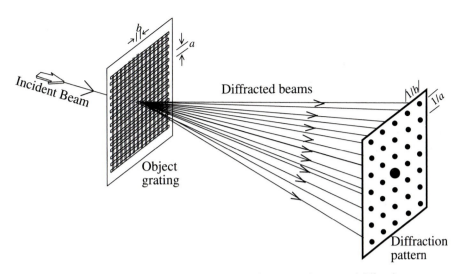

Figure 3.3. Diffraction from a two-dimensional grating and the formation of a rectangular array of diffraction spots.

through which the incident beam is diffracted, λ is the wavelength of the radiation and h is an integer termed the order of diffraction. There will be a number of orders $h = 1, 2, 3,..$ of diffracted beams and the resultant diffraction pattern will appear as an array of spots on a screen placed at some distance L from the object. From simple trigonometry it can be shown that the spacing d between the spots in the diffraction pattern is reciprocally related to the spacing of the slits:

$$d = L \tan \phi$$

$$L = \lambda \, \frac{\sin \phi}{\cos \phi}$$

$$d = \frac{L\lambda}{a \cos \phi} \qquad (3.2)$$

The conclusions we can draw from this simple experiment are as follows.

1. For a diffraction pattern to be formed, a cannot be smaller than λ, otherwise $\sin \phi$ becomes greater than unity. Also if a is very much larger than λ, the diffracted beams will be too close to the direct beam to be seen. (The requirement that $\lambda \sim a$ was well recognized in the study of optics at the turn of the century but its implications for the determination of atomic structure had to await the experiment, by Friedrich and Knipping in 1912, in which X-rays which have a similar wavelength to interatomic distances in crystals were diffracted by a crystal of copper sulphate.)

2. The spacing in the diffraction pattern is *reciprocally related* to the spacing in the object.

3. The direction of the row of spots in the diffraction pattern is normal to the lines of the grating.

We now extend this same treatment to a two-dimensional grating which we can consider as the superposition of two one-dimensional gratings, one with spacing a, the other with spacing b. Both diffraction equations:

$$a \sin \phi_1 = h\lambda$$
$$b \sin \phi_2 = k\lambda$$

(where h and k are both integers, positive or negative) must now be simultaneously satisfied for a diffracted beam to occur. The solutions produce a two-dimensional array of spots on a screen (Figure 3.3) each spot associated with a particular pair of h,k values defining the order of diffraction. The row of spots in the diffraction pattern normal to the a grating has the value of $k = 0$; similarly the row of spots normal to the b grating has $h = 0$. The direct beam at the origin of the diffraction pattern is labelled (0 0). The production of the other spots in the two-dimensional diffraction pattern can be thought of as originating from the other periodicities generated in a two-dimensional grating, as shown in Figure 3.4. For example, using the Miller index notation in two dimensions, the set of spacings shown dashed in Figure 3.4(a) and labelled (1 1) gives rise in the diffraction pattern to a row of spots

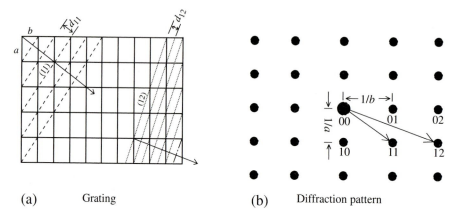

(a)   Grating   (b)   Diffraction pattern

Figure 3.4. (a) A rectangular grating and (b) its associated diffraction pattern, illustrating the relationship between spacings of planes in the grating and the position of the diffraction spots. Each diffraction spot is related to a specific set of planes in the grating.

which is normal to this (1 1) grating and has a spacing reciprocally related to that in the grating. Similarly, a set of grating spacings labelled (1 2) in Figure 3.4(b) is associated with a spot in the diffraction pattern labelled (1 2), in that a vector from the origin (0 0) to the point (1 2) in the diffraction pattern has a length reciprocal to the spacing of the (1 2) grating and a direction normal to it.

Thus a two-dimensional periodic object has a diffraction pattern which is a two-dimensional array of spots with the following properties.

(i)   Each spot in the diffraction pattern refers to a particular periodicity in the object.
(ii) The length of the vector from the origin to any spot in the diffraction pattern is reciprocal to the object spacing to which it refers, and
(iii) The direction of the vector from the origin to a spot in the diffraction pattern is normal to the spacing to which it refers.

Due to the reciprocal relationship between the diffraction pattern and the periodic object, we will use the terminology that the solutions to the diffraction equations, represented by the array of spots in the diffraction pattern, define a *reciprocal lattice*. Just as the diffraction pattern has an origin defined by the direct beam, the reciprocal lattice has a unique origin, and the position of any point hk in the reciprocal lattice defines the spacing and orientation of planes with Miller indices (hk) in the object. The reciprocal lattice concept is particularly important in any discussion of diffraction and imaging as well as in many other aspects of mineral physics. We will return to it in the next section.

Figure 3.5. (a) A non-periodic object (mouse) and (b) its optical diffraction pattern. This is one of many optical transforms in Harburn, Taylor and Welberry (1975).

The term reciprocal *lattice* can be misleading when compared to the definition of a real lattice, in that apart from having equally spaced points, the reciprocal lattice is not repetitive and each point has its own identity defined in relation to a fixed origin. In the real lattice each point is identical and any lattice point can be defined as the origin.

In terms of the notion that the diffracted beams transfer spatial information from the object to the image, it should be clear from the above that the most widely diffracted beams refer to the finest spacings in the object. The same applies to non-periodic objects, in which case the diffraction pattern is also non-periodic (Figure 3.5). In the imaging process therefore, achieving a high resolution depends on having an objective lens with a wide enough aperture to be able to collect the beams diffracted through large angles. The lens quality is also very important, as we shall see in the discussion on high resolution electron microscopy (Section 3.13).

3.3 Diffraction from a three-dimensional object – the reciprocal lattice

The extension of the above ideas to three dimensions is intuitively quite straightforward, although it needs some caution in its literal interpretation. The development we have followed from one dimension, where the diffraction pattern was specified by one integer h, to two dimensions where two integers h and k were needed to define the two-dimensional diffraction pattern, leads naturally to three dimensions, where we now need three integers h, k, and l to specify each diffracted beam. However, in one and two dimensions all of the diffracted beams are produced simultaneously, irrespective of the orientation of the incident beam relative to the object grating. In three dimensions, the simultaneous solution of three diffraction equations requires very specific geometrical conditions to be satisfied before any diffracted beams are produced, and each beam is produced separately. Thus the result is not a three-dimensional diffraction pattern, although it is very useful to extend the reciprocal lattice concept to three dimensions.

Just as the solutions for two-dimensional diffraction can be considered as lying on a two-dimensional reciprocal lattice, we define a three-dimensional reciprocal lattice which describes all the *possible* solutions to the diffraction equations in three dimensions, noting that of all the solutions only a few may operate in any specific geometrical case. In the three-dimensional reciprocal lattice we define an origin, 000, and an array of points described by three coordinates h,k,l. Each point hkl in the reciprocal lattice refers to a set of planes (hkl) in the object, and the position of the reciprocal lattice point defines the interplanar spacing and the orientation of these planes. The direction of the reciprocal lattice vector from 000 to hkl is normal to the (hkl) planes and its length is reciprocal to the interplanar spacing d_{hkl}. The terminology used is that the reciprocal lattice occupies *reciprocal space*, just as the direct lattice of the crystal is in '*real* (or direct) space'. Although somewhat abstract, the concept of reciprocal space provides a practical tool, as well as a unifying common basis, for the interpretation of all of the different experimental set-ups for diffraction and imaging.

3.3.1 From real lattice to reciprocal lattice – an example

The example which follows demonstrates how to construct any reciprocal lattice from a given real lattice. We will restrict ourselves to two-dimensional sections of real and reciprocal lattices. In a real lattice the lattice parameters (the magnitudes of the unit translation vectors) along the three axes are denoted a, b, c; the unit translation vectors of the reciprocal lattice are defined as $\mathbf{a^*}, \mathbf{b^*}, \mathbf{c^*}$. As justified in the discussion above,

$\mathbf{a^*}$ is perpendicular to the (100) planes of the real lattice and has a length equal to $1/d_{100}$.

$\mathbf{b^*}$ is perpendicular to the (010) planes of the real lattice and has a length equal to $1/d_{010}$.

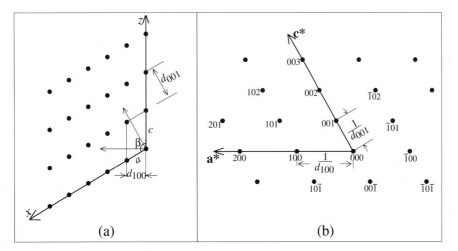

Figure 3.6. An example of a reciprocal lattice construction from a real lattice. (a) An *ac* section of a monoclinic lattice showing the *d* spacing of (100) and (001) planes, and the normals to these planes. (b) The $\mathbf{a^*c^*}$ reciprocal lattice section in the same orientation as (a).

c* is perpendicular to the (001) planes of the real lattice and has a length equal to $1/d_{001}$.

Consider the *ac* section of a real monoclinic lattice shown in Figure 3.6(a). To construct the reciprocal lattice section, we draw a line from the origin perpendicular to (100) planes – this is the **a*** direction of the reciprocal lattice. The spacing of reciprocal lattice points along this axis is $1/d_{100}$ and these points are labelled 100, 200, 300 etc as shown. The distance from the origin to the reciprocal lattice point 200 is therefore the reciprocal of the (200) interplanar spacing etc. The same procedure is used to construct the **c*** axis normal to (001) planes and the reciprocal lattice points 001, 002, 003 etc.(Figure 3.6(b)). The rest of the points in the **a*c*** reciprocal lattice section are simply generated by adding pairs of vectors and their indices as shown in Figure 3.6(b).

The reciprocal lattice section we have generated has the same properties as the diffraction pattern in Figure 3.4. The vector from the origin to the reciprocal lattice point 101 for example, has a length reciprocal to the (101) interplanar spacing and a direction normal to (101) planes. Every reciprocal lattice point has this property. In the third dimension of the reciprocal lattice, **b*** will be vertical and the

point 010 will therefore lie directly above the origin at a distance $1/d_{010}$ from it. The point 111 will be at a vertical height $1/d_{010}$ above the point 101 etc. In this way we can visualize a three-dimensional reciprocal lattice associated with every real lattice, each reciprocal lattice point recording the interplanar spacing and its orientation. In other words, all of the information about the real lattice is contained in the reciprocal lattice which, as we have already noted, is identical to diffraction space.

3.4 Systematic absences in the reciprocal lattice

The reciprocal lattice of a crystal is determined solely by its interplanar spacings and their orientation, and is independent of the choice of unit cell in the real lattice. We saw in Chapter 1 that there are many different ways of defining a unit cell, and in some cases it was convenient to define a non-primitive unit cell. We noted then, that in a non-primitive unit cell, the unit repeat is larger than if a primitive cell had been chosen and that this would have to be taken into account when diffraction was considered.

The way this is done is as follows. When a larger non-primitive unit cell is defined in the real lattice,

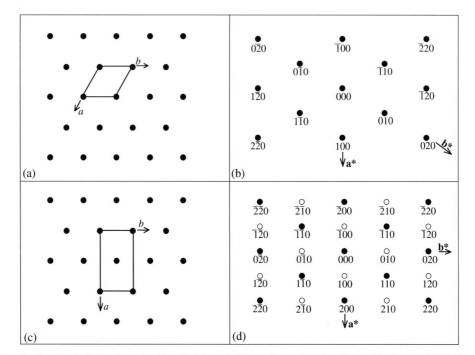

Figure 3.7. Systematic absences in the reciprocal lattice. In (a) a primitive unit cell describes the real lattice, and (b) is the reciprocal lattice section in the same orientation. In (c) a centred unit cell is used to describe the same lattice as in (a). In the reciprocal lattice based on this non-primitive cell, shown in (d), spots with indices $h+k$ odd (open circles) must be removed.

the construction of the associated reciprocal lattice, outlined above, would lead to a smaller reciprocal unit cell than would have been the case if a primitive cell had been chosen. Clearly, the reciprocal lattice, which defines all of the possible solutions to the diffraction equations, cannot be dependent on how we might choose to define the unit cell. In constructing the reciprocal lattice for a non-primitive real cell using the above method, we must remove some of the reciprocal lattice points, so that the end result is the same as that which would be produced from the equivalent real lattice with a *primitive* unit cell.

A simple example will illustrate the principle: Figure 3.7(a) shows a primitive unit cell and the associated reciprocal lattice (Figure 3.7(b)) drawn out and labelled according to the method above. Figure 3.7(c) shows the same real lattice, but described with a non-primitive unit cell. This larger unit cell is used to generate the reciprocal lattice in Figure 3.7(d). To make both reciprocal lattices the same, those marked with open circles in Figure 3.7(d) must be removed. Thus when the reciprocal lattice is defined in terms of a non-primitive cell in the real lattice, there are certain points missing from the reciprocal lattice. In this case those for which $h+k$ is odd are absent, and these are termed *systematic absences*.

Using the same argument it can be shown that the following absences apply to the non-primitive Bravais lattices:

F lattice: all points absent for which h,k,l, are mixed even and odd integers (neither all odd nor all even).

I lattice: all points absent for which $h+k+l$ is odd.

C lattice: all points absent for which $h+k$ is odd.

3.5 Diffraction by a crystal

In 1912 Max von Laue realized that a crystal will behave as a three-dimensional diffraction grating for incident radiation with a wavelength of the same order as the interatomic distance. His interpretation of the experimental observation of X-ray diffraction by a crystal of copper sulphate was the beginning of the study of the internal structure of crystals by diffraction, and X-ray diffraction remains the principal technique for the determination of atomic structure. Beams of neutrons and electrons of suitable

energies are also diffracted by crystals, and are governed by the same geometric laws.

In Section 3.7 we will return to the discussion of how the reciprocal lattice concept is applied in principle to the interpretation of diffraction experiments on crystals. First, however, some aspects of the diffraction process, whereby atoms can act as scattering centres for incident radiation, need to be briefly described.

X-rays are very high frequency ($\sim 10^{18}$ Hz) electromagnetic waves. The oscillating electric field of the X-rays interacts with the charged particles of an atom, causing these to oscillate with the same frequency. The protons, being much heavier than the electrons, are not affected very much and it is the extra-nuclear electron cloud which oscillates and hence becomes a source of secondary X-rays. This is the main cause of scattering of X-rays by matter. In general, heavier atoms with more electrons scatter more strongly than lighter atoms, and so X-rays are not successful at locating the position of very light atoms such as hydrogen. This proportionality would only be a simple one if the size of atoms was very small relative to the X-ray wavelength. However, heavier atoms have dimensions of the same order as the X-ray wavelength. The effect of this is that X-rays scattered from different points in the electron cloud of the same atom interfere with one another resulting in a fall-off in the scattered intensity with the angle through which the incident beam has been scattered (Figure 3.8).

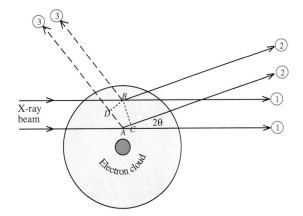

Figure 3.8. X-rays scattered from two different points A and B of the electron cloud have path differences which depend on the scattering angle, 2θ. In case 1, the scattering angle, and the path difference are zero; in case 2, the path difference is given by AC; in case 3, the path difference is AD. The interference between X-rays with increasing path difference results in a fall-off of the scattered intensity with the scattering angle.

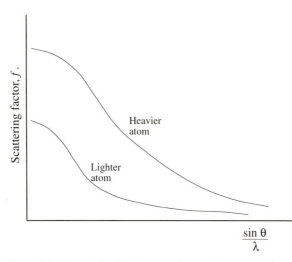

Figure 3.9. The relationship between the atomic scattering factor and the term $\sin\theta/\lambda$ where 2θ is the scattering angle and λ is the X-ray wavelength. When $\sin\theta/\lambda$ is zero, the scattering factor f is proportional to the number of extranuclear electrons.

Figure 3.9 summarizes the effect of scattering angle and incident wavelength on the *atomic scattering factor*, as it is known, and shows that at small scattering angles the scattering is proportional to the atomic number, and falls off with angle more slowly for heavy atoms than for light atoms. A knowledge of these scattering curves is the starting point for working out the intensity of the diffracted beams from an assembly of atoms.

Next we need to consider how the scattering from each atom is related to every other atom in the diffracting material. If the material was amorphous, the total scattered intensity would be spread in all directions. In a crystalline material the periodicity in the atomic distribution results in the diffracted intensity being restricted to certain directions, in the same way as for optical diffraction discussed earlier. To calculate the intensity of the various orders of diffraction for a three-dimensional array of atoms, we need to know the positions of the atoms in the unit cell and hence the relative phase of the scattered intensity from each atom. When the first X-ray diffraction experiments were done in 1912, the periodicity of the internal structure of crystals had long been inferred from their symmetry and morphology, but the positions of atoms within the unit cell were not known. The interpretation of the complex diffraction patterns from crystals was therefore not straightforward and two separate ways of treating crystal diffraction were devised. The first was by Laue, the second by W.L. Bragg in 1913.

3.5.1 The Laue equations

Laue's attempt to explain diffraction from a crystal was to consider the crystal as a three-dimensional grating with scattering from an array of lattice points, acting as scattering centres separated by spacings a, b, and c along the three crystallographic axes. Maxima of diffracted intensity only occur if the scattering from all atoms, or groups of atoms associated with each lattice point, is in phase. Laue's approach is analogous to that for optical diffraction from a grating, although in a crystal there is no grating plane as such to act as a reference for measuring the angle of diffraction, and Laue's theory is thus more general.

Consider a row of lattice points, spacing a, along the x axis (Figure 3.10(a)). For an incident beam of X-rays at an angle α_1 to the lattice row, the condition for reinforcement of the scattered radiation, i.e. for a diffracted beam, is that the path difference between adjacent rays 1 and 2 is an integral number of wavelengths, i.e.

$$a(\cos \alpha_2 - \cos \alpha_1) = h\lambda \quad \text{(where h is an integer)}$$
$$(3.3)$$

For a given value of α_1, this equation is satisfied by any direction making an angle α_2 with the x axis. So for a given integer h the diffracted beams are confined to the surface of a cone with semi-angle α_2. For values of $h = 0, \pm1, \pm2$ etc. the diffracted beams lie on a set of cones coaxial with the x axis (Figure 3.10(b)).

For the special case where $\alpha_1 = 90°$,

$$a \cos \alpha_2 = h\lambda \ (h \text{ integer}).$$

Compare this with eqn (3.1).

For diffraction to take place in a three-dimensional crystal, three such equations must be simultaneously satisfied:

$$\left.\begin{array}{l} a(\cos \alpha_2 - \cos \alpha_1) = h\lambda \\ b(\cos \beta_2 - \cos \beta_1) = k\lambda \\ c(\cos \gamma_2 - \cos \gamma_1) = l\lambda \end{array}\right\} \quad (3.4)$$

where h, k, l are integers and the angles β and γ are equivalent in meaning to α in Figure 3.10(a).

Thus for a diffracted beam to be produced in a particular direction, three sets of cones representing three possible solutions to these *Laue equations* must intersect along that direction. In general this will not happen. Even when the three Laue equa-

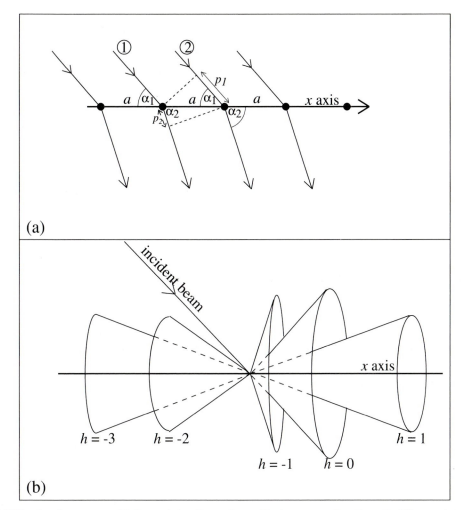

Figure 3.10. (a) Diffraction from a row of lattice points with spacing a. Maxima occur when the path difference between two adjacent beams is an integral number of wavelengths. (b) The diffraction conditions are satisfied by a family of cones coaxial with the lattice row, each cone described by an integer h, shown here with $-3 < h < 1$.

tions are satisifed, the intersection of three cones is difficult to visualize, and it is also difficult to apply, except in certain specific geometric cases. Thus although the Laue theory is correct, an alternative formulation of it was sought by W.L. Bragg who, with his father W.H. Bragg, was the first to use X-ray diffraction to solve crystal structures.

3.5.2 The Bragg equation

The Bragg approach is an alternative, equally valid and completely equivalent way of expressing the diffraction condition, that scattering from all 'scattering units' must be in phase for maxima to occur, and is easier to apply than the Laue equations in most diffraction experiments. Bragg realized that

X-rays scattered by all the lattice points in a plane (hkl) must be in phase for the Laue equations to be satisfied, and further, that scattering from successive (hkl) planes must also be in phase. For a zero phase difference, the laws of simple reflection must hold for a single plane, and the path difference for 'reflections' from successive planes must be an integral number of wavelengths. Thus although we are dealing with a diffraction phenomenon, the geometry is similar to that for reflection from (hkl) planes.

Figure 3.11(a) shows the diffraction geometry. An incident beam is diffracted from two successive (hkl) planes with interplanar spacing d_{hkl}. The path difference for beams from successive planes is given by $AB+BC = 2d_{hkl} \sin \theta$, and hence the condition for diffraction maxima is that

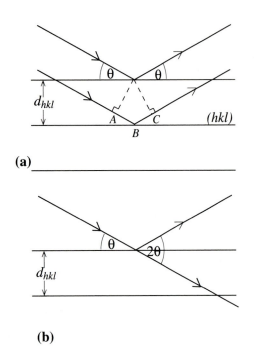

(a)

(b)

Figure 3.11. The condition for Bragg reflection from planes *hkl* with spacing d_{hkl}. (a) The path difference between X-rays 'reflected' from successive planes is $AB+BC$. (b) X-rays incident on a set of planes at the Bragg angle θ are diffracted through an angle 2θ.

$$2d_{hkl} \sin \theta = n\lambda$$

In practice it is more convenient to treat n^{th} order reflection from (*hkl*) planes as 1st order reflection from planes (*nh,nk,nl*). Thus the second order reflection from (100) planes is equivalent to the first order reflection from (200) planes, i.e.

$2d_{100} \sin \theta = 2\lambda$ is equivalent to $2d_{200} \sin \theta = \lambda$

and we write the Bragg Equation as:

$$\lambda = 2d_{hkl} \sin \theta \qquad (3.5)$$

Although the terms 'Bragg reflection' or '*hkl* reflection planes' are commonly used in the description of diffraction in crystals, the terms suggest a conceptual method used to interpret the diffraction pattern rather than a description of the scattering process. In other words, Bragg found that the diffraction patterns could be explained *as if* they were produced by reflection of X-rays from (*hkl*) planes, but only when the Bragg equation is satisfied. As shown in Figure 3.11(b), X-rays incident on a set of planes (*hkl*) at the Bragg angle, θ, are deviated by an angle 2θ from the undiffracted beam which passes straight through the crystal.

The most common application of the Bragg equation is in the interpretation of X-ray diffraction patterns from powdered crystals. X-ray powder diffraction is a standard method for identifying crystalline minerals, determining their lattice parameters, and in some cases determining crystal structures from the diffracted intensities. Before describing powder diffraction it is worth making a preliminary note on the relative amount of information obtained from a powdered sample relative to a single crystal in a diffraction experiment.

Consider a beam of monochromatic X-rays, wavelength λ, incident on a fixed single crystal. Unless the crystal is rotated to set up all the different (*hkl*) planes to the appropriate Bragg angle in turn, the chances of any diffracted beams being produced at all are not great. We will come back to this point later, but note here, that by crushing up the crystal into a powder, all the different (*hkl*) planes in the crystal are completely randomly oriented, such that we can assume that there are sufficient at the correct Bragg angle for each (*hkl*) plane to be represented in the diffraction pattern.

3.6 The interpretation of X-ray powder diffraction patterns

There are many ways of detecting and recording X-ray powder diffraction patterns, but the geometry of the experimental set-up is the same in each case, and is well illustrated by a simple powder diffraction camera, shown schematically in Figure 3.12. A monochromatic beam of X-rays passes through a collimator into a light-tight metal cylinder, at the centre of which is a capillary containing the

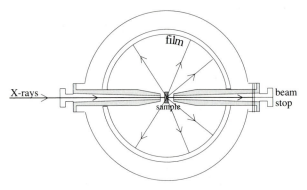

Figure 3.12. Schematic diagram of an X-ray powder diffraction camera. The X-rays enter a cylindrical metal chamber with the sample held at the centre and a strip of film around the inside wall.

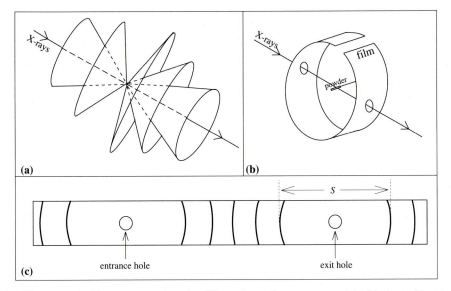

Figure 3.13. (a) When X-rays are incident on a powder, the diffracted rays form cones coaxial with the incident beam. Each cone of diffracted rays represents Bragg reflection from a set of lattice planes in the sample. (b) In a powder camera a cylindrical strip of film detects the diffracted beams. (c) When the strip of film is laid flat, the intersections of the cones of diffraction with the film form pairs of arcs around the exit and entrance holes through which the X-rays pass.

powdered sample. The diffraction pattern is recorded on a thin strip of photographic film on the inside wall of the cylinder.

For each set of planes (*hkl*) in the sample there will be sufficient at the Bragg angle θ to satisfy the Bragg equation $\lambda = 2d_{hkl} \sin \theta$. In a random sample these planes will be present in every angular orientation around the incident beam axis, and hence the diffracted beams will form a cone from each set of (*hkl*) planes. Since the angle between the diffracted and undiffracted beams is 2θ, the cone angle is 4θ. Planes with the simplest (*hkl*) indices have the largest interplanar *d* spacings and hence the smallest values of 2θ. The diffraction from the powdered sample therefore, consists of a set of cones around the direction of the incident beam (Figure 3.13(a)). The intersection of each cone of diffracted beams with the photographic plate (Figure 3.13(b)) produces two short arcs which are symmetrical about the entrance and exit holes in the film shown opened out in Figure 3.13(c). Thus although most of the crystals in a powdered sample may not be diffracting at all, there are usually sufficient in the correct position to contribute to the arc of diffraction seen on the film. If there are insufficient crystals, or they are too coarsely ground, the arcs will appear speckled.

To obtain the *d*-spacings of the crystal from such a

film, the separation S between pairs of corresponding arcs (Figure 3.13(c)) around the exit hole is measured, and given the camera (and hence film) radius to be R, then

$$\frac{S}{2\pi R} = \frac{4\theta}{360}$$

from which 2θ values (in degrees) can be tabulated for each pair of arcs. The *d*-spacings are then calculated from the Bragg equation, $d_{hkl} = \lambda/(2\sin \theta)$. Since the set of *d*-spacings is characteristic of a specific set of lattice parameters $a,b,c,\ \alpha,\beta,\gamma$ of the sample, it is the basis for the identification of unknown materials. The general relationship between the *d*-spacing of any set of planes, d_{hkl}, and the lattice parameters $a,b,c,\ \alpha,\beta,\gamma$ is quite complicated, especially for non-orthogonal crystal systems, and in most cases assigning *hkl* values to the diffraction lines is carried out with reference to the *Powder Diffraction File*. This lists the Miller indices (*hkl*) alongside the *d*-spacings and relative intensities of (*hkl*) reflections of most known materials, with which the experimental data can be compared.

When the powder pattern is indexed, i.e. values of (*hkl*) assigned to each pair of arcs on the film, lattice parameters can be determined from their *d*-spacings. The principle of the method is illustrated for a cubic crystal in the example below.

3.6.1 Example: The powder diffraction pattern from a cubic crystal

Figure 3.14 is an X-ray powder diffraction pattern of a cubic crystal with a primitive unit cell. In the cubic system the relationship between the d-spacing and the lattice parameter a is given by

$$a = d_{hkl}\sqrt{h^2 + k^2 + l^2} \qquad (3.6)$$

If we replace $h^2 + k^2 + l^2$ by N, which must be an integer, then $d_{hkl} = a/\sqrt{N}$, with N taking values 1,2,3,4 etc. (but not N=7 or 15, which cannot be expressed as the sum of three squares). Thus decreasing values of d_{hkl} are represented by increasing N values.

The diffraction lines in Figure 3.14 have been indexed by inspection, beginning with the line with the smallest 2θ value and N=1, which is indexed as {100}, i.e all planes of the form {100} will diffract into this line. The rest of the lines are indexed with N = 2,3,4,5,6,8. The lattice parameter is usually determined by obtaining the best fit of the measured d-spacings of these indexed reflections to a value of a. If only a single line is used in the determination of a, the maximum accuracy is achieved by measuring the line with the largest θ value. This can be seen from the fact that differentiating the Bragg equation with respect to θ and d gives

$$2\sin\theta \, \mathrm{d}d + 2d\cos\theta \, \mathrm{d}\theta = 0$$

Thus $\mathrm{d}d/\mathrm{d}\theta = -d\cot\theta$, and hence for a fixed error in θ, the error in d will be minimum as $\cot\theta$ approaches 0, i.e. as θ approaches 90°.

X-ray powder diffraction patterns from crystals with non-primitive lattices will have systematic absences for precisely the reasons described in Section 3.4. This will become even clearer after the discussion in the next section. Figures 3.14(b) and (c) show indexed powder diffraction patterns from cubic materials with F and I lattices respectively. Note that for an F lattice $h, k,$ and l must be either all even or all odd for a reflection to be present. Thus only lines with N= 3,4,8,11,12,16,.. will be present. For an I lattice only reflections for which $h+k+l$ are even are present, i.e. those for which N= 2,4,6,8,10 . . .

Most minerals have powder patterns which are considerably more complex than the schematic diagrams in Figure 3.14. Larger lattice parameters and decreased symmetry results in many more lines. Figure 3.15 is an X-ray powder of magnetite which has a cubic F cell with a lattice parameter 8.4Å. The 440 line has been marked so that this figure provides a simple exercise in the use of the Bragg equation in this context. The angular scale of the powder pattern can be determined from the fact that the angle between the entrance and exit holes is 180°. A further point worth mentioning is that the highest angle line (the one nearest the entrance hole) con-

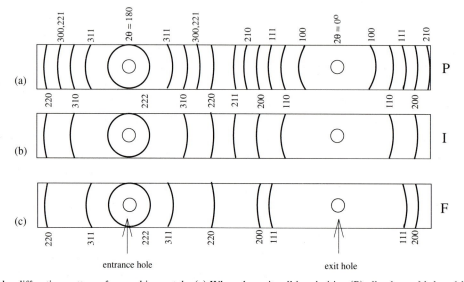

Figure 3.14. Powder diffraction patterns from cubic crystals. (a) When the unit cell is primitive (P) all values of h, k and l are allowed; (b) for a body-centred (I) unit cell $h+k+l$ must be even, and (c) for a face-centred (F) unit cell only reflections for which h, k and l are all even or all odd are allowed. The diffraction patterns in this figure have been calculated for X-rays of wavelength $\lambda = 1.54$Å incident onto a material with a cubic cell with lattice parameter $a = 2.7$Å. This is unrealistically small for a real mineral structure. For a larger lattice parameter, more lines out to higher values of $h^2+k^2+l^2$, will be present in the powder patterns.

Figure 3.15. An X-ray powder diffraction pattern of magnetite, Fe_3O_4, which has a cubic F cell with lattice parameter $a = 8.4$Å. The radiation used was Fe K_α, wavelengths Fe $K_{\alpha 1}$ 1.936Å , Fe $K_{\alpha 2}$ 1.940Å.

sists of a closely spaced pair. This is due to the small difference in the wavelength of $K_{\alpha 1}$ and $K_{\alpha 2}$ radiation which is resolved at these high angles.

3.7 The Ewald method

We have seen how the Laue equations and the Bragg equation represent two different but equivalent (although we have not proved this) ways of interpreting diffraction patterns. While the Bragg equation is simple to use for cases such as powder diffraction, it is not easy to visualize how a single crystal will diffract a particular incident beam, given the large number of possible (hkl) planes and their various orientations. The Ewald method, which uses the reciprocal lattice concept introduced at the beginning of this chapter, is the most general approach and provides a framework for understanding all diffraction experiments. In Section 3.3 we showed that the orientation and spacing of all planes of a crystal were represented by its reciprocal lattice, and furthermore, suggested that the reciprocal lattice represented all possible solutions to the diffraction equations. We left it at that point, noting that only some of the solutions applied in any particular geometrical conditions. We now continue with that theme.

In Figure 3.16(a) an X-ray beam, wavelength λ is incident along the x axis of an orthorhombic single crystal with a primitive lattice with parameters a,b,c. Consider the reciprocal lattice of this crystal. Figure 3.16(b) is the **a*b*** section of the reciprocal lattice in the same orientation as the real crystal lattice, i.e. with **a*** normal to (100) planes and **b*** normal to (010) planes of the real lattice. We now add the direction of the X-ray beam such that it passes through the origin of this reciprocal lattice, and then draw a sphere, radius 1/λ, with the X-ray beam direction as diameter and also passing through the origin. Figure 3.16(c) shows the two-dimensional section of this *Ewald sphere* (also known as the

reflecting sphere) as a circle. Now if any point hkl of the reciprocal lattice lies on the surface of this sphere, such as the point 320 in our example, then from simple geometry, this point satisfies the equation $\lambda = 2d_{hkl} \sin\theta$, i.e. the Bragg equation is satisfied for the particular (hkl) planes.

To bring a different set of planes into the diffraction condition the crystal must be rotated, and so the reciprocal lattice is rotated in the same way. In Figure 3.16(d) the reciprocal lattice has been rotated clockwise about the **c*** axis through an angle of 10° so that the (130) planes are diffracting, i.e. the reciprocal lattice point 130 now lies on the Ewald sphere, and $\lambda = 2d_{130} \sin\theta$. (Note that the angle θ has also changed).

Thus whenever a reciprocal lattice point hkl lies on the surface of an Ewald sphere, radius 1/λ, drawn through the origin of the reciprocal lattice, with the incident beam direction as diameter, the set of planes (hkl) are diffracting. Furthermore, as can be seen from Figure 3.17, in which the orientation of the real reflecting plane has been added, the direction from the centre of the sphere through the reciprocal lattice point represents the direction of the diffracted beam. We have only taken a two-dimensional section of the Ewald sphere but note that when a reciprocal lattice point above or below the page lies on the surface of the sphere, a diffracted beam will be produced from the centre of the sphere through this point.

Although the Ewald sphere construction is simply a geometrical description of the Bragg equation in reciprocal space, it has some fundamental implications for the way in which we think about a diffraction experiment. As stated earlier, the reciprocal lattice contains all the information about the crystal. So far, we have mainly considered the position of the reciprocal lattice points which are determined by the crystal lattice. However, we can associate an intensity with each reciprocal lattice point hkl, equivalent to the intensity of the diffracted beam from the equivalent set of (hkl) planes. This intensity depends

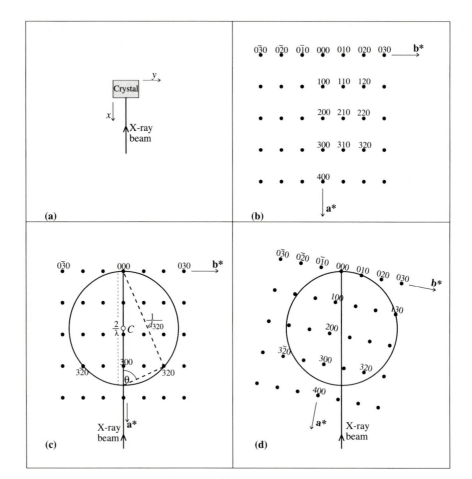

Figure 3.16. The Ewald sphere method for describing diffraction. (a) X-rays incident on a single crystal with the *xy* plane in the plane of the page. (b) The **a*b*** reciprocal lattice section of the crystal in the same orientation as in (a). (c) The Ewald sphere, with radius 1/λ and diameter parallel to the X-ray beam direction, is drawn through the origin of the reciprocal lattice. Any reciprocal lattice point which lies on the surface of the sphere (such as 320 in this case) represents a set of lattice planes satisfying the Bragg equation. (d) If the crystal is rotated through a small angle, a different set of planes (e.g. 130) can be put into the Bragg condition.

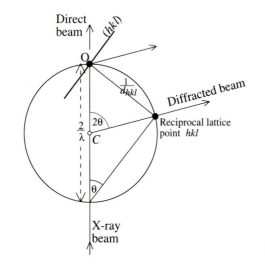

Figure 3.17. The Ewald sphere construction showing the direction of the incident X-ray beam, the orientation of the diffracting planes (*hkl*), the associated reciprocal lattice point *hkl* and the direction of the diffracted beam, from the centre of the sphere through the reciprocal lattice point on its surface.

on the scattering factors and positions of the atoms in the unit cell, i.e. the atomic structure.

Thus the reciprocal lattice, both in the positions and intensities of the points, represents the crystal. The Ewald sphere, in its size and orientation relative to the crystal, represents the probe which we use to extract the structural information from the crystal. This structural information takes the form of a list of positions and intensities of reciprocal lattice points, or in real space language, a list of *d*-spacings of (*hkl*) planes and the intensity of diffraction from them. We sample the reciprocal lattice with the Ewald sphere by setting up the geometry so that as many reciprocal lattice points as possible are, in turn, made to lie on the surface of the sphere.

At this point it is worth being reminded of the optical analogy with which we began this chapter. Imaging, in Abbe theory, is a two step process. First the incident radiation interacts with the object and is scattered. A lens collects these diffracted beams, which contain the information that is to be transmitted to the image, and focuses them to produce the diffraction pattern. The recombination of the diffracted beams is the second stage which leads to the image of the object. So far we are only discussing the first stage, the diffraction process, but note that just as in the optical analogy, the resolution of the final image depends on being able to collect beams diffracted through large angles, i.e. beams diffracted from planes with small *d*-spacings, or in our preferred terminology, being able to sample points far out in reciprocal space.

3.8 The reciprocal lattice and structure factors, F(*hkl*)

As we have seen, every X-ray reflection is characterised by a unique index *hkl*, and for every such reflection there exists a point in reciprocal space whose coordinates (measured in units of **a***, **b*** and **c*** respectively) are also *h*, *k* and *l*. Thus the reciprocal lattice points are a convenient device for filing information about any reflection *hkl*.

We can take this one step further by noting that each diffracted beam is a wave with a characteristic amplitude and phase. This combined quantity is represented by a term called the *structure factor*, F(*hkl*) for each diffracted beam. The structure factor is derived by summing the waves diffracted by each of the atoms in the unit cell, for the conditions specified by the Bragg equation, i.e. for each *hkl* reflection. The amplitude of the resultant wave is termed the *structure amplitude* |F(*hkl*)| and the intensity of a diffracted beam is defined as the square of the structure amplitude |F(*hkl*)|^2. Both the amplitude and phase of the diffracted beam depend on the scattering factors of the atoms, and on the position of each atom in the unit cell.

If we attribute to each reciprocal lattice point, not only the coordinates *h*, *k* and *l* but also the structure factor F(*hkl*), we include information about the arrangement of atoms in the cell as well as its cell parameters. A reciprocal lattice drawn with the sizes of the spots proportional to the relative intensities is termed a *weighted reciprocal lattice*.

3.9 Diffraction experiments – sampling the reciprocal lattice

In this section we will outline some of the principal experimental methods of obtaining diffraction patterns from crystals, and relate them to the Ewald method, in which the aim is to set up an experimental situation so that as many reciprocal lattice points as possible are made to touch the surface of the Ewald sphere.

The geometry described above applies to X-ray, neutron and electron diffraction, but there are aspects of each type of radiation which provide different information about the crystal as well as needing different experimental arrangements.

1. X-rays are produced when a high energy electron beam strikes a metal target. The resultant X-rays emitted from an X-ray tube consist of a continuous 'white' spectrum, as well as lines characteristic of K-shell electronic transitions induced in the target metal. A filter is used to isolate the K$_\alpha$ wavelength. Values of the K$_\alpha$ wavelengths for the commonly used targets are: 0.71 Å for Mo, 1.54 Å for Cu, 1.79 Å for Co and 2.29 Å for Cr. The energy of X-rays of wavelength 1Å is ~ 12keV, and they can penetrate crystals up to 0.5 mm thick, depending on their molecular weight (heavy elements are more strongly absorbant). X-rays are scattered by the extranuclear electron cloud and the scattering factor for each element varies systematically with atomic number (see section 3.5). X-rays are detected on a photographic film, or by electronic counters.

2. Electron diffraction is generally carried out in a transmission electron microscope in which high energy electrons are accelerated from a heated filament by a high voltage field (100kV–1000kV). The wavelength, which is proportional to 1/ $\sqrt{}$(applied voltage) is 0.04Å for 100 kV electrons. Because electrons are charged particles, they are scattered by electric fields in the crystal, which in turn depend on the atomic positions. Electrons interact very strongly with matter and can only penetrate a few hundred ångströms into a crystal before they begin to lose their energy through multiple and inelastic collisions with atoms. Samples for electron diffraction must therefore be prepared as thin films, using special thinning techniques. The multiple scattering means that the intensity of electron diffracted beams is not simply related to the atomic structure and electrons cannot compete with X-rays as probes for the study of bulk crystal structure. However, the power of electron diffraction as a probe lies in the fact that electrons can be focussed in a microscope, a topic dealt with in some detail later in this chapter.

3. Neutrons are generated in a reactor which produces a range of neutron wavelengths, the desired wavelength being selected by a monochromator, which is a crystal set at the Bragg angle for the selected value of λ. Neutrons are scattered by nuclei, but there is no systematic dependence on the atomic number. This can often be an advantage in distinguishing elements of similar atomic number which will have very similar scattering factors for X-rays, but may scatter neutrons to different extents. Also, neutrons scatter the light elements just as efficiently, in contrast to X-rays. Large sample volumes (up to 10 cm^3) are needed for neutron diffraction due to the low intensity of the incident neutron beam although there is no problem with absorption as neutrons can typically penetrate through metres of material.

Two further applications of neutron scattering are of interest. First, in magnetic materials, unpaired electron spins interact with a neutron beam, thus providing information about the distribution of magnetic moments in a crystal. Second, the energy of neutrons with a wavelength of 1Å is only about 0.08 eV. This is similar to the energy of the thermal vibrations of atoms in the crystal, and so, as well as the elastic collisions used for diffraction experiments, neutrons also undergo *inelastic scattering* in which the energy of the incident neutrons is changed by interacting with the atomic vibrations. The change in neutron energy in such collisions is large and easily measured, and provides information about the lattice vibrations in a crystal (see Chapter 4).

X-ray and neutron diffraction experiments have many features in common, even though the apparatus involved is very different. Of the very many experimental arrangements used to obtain diffraction patterns only a few will be mentioned here to illustrate the principles they have in common. As we noted in Section 3.7 and Figure 3.16, monochromatic X-rays incident on a crystal at some arbitrary angle will not usually be diffracted, so either X-rays with a range of wavelengths must be used, or the

crystal (and in many cases, the detector) must be rotated in such a way as to place as many reflecting planes as possible into the Bragg condition, i.e. to make as many reciprocal lattice points as possible intersect the surface of the Ewald sphere.

3.9.1. 'White' X-rays and a fixed single crystal – Laue photographs

The experimental arrangement is very simple with unfiltered X-rays (i.e. with a broad band of wavelengths) directed onto a single crystal, mounted on a goniometer (Figure 3.18). A flat photographic plate placed in the forward (small 2θ) position or the back-reflection position (large 2θ) collects the diffraction pattern. As shown in Figure 3.19, all reciprocal lattice points in the shaded area between the

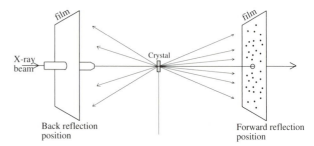

Figure 3.18. A schematic diagram illustrating Laue diffraction on a flat plate film, in either the forward reflection position (collecting X-ray beams diffracted through small angles) or the back reflection position (diffraction through large angles).

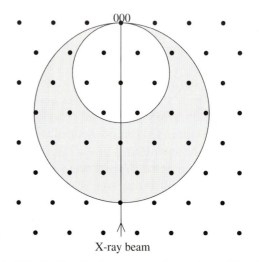

Figure 3.19. The Ewald sphere construction for Laue diffraction, in which the incident beam with a range of wavelengths is represented by the shaded region between the minimum value of λ (largest sphere) and the maximum value of λ (smallest sphere). All reciprocal lattice points in the shaded region represent diffracting planes.

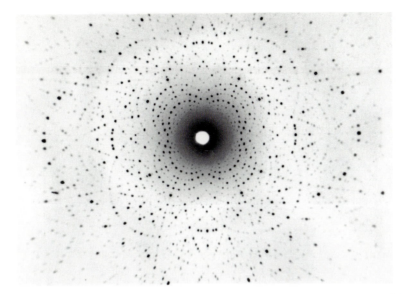

Figure 3.20. Photograph of a Laue diffraction pattern of zircon, $ZrSiO_4$, taken with the X-ray beam parallel to the c axis. Zircon is tetragonal, point group $4/mmm$. The 4-fold symmetry axis and the mirror planes are evident in the symmetry of the diffraction pattern.

maximum and minimum diameter Ewald spheres (i.e. representing the smallest and largest values of λ respectively), will be sampled. The diffracted beams directed towards the photographic plate produce a Laue diffraction pattern (Figure 3.20). Clearly, it is very difficult to assign indices to the spots in the pattern as only the 2θ angle can be determined without fixing λ or the d-spacing. Note however, the symmetry in the distribution of the diffraction spots in the photograph. Laue photographs are used to provide information on the symmetry and orientation of the crystal.

3.9.2. Monochromatic X-rays and a rotating crystal – rotation photographs

As suggested by Figure 3.16, rotating the crystal about a fixed axis normal to the incident beam is equivalent to rotating the reciprocal lattice about an axis through the origin. This sweeps the reciprocal lattice through the Ewald sphere. Each time a reciprocal lattice point intersects the sphere, a diffracted beam passes from the centre of the sphere through that point. If the rotation axis is normal to the page as in Figure 3.16, those reciprocal lattice points in the plane of the page (the zero layer), produce diffracted beams also in the plane of the page. On a cylindrical photographic plate around the crystal (Figure 3.21(a)), the zero layer forms a row of diffraction spots through the centre of the film.

The reciprocal lattice points which lie above and below the zero layer produce rows of spots on parallel *layer lines* (Figure 3.21(b)). These layer lines are therefore "edge-on" projections of the layers of reciprocal lattice points.

In the example shown in Figure 3.16, the rotation axis was [001], and so the zero layer of the reciprocal lattice contains spots indexed as $hk0$. In general, if the rotation axis is [UVW], the zero layer hkl contains spots which lie in the [UVW] zone, such that $hU+kV+lW = 0$. The first layers above and below the example in Figure 3.16 contain reciprocal lattice points $hk1$ and $hk\bar{1}$ respectively, as do the spots on the first layer in the diffraction pattern. The second layer contains reflections from $hk2$ planes etc. If the crystal is then reset so that the rotation axis was [010], the zero layer would contain reflections $h0l$, the first layer $h1l$, the second layer $h2l$ etc.

The spacing between the layer lines on the rotation photograph (Figure 3.21(b)) is simply related to the distance between the layers of reciprocal lattice points and the cylindrical camera radius. From the definition of the reciprocal lattice, the spacing between these *layers* of reciprocal lattice points is reciprocally related to the distance, along the rotation axis, between lattice points in real space. This technique is therefore applicable to the determination of real lattice spacings, and hence cell dimensions, in single crystals.

Thus rotating about [001] for example, gives layer

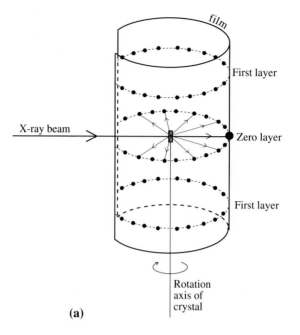

(a)

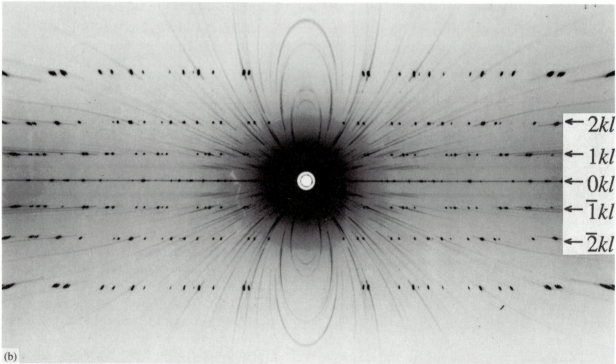

(b)

Figure 3.21. (a) Schematic representation of diffraction in a rotation camera in which a monochromatic X-ray beam is incident on a single crystal which is rotated about an axis normal to the beam. The diffracted beams form on a series of 'layer lines' which can be thought of as 'edge-on' projections of layers of the reciprocal lattice. (b) An X-ray rotation photograph of zircon, with the *a* axis as rotation axis. The spacing between layer lines is reciprocally related to the distance between lattice points along the *a* axis of zircon.

lines which can be used to calculate the lattice repeat along this direction, although we will not describe the method of calculation here.

To determine the amount of reciprocal space which can be sampled in such an experiment it is easier to visualize the reciprocal lattice as stationary and to rotate the Ewald sphere about 000. Rotating a sphere produces a torus (Figure 3.22) which contains all the reciprocal lattice points which could possibly be sampled for a single rotation axis of the

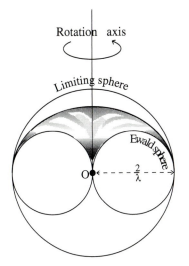

Figure 3.22. The volume of reciprocal space sampled in a rotation photograph is described by rotating the Ewald sphere around the rotation axis. By choosing different rotation axes, the maximum volume which can be sampled is a limiting sphere, centred on the origin of the reciprocal lattice and having a radius $2/\lambda$.

crystal. The only way to sample more of the reciprocal lattice is to change the rotation axis of the crystal and repeat the experiment. The torus lies within a sphere of radius $2/\lambda$, called the limiting sphere, (Figure 3.22) which contains all the reciprocal lattice points which can be sampled using a variety of crystal rotation axes. As with all X-ray diffraction experiments, more reciprocal lattice points can be sampled if the Ewald sphere is larger, i.e. the wavelength is shorter.

While useful in determining layer line spacings, rotation photographs are not a very powerful method of studying single crystals.

3.9.3. Moving film, moving crystal – Weissenberg and Precession photographs

Weissenberg and Precession methods are more powerful techniques for single crystal studies as each reflection can be measured separately, without the overlapping spots which frequently occur on layer lines in rotation photographs. We mention them briefly here, merely to illustrate the sort of approach needed to record single crystal diffraction patterns on a photographic plate. Since we cannot record a three-dimensional diffraction pattern on a two-dimensional film, we attempt to reduce the pattern to two dimensions.

The *Weissenberg* method is similar to the rotation method except that by using a metal screen between the crystal and film the layers are selected one at a time. Since we know one index of the spots on each layer line (e.g. the *l* index if the rotation axis is [001]) we need to determine the other two indices in each layer line. The diffraction pattern is photographed on a film which moves up and down the rotation axis so that each two-dimensional layer is recorded in two dimensions. This involves a complicated coupled motion between the oscillating crystal and the film but produces a record of a large number of individual *hkl* reciprocal lattice positions.

The *precession* method also involves a coupled motion of the crystal and the film, but in this case the crystal precesses about the X-ray beam direction while the diffraction pattern is recorded on a film undergoing a coupled and parallel precession. The advantage of this method is that an undistorted layer of the reciprocal lattice can be recorded on a flat plate film (Figure 3.23). By altering the film position other parallel reciprocal lattice layers can be recorded.

3.9.4. Automated four-circle single crystal diffractometers

The ultimate instrument for sampling the reciprocal lattice is a computer operated four-circle diffractometer. The crystal is mounted on a holder which is able to rotate the crystal around three separate rotation axes (Figure 3.24), and the photographic plate is replaced by a sensitive electronic detector such as a Geiger counter, which is also able to rotate in a horizontal circle. The instrument is programmed to rotate the crystal, while the detector scans its 2θ circle at twice the angular velocity of the crystal in order to keep the Bragg condition satisfied. Once a few reflections have been located, a consistent unit cell can be identified and the instrument subsequently set to systematically collect a set of *hkl* reflections and measure their intensities. Typically, several thousand reflections are collected and measured in this way.

3.9.5. Powder diffraction

As we have already discussed in Section 3.6, another way of obtaining more diffraction information about a crystal is to grind it to a randomly oriented powder so that the Bragg equation is simultaneously satisfied

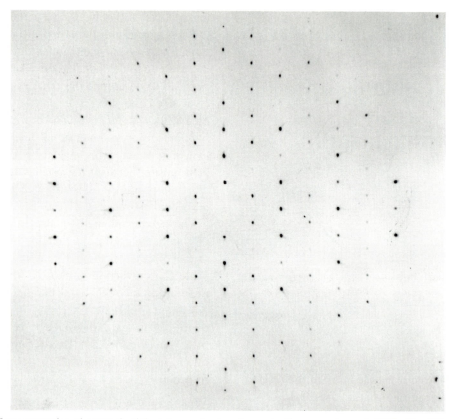

Figure 3.23. An X-ray precession photograph of the tetragonal mineral vesuvianite, precession axis [001]. The diffraction pattern is an undistorted **a*c*** section of the reciprocal lattice.

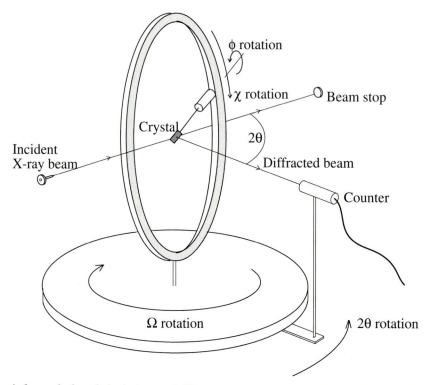

Figure 3.24. A schematic figure of a four-circle single crystal diffractometer, in which a crystal, centrally mounted in a circular ring can be automatically rotated around three different axes: the φ, χ and Ω rotations, while the detector moves on the 2θ rotation. This instrument enables the positions and intensities of several thousand *hkl* reflections to be recorded automatically.

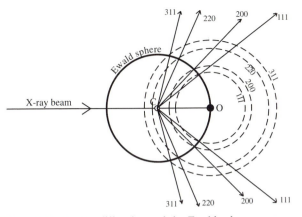

Figure 3.25. Powder diffraction and the Ewald sphere construction. The reciprocal lattices of the powdered crystals coalesce into spheres around the origin of reciprocal space, the radius of each sphere being the reciprocal of a d spacing in the crystal. (In this case the sequence of hkl reflections indicates a face-centred lattice.) When the Ewald sphere is drawn through the origin, its intersection with the reciprocal lattice sphere defines a set of cones, in the same way as shown in Fig. 3.13.

for a large number of different lattice planes. In reciprocal lattice terms, a powder specimen is equivalent to many different orientations of the reciprocal lattice sharing a common origin. Thus all the reciprocal lattice spacings coalesce into concentric spheres about the origin, the radii being all the possible reciprocal lattice vectors $1/d_{hkl}$ of the crystal (Figure 3.25).

If we now put the Ewald sphere into this reciprocal space passing through the origin, each reciprocal lattice sphere intersects the Ewald sphere along a circle, so that cones of diffracted rays emanate from the specimen. These diffraction cones are concentric about the X-ray beam and intersect a strip of film in arc segments as illustrated in Figure 3.13. Alternatively a counter can be used to detect the diffracted beams and the diffraction pattern recorded on a chart plotting the diffracted intensity versus 2θ.

3.10 Electron diffraction

The interpretation of electron diffraction patterns is the most vivid demonstration of the reciprocal lattice concept, since the pattern is identical to a two-dimensional section of the reciprocal lattice. In this it is similar to precession photographs, but is obtained without any motion of the crystal or the film. Electron diffraction is also the closest analogy to the optical diffraction process described at the beginning of this chapter. As mentioned above,

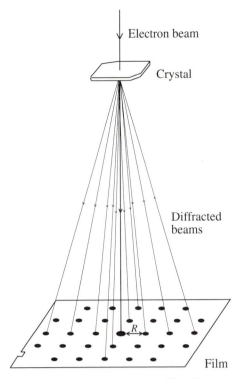

Figure 3.26. The formation of an electron diffraction pattern from a thin crystal. The diffraction pattern is a projection of the reciprocal lattice section in the plane of the crystal normal to the electron beam. The beam divergences are greatly exaggerated.

electron diffraction is carried out in an electron microscope, the details of which will be discussed later. However, the essential features of electron diffraction are shown schematically in Figure 3.26, i.e. an electron beam passes through a very thin flake or thin film (see Section 3.9) of the crystal, and the diffraction pattern is recorded on a flat photographic plate producing a simple geometric array of spots. Before we can adequately describe why this array is a simple section of the reciprocal lattice we need to introduce one extra concept into our description of the reciprocal lattice.

Throughout this chapter we have emphasised that all of the information about the crystal is contained in its reciprocal lattice. From the point of view of imaging, as in the optical imaging system of the Abbe theory (Section 3.1) the diffraction pattern must also contain information about the *dimensions* of the crystal as well as its internal structure. While the information about internal structure is contained in the *position and intensity* of the reciprocal lattice points, information about the shape and size of the crystal is contained in the *shape* of the reciprocal lattice points. As expected, this information is recip-

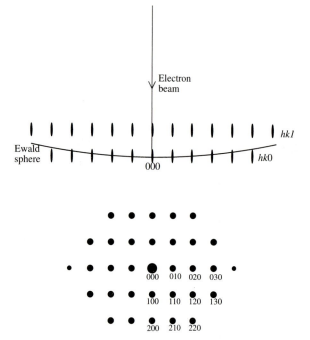

Figure 3.27. The formation of an electron diffraction pattern. The upper diagram shows an electron beam along [001] of a crystal, and a section through the reciprocal lattice planes *hk*0 and *hk*1. The Ewald sphere, only part of which is shown, is large in comparison with the reciprocal lattice spacings and intersects a large number of reciprocal lattice spots in the *hk*0 layer. The reciprocal lattice spots are cigar-shaped because they are elongated normal to the thin crystal film. The lower diagram shows the resulting **a*c*** diffraction pattern.

rocally related to real space – a very small spherical crystal will have reciprocal lattice spots which are also spherical but with a diameter reciprocally related to the diameter of the crystal. Clearly, the size of the crystal is generally very large compared with lattice plane spacings and so the spread of intensity at the reciprocal lattice point is small compared with the distance between reciprocal lattice points.

In the case of electron diffraction however, the spread of intensity at each reciprocal lattice point does become significant, since the crystal which must be used for electron diffraction approximates to a thin film or flake of material. For a thin film the reciprocal lattice points become spikes, the narrow dimension corresponding to the reciprocal of the width of the crystal and the long dimension corresponding to the reciprocal of the thickness.

This feature of the reciprocal lattice of thin flakes, coupled with the fact that the Ewald sphere for electrons is very large compared with reciprocal lattice spacings (the wavelength of 100keV electrons is only 0.04Å) produces a situation in which the Ewald sphere may intersect a large number of reciprocal lattice spikes in one plane simultaneously. Figure 3.27 shows the Ewald sphere construction

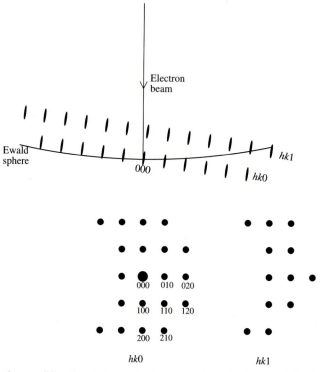

Figure 3.28. The formation of an electron diffraction pattern as in Figure 3.27, but with the crystal tilted so that the Ewald sphere intersects a smaller part of the *hk*0 layer, but also part of the *hk*1 layer. The resulting diffraction pattern shows parts of each reciprocal lattice layer.

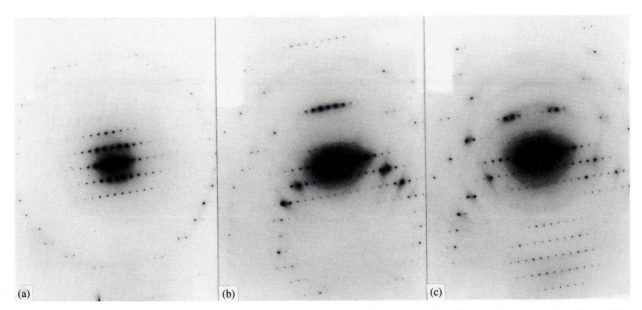

Figure 3.29. A sequence of electron diffraction patterns showing the effect of tilting the crystal, as illustrated schematically in Figure 3.28. (a) Electron beam normal to the reciprocal lattice layer (b), slight tilt of the crystal and (c) further tilting showing the intersection of the Ewald sphere with higher layers of the reciprocal lattice.

which leads to an electron diffraction pattern which is a slice, through the origin, of a plane of reciprocal space normal to the electron beam direction. As the 2θ angle is very small ($< 0.5°$ for $d_{hkl} = 5$Å) the small divergence of the diffracted beams allows an essentially undistorted two-dimensional section of the reciprocal lattice to be recorded on a flat screen (Figure 3.29(a)). If the reciprocal lattice of the crystal is sufficiently small, an upper layer of the reciprocal lattice may also be intersected forming a ring of spots around the zero layer (Figure 3.29(b)). Slight tilting of the crystal relative to the electron beam direction produces diffraction patterns in which the origin 000 (i.e. the undiffracted beam) is not at the centre of the array of diffracted beams (see Figures 3.28 and 3.29(c)).

As can be seen from Figure 3.26, the size of the electron diffraction pattern on the photographic plate depends on the distance L between the crystal and the film. Figure 3.30 shows the simple geometry which relates the real space experiment to the reciprocal lattice construction and derives the lattice plane spacing d in terms of the distance R on the experimental diffraction pattern:

$d = \lambda L / R$ λL is known as the camera constant.

When the unit cell of the crystal is known, indexing electron diffraction patterns is relatively straightfor-

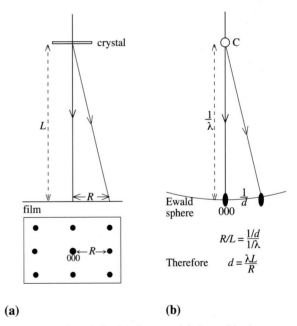

Figure 3.30. The relationship between (a) the real lattice geometry of electron diffraction, and (b) the equivalent reciprocal lattice geometry.

ward from measurements of the reciprocal d spacings. All points hkl in zero layer of the electron diffraction pattern must satisfy the equation $h\mathrm{U}+k\mathrm{V}+l\mathrm{W} = 0$, where [UVW] is the electron beam direction (i.e. the zone equation, see Appendix 1B).

As an example, consider the determination of the electron diffraction pattern formed when an electron beam is incident along the [1$\bar{1}$0] direction of a cubic crystal with lattice parameter $a = 4\text{Å}$. This diffraction pattern is therefore a two-dimensional section of the reciprocal lattice passing through the origin 000 and normal to [1$\bar{1}$0]. All planes hkl for which $h = k$, lie in the [1$\bar{1}$0] zone and so the diffraction pattern will contain reflections 001 and 110. These two specific reflections have been chosen because the angle between the planes is known (they are perpendicular).

Since $1/d_{001} = 1/a = 0.25\text{Å}^{-1}$ and $1/d_{110} = \sqrt{2}/a = 0.35\text{Å}^{-1}$ the reciprocal lattice points 001 and 110 can be readily plotted, after which the rest of the reciprocal lattice plane can be filled in by vector addition. Figure 3.31(a) shows the indexed [1$\bar{1}$0] axis diffraction pattern from a cubic crystal with a primitive cell. If the 4Å cell was face-centred F, for example, only reflections for which h, k and l were all even or all odd would be present (Section 3.4) and the diffraction pattern would be as shown in Figure 3.31(b). For a body-centred, I, unit cell only reflections for which $h+k+l$ is even are present (Figure 3.31(c)).

3.11 Imaging – some introductory remarks

In the rest of this chapter we take up a theme raised at the beginning, and that is, that the formation of an image is a two-stage process (see Abbe theory, Section 3.1). The first involves the formation of the diffraction pattern, and we have emphasised throughout that the diffraction pattern contains all the information about the object which can be used to form an image. In the optical analogy (Section 3.1) the recombination of the diffracted beams to form the image is carried out by a lens, which collects the diffracted beams. The extent to which the image represents a true description of the object depends on two main factors. First, the recombination process must preserve the relative amplitudes and phases of the diffracted beams. Any relative changes introduced represent a degradation of the image quality. Second, the resolution of the image depends on being able to collect and recombine beams diffracted through large angles, as these carry the information about the finest details in the object. In terms of the reciprocal lattice of a crystal, points far out in reciprocal space must be sampled to obtain high resolution information about interatomic distances.

Apart from the principles involved, the Abbe theory may appear to have little relevance to the process of 'imaging' with X-rays or neutrons, that is, the process of structure determination by X-ray and neutron diffraction. Firstly, X-rays and neutrons cannot be focussed by lenses, and secondly, even if X-rays could be focussed, an optical system could not be devised unless many diffraction maxima could be produced simultaneously. As we have seen, only limited X-ray diffraction occurs with monochromatic radiation and a fixed single crystal. Thus with X-rays and neutrons, only the first stage of the imaging process is achieved in the experiment, namely the formation and collection of the diffraction pattern. For electrons however, the Abbe theory is directly applicable. First, because electrons can be focussed by electrostatic lenses in an electron microscope, and second, because a large number of diffracted beams are produced simultaneously when an incident electron beam passes through a thin crystal flake.

3.12 Crystal structure determination and the phase problem

In terms of 'image formation' crystal structure determination must involve a recombination of the dif-

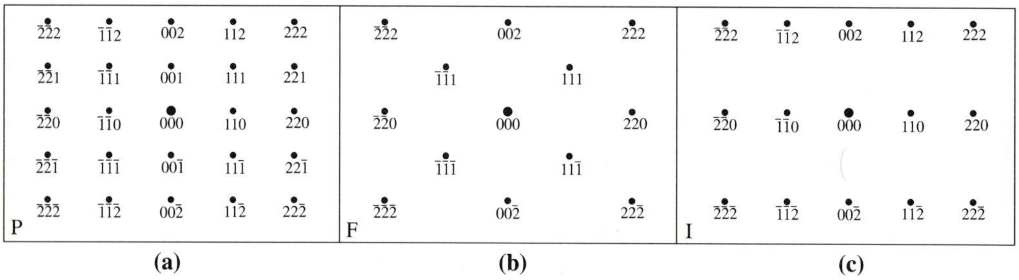

Figure 3.31. Electron diffraction patterns of cubic crystals with lattice types P, F and I. In each case the electron beam direction is [1$\bar{1}$0] and the differences are due to systematic absences. All points hkl lie in the [1$\bar{1}$0] zone, so that $h = k$.

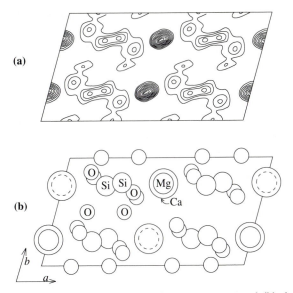

Figure 3.32. (a) An electron density contour map and (b) the crystal structure of diopside, both projected onto (010). The correspondence between regions of high electron density and the projection of atomic positions is evident.

fracted beams, with correct amplitudes (intensities) and relative phases. The image formed then portrays the density of the scattering material in the crystal. For X-rays this is the electron density, and the image is an *electron density map*, which is a projection of the electron density onto a given plane (Figure 3.32). The map is a set of contours of electron density with the peaks representing atomic positions in the unit cell.

As we cannot recombine diffracted X-rays experimentally, this must be done mathematically by a process termed *Fourier transformation*. *Fourier analysis* is a mathematical method of expressing any periodic function with wavelength λ, as a sum of sinusoidal functions whose wavelengths are integral fractions of λ (i.e. λ, λ/2, λ/3, etc.). In general, these sine curves do not share a common origin and can be displaced relative to one another i.e. have different phases. The amplitude and phase of each term in a Fourier series is the Fourier coefficient. *Fourier synthesis* is the process of adding up all these terms in the series to obtain the original periodic function. As Fourier analysis is an approximation (unless the number of terms is infinite), the original function cannot be reproduced precisely.

The periodic distribution of electron density in a crystal can also be expressed as a three-dimensional Fourier series, and W.H. Bragg realized that the process of X-ray diffraction in a crystal was essentially an experimental Fourier analysis process. Each term is represented by a point in the reciprocal lattice, points further from the origin representing the smaller wavelength terms. The amplitude and phase i.e. the structure factor $F(hkl)$ at each reciprocal lattice point is the Fourier coefficient, and hence by carrying out a Fourier synthesis the original electron density can be determined. To obtain a high resolution electron density map, we must include the small wavelength terms in the summation, i.e. points far out in reciprocal space. In the optical analogy, the glass lens is the Fourier synthesiser, a job which it does particularly well, preserving the relative amplitudes and phases of the diffracted beams to reproduce an image of the scattering object.

In an X-ray diffraction experiment, the Fourier synthesis must be done mathematically. However, while we can measure the amplitudes of the diffracted beams, *there is no routine experimental way of determining their relative phases*, and so we only have half of the information we need. Furthermore, the general features of a Fourier synthesis (and hence the resultant electron density map) depend more on the relative phases than on the amplitudes. In crystallography this is called the *phase problem*, and methods of structure determination are largely concerned with ways of getting around it. It is beyond the scope of this book to go into this topic in depth, but we will outline some of the basic principles.

Early structure determinations were made by trial-and-error methods: a structure was assumed and the amplitudes of diffracted beams calculated from the positions of the atoms in the unit cell and their scattering factors. A comparison of calculated and experimentally measured amplitudes is a measure of the success of the initial trial structure. It is quoted as an *R*-factor given by

$$R = \frac{\Sigma|\Delta|}{\Sigma|F_o|}$$

where $\Delta = |F_o| - |F_c|$, F_o is the observed amplitude, and F_c is the calculated amplitude for the trial structure. The values of Δ and F_o are summed for a number of observed and calculated amplitudes. The lower the value of R, the better the fit (e.g. $R = 0.1$ is small enough to indicate that the positions of the atoms are essentially correct). Trial structures can be derived in a number of ways.

1. For simple structures, consideration of close-packing and cation coordination (see Chapter 5), as well as comparison with other known structures etc. can lead to a trial structure, although there is always a problem of knowing whether a promisingly small R-factor represents an approximation to the truth, or merely an approximation to a good fit. The positions of atoms in a unit cell are constrained by the space group, as discussed in Chapter 1. The determination of the most likely space group is made in the earlier stages, from the distribution of systematic absences in the diffraction pattern.

2. If in the Fourier synthesis the summation is made using only the measured intensities, (i.e. $|F(hkl)|^2$ values), it was shown theoretically by Patterson in 1935 that the resulting function is a map, not of the electron density, but of interatomic distances. A peak at some position (X, Y, Z) means that there are atoms whose coordinates differ by these values i.e., atoms at positions $x_1 y_1 z_1$ and $x_2 y_2 z_2$ such that $x_2 - x_1 = X$, $y_2 - y_1 = Y$, and $z_2 - z_1 = Z$. A strong peak implies that either the atoms are strong scatterers, or there are several atoms related by this interatomic distance. This method is called *Patterson synthesis* and the resulting map is termed 'the *Patterson*'. The Patterson produces a set of interatomic distances between the atoms of a structure, without providing any information about how they are arranged in the unit cell. Unravelling the map to find a trial structure is sometimes possible but not always easy given the fact that there are $n(n-1)$ possible interatomic distances in a cell containing n atoms. However, a Patterson can be very useful when combined with the *heavy atom method*, described next.

3. If the unit cell of a structure contains a fairly uniform distribution of light elements and a few heavy atoms, X-ray scattering will be dominated by the heavy atoms. A Patterson map can be used to locate the heavy atom, and the assumption is made that the phase for each diffracted beam due to the whole structure is approximately the same as that due to the heavy atom. This phase can be readily calculated from the atomic position and a Fourier synthesis carried out to obtain a first approximation to the electron density map. Peaks in the electron density map, together with known features of cation–anion coordination etc. are then used in a calculation of the relative phases for a second improved approximate electron density map. The

process is continued until all the atoms have been located. A drawback of this approach is that under conditions where this heavy atom method works, i.e. when the heavy atom is the dominant scatterer, the lighter atoms become difficult to locate with any precision.

4. *Isomorphous replacement* is a direct method of obtaining phases which works when there are two compounds with the same structure and similar unit cell dimensions, but with an atomic position into which different atoms can be substituted. For example, $KAl(SO_4)_2.12H_2O$ and $KCr(SO_4)_2.12H_2O$ have the same structure. The position of the two interchangeable atoms is determined from a Patterson map and the difference between their contribution to the phase of each diffracted beam can be calculated. By experimentally measuring the difference between the intensities of each hkl reflection for the two compounds, and knowing the calculated phase difference, a value of the phase for each reflection can be determined algebraically.

5. Finally, *direct methods* of determining phases exist without the need for heavy atoms or any initial assumptions about trial structures. These are analytical methods whose success depends on the availability of high-speed computers. Briefly, if phases are assigned to the reflections at random, a Fourier synthesis would probably be meaningless and likely to show regions of negative electron density. As this is impossible, those combinations of assigned phases which give negative electron density are discarded. It was found that the condition that the electron density must be positive, as well as approximately spherical about atomic positions, imposes constraints on the relationships between the phases of the strong reflections. From such relationships, phases can be directly assigned to strong reflections and an approximate electron density map produced. Successive iterations of this process gradually improve the fit between the experimental and calculated intensities for successive trial structures.

The overall structures of most of the important mineral groups are well known, and at the present time structure determinations are essentially processes of refining these known structures, and determining in considerable detail, the way a structure changes with temperature, pressure and composition. Refining a structure involves systematically varying the atomic positions to give the best possible agreement between the observed and calculated intensities. This usually takes successive cycles until the structure refinement converges to a point where no further improvement is possible. As there may be many thousands of reflections measured in a precise refinement, computer programs, based on statistical methods (least-squares) of fitting calculated to observed data, have been written to carry out this task. In a precise structure determination positional coordinates of atoms are quoted to the fourth decimal place in ångström units.

3.13 Direct high-resolution imaging in a transmission electron microscope

High resolution imaging in a transmission electron microscope (TEM) is analogous to optical imaging and the Abbe theory (Section 3.1) can be directly applied. The electron diffraction pattern from a thin slice of a crystal can be regarded, in the first instance, as that obtained from a two-dimensional diffraction grating, and represents a relatively undistorted slice of the reciprocal lattice (Section 3.10). The fact that electrons can be focussed by electromagnetic lenses means that the incident electron beam can illuminate a very small area of the crystal, and furthermore, the diffracted beams can be recombined to form an image. The image is then a two-dimensional projection of the density of the scattering material in the crystal. As electrons are scattered by electric fields in the crystal, the differences in the distribution of electrostatic potential due to atomic positions can be imaged directly.

Given that the Abbe theory predicts that the resolution in the image is defined by the wavelength of the radiation and the ability of the objective lens to collect widely diffracted beams, the very small wavelength of electrons (0.04Å at 100kV and 0.01Å at 1000kV) raises the possibility that an electron microscope could reveal sub-ångström detail in an image. Unlike optical lenses however, electron lenses have unavoidable aberrations which affect the relative phases of the diffracted beams. This limits the effective image resolution of the best modern microscopes to around 1.5Å. In this section, we will briefly describe a transmission electron microscope, and the process of image formation.

Figure 3.33 is a schematic diagram of the main components of a transmission electron microscope. The electrons generated in the heated filament are accelerated at voltages between 100kV and 1000kV

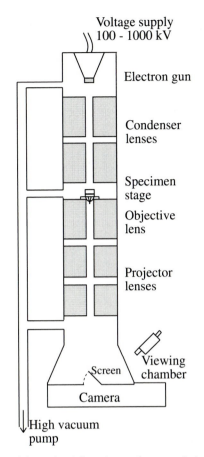

Figure 3.33. Schematic of the column of a transmission electron microscope.

(depending on the microscope) and focussed by the condenser lenses onto the specimen. The objective lens, immediately below the specimen, collects and focuses the diffraction pattern. Apertures of various sizes inserted into the back focal plane control how much of the diffraction pattern passes on to contribute to the image. As we have stressed throughout this chapter, the widely diffracted beams carry the information about the fine detail of the scattering matter in the object, and hence must be collected to achieve a high resolution in the image. The projector lenses control the magnification of the image formed on the fluorescent viewing screen. By changing the strength of these lenses (altering the lens current) either the diffraction pattern or the image can be focussed on the screen. Figure 3.34 shows an electron diffraction pattern with the position of the circular objective aperture shown by the circle. This aperture limits the number of diffracted beams which are allowed to contribute to the image.

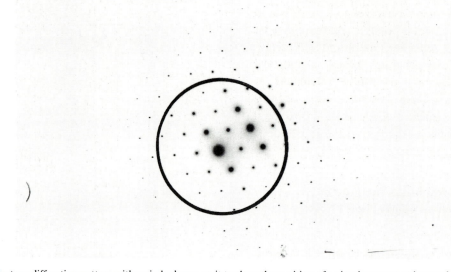

Figure 3.34. Electron diffraction pattern with a circle drawn on it to show the position of a circular aperture inserted to limit the number of diffracted beams which contribute to the high resolution image. All diffracted beams outside the circle are blocked by the aperture. (The uneven distribution of intensity is due to a slight misorientation of the crystal relative to the zone axis.)

Figure 3.35(a) is a high resolution image of the structure of cordierite. The contrast observed in this image arises from the differences in the electrostatic potential in the crystal, and an ideal image would be a true representation of the projected electrostatic potential. Dense rows of heavy atoms are regions of high electrostatic potential; regions of low average atomic number have a low projected potential. In general, projections of rows of atoms appear as dark spots on a photographic plate, although the dark–light contrast can be reversed by instrumental factors. Figure 3.35(b) is also a high resolution image of the same crystal of cordierite but under slightly different instrumental conditions, illustrating the image variation possible. Although the basic features of the image are the same, in Figure 3.35(b) the projected regions of low atomic density appear as white spots, the reverse of that in Figure 3.35(a). The insert in Figure 3.36 shows the projection of the cordierite structure superimposed on a magnified image of Fig.3.35(a). Although the electron micrograph is a faithful representation of the cordierite structure, it is still far from being able to resolve individual atoms in this material. (Further details of the structure of cordierite may be found in Section 6.8.4.)

The limitation in image resolution arises from a number of different factors. The most important is the fact that the electromagnetic objective lens can-

not be designed without some spherical aberration which introduces phase shifts in the diffracted beams. To obtain a true image of the object the lens must recombine the diffracted beams while preserving their relative amplitudes and phases. In a high-quality lens the phases are able to be preserved but only for beams near the origin and hence diffracted through small angles, thus effectively limiting the interpretable resolution. If beams diffracted through larger angles are included in the image forming process the extra apparent resolution is an artefact and may not relate to any fine details in the object. Defocussing the objective lens, such that the image is viewed in a slightly different focal plane than that for exact focus, also changes the relative phases and can partially correct lens aberrations. There is an optimal defocus which, combined with the lens aberrations, produces the best image. The objective aperture limits the collection of diffracted beams to those within the interpretable resolution.

Other instrumental effects which determine the resolution in the image are chromatic effects, which define the energy spread and hence wavelength, and the spatial coherence which defines the angular spread of the incident electron beam. Higher voltage electron microscopes have a shorter electron wavelength and therefore a potentially better resolution. They can also penetrate a thicker specimen. These advantages are offset to some extent by increased

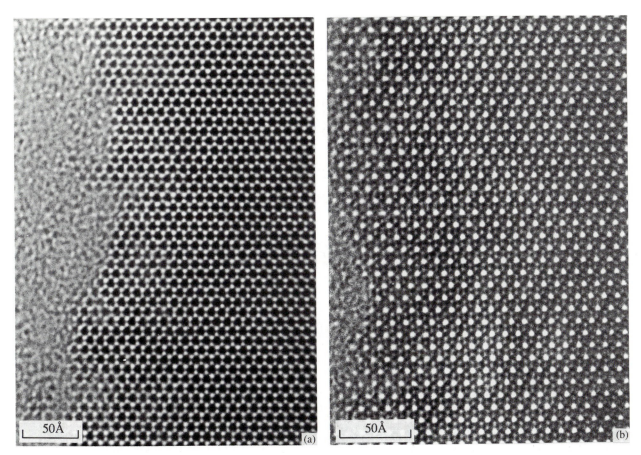

Figure 3.35. High resolution transmission electron micrograph of cordierite along [001]. In (a) the large dark areas represent projections of the channels in the structure along [001], and hence the light areas are projections of high electrostatic potential. In (b) the contrast is reversed with the light areas representing the low electrostatic potential (i.e. the channels).

chromatic aberration at higher voltages, and by increased costs.

Another difficulty in image interpretation arises from the fact that unlike X-rays, electrons are very strongly scattered by a crystal. Even in a thin flake, electrons can experience several scattering events. Multiple (or dynamical) diffraction affects the relative amplitudes and phases of the diffracted beams emerging from the specimen, so that the diffraction pattern does not represent a simple Fourier transform of the specimen. This becomes a more serious problem for thicker specimens and for those containing heavy atoms which are strong scatterers. In a thin specimen, if the experimental parameters are carefully controlled, a high resolution TEM image can be interpreted as a direct representation of the crystal structure. Experimental images are commonly compared with calculated images from computer simulations which can reproduce the multiple scattering of electron beams through a given crystal

thickness and their recombination by the objective lens.

The best attainable resolution of an electron microscope is at best around 1.5Å, and hence it cannot compete with X-ray diffraction in terms of defining precise atomic positions. There are some cases however, when only very small amounts of a specimen are available, insufficient for X-ray diffraction, and electron microscope imaging can contribute significantly to determining an unknown structure. The most powerful application of such high resolution imaging is in the study of defects in mineral structures, discussed in more detail in Chapter 7.

3.14 Other TEM imaging modes I: Lattice fringe images

From the above discussion it should be apparent that the nature of the image formed by recombining

10Å

Figure 3.36. A magnified image of Figure 3.35(a), with a projection of the cordierite structure superimposed on it. The structure is made up from 6-fold rings of tetrahedra which are stacked above one another forming channels parallel to the *c* axis, here shown as the dark areas. Each channel is surrounded by columns of Mg atoms, imaged as the white areas.

diffracted beams depends on how many beams are allowed to pass on through the objective aperture to the image forming process. If for example, a diffraction pattern such as shown in Figure 3.37(a) was obtained from a specimen, and a small objective

aperture placed in the beam path to block out all beams except the central undiffracted beam 000 and the two diffracted beams on either side, labelled 100 and $1\bar{0}0$, the resultant image would only contain information about the spacing of (100) planes in the

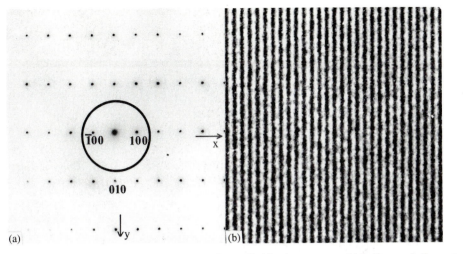

Figure 3.37. (a) Electron diffraction pattern showing the position of a small objective aperture which allows only the central beam and two diffracted beams, 100 and $\overline{1}00$ on either side to contribute to the image. (b) The resulting image shows only the (100) lattice planes.

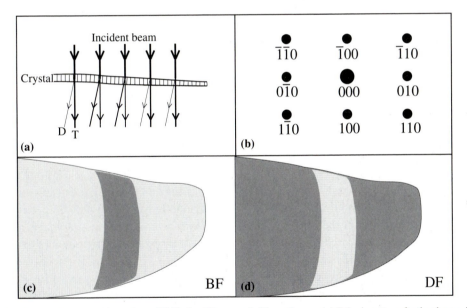

Figure 3.38. (a) An electron beam incident along [001] of a thin crystal. Since the crystal flake is not perfectly planar, in some parts the diffracting lattice planes, e.g. (010), will be strongly diffracting, while in other parts diffraction from the same planes will be weak. Hence the intensity of the diffracted beam (D) relative to the transmitted beam (T) varies from place to place (as indicated by the thickness of the line). (b) The **a*b*** electron diffraction pattern. (c) A bright field image using only 000 shows a bend contour of depleted intensity in the image in those parts which are strongly diffracting. (d) A dark field image using only 010 shows a brighter contour where 010 planes are strongly diffracting.

object (Figure 3.37(b)). The information about the rest of the structure has been filtered out by the objective aperture.

In many cases, lattice fringe images are sufficient to study defects and intergrowths in crystals, and are both easier to obtain and interpret than higher resolution images.

3.15 Other TEM imaging modes II: diffraction contrast imaging

Transmission electron microscopy need not involve the formation of high resolution images, and in many problems encountered in mineralogy *diffraction contrast imaging* is more relevant. The origin of diffraction contrast can be described by considering the nature of an image produced when only a single beam (the undiffracted beam or one of the diffracted beams) is allowed to pass through the objective aperture. All details of atomic structure are completely removed from such an image, and the resolution, defined by the size of the objective aperture, is about 10Å–20Å. The information contained in this type of image must relate to the information carried by any single beam, namely

(i) the shape of the electron diffraction spot defines the shape and size of the crystal flake being illuminated, and

(ii) the intensity of the spot depends on the extent of diffraction of the incident beam.

Consider a perfectly planar thin film with no defects, grain boundaries or intergrowths, uniformly illuminated by a parallel electron beam. The extent to which it diffracts and hence the intensity of the diffracted beams depends on its exact orientation relative to the incident beam, as discussed in Section 3.10. If the crystal film is perfect and planar, all parts of it diffract to the same extent. Every spot hkl in the diffraction pattern is the sum of the diffracted beams contributed equally from (hkl) planes throughout the film. The intensity of the undiffracted beam, 000, depends on how much of the incident beam intensity has been subtracted by the diffracted beams. The brightness of the image of the crystal film, formed using a single beam, will depend on the intensity of that beam. If all parts of the illuminated crystal have contributed equally to this beam, the image will have a uniform brightness throughout and be featureless.

However, even a perfect crystal film is never completely planar, and will have some degree of bending. As the Bragg angle for electron diffraction is so small ($<1°$), even small variations in the orientation of the film relative to the incident beam will strongly affect the extent of diffraction throughout the film. As shown schematically in Figure 3.38(a) some parts of a bent crystal film or flake will diffract more strongly and deplete the direct beam intensity more than other parts of the crystal. Assume that the incident beam in Figure 3.38(a) is along the [001] direction of an orthorhombic crystal. The electron diffraction pattern will be the **a*b*** orientation (Figure 3.38(b)) but different parts of the crystal, being in slightly different orientations, will have contributed unequally to the intensity of

each diffraction spot. For example, there may be a locus of points in the crystal film which are in the exact orientation for strong diffraction from (010) planes but only weak diffraction from other ($hk0$) planes. In this situation, an image of the crystal flake using only the undiffracted beam would show a dark (depleted intensity) contour corresponding to this locus of points where diffraction is strongest (Figure 3.38(c)). Conversely, an image using only the 010 diffraction spot would only be bright in those parts strongly diffracting into 010 (Figure 3.38(d)).

A diffraction contrast image therefore is a low resolution, but high magnification map of the extent of diffraction in different parts of a thin crystal flake. The power of the technique lies in the ability to be able to relate each spot in the diffraction pattern to some part of the image. The example given above described the formation of a *bend contour*, which defines that part of the crystal diffracting into a particular diffraction spot. In practice, when a diffraction pattern contains many strong spots, there will be many bend contours, each associated with diffraction by a particular set of lattice planes (Figure 3.39(a)).

A **bright field (BF) image** is one using only the undiffracted (often termed transmitted) beam to form the image.

A **dark field (DF) image** uses one of the diffracted beams to form the image. Each beam can be selected individually by the objective aperture (Figure 3.40) to form a dark field image which illuminates those parts of the image diffracting into this reflection.

Figure 3.39(a) is a bright field image showing complex bend contours indicating strong diffraction from a number of different sets of lattice planes. Figs.3.39(b) and (c) are two dark field images formed using, in turn, two different diffracted beams, say 100 and 010 respectively. The parts of the crystal where 100 planes are diffracting strongly are illuminated in Fig 3.39(b); those parts in which 010 planes are strongly diffracting are shown in Fig.3.39(c). In this example the thin foil of crystal is quite strongly buckled, giving well-defined bend contours. In practice, electron micrographs generally show softer, more diffuse diffraction contrast, either because they are less buckled, or because they are tilted to reduce the intensity of the diffracted beams.

The application of this method of imaging is not in studying the bend contours themselves, but in the way that the diffraction contrast is modified by

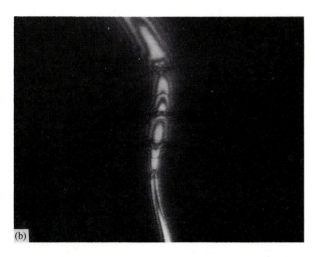

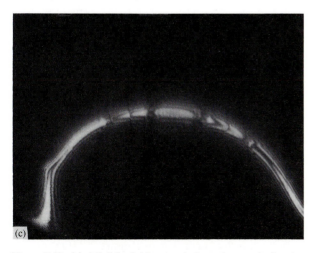

Figure 3.39. (a) A bright field transmission micrograph showing bend contours (dark) along which strong diffraction is taking place. (b) and (c) are two dark field images taken using two different diffracted beams. These show how each bend contour can be related to a specific diffracted beam, i.e. to diffraction from a particular set of lattice planes.

intergrowths and other defects. Figure 3.41 shows a bright field image of a thin polycrystalline specimen. The contrast here is due to the fact that the individual crystals are in different orientations and so diffract differently. The grain boundaries become visible because of the difference in diffraction contrast on either side.

We can illustrate the principle of bright field – dark field imaging using a relatively simple schematic example. In Figure 3.42(a) the bright field image of a thin edge of a crystal flake shows a microstructure consisting of parallel lamellae. At this stage we are not concerned with the nature of this lamellar intergrowth, except to point out that the different contrast in alternate lamellae is due to their slightly different crystallographic orientation. Figures 3.42(c) and (d) show electron diffraction patterns taken from each lamella separately i.e. from the areas A and B respectively. The diffraction pattern from the whole crystal (Figure 3.42(e)) is the superposition of these two diffraction patterns, one from each set of lamellae. By isolating a diffraction spot from one set of lamellae, and making a dark field image with this reflection, only this set will be visible (Figure 3.42(b)).

Very often mineral intergrowths are very fine-scaled and it is not possible to obtain separate diffraction patterns from each phase. The ability to relate each diffracted beam in the diffraction pattern to individual domains in an intergrowth makes diffraction contrast electron microscopy a very important method of studying internal microstructure in minerals.

Figure 3.43 shows a bright field – dark field pair of electron micrographs taken over a complex intergrowth in alkali feldspar. In the bright field image the diamond shaped regions (Na-feldspar, albite) are darker and hence are diffracting more strongly than the matrix between them (K-feldspar). A dark field image using one of these strongly diffracting beams 'lights-up' these albite regions. This micrograph has the added complication that the albite regions themselves are finely twinned and each twin diffracts slightly differently giving the striped appearance in the image. The origin of such intergrowths will be discussed in Section 11.3.4.

In describing defects, phase transformations and the origin of microstructure later in the book, we will make considerable use of bright field – dark field micrographs when further details of their interpretation will be discussed.

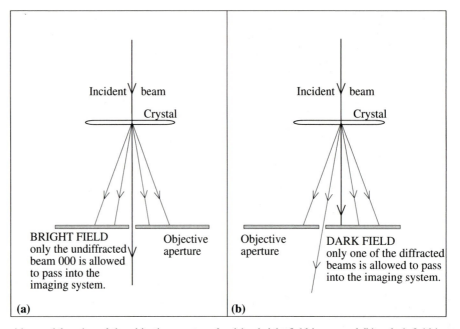

Figure 3.40. The position and function of the objective aperture for (a) a bright field image and (b) a dark field image. In practice the incident beam is tilted to obtain a dark field image, rather than the aperture off-centred, so that the selected diffracted beam is parallel to the microscope column.

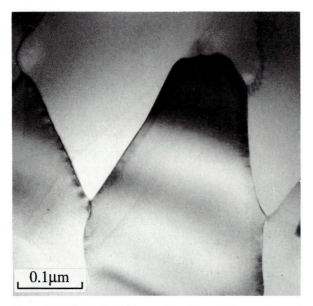

Figure 3.41. Bright field image of a polycrystalline specimen of cordierite, in which the different crystals are in different orientations, and hence diffract differently. This shows up the grain boundaries.

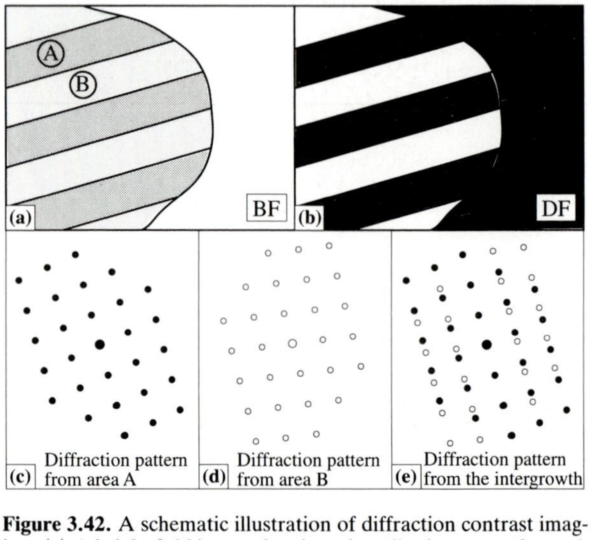

Figure 3.42. A schematic illustration of diffraction contrast imaging. (a) A bright field image showing a lamellar intergrowth, set A diffracting more strongly than B. (c), (d) and (e) are diffraction patterns from individual lamellae and over the whole intergrowth respectively. The dark field image in (b) is formed using one of the diffracted beams from set B, i.e. one of the spots drawn as open circles.

Another contrast feature of a single crystal flake arises due to thickness variations. Most commonly, in the preparation of thin specimens for transmission electron microscopy, the specimen is slightly wedge-shaped, thinner at the edge than in the bulk. This gives rise to *thickness fringes* (Figure 3.44(a)) due to the changing path difference for the transmitted and diffracted beams. Again, the way in which this diffraction contrast is modified by defects and intergrowths (Figure 3.44(b)) allows their observation

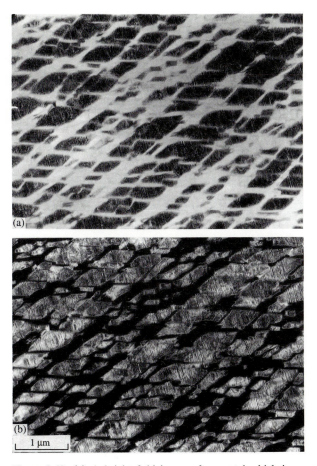

Figure 3.44. (a) Bright field micrograph showing thickness fringes parallel to the edge of a wedge shaped flake of a crystal. The fringes represent contours of equal thickness. (b) The diffraction contrast is modified by the presence of planar defects which diffract differently.

Figure 3.43. (a) A bright field image of a crystal which is an intergrowth of two phases: the darker diamond shaped 'islands' are diffracting more strongly than the pale matrix in between. An electron diffraction pattern of the intergrowth consists of a superposition of two diffraction patterns, one from the islands and one from the matrix. (b) By choosing a strongly diffracted beam to form a dark field image only the islands light up within a black matrix. This indicates that the chosen beam comes only from these islands. This micrograph has the added complication that the islands themselves are finely twinned and each twin diffracts slightly differently giving the striped appearance in the image. (Micrographs courtesy of P.E. Champness).

and analysis. The fine parallel lines cutting across the thickness fringes are planar defects, discussed in more detail in Section 7.3.1.

Figure 3.45 shows another example of diffraction contrast imaging. The two parallel lamellae are twins, in a different crystallographic orientation relative to the matrix. In a bright field image, the difference in the way the twins diffract relative to the matrix has the effect of displacing the diffraction contours.

In comparing diffraction contrast imaging with high-resolution imaging, it is worth pointing out that although the resolution in diffraction contrast imag-

Figure 3.45. Bright field image showing two parallel twin lamellae. Since the lamellae diffract differently from the matrix, the diffraction contours are displaced as they cross twin boundaries.

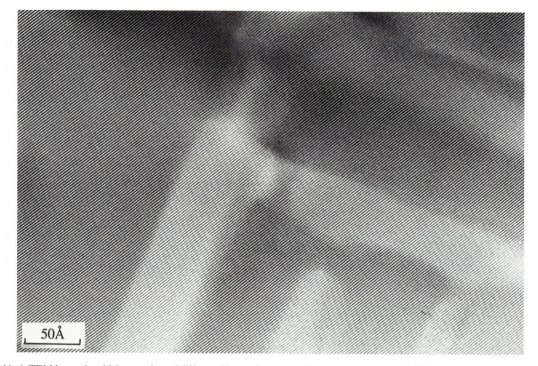

Figure 3.46. A TEM image in which a number of diffracted beams have contributed to the image which therefore shows lattice fringes, but also shows differences in contrast across a domain structure. The domains are visible because the domains and matrix diffract differently and some of these diffracted beams have been excluded by the objective aperture. Note how the lattice fringes bend slightly on passing across domain boundaries, a feature seen more clearly if the fringes are viewed at a low angle along their length.

ing is low, it can be a much more sensitive technique for studying subtle effects in a crystal. For example, strain in a crystal originating from very small displacements of atoms ($\ll 1$Å) can arise during structural transformations and around defects. Such displacements cannot be resolved by high-resolution imaging, but do result in local strains which have a considerable effect on the diffraction contrast. A small strain in a crystal has the effect of locally compressing or dilating lattice plane spacings, which in turn must slightly change the Bragg angle. This broadens the intensity distribution at a reciprocal lattice point. Diffraction contrast imaging, by using only a single reflection to form an image, isolates this effect from the information in the rest of the diffraction pattern. High-resolution imaging, by using all the diffracted information out to the resolution limit, is not able to image any spatial differences in the strength of diffraction into individual reflections.

We have treated these two imaging modes separately to draw the distinction between the two basic mechanisms of producing contrast in a transmission electron microscope. However, it is possible to include both contrast mechanisms in the image by using an intermediate objective aperture size which includes enough diffracted beams to be able to image the unit cell spacing, but excludes the more widely diffracted beams. Diffraction contrast will then arise if different parts of a specimen diffract differently into these outlying diffraction spots. Figure 3.46 shows an example.

3.16 Scanning electron microscopy (SEM)

To complete this chapter, we will briefly describe the scanning electron microscope (SEM) although the principles of its operation have little in common with transmission electron microscopy apart from the use of an electron source and a condenser lens to produce a focussed electron beam. This will be apparent from Figure 3.47.

Figure 3.48 shows the basic components of an SEM. A very fine 'probe' of electrons with energies up to about 30kV is focussed onto the surface of a specimen and scanned across it in a 'raster pattern of parallel lines. A number of phenomena occur at the surface under electron impact; the most important for scanning microscopy is the emission of *secondary*

Figure 3.47. An SEM image of an aphid examining a crystal of anglesite, PbSO$_4$, on a leaf.

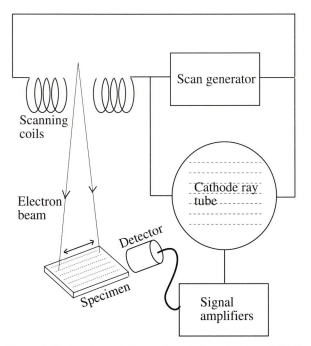

Figure 3.48. A schematic illustration of the principles of SEM imaging.

electrons with energies of a few tens of eV, and high energy *backscattered electrons*. Secondary electrons are emitted from the specimen due to collision with the high energy incident beam. The backscattered electrons are electrons from the incident beam which have interacted with atoms in the specimen and reflected out again. The intensity of both emissions is very sensitive to the angle at which the electron beam strikes the surface i.e. the surface topography.

The emitted electron current is collected and amplified and this signal is used to vary the brightness of the trace of a cathode ray tube being scanned in synchronism with the electron probe. The SEM image is thus a high magnification view of the topography of the specimen, analogous to that provided by a reflected light microscope. Suitable detectors which discriminate between different electron energies are available so that either a secondary electron or backscattered image can be produced.

The intensity of emission of the backscattered electrons is dependent on the average atomic num-

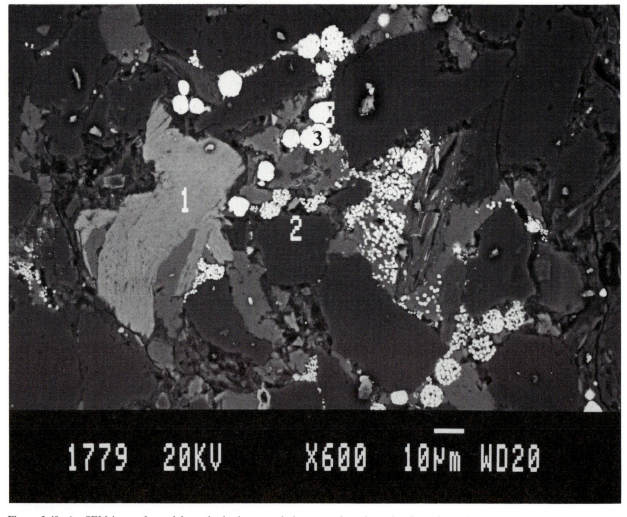

1779 20KV   X600   10μm WD20

Figure 3.49. An SEM image formed from the backscattered electrons whose intensity depends on the average atomic number of the specimen being bombarded. The specimen shown here is a polished rock slice of shale. The brightest grains are pyrite, FeS_2, the grey grains are ferromagnesian silicates such as mica and pyroxene, and the darkest contrast grains are quartz SiO_2. The numbered labels refer to crystals whose X-ray spectra are shown in Figure 3.50.

ber of the specimen, as well as surface topography. Heavy atoms produce many more backscattered electrons. Therefore, local variations in average atomic number result in variations in the contrast of the image. A backscattered SEM image of a polished slice of rock for example, would reveal the different minerals by their contrast differences (Figure 3.49).

A further important emission from a specimen when it is bombarded with electrons in either an SEM or TEM, are X-ray photons with wavelength and energy characteristic of the elements in the specimen. The spectrum of the X-radiation emitted

from a mineral (e.g. Figure 3.50) can be used for quantitative chemical microanalysis. In an SEM, by using a suitable detector, an image can be formed using the emitted X-rays using the same principle as in secondary electron imaging. The image is then an element distribution map in which the contrast variation reflects the concentration of any chosen element (Figure 3.51). X-ray emission from minerals is discussed again in Section 4.6.2.

Finally, if the scanned specimen is very thin, as in TEM, much of the intensity of the scanning electron beam will be transmitted and a detector placed under the specimen can be used to form an image in

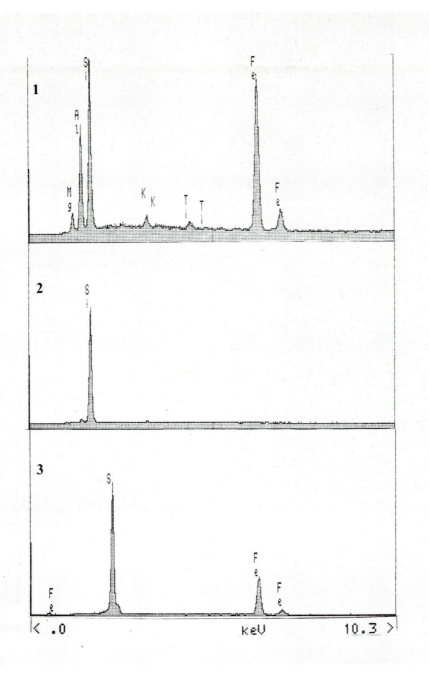

Figure 3.50. X-ray spectra taken from the numbered crystals in the shale shown in Figure 3.49. Spectrum 1 is from mica, a ferromagnesian aluminosilicate containing some potassium, spectrum 2 is from quartz, SiO_2 and spectrum 3 from pyrite, FeS_2. Note that elements with low atomic number such as O, do not appear in the spectra since their X-ray emission is too weak to be detected without special techniques.

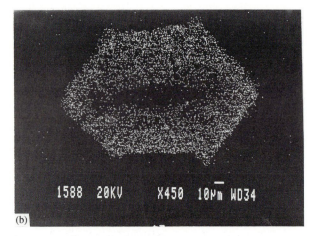

Figure 3.51. (a) A backscattered electron SEM image of a polished section through a crystal of $(Ba,Sr)SO_4$, with a core which is Sr-rich, and a Ba-rich boundary. A map of the distribution of each element can be made by forming an image using the characteristic X-ray emission. In (b) the image has been formed only by the Ba peaks in the spectrum, and hence shows the Ba distribution in the crystal. In (c) the Sr peak has been chosen for the image formation, and shows that the Sr is concentrated in the centre of the crystal.

the same way as in an SEM. This is the basis for a scanning transmission electron microscope (STEM). The intensity of the transmitted beam will vary in different parts of the specimen depending on the extent of diffraction, and so the interpretation of images is similar to that for diffraction contrast imaging in a TEM.

Bibliography

BUERGER, M.J. (1970). *Contemporary crystallography*. McGraw-Hill Book Co.

BUSECK, P.R. AND IIJIMA, S. (1974). High resolution electron microscopy of silicates. *Amer. Mineral.* **59**, 1–21.

DENT-GLASSER, L.S. (1977). *Crystallography and its applications*. Van Nostrand Reinhold.

GLUSKER, J.P. AND TRUEBLOOD, K.N. (1985). *Crystal structure analysis – a primer*. Oxford University Press.

HARBURN, G., TAYLOR, C.A. AND WELBERRY, T.R. (1975). Atlas of optical transforms. G.Bell and Sons.

HECHT, E. (1987). *Optics*. Addison-Wesley.

McLAREN, A.C. (1991). *Transmission electron microscopy of minerals and rocks*. Cambridge University Press.

VEBLEN, D.R. (1985). Direct imaging of complex structures and defects in silicates. *Ann. Rev. Earth Planet. Sci.* **13**, 119–146.

WENK, H.-R. (ed.). (1976). *Electron microscopy in mineralogy* Springer-Verlag.

4 Spectroscopic methods

The main aim of this chapter is to introduce spectroscopic methods as complementary techniques to diffraction in the study of the structure of minerals. In terms of the interaction of radiation with matter, diffraction involves a change in direction of the incident radiation, without any change in its energy. The determination of structure by diffraction depends on the periodicity of the structure and therefore produces a long-range or average picture. Any local disorder such as defects, impurities etc. appears in the diffraction pattern as diffuse scattering in the background, and in X-ray structure analysis in which only the Bragg reflections are used, this information is generally not used. In high resolution electron microscopy, the background scattering is automatically added into the Fourier synthesis carried out by the lens, and so the image can contain information about local disorder, although the resolution is limited, as explained in the previous chapter. Spectroscopic methods, on the other hand, provide information about local structure – site symmetry, coordination number, local chemical and crystallographic environment etc. and the methods do not depend on long-range periodicity or crystallinity. In many cases amorphous materials can be studied equally well.

Although there are very many different spectroscopic methods they all work on the same basic principle. Under some conditions, an incident beam of radiation can be absorbed by matter, or alternatively, can cause the emission of radiation from the material. The absorption and emission of energy arises when incident radiation of the appropriate frequency induces changes in the energy levels in the material. A wide range of phenomena contribute to the energy of a material at temperatures above absolute zero, and it is usual to describe these individual phenomena as if they possessed distinct reservoirs of energy. Each phenomenon is then associated with a range of energy – from the zero energy or *ground state* to higher energy *excited states* which it can occupy by absorbing some of the incident radiation. The energy range is expressed in terms of frequency or wavelength, related by the equation

$$E = h\nu = hc/\lambda$$

where h is Planck's constant (6.6×10^{-34} J s), c is the velocity of light (3×10^8 m s^{-1}), ν is the frequency (Hz or cycles/s) and λ is the wavelength (cms). E is the energy in joules.

The energy differences between ground and excited states cover the entire range of the electromagnetic spectrum from radio frequency ($\sim 10^6$Hz) to X-ray and γ-ray frequencies (up to $\sim 10^{20}$Hz). For example, in the lowest energy range, nuclei have energy levels associated with their spin. The differences between these energy levels is very small, in the radio frequency range. Electrons also have spin, and transitions between spin energy levels have energy in the microwave region. Rotations of molecular groups (e.g CO_3 groups in carbonates) are in the far infra-red range. Molecular vibrations, which involve stretching and bending of atomic bonds, have a higher energy, in the infra-red range. At still higher energies, electronic transitions involving valence electrons are associated with frequencies in the visible range, and those involving inner shell electrons at still higher energies, in the X-ray range. Each of these phenomena in a material can be studied by using incident radiation within a limited frequency range appropriate to the energy of the transitions (Table 4.1).

Spectroscopy measures the difference between energy levels in a material by measuring the energy of the absorbed or emitted radiation when the material is excited to a higher energy state, or as it decays back to the ground state. The intensity of the absorption or emission depends on the number of molecules, or atoms, or electrons (depending on the phenomenon being observed) moving between

Table 4.1. *The electromagnetic spectrum and phenomena causing absorption*

Nuclear spin resonance		Electron spin resonance	Molecular rotations – vibrations	Valency electron transitions	Core electron transitions	Nuclear transitions
radio frequency		microwave	far infrared / infrared	visible / ultraviolet	X-rays	γ-rays

Energy, log E (eV):
$-10 \quad -9 \quad -8 \quad -7 \quad -6 \quad -5 \quad -4 \quad -3 \quad -2 \quad -1 \quad 0 \quad 1 \quad 2 \quad 3 \quad 4 \quad 5$

Wavelength, log λ (m):
$4 \quad 3 \quad 2 \quad 1 \quad 0 \quad -1 \quad -2 \quad -3 \quad -4 \quad -5 \quad -6 \quad -7 \quad -8 \quad -9 \quad -10 \quad -11$

Frequency, log ν (Hz):
$4 \quad 5 \quad 6 \quad 7 \quad 8 \quad 9 \quad 10 \quad 11 \quad 12 \quad 13 \quad 14 \quad 15 \quad 16 \quad 17 \quad 18 \quad 19 \quad 20$

energy levels. The application of spectroscopy to the study of mineral structure depends on the fact that the differences between energy levels are often very sensitively determined by the local structural and chemical environment of the molecule or atom. Thus the incident radiation is a probe of *local* structure, measured by the frequency (energy) at which absorption or emission takes place. The local nature of the information derived from a spectroscopic method will be important when we compare diffraction and spectroscopy for studying phase transformations in minerals in later chapters.

Since the energy absorption and emission from matter is quantised, corresponding to the discrete energy levels which can be occupied, the theory of spectroscopy is closely linked with quantum theory and many of the developments of quantum theory arose from early spectroscopic observations. It would not be appropriate to attempt to cover this theoretical background even in a simplified way in a mineralogy book, and the aim here is to provide a brief description of each phenomenon and then describe some applications of each spectroscopic method. In later chapters when minerals are discussed in more detail, spectroscopic data will be used to illustrate certain aspects of their structure.

4.1 Nuclear magnetic resonance (NMR) spectroscopy

NMR spectroscopy is concerned with the energy differences between allowed spin states of atomic nuclei. The spin of a nucleus is described by I, which is called the spin quantum number. Nuclei with odd mass numbers have $I = n/2$, where n is an integer, e.g. ^1H ($I = 1/2$), ^{17}O ($I = 5/2$), ^{23}Na ($I = 3/2$), ^{27}Al ($I = 5/2$), ^{29}Si ($I = 1/2$). Nuclei with even mass numbers but odd atomic numbers have integral values of I, e.g. ^{14}N ($I = 1$). If both mass and atomic numbers are even, $I = 0$ (e.g. ^{16}O).

All nuclei for which the spin $I \neq 0$ have a magnetic moment. In the absence of an external magnetic field the spin states of the nucleus all have the same energy – the energy levels are said to be *degenerate*. When a strong external magnetic field is applied, it interacts with the magnetic moment of the nucleus, splitting the energy levels into groups. When a nucleus with spin I is placed in an applied magnetic field H_o, $2I+1$ energy levels exist, spaced by I, from $-I$ to $+I$. Thus when $I = 1/2$ the nucleus has two energy levels $-1/2$ and $+1/2$ and behaves as a magnetic dipole (Figure 4.1); when $I \geq 1$, the nucleus has more than two energy levels and behaves as a magnetic multipole.

Here we will mainly consider the simplest case when $I = \pm 1/2$, which includes the most important nucleus for mineral studies, namely ^{29}Si. The differences in the energy between nuclear energy levels are in the radio frequency range, and transitions between the spin states can be induced by applying a radio frequency field to the sample in a large static magnetic field H_o. As shown in Figure 4.1, increas-

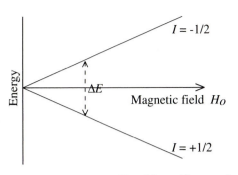

Figure 4.1. Splitting of the $I = -1/2$ and $I = +1/2$ energy levels in a magnetic field H_o.

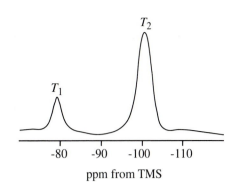

Figure 4.2. ^{29}Si NMR spectrum of cordierite representing Si atoms in two different environments, denoted T_1 and T_2. Since the area under the T_2 peak is four times that of the smaller peak, there are four times as many Si atoms in T_2 sites as there are in T_1 sites. (After Putnis and Angel, 1985.)

ing the magnetic field H_o increases the energy difference between the spin states. Magnetic resonance occurs when the applied radio frequency v is equal to $\Delta E/h$, where ΔE is the difference in the spin energy levels for the nucleus being studied, h is Planck's constant.

All nuclei of the same isotope in the same static magnetic field H_o should have the same resonance frequency. However, the *exact* resonance frequency depends on the local electronic environment of the particular nucleus, because electrons in the vicinity of the nucleus shield it to varying degrees from the applied magnetic field. Hence the static field experienced by the nucleus depends on the local chemical and crystallographic environment, which can be studied using this nuclear method. The 'environment' in this case means the nearest and next nearest atomic neighbours which are primarily responsible for the shielding. NMR spectroscopy is a technique for probing the environment of any specific isotopic nucleus which has a magnetic moment.

4.1.1 Chemical shift

The resonance frequencies in NMR spectroscopy are not expressed in energy terms, but as *chemical shifts*, which are differences in parts per million (ppm) between the resonance frequency of the nucleus in the sample to the same nucleus in a standard. The chemical shift δ is then defined as

$$\delta = \frac{v_{\text{sample}} - v_{\text{standard}}}{v_{\text{standard}}} \times 10^6$$

For example, in ^{29}Si NMR spectroscopy the standard used is tetramethyl silane (TMS), $(CH_3)_4Si$, and the overall range of chemical shifts for ^{29}Si in various silicate mineral structures is from -120ppm

to -60ppm. The negative values indicate greater shielding of the external field, less negative values indicating less shielding relative to the shielding in the standard. Less shielding inplies that the nuclei experience a larger value of the external field and resonate at a higher frequency. Figure 4.2 shows a ^{29}Si NMR spectrum with two peaks. Each peak represents Si in a specific environment, and the intensity of each peak is proportional to the number of atoms in that environment.

4.1.2 Spin relaxation times

In an NMR experiment a high power radio frequency pulse is applied to the sample slightly increasing the number of nuclei in the high energy spin state. The relaxation time is a measure of the rate at which spins in excited states decay back to thermal equilibrium i.e. to the ground state at the particular temperature. Due to the very small energies involved, this relaxation is not spontaneous as in visible or infra-red spectroscopy, and the emission of energy from the spin system must be stimulated by local atomic or molecular motions. The fluctuating magnetic fields due to these motions allow the magnetic energy to be dissipated. Thus the relaxation time itself depends on the immediate chemical surroundings of the nucleus and can be an important indicator of structural environment. In silicates these relaxation times can vary from microseconds to several thousand seconds. A small amount of paramagnetic impurity (e.g. $< 1\%$ Fe^{2+}) significantly reduces the relaxation time. A long relaxation time can be a time-consuming problem as spectra are collected from a large number of pulse–decay cycles, and quantitative information can only be extracted from the spectra if the spin system is allowed to relax between cycles.

The decay of the spin system between each radio frequency pulse is measured in the NMR experiment as a function of time. This *time domain* information is transformed via an interfaced computer into the *frequency domain* to produce the NMR spectrum. The time domain measurement uses a pulse with a sufficiently wide radio frequency range to span the entire range of likely resonances and hence the decay of all spectral lines is recorded simultaneously.

4.1.3 NMR of solids and magic angle sample spinning

As with many of the spectroscopic techniques, NMR was developed for the study of molecules in liquids and gases, and the application to solids came much later after further theoretical and experimental developments. The problem with spectroscopy of solids is that we are not dealing with free atoms and molecules, and interactions between atoms can seriously complicate the spectra. In the case of solid state NMR there are a number of factors which cause the broadening of resonance peaks which smears out the structural details.

1. For spin $I = 1/2$, dipole–dipole interactions between neighbouring nuclei results in a range of effective magnetic field strengths at individual nuclei producing a variation in the chemical shift which greatly broadens the NMR spectrum. For nuclei with $I \geq 1$, an additional effect due to the interactions of the nuclear magnetic moments with the local electric field gradients in a solid introduces further broadening of the resonance peaks.

2. If the shielding of the applied field for a particular nucleus in a site is markedly anisotropic, there will also be a variation in the chemical shift with direction of the crystal relative to H_o. This again results in a range of effective magnetic field strengths at individual nuclei and is termed the *chemical shift anisotropy* (CSA).

In liquids, rapid molecular tumbling averages these interactions to zero and sharp spectral lines are readily available, making NMR spectroscopy a principal technique for the study of species in solution.

In solids, there are a number of strategies for

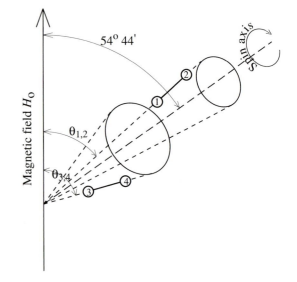

Figure 4.3. Magic angle spinning NMR in which the sample is spun at an angle of 54°44′ to the magnetic field direction. Although the vectors between nuclei 1 and 2 are at a different angle to the field than nuclei 3 and 4, spinning causes the time-averaged direction of these internuclear vectors to lie along the spin axis.

improving the resolution of the spectrum by removing these interactions. One approach is to use elaborate sequences of radio frequency pulses designed to reduce the interaction times between nuclei by averaging the spins. However, the approach which has made the greatest impact in mineralogy is *magic angle sample spinning* (MAS NMR) which involves physically spinning the sample at kilohertz frequencies about an axis oriented at 54.7° (the magic angle) to the applied static field H_o. The theoretical basis for this experimental technique lies in the fact that dipole–dipole interactions and chemical shift anisotropy, which are the major causes of line broadening, are described by equations which contain $(3\cos^2\theta - 1)$ terms, where θ is the angle between the applied field H_o and the vector joining the two interacting nuclei. When $\theta = 54°44′$, $(3\cos^2\theta - 1) = 0$ and the interaction terms which cause the line broadening also equal zero.

Although the different internuclear vectors in the sample have different values of θ and hence are rotated on different cones, by fast rotation their time-averaged direction lies on the spinning axis (Figure 4.3). The spinning frequencies must be comparable to the resonance frequency to be measured and are typically between about 2 and 10 kHz.

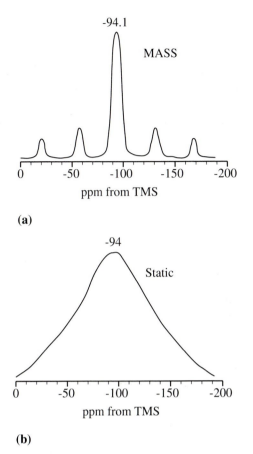

Figure 4.4. The effect of magic angle spinning on the NMR spectrum. (a) The ^{29}Si MAS NMR spectrum of a clay sample, showing a narrow peak at -94.1ppm, flanked by two sets of spinning sidebands. (b) The ^{29}Si NMR spectrum of the same sample without sample spinning. (after Kirkpatrick,1988)

and have quadrupole moments due to the non-spherical distribution of charge on the nucleus. There are now more possible energy levels and hence more transitions between spin states at each nucleus. The peak-broadening interactions between the quadrupole moment and the local electric field gradients are more complex and cannot be easily removed by magic angle spinning. In ^{27}Al NMR spectra for example, usually only the central $+1/2, -1/2$ transition is observed as this is free of strong first-order quadrupole effects, and is only broadened by much weaker second-order interactions. These interactions are diminished and modified by magic angle spinning but not entirely averaged to zero. The extent of interaction depends on the symmetry of the site occupied by the nucleus. The peak-broadening, distortion of the peak shape, and the displacement of the peak from the true chemical shift value all make quantitative interpretation of spectra of nuclei with spin ≥ 1 more difficult.

^{29}Si is the most widely investigated nuclide in minerals and, having spin 1/2, high resolution spectra of most of the common minerals have been obtained. The natural abundance of ^{29}Si is only 4.7% which has the disadvantage that collecting the spectrum takes longer, but the advantage that interactions between nuclei at such a low dilution are reduced. ^{27}Al has also been fairly widely studied and is 100% abundant, but has $I = 5/2$ with its associated problems. ^{17}O NMR spectra are potentially very important in mineralogy but ^{17}O has only 0.037% abundance and experiments have to be carried out on isotope enriched samples.

4.1.4 Assigning the peaks in an NMR spectrum

There is no direct way of assigning peaks to specific sites and extracting details of the local chemical and structural environment of a nucleus from the value of the chemical shift, as the theory has yet to be fully developed. The experimental developments in the 1980s enabled ^{29}Si NMR spectra to be obtained for a large number of silicate minerals whose structure was already well known and this formed the beginning of the application of NMR to mineralogy. A data base was built up from which ^{29}Si chemical shifts could be correlated with different structure types and empirical rules drawn up to relate them.

In virtually all silicates Si occupies a tetrahedral

One side-effect of the spinning is that it can produce a modulation of the radio frequency signal producing beats which appear in the NMR spectrum as *spinning sidebands* on either side of each resonance peak. They can be distinguished from the resonance peaks because they change position with changing spinning frequency. Figure 4.4 shows the ^{29}Si NMR spectrum of a clay sample illustrating the effect of spinning on the peak width, and also the appearance of spinning sidebands.

For nuclei with spin 1/2, MAS NMR can result in a spectrum with well-resolved peaks, almost as narrow as in solutions. Each peak represents a separate chemical and/or crystallographic environment for the particular nuclide being probed. By re-tuning the radio frequency, spectra can be obtained for different nuclides.

Extra complications arise however for nuclei with spin $I \geq 1$ which no longer behave as simple dipoles

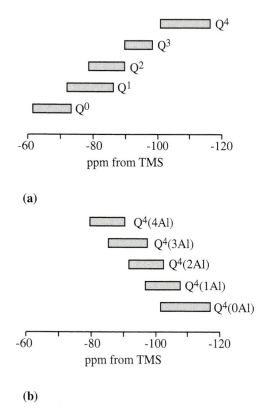

(a)

(b)

Figure 4.5. (a) Variation of the ^{29}Si NMR chemical shift for Si in tetrahedral coordination with the degree of polymerisation of the tetrahedra. Q^0 represents isolated SiO_4 tetrahedra in a structure, Q^1 are structures with pairs of tetrahedra sharing corners, Q^2 are structures with chains of tetrahedra, Q^3 are sheet silicates and Q^4 are framework silicates. When Al is also present in these structures there is greater overlap between the chemical shift ranges. (b) The correlation between the ^{29}Si chemical shift for framework aluminosilicates and the number of Al atoms in adjacent tetrahedral sites.

site, and silicate minerals are classified according to the degree of polymerisation of the tetrahedra, from structures in which the tetrahedra are isolated from each other, termed Q^o, to framework structures in which all tetrahedral corners are shared with other tetrahedra (Q^4) (see Chapter 6 for a more detailed discussion of silicate structures). The Q^n notation represents the number of bridging oxygens per tetrahedron; thus Q^o (isolated tetrahedra), Q^1 (dimers), Q^2 (chains), Q^3 (sheets) and Q^4 (frameworks).

The most useful correlations of ^{29}Si chemical shift are with the degree of polymerisation and with the number of Al nearest neighbours in the polyhedra adjacent to the Si tetrahedron. Increasing the degree of polymerisation from Q^o to Q^4 increases the degree of shielding and leads to more negative chemical shifts (Figure 4.5(a)). The presence of Al

in adjacent sites has the opposite effect. In framework aluminosilicates for example, each Si tetrahedron is connected to four other tetrahedra which may contain Si or Al (Figure 4.6). This is usually written Si(nAl) where n is the number of AlO_4 tetrahedra linked to the SiO_4 tetrahedron. The effect on the NMR spectrum is to shift the position of the Si peak about +5ppm for each Al atom added to the local environment (Figure 4.5(b)).

Another important correlation is that within any one structural group, e.g. Q^4(0Al) which are framework silicates with no aluminium, the variation in ^{29}Si chemical shift correlates well with the mean Si–O–Si bond angle (Figure 4.7).

The assignment of each peak in the ^{29}Si NMR spectrum to an Si site in the crystal structure is made with reference to the known average structure, although even this first step is not always straightforward, especially if there is a large number of sites with only slightly different environments. The empirical relationships discussed above can then be used for more detailed structural studies. Particularly useful in framework aluminosilicates is the correlation of chemical shift with Al nearest neighbour occupancy, Si(nAl), as it enables the local Al,Si distribution to be determined if the individual Si sites can be resolved. Understanding the Al,Si distribution (i.e. the degree of order of the Al,Si atoms in the tetrahedra) is important in describing the energy of aluminosilicates and is difficult to determine using X-rays due to the very similar scattering factors of Al and Si. Furthermore, diffraction methods determine the average distribution which need not be the same as the local distribution determined by NMR. This difference between local structure and average structure is very important in mineralogy and we will return to it in later chapters.

^{27}Al is the next most widely studied nuclide in minerals, but for the reasons described above, its NMR spectra are more difficult to interpret quantitatively to the same detail as ^{29}Si spectra. One unambiguous application of ^{27}Al NMR spectra is in determining the coordination of Al. Whereas Si in minerals is generally in tetrahedral sites, Al may occupy both AlO_4 tetrahedra and AlO_6 octahedra. NMR spectra provide a clear distinction between these coordinations with well separated chemical shift ranges. For AlO_4 tetrahedra ^{27}Al chemical shifts range from +50ppm to +80ppm; for AlO_6 octahedra the range is between $-$10ppm and

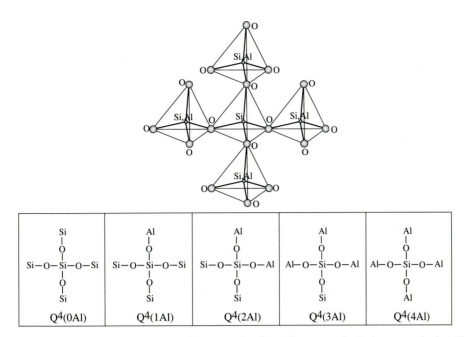

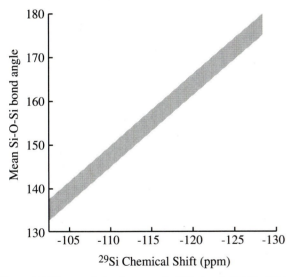

Figure 4.6. Each Si tetrahedron in a framework structure is connected to four adjacent tetrahedra by corner sharing. These neighbouring tetrahedra may contain either Si or Al atoms giving rise to the nomenclature Si(nAl), where n may be 0,1,2,3 or 4.

Figure 4.7. The correlation between the ^{29}Si chemical shift of Si in framework silicates and the average Si–O–Si bond angle between the Si atom and its four neighbouring Si atoms. Data points lie within the dark band.

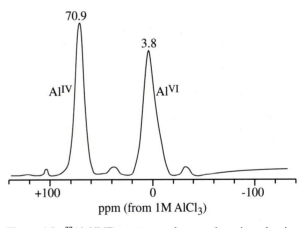

Figure 4.8. ^{27}Al NMR spectrum of muscovite mica, showing peaks for Al in tetrahedral (70.9ppm) and octahedral (3.8ppm) sites. The smaller peaks are spinning sidebands. (After Kirkpatrick, 1988.)

+20ppm. Figure 4.8 shows the ^{27}Al NMR spectrum of muscovite mica, showing Al in both tetrahedral and octahedral sites.

When all of the Al is in tetrahedral sites, as in the framework aluminosilicates, the ^{27}Al chemical shift becomes more shielded (more negative) with increasing mean Al–O–Si bond angle just as in ^{29}Si.

4.1.5 Applications

More detailed examples of applications will be discussed later in the book in the context of particular mineralogical problems, and here we merely summarize some of the important structural information which can be obtained from ^{29}Si and ^{27}Al NMR spectra of minerals.

1. Each line in the NMR spectrum represents a distinct site for the nucleus under investigation and the peak intensities are dependent on the relative proportions of Si and Al in each site. Determining the number of structurally different environments obviously depends on being able to resolve the chemical shift differences between them. In glasses for example, the local environments of Si and Al are not very different from crystals, but there is a wide variety of small variations possible between environments when there is no long-range crystallographic constraint on atomic positions. This has the effect of producing many more local environments with small differences which cannot be resolved, thus producing a broadening of the NMR lines in glasses relative to crystals.

2. The coordination of Si and Al can be readily determined in crystalline and amorphous materials.

3. The local environment of SiO_4 tetrahedra can be determined from the correlations of chemical shift with degree of polymerisation. This can be especially useful in studying silicate glasses and the speciation of silica in solution.

4. In framework aluminosilicates, the correlation of chemical shift with local $Si(nAl)$ linkages can lead to a measure of the local degree of Al,Si order.

5. Finally, one limitation of using NMR spectroscopy on natural minerals is that the presence of paramagnetic impurities (mainly Fe or Mn) can cause severe line-broadening effects due to the strong interaction of the unpaired d electrons with the applied magnetic field. The magnetic moment of the electron is about 10^3 times greater than the nuclear magnetic moment, and even though the electron spin has a very fast relaxation time, the nucleus senses a time averaged value of this moment which causes a shift of the nuclear resonance line. More than a few % Fe oxide can destroy the resolution of an NMR spectrum.

6. Many nuclei are suitable for NMR spectroscopy and so there are many possible applications not mentioned here, e.g. ^1H NMR spectra to study the structural environment of water and OH groups in minerals, and ^{13}C NMR in the study of coal and other hydrocarbon deposits. Both of these elements have been in standard use in NMR spectroscopy in organic chemistry for many years.

4.2 Electron spin resonance (ESR) spectroscopy

Electron spin resonance (ESR), sometimes referred to as electron paramagnetic resonance (EPR), is similar in principle to NMR. An electron has a spin S of 1/2 and an associated magnetic moment. In a magnetic field the two spin states $S = +1/2$ and $S = -1/2$ have different energies, just as in the analogous case for nuclei (Figure 4.1), although for electrons the energy of the $S = +1/2$ and $S = -1/2$ spin states are reversed (the spin and magnetic vectors are opposed for the electron, making the $-1/2$ state the lower energy). The energy difference for electron spin is in the microwave region (Table 4.1) and hence transitions between these spin states can be induced by applying microwave radiation to the sample within a magnetic field.

For technical reasons, ESR spectra are obtained by keeping the microwave frequency fixed and varying the magnetic field until the resonance is reached. The collection of the spectrum is by cycling the magnetic field strength over the resonance position. At resonance, the energy difference between the two states is given by

$$\Delta E = h\nu = g\beta_e H_o$$

where β_e is the Bohr magneton (magnetic moment of a free electron), H_o is the applied magnetic field, ν is the frequency and g is a proportionality constant termed the *g-factor*.

For an isolated free electron $g = 2.0023$, but may be very different in the environment of a solid. The value of g depends on the local environment of the electron, and hence on the oxidation state, coordination number and local distortions around the ion carrying the free electron. The position of the absorption peak is determined by the value of g and in this sense g in ESR is analogous to the chemical shift in NMR. The *g-factor* is strictly a tensor and its effective value may vary with direction depending on the symmetry of the paramagnetic centre. In single crystal studies the symmetry may be determined by observing the variation of ESR transition frequency with changing orientation of the crystal relative to the magnetic field.

An important difference to note between nuclear and electron resonance is that in electron resonance, transitions can only be induced by microwave radiation when there is an *unpaired* electron present i.e. a paramagnetic centre associated with an ion or defect. Most electrons occur in pairs with opposed spins as required by the Pauli exclusion principle, but unpaired electrons are present in transition metal ions, organic radicals, and certain types of defects. In mineralogy, ESR is primarily applied to the study of transition metal ions and their environment.

One problem with ESR spectroscopy in minerals is that if there is a high concentration of ions with unpaired electrons, spin – spin interactions between neighbouring electrons can greatly broaden the absorption peaks. This limits ESR studies to minerals with a low natural concentration of unpaired electrons, such as on defects or trace elements, or on doped minerals in which a low concentration (0.1% – 1%) of ions with unpaired electrons is substituted into the diamagnetic host structure. The assumption is then made that the site symmetry of the dopant is the same as that of the ion it replaces. ESR is very much more sensitive than NMR in detecting low spin concentrations, and the detection limit can be as low as 10^{11} spins (i.e. 10^{-11} molar concentration). Concentrations in the parts per billion range can be theoretically detected. Sensitivity is improved by keeping the sample at low temperature (e.g. liquid helium temperature, 4.2K), which increases the population difference between the energy levels.

ESR spectra are usually presented as the first derivative of the absorption versus the magnetic field, rather than the absorption peak itself (Figure 4.9). A negative slope in the derivative indicates a peak or shoulder in the absorption peak. A maximum in the absorption peak occurs when the axis of

the derivative is crossed with a negative slope, as at the centre of Figure 4.9(b), while a minimum in the absorption is equivalent to a positive crossing of the derivative axis. Figure 4.9 is the simplest spectrum obtained from a $(+1/2, -1/2)$ transition of a single unpaired electron. For the present we will limit the discussion to cases where there is only one unpaired electron per ion.

4.2.1 Hyperfine splitting in the ESR spectrum – the effect of the nucleus

An interesting feature frequently arises in the ESR spectrum when the unpaired electron senses the magnetic field associated with the spinning nucleus of the same atom, when its nucleus has non-zero spin, $I > 0$. This has the effect of modifying the external magnetic field experienced by the electron during the transitions between energy levels, producing extra resonance lines at equally spaced intervals along the magnetic field axis.

The extra energy contribution due to the interaction between the electron and the nucleus is proportional to the term AIS where A is the *hyperfine coupling tensor*, I is the nuclear spin number and S is the electron spin number. Transitions between energy states are always governed by *selection rules* which only allow certain transitions to take place. For this case the selection rules stipulate that for an electron spin transition, $\Delta S = \pm 1$ and $\Delta I = 0$. For example, if an electron with $S = 1/2$ interacts with a nucleus with $I = 1/2$, there will be two peaks in the ESR spectrum – one associated with the $+1/2$ to $-1/2$ electron transition interacting with nuclear spin $I = +1/2$, and the other with the same electron transition interacting with nuclear spin $I = -1/2$. If $I = 1$ there will be three peaks, the $\pm 1/2$ electron transition interacting with nuclear spins $I = -1$, 0 and $+1$.

In general, a magnetic nucleus of spin I will split the electron resonance into $(2I+1)$ lines separated by a distance which depends on the hyperfine coupling tensor A. The value of A defines the magnitude and direction of the interaction between the electron magnetic moment and the nuclear moment. Its magnitude is dependent on the nature of the absorbing paramagnetic centre and its immediate environment.

Figure 4.10 shows the effect of hyperfine splitting on an ESR spectrum. Analysis of this hyperfine

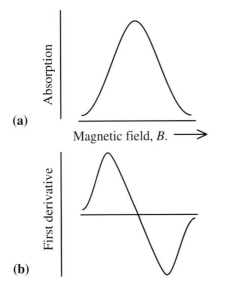

Figure 4.9. An electron resonance absorption peak and its first derivative. ESR spectra are normally displayed as the first derivative, and this curve is the simplest spectrum for a transition of a single unpaired electron.

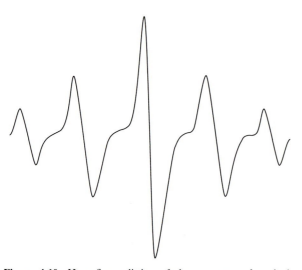

Figure 4.10. Hyperfine splitting of the spectrum of a single unpaired electron, due to the interaction between the electron spin transition and the nuclear spin.

splitting can provide information on the symmetry of the electron distribution around the paramagnetic centre and hence on the nature of the bonding between it and the surrounding structure.

In some cases there is an additional 'superhyperfine' splitting due to the interaction of the unpaired electron with the nuclear moment of a nucleus in a neighbouring ion.

4.2.2 Fine structure in an ESR spectrum when $S > 1/2$

So far we have assumed that the electron had only two spin states $+1/2$ and $-1/2$. This is only true if the paramagnetic ion has only one unpaired electron and the interpretation of ESR spectra is greatly simplified when this is the case. In many transition metal ions in minerals however there is more than one unpaired electron, e.g. Mn^{2+} has five unpaired $3d$ electrons in which case the total spin is $S = 5/2$. In general $S = n/2$ where n is the number of unpaired electrons, with $2S+1$ spin energy levels associated with different directions of S in a magnetic field. Transitions within these $2S+1$ levels are governed by the selection rule that $\Delta S = \pm 1$, which results in a spectrum with $2S$ lines, referred to as the *fine structure*.

In a spherically symmetric electric field, as in a free atom, all of these $2S+1$ energy levels are degenerate, and the application of an external magnetic field, H_o, splits the levels such that they are

equally spaced (Fig. 4.11(a)). For a given microwave frequency, the $\Delta S = \pm 1$ transitions will all occur at the same value of the magnetic field H_o and so only one line will be seen in the ESR spectrum. However, in a crystal structure the electric field is generally distorted, depending on the coordination of the ion, and therefore the effect of the unpaired electrons on each other is direction dependent, and will be different for each energy level. This results in a certain amount of splitting of the spin energy levels even in the absence of an applied magnetic field. This *zero field splitting* depends on the degree of distortion of the coordination polyhedron around the paramagnetic ion. A result of zero field splitting is that in a given applied magnetic field the $(2S+1)$ levels are usually unequally spaced in energy, and hence the resonances do not occur at the same value of the applied field H_o, as shown in Figure 4.11(b).

Complications can arise when both fine structure, and hyperfine splitting (and even superhyperfine splitting) exist and are superimposed in the ESR spectrum. As an illustration, Figure 4.12 shows the ESR spectrum due to trace amounts of Mn^{2+} in an amphibole mineral.

4.2.3 Applications

ESR spectroscopy is a very sensitive tool for detecting unpaired electrons, which in minerals are mainly due to the presence of transition metal ions and point defects. When these paramagnetic centres are present in dilute concentrations the spectra can be interpreted to provide information about the local structural and electronic environment. This can be applied to a wide variety of problems either in directly studying the local structure, or in using the paramagnetic centre as a probe to monitor processes and reactions in minerals. Some general applications are listed below.

1. The site in a mineral structure occupied by trace amounts of transition metals, and the local distortions around the site, can be determined. When the transition metal ion substitutes for a major element it can be used as a probe to study the local structure if the size mismatch between host and doping ion is not too great to alter the site geometry. This approach can be applied to both crystalline and non-crystalline materials.

2. The oxidation state of an element can be determined from the number of unpaired electrons.

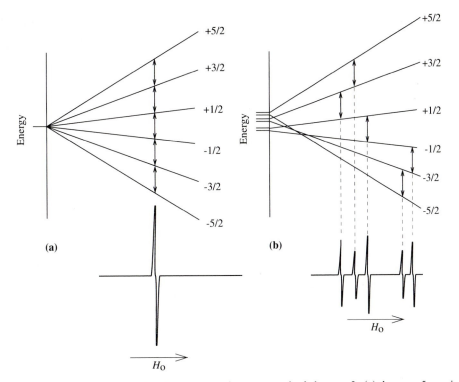

Figure 4.11. Fine structure in an ESR spectrum when there is more than one unpaired electron. In (a) there are 5 unpaired electrons and $S = 5/2$, as in Mn^{2+}. If the ion is in a spherically symmetric field, application of an external field H_o splits the 6 energy levels so that they are equally spaced. Transitions between the levels therefore occur at the same value of H_o, resulting in a single peak in the ESR spectrum. In (b) the ion is in an axially distorted field and zero field splitting occurs when H_o is zero. As H_o increases the resonance position occurs at different values resulting in 5 separate peaks.

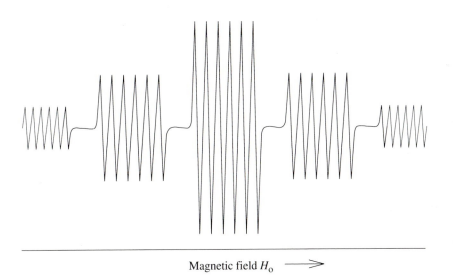

Magnetic field H_o ⟶

Figure 4.12. The ESR spectrum of Mn^{2+} in an amphibole, showing a combination of fine structure (5 groups of peaks) and hyperfine splitting (each group is further split into 6 peaks). (After Golding *et al.*,1972.)

In Fe^{2+} for example there are 4 unpaired $3d$ electrons, while in Fe^{3+} there are 5 unpaired electrons.

3. Changes in local structure as a mineral undergoes a phase transition from one structure to another as a function of temperature can be studied by

observing the temperature dependence of the ESR spectrum of a trace transition element.

4. Electrons or positive holes are formed in minerals by high energy radiation, and these are trapped at structural or chemical defects. The defects can then be studied by ESR. The role of point defects in minerals is discussed in Chapter 7.

5. ESR dating is a technique applicable to relatively recent ($<10^6$ years) minerals and makes use of the assumption that the defect concentration produced by irradiation from radioisotopes in a rock increases linearly with time. A calibration experiment is used to determine the relationship between the ESR signal intensity and the defect population induced by an artificial dose of radiation.

4.3 Vibrational spectroscopies – infra-red and Raman

Atoms bonded in molecular systems vibrate at frequencies between 10^{12} and 10^{14} Hz. The interatomic forces between bonded atoms control the exact frequency of the vibrations, in much the same way as a spring connecting two vibrating masses. Pairs of atoms or molecular groupings will have a number of fundamental vibrational frequencies, or vibrational modes, with which they can vibrate and these correspond to specific energy states of the molecule. These energies are in the infra-red range, and hence transitions between vibrational energy levels of a molecule can be induced by incident infra-red radiation, which will be absorbed at the frequency corresponding to the vibrational modes in the sample. The vibrational modes may also be studied by a light scattering experiment in which an incident light beam has its energy changed by an inelastic collision with the vibrating molecular groups. This is termed Raman scattering.

Both infra-red and Raman spectroscopies are plots of the intensity of absorption (in the case of IR) or scattering (in Raman) as a function of the frequency or wavenumber. Each peak in the vibrational spectrum corresponds to the energy of a particular vibrational mode within the sample.

4.3.1 Normal vibrational modes

The concept of a vibrational mode can be illustrated by considering the vibrations of atoms in a water molecule. In the classical description of atomic vibrations we can think of the vibrational motions of a molecule in terms of a model where the atoms are connected by springs whose stiffness represents the interatomic forces. If the springs obey Hooke's Law (i.e. the restoring force is proportional to the displacement from the equilibrium position) the model is harmonic. In Figure 4.13 the weaker force between the hydrogen nuclei is represented by a weaker spring which controls the increase or decrease of the H–O–H angle. Stronger bonds control the O–H bond length stretching, and so the bond stretching mode will vibrate at a higher frequency than the bond bending mode.

If one of the atoms is suddenly displaced relative to the others, the whole molecule will undergo some complicated vibration which will be some combination of bond bending and bond stretching. This complex motion can always be reduced to different proportions of *normal vibrations* of the system. A normal vibration is one in which all the atoms vibrate with the same frequency and move in phase, although they may have different amplitudes. The three normal vibrational modes of an H_2O molecule are shown in Figure 4.13. In general, a molecule containing N atoms will have $3N–6$ vibrational modes ($3N–5$ for a linear molecule such as OH).

If we know the bond lengths, bond angles and the force constants for bond stretching and bond bending (i.e. the strengths of the springs), the frequencies of the normal modes can be calculated, essentially by solving the equations of motion for the assembly of atoms. This is more difficult than it sounds, and we will return to this theme in a later section.

When a quantum mechanical treatment is applied

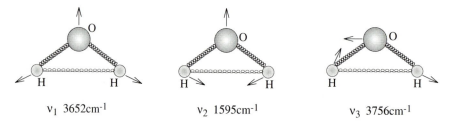

$$\nu_1 \ 3652 \text{cm}^{-1} \qquad \nu_2 \ 1595 \text{cm}^{-1} \qquad \nu_3 \ 3756 \text{cm}^{-1}$$

Figure 4.13. Representation of the three normal vibrational modes of a water molecule, the stronger O–H bonds represented by stronger springs, and the weaker H–H bonds by weaker springs. ν_1 and ν_3 are the symmetric and asymmetric stretch modes and ν_2 is the bending mode.

to vibrational modes the result is a quantised set of vibrational energy levels and so the vibrational spectra are distinct lines rather than a continuous range of energy. Another outcome of this treatment is the derivation of *selection rules* which explains why not all vibrational modes of a molecular group can be observed by IR or Raman spectroscopy. These selection rules define which vibrational modes can interact with the incident radiation. For a vibrational mode to be *infra-red active*, the oscillating electric field of the incident IR radiation must be able to interact directly with a changing molecular dipole moment associated with the vibration. If there is no change in the dipole moment (e.g. if the vibrational mode has a centre of symmetry) the mode is IR *inactive*. In general asymmetric vibrations give the strongest IR absorption peaks.

In Raman scattering, the incident light is responsible for *inducing* an instantaneous dipole moment by deforming the electron cloud around the molecule. If the displacement of the electron cloud corresponds to that of a vibrational mode, it is *Raman active*. The magnitude of this instantaneous dipole moment depends on how easily the electron cloud can be deformed, a property measured by the polarizability. In contrast to IR spectra, symmetric modes give the strongest Raman peaks, as they are associated with the greatest change in polarizability. The IR and Raman spectra of a material can be quite different and both methods are complementary. A more detailed discussion of the selection rules involves a treatment of the symmetries of molecular vibrations using group theory. This is beyond the scope of this book but some references may be found in the bibliography.

4.3.2 IR and Raman spectra

In an infra-red absorption experiment, infra-red radiation with a range of frequency is passed through a sample, which may be a powder or a single crystal, and the intensity of the transmitted beam is measured as a function of frequency. Since the vibrational frequencies of molecules are so high, the usual frequency units such as hertz (cycles/s) are not convenient. The most common unit is the *wavenumber* with units of cm^{-1}, which is defined as the inverse of the wavelength (in vacuum) of the radiation in cm. In terms of the more familiar frequency units, $1\ cm^{-1} = 3 \times 10^{10}$ Hz. Expressed as energy

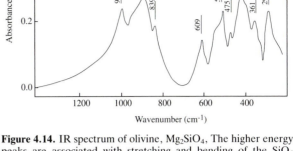

Figure 4.14. IR spectrum of olivine, Mg_2SiO_4, The higher energy peaks are associated with stretching and bending of the SiO_4 tetrahedra, while the lower energy peaks are due mainly to the MgO_6 octahedral vibrations. (After Jeanloz, 1980.)

units, $1\ cm^{-1} = 1.24 \times 10^{-4}$ eV (electron volts) $\equiv 11.97\ Jmol^{-1}$.

The IR absorption spectrum represents the set of IR active vibrational modes in the sample. The interpretation of a spectrum of a relatively simple mineral such as olivine, Mg_2SiO_4, shown in Figure 4.14 is quite complex. In general the higher energy peaks between about $500 cm^{-1}$ and $1000 cm^{-1}$ can be identified as stretching and bending vibrations of the SiO_4 tetrahedra, while the lower energy peaks are associated with the MgO_6 octahedral vibrations. However, in a crystal these vibrations cannot always be regarded as independent of each other, and in most cases there is coupling between various vibration modes in a crystal, so that a particular peak cannot be simply related to a molecular group as if it were isolated.

In Raman spectroscopy the sample is illuminated with monochromatic light, and as Raman scattering is a weak effect, this is usually from an intense source such as a laser. Some of the incident beam may be absorbed by transitions corresponding to the incident frequency, but most of the light is transmitted or scattered without loss of energy, a process termed Rayleigh scattering. A small fraction of the incident light however interacts with the vibrational modes in such a way that the vibrational state is raised or lowered in energy as shown in Figure 4.15. The Raman scattered light therefore loses or gains a small increment of energy which corresponds to the energy of the vibrational mode (i.e. $n = 0$ to $n = 1$ transition in Figure 4.15).

When energy is lost by the incident beam it is referred to as Stokes Raman scattering, and anti-

Stokes when the energy is gained. The energy loss process is the most probable since at normal temperatures the initial population of the $n = 1$ excited

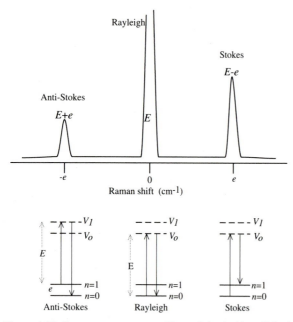

Figure 4.15. Raman spectroscopy. Most of the incident light is scattered with no energy change (Rayleigh scattering), but some of the incident photons may gain or lose a small amount of energy e from the vibrational modes in the crystal, giving peaks at $E+e$ (anti-Stokes scattering) and $E-e$ (Stokes scattering). The energy e is the energy of the vibrational modes in the crystal, and the Raman spectrum displays these modes as a shift from the central peak. (b) $n = 0$ and $n = 1$ are two vibrational energy levels in the crystal raised to V_o and V_1 respectively, by incident photons. These short lived 'virtual states' decay with the release of a photon which may have energy $E+e$, E or $E-e$. (After McMillan and Hofmeister, 1988.)

vibrational state may be very low. Usually only the more intense Stokes side of the Raman spectrum is analysed.

The Raman scattered light is detected by a spectrometer at right angles to the incident beam, and the spectrum plotted as intensity versus the energy shift (in cm^{-1}). The spectrum consists therefore of a strong central Rayleigh peak (at Raman shift 0) with additional weak peaks on either side at energy intervals corresponding to the Raman active vibrational modes in the sample. Figure 4.16 shows the Raman spectrum of olivine, Mg_2SiO_4, in the range $0–1200$ cm^{-1}.

4.3.3 Material identification and molecular groups

Since the set of vibrational modes is a characteristic 'fingerprint' of the chemistry and structure of the molecular groups in a sample, IR and Raman spectroscopies are widely applied for routine identification of materials, without necessarily knowing the structural origin of the individual spectral peaks. In some cases there are significant advantages over powder diffraction methods of identification, as the spectra are easy to obtain from solids, liquids and gases, and from materials with low atomic weight elements. Identification has to be made by comparison with a file of reference spectra, which are most widely available for organic materials.

In minerals, IR spectroscopy is commonly used to identify specific molecular groups such as hydroxyl

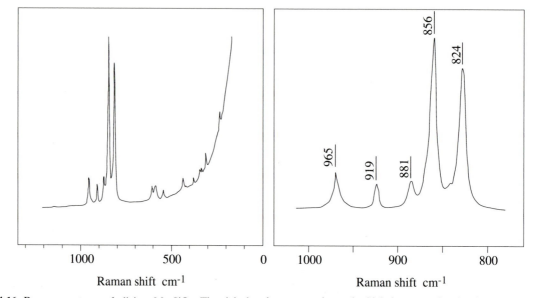

Figure 4.16. Raman spectrum of olivine, Mg_2SiO_4. The right hand spectrum shows the high frequency bands, due to SiO_4 tetrahedral vibrations, in more detail. (After Piriou and McMillan, 1983.)

groups, water molecules, and oxyanions such as CO_3, SO_4, NO_3, etc., all of which give intense IR and Raman peaks. Many minerals contain hydrous components, and often nominally anhydrous minerals such as quartz contain trace amounts of OH and H_2O which can play a very important role in modifying physical and chemical properties.

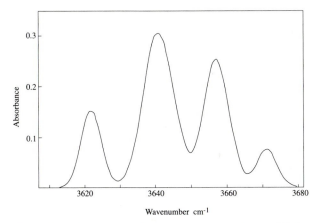

Figure 4.17. The OH-stretching frequencies in the amphibole mineral actinolite, $Ca_2(Mg,Fe)_5Si_8O_{22}(OH)_2$. The four peaks indicate that there are four different structural environments for the OH group in this sample. These different environments arise due to the distribution of cations in adjacent sites. (After Burns and Greaves, 1971.)

The OH stretching motion vibration for free ions occurs at around 3735 cm^{-1}, but in a mineral structure the exact frequency depends on the strength of the hydrogen bonding to the other surrounding anions. Figure 4.17 shows the IR spectrum in the OH stretching region of the amphibole mineral, actinolite. The four absorption features are due to the effects of different local cation environments around the OH ions. When the OH ion is strongly bonded in a structure this weakens the O–H bond and the OH stretch vibration occurs at lower energies. In single crystal spectra, the orientation of the OH bond can also be determined by using polarized incident radiation and observing the change in the absorption for different orientations of the crystal. As the strongly polar OH groups are efficient absorbers, this orientational dependence can be very pronounced. In this way the structural environment of both OH and H_2O groups has been studied.

For H_2O molecules, the symmetric and asymmetric vibrations occur in the 3800–3600 cm^{-1} region, while the bending motion is at ~1600 cm^{-1}. This can be seen in the spectrum of hydrated minerals such as gypsum, $CaSO_4.2H_2O$ (Figure 4.18). The dehydration of such materials can be observed directly in

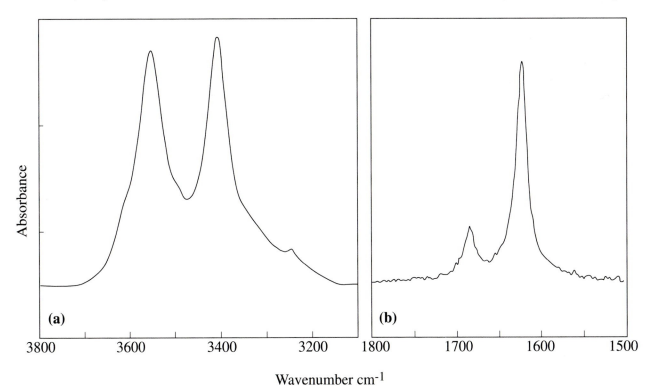

Figure 4.18. Part of the IR spectrum of gypsum, $CaSO_4.2H_2O$. (a) The OH stretch vibrations in the H_2O molecule. (b) The OH bending modes in the H_2O molecule. Note the broadness of the peaks from H_2O molecules compared to those in Figure 4.17 for the hydroxyl group vibrations. (After Putnis *et al.*, 1990.)

spectrometers equipped with a sample heating device.

In complex mineral structures such as silicates, 'molecular groups' are not always readily identifiable, especially where the structure is strongly bonded in three dimensions. In silicates with isolated SiO_4 tetrahedra, such as olivine Mg_2SiO_4 (see Chapter 6), it is a useful starting point to consider the vibrational modes of an SiO_4 tetrahedron. The Si–O stretching modes are in the high frequency range ($1100 cm^{-1}$–800 cm^{-1}), while the Si–O bending modes have a lower frequency ($650 – 475$ cm^{-1}). At even lower frequencies are the vibrations of Mg–O bonds (Figure 4.14). These rather generalized assignments are made by observing the effects of Si and Mg isotopic substitutions on the vibrational spectrum, although most of the modes cannot be related to simple molecular vibrations. The situation becomes more difficult in more highly polymerized silicates. Even a superficial understanding of complex spectra requires a more detailed treatment of vibrations in crystals rather than simply molecules. Some aspects of this will be discussed in the next section.

Therefore, apart from the study of specific molecules such as OH etc. in minerals, IR and Raman spectra are not, in a simple analysis, assigned to specific vibrational modes. Nevertheless, the information they contain is more than just an identifiable fingerprint of the sample. Two aspects will be mentioned here and developed further in later chapters.

1. When a mineral changes its structure as a result of changing temperature or pressure, the vibrational spectrum invariably changes in some way. A particular vibrational mode may change its frequency or intensity with temperature and this can be used to monitor structural changes in a much more sensitive way than by X-ray diffraction techniques, especially lattice parameter changes.

2. The vibrations in a crystal are the result of thermal excitation, and the way a mineral structure responds to changes in temperature (and pressure) is largely by changing its vibrational states. Thus most of the energy changes which define the thermodynamic properties of a mineral and hence its stability in various environments in the Earth are recorded in the vibrational spectrum. This link between macroscopic behaviour and microscopic properties is one of the central themes in mineralogy and has given

the study of vibrational spectra particular relevance. We shall return to this topic in Chapter 8.

4.4 More on vibrations in crystals – inelastic neutron scattering

So far we have discussed some aspects of vibrational modes in crystals from the point of view of their internal molecular vibrations. While in some cases it is possible to identify tightly bound molecular clusters in a complex structure, a crystal does not have the same degrees of freedom as isolated molecules and additional concepts relating to the rigidity and periodicity of the crystal lattice need to be introduced.

In a crystal, when a vibrating atom or molecular group is displaced from its mean position the vibrational motion of its neighbours is also affected. The result is that the individual vibrations of the atoms are correlated in such a way that the *internal modes* undergo a collective motion producing travelling waves through the crystal. These waves are a consequence of the periodicity of the crystal lattice, and are termed *lattice modes*.

We can illustrate some of the basic ideas about lattice modes by considering a simple example of a 'one-dimensional' crystal containing 2 atoms in the unit cell repeat a as illustrated in Figure 4.19(a). With the atoms vibrating normal to the chain length there are two different lattice modes of vibration

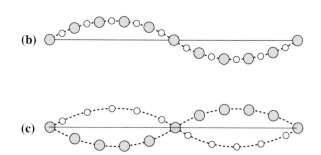

Figure 4.19. The sequence of atomic displacements involved in the transmission of a transverse wave along a diatomic chain. (a) Diatomic row of atoms with unit cell repeat a. (b) The acoustic mode, in which both atoms of the unit cell move in the same direction. (c) The optic mode, in which the two atoms move in opposite directions. The wavelength of the transverse vibrations is $10a$ in each case.

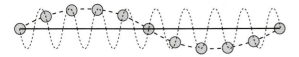

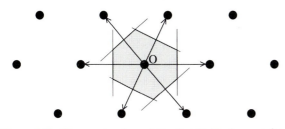

Figure 4.20. A transverse wave which can be described by a wave with $\lambda < 2a$, where a is the lattice repeat, can always be described equally well by a longer wavelength, in this case $\lambda = 10a$.

Figure 4.21. The construction of the first Brillouin zone for a two-dimensional lattice. The spots represent a part of the reciprocal lattice, and lines bisecting the six shortest reciprocal lattice vectors from the origin O define the shaded Brillouin zone.

possible: first with both atoms of the unit cell moving in the same direction (Figure 4.19(b)) or moving in opposite directions (Figure 4.19(c)). In both cases the oscillations are correlated in such a way as to generate transverse travelling waves along the chain. The first case (Figure 4.19(b)) is termed an *acoustic mode*, as these are the kinds of waves responsible for the propagation of sound waves through a crystal. The second (Figure 4.19(c)) is termed an *optic mode*, as in an ionic crystal it generates a dipole moment and hence can interact with light.

In our example the displacements are repeated at intervals of 10 unit cells and hence the lattice mode has wavelength $\lambda = 10a$. The long wavelength limit tends to infinity (i.e. the size of the crystal), which in the case of an acoustic mode would correspond to translations of the entire crystal, without individual atomic oscillations. In an optic mode the individual atoms still vibrate relative to each other even in the long wavelength limit. We will return to this point below.

The shortest wavelength possible in such a chain has $\lambda = 2a$, with adjacent unit cells vibrating in antiphase. This wave is now a standing wave, rather than a travelling wave. All possible patterns of displacement and all possible frequencies of lattice modes are described by waves between these two limits of wavelength, $2a \leq \lambda \leq \infty$. This is further illustrated in Figure 4.20 which shows that the motion of lattice points described by a wave with $\lambda < 2a$ can equally well be described by a longer wave within the above wavelength range.

The same general description applies to longitudinal waves in which the atomic displacements are parallel to the direction of propagation of the wave. Thus there are longitudinal acoustic (LA) and longitudinal optic (LO) modes as well as the transverse acoustic (TA) and transverse optic (TO) modes described here. There are always twice as many transverse modes as longitudinal modes.

4.4.1 Waves in reciprocal space

In a three-dimensional crystal the vibrations produce a complex situation with waves of various amplitudes, wavelengths and phases travelling in all possible directions. As was the case in diffraction (Chapter 3) it is more convenient to describe lattice modes in reciprocal space. Each lattice vibration is described by a wavevector **k**, where **k** is the direction of propagation of the wave. The magnitude of **k** is $2\pi/\lambda$, where λ is the wavelength of the lattice mode. Hence **k** is a vector in reciprocal space, also often referred to as **k**-space.

[For historical reasons, reciprocal space on this branch of the subject is multiplied by 2π, relative to that used in diffraction theory, and hence the reciprocal lattice vector normal to a lattice plane spacing a has a length $2\pi/a$ rather than $1/a$ as described in Chapter 3.]

The infinite wavelength limit of vibrational waves is now described by $\mathbf{k} = 0$, i.e. the origin of reciprocal space, while in the one-dimensional example in Figure 4.20 the short wavelength limit is described by $\mathbf{k} = 2\pi/2\lambda = \pm \pi/a$ where a is the real lattice repeat. The positive and negative signs represent waves travelling in opposite directions. All possible wavevectors describing lattice modes lie between $\mathbf{k} = 0$ and $\mathbf{k} = \pm\pi/a$ in reciprocal space. The vector $\mathbf{k} = \pi/a$ is one half of the distance to the reciprocal lattice point (at $2\pi/a$) in this direction and lies at the boundary of a volume of reciprocal space, centred around the origin, termed the *Brillouin zone*.

In three dimensions the Brillouin zone defines the short wavelength limits of all wavevectors in the crystal. The Brillouin zone is the smallest volume of reciprocal space entirely enclosed by planes that are perpendicular bisectors of the reciprocal lattice vectors drawn from the origin. Figure 4.21 shows the construction of the Brillouin zone for a general two-dimensional lattice.

The complete description of lattice vibrations of a crystal is given by the relationship between the frequency of the normal vibrational modes and the wavelength of their propagation through the crystal for the whole range of wavelengths corresponding to wavevectors **k** in the Brillouin zone. This relationship is called the *dispersion relation* and is described by a plot of frequency against the wavevector **k** for each normal mode.

4.4.2 Dispersion curves for normal vibrational modes

We can now bring together the ideas outlined so far in this section to describe the dispersion relations in crystals. In the simple diatomic chain example (Figure 4.19) we defined the difference between acoustic and optic modes, and pointed out that in the long wavelength limit the frequency of acoustic modes is zero, while this was not the case for optic modes. Figure 4.22 shows the general form of the acoustical and optical branches, as they are known, of the dispersion relation for a diatomic chain. The limiting wavelengths are at **k** = 0, termed the zone centre, and **k** = π/a, at the zone boundary.

In three dimensions the general form of the dispersion curves is similar and they are usually described for a number of different directions in reciprocal space. It can be shown that in a crystal with N atoms in the primitive unit cell there are $3N$ normal modes: 3 acoustic modes in which the atoms vibrate in phase and $3N-3$ optic modes in which adjacent atoms vibrate with a phase difference. For

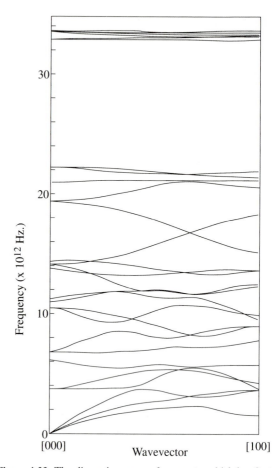

Figure 4.23. The dispersion curves for quartz, which has 9 atoms per unit cell, and hence has 3 acoustic modes (frequency zero at **k** = 0), and 24 optic modes. Each mode of vibration is represented as a separate branch on the diagram. The six optic modes isolated at high frequencies are due to the internal Si–O stretching modes within the SiO_4 tetrahedra, while the lower frequency modes involve motion of Si atoms in adjacent tetrahedra relative to one another, usually referred to as Si–Si stretching modes. (after Kieffer, 1985)

any real mineral structure therefore, the dispersion diagram is very complex, but will have a general form similar to that in Figure 4.23 which shows a typical set of dispersion curves for quartz, SiO_2, with 9 atoms per unit cell.

4.4.3 Phonons – quantum units of vibrational energy

In a rigorous treatment of lattice vibrations, the energy of the whole crystal, which is made up of the superposition of all the individual atomic vibrations, must obey the rules of quantum mechanics. Its energy can only increase or decrease in discrete steps, and the lattice vibrations have a dual wave–particle nature, just as electromagnetic radiation or

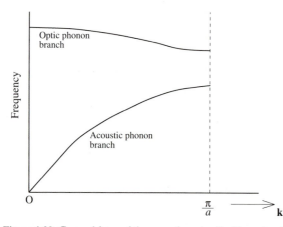

Figure 4.22. General form of the acoustic and optical branches for a diatomic chain, as a function of the wavevector **k**, from the long wavelength limit (**k** = 0, λ infinite) to the boundary of the Brillouin zone (**k** = π/a, λ = 2a). Such curves are termed dispersion curves.

atomic particles. The particle aspect of lattice vibrations is called a *phonon*, by analogy with the photon of the electromagnetic wave. The phonon is the quantum unit of vibrational energy of the crystal.

The transmission of vibrational waves, such as heat and sound, through a crystal is then described as the movement of phonons, each carrying a quantum of energy. If the temperature is raised, the amplitude of the atomic vibrations increases, and this corresponds to an increase in the number of phonons in the crystal.

One of the important properties of a crystal is its specific heat, and this is determined by the frequency distribution of the phonons. The terminology used is the *phonon density of states*, which is the number of discrete states per unit energy (or frequency or wavevector) interval.

4.4.4 Experimentally determining dispersion curves

In the previous section we briefly described infra-red absorption and Raman scattering as spectroscopic methods of studying the vibrational modes in crystals. For light to interact with a lattice vibration they both must have comparable wavelengths. Typically IR radiation has a wavelength around 10^5Å, while the laser light used in Raman spectroscopy is in the visible region with wavelength around 5000Å. These waves therefore can only interact with long wavelength lattice modes. Thus both IR and Raman spectroscopy are only capable of determining the frequency of normal modes at near the centre of the Brillouin zone, $\mathbf{k} = 0$. No information about the dispersion relation in the rest of the Brillouin zone can be determined with these methods.

In terms of the phonon description of lattice modes, Raman scattering involves either the creation or annihilation of phonons by the incident photons which themselves either lose (Stokes shift) or gain (anti-Stokes shift) energy in the process. Raman scattering involves optic phonons. A related technique which involves the creation or annihilation of acoustic phonons by incident photons is called *Brillouin scattering*.

To probe the rest of the Brillouin zone and determine the dispersion relation for the normal modes, the incident radiation must have a similar frequency to the vibrational mode as well as a wavelength comparable to interatomic spacings in crystals. Both of these requirements are satisfied by thermal neutrons which have a wavelength between 1Å and 5Å and frequency in the range 10^{12}–10^{13} Hz. Thermal neutrons are produced by slowing down high energy neutrons from a reactor or pulsed source by passing them through a moderating material. Such neutrons can scatter elastically and be used in diffraction experiments, or inelastically in which case some energy is lost or gained by the scattering process. Since the energy of a neutron is comparable to the phonon energies in solids, the creation or annihilation of a phonon means that the change in the neutron energy and direction is large and readily measured. In principle, inelastic scattering of X-rays could also be used to study phonons in crystals but because the energy change associated with creating or annihilating a phonon is so much smaller than the energy of the incident X-rays, it is very difficult to detect.

In an inelastic neutron scattering experiment the wavelength and direction of the incident beam can be chosen by using a suitable monochromator and the energy gain or loss of the scattered neutrons measured as a function of the scattering direction. The experiment involves very large scale apparatus, available in less than a dozen laboratories throughout the world, and a full determination of the dispersion relations is a complex procedure. The low intensity of neutron beams means that experiments take a long time and need large single crystal samples, which are not always readily available. Although the information obtained from IR and Raman spectroscopy is much more restricted, the spectra are very easy to obtain from very small single crystal or polycrystalline samples.

4.4.5 Applications

The ultimate aim of most studies of vibrational behaviour in minerals is to gain a better understanding of their behaviour and properties from first principles. The vibrational spectrum is determined by the strength and nature of the interatomic bonding forces which hold a crystal together, and is also directly related to the bulk thermodynamic properties. As well as the experimental methods for determining vibrational modes and phonon dispersion curves, considerable efforts are being made to develop methods of calculating these vibrational properties from computer simulations of the crystal structures and models of the interatomic forces.

These models are then used to calculate heat capacities, entropies and thermal expansion coefficients, and ultimately to predict the relative thermodynamic stability of minerals in a wide variety of physical environments. Although calculations of the thermodynamic properties from computer simulations is still in its early stages, the possibility of obtaining these quantities under conditions where experiments are very difficult (e.g. mantle conditions) will continue to stimulate improvements in these methods.

4.5 Optical spectroscopy – visible and ultraviolet

Optical spectroscopy is concerned with transitions of electrons between outermost energy levels, which have energy differences in the range from near infra-red, through visible, to ultra-violet (Table 4.1). Such optical transitions are often responsible for the colour of minerals, produced when absorption occurs in some part of the visible spectrum. A detailed analysis of the optical spectrum provides information about electronic energy levels of the absorbing ion, and how they are affected by local structure such as the type of coordination, its symmetry and distortions, bond types, as well as interactions with neighbouring anions and cations.

The basic nomenclature of electronic structure and quantum numbers is reviewed in Appendix A.

There are various types of electronic transitions which can be observed in the UV to IR spectral range and those of most importance in minerals are:

1. Transitions in which an electron from one orbital is promoted to a higher energy orbital in the same atom. Usually this involves electrons in the unfilled *d*-orbitals of ions in the first row transition elements such as Cr^{3+}, Mn^{2+}, Mn^{3+}, Fe^{2+} and Fe^{3+}. These transitions involve absorption in the visible to near infra-red and are the major cause of colour in minerals. They are often called crystal field spectra for reasons which will be explained below.

2. Transitions in which an electron from an orbital on one atom is promoted to a higher energy orbital on an adjacent atom. There are two ways in which this can occur.

(i) Electron transfer from an anion to a cation e.g. transfer from a filled oxygen *p* orbital to a partially occupied Fe^{3+} *d* orbital. These transitions occur in the ultra-violet region with a high probability and produce a very intense

absorption, up to 10^3 to 10^4 times higher than those of crystal–field transitions.

(ii) Electron transfer from one cation to a higher energy orbital in an adjacent cation. The cations must have different oxidation states and the transfer generally occurs between cations in adjacent edge-shared coordination polyhedra, especially when metal–metal interatomic distances are small. Absorption bands for these *intervalence charge transitions* (IVCT) occur principally in the visible region with intensities between 10^2 and 10^3 times stronger than transitions between *3d* orbital energy levels.

3. Of less importance in minerals are electron transitions from a localized orbital on an atom to the conduction band which is a delocalized energy band characteristic of the whole solid. Such transitions occur in sulphide minerals with absorption in the visible/ultra-violet range e.g. the red colour of cinnabar, HgS, is due to the absorption of wavelengths shorter than 600nm. In silicates this absorption takes place at higher energies, far into the ultra-violet.

The principal features of a typical absorption spectrum in the UV to IR spectral range are shown in Figure 4.24. Energy units used in optical spectroscopy are often quoted in terms of wavelength, as well as wavenumber. At the UV end of the spectrum an absorption edge occurs beyond which the transmittance of the sample virtually drops to zero. This feature is due to oxygen $\Rightarrow$ metal charge transfer, and in some cases due to electron transfer to the conduction band. The energy at which oxygen $\Rightarrow$ metal transfer occurs depends on the cation and the symmetry of its coordination site. For octahedrally coordinated cations the energies decrease in order from $Cr^{3+} > Ti^{3+} > Fe^{2+} > Ti^{4+} > Fe^{3+}$. For Fe^{3+} in tetrahedral coordination the charge transfer occurs at lower energies, and may extend well into the visible region.

At frequencies below the absorption edge a number of broad peaks or absorption bands occur. Strong absorption bands in this region may be the result of intervalence transitions, and the weaker features are due to the intra-electronic *d–d* crystal field transitions. Most mineralogical studies have concentrated on this region, both in studies of local structural environment around transition metal cations as well as for the determination of the origin of colour and pleochroism in minerals.

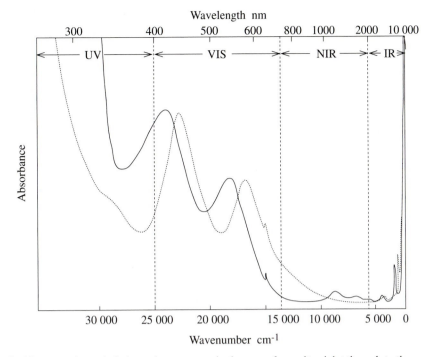

Figure 4.24. The principal features of a typical absorption spectrum in the range from ultraviolet through to the near infra-red, shown for two different garnet samples, one with 5.5 wt% Cr_2O_3 (solid line), the other with 17 wt% Cr_2O_3 (dotted line). (After Langer, 1987.)

Optical spectra are obtained by passing light through a polished crystal slice which must be thin enough to transmit. Spectrometers can also be fitted to microscopes allowing work on samples a few tens of micrometres in diameter.

4.5.1 Crystal field spectra

Understanding the origin of crystal field spectra involves a more detailed description of the nature of d-orbitals and their energy when the ion is in various crystallographic environments. Crystal field theory provides a simple illustration of the principles which define how electron energy levels are affected by their environment, and the description of the inter-action of the symmetry of the various energy states with the symmetry of the environment are applicable in a general way to other spectroscopic methods.

The d electrons are divided into 5 orbitals, with shapes represented by the angular distribution pro-babilities for electrons in each orbital (Figure 4.25). On the basis of this angular distribution the 5 orbitals are divided into 2 groups, termed t_{2g} and e_g (the terminology comes from the symmetry of the wave functions of the electrons, but need not con-cern us here). Each orbital can accommodate 2

electrons spinning in opposite directions, $m_s = \pm \frac{1}{2}$ and so 10 electrons can occupy the d-orbitals. Partly filled d-orbitals are occupied with electrons spread-ing out over as many orbitals as possible, one electron in each before any pairing of spins takes place. This distribution is termed the high-spin state and minimizes inter-electron repulsion (Hund's rule). The alternative arrangement where spins are paired where possible is termed the low-spin state.

In a free ion, or an ion in a spherical field, all of the $3d$ electrons have equal probability of being in any of the $3d$ orbitals, since they all have the same energy. When such an ion is in a crystal structure the relative energies of the $3d$ orbitals are controlled by the different repulsive energies of the anions coordi-nated to it. This is where the symmetry of the orbitals is important, for if the lobes of the electron distributions of the different orbitals (Figure 4.25) point towards the anions, electrons in these orbitals are repelled to a greater extent and their energy is raised relative to orbitals which project between the anions. This can be illustrated by considering a transition metal ion in octahedral coordination.

Figure 4.26 shows an octahedron with anions at the corners, and a transition metal cation at its centre. By comparing this figure with Figure 4.25,

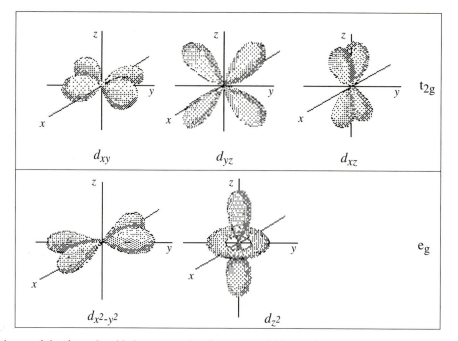

Figure 4.25. The shapes of the d atomic orbitals, representing the regions which contain most of the electron density. The orbitals are classified into two groups, t_{2g} and e_g on the basis of their orientation relative to the x,y and z axes.

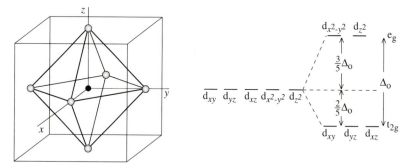

Figure 4.26. Crystal field splitting in octahedral coordination. When the transition metal cation (black sphere) is octahedrally coordinated by 6 anions (shaded spheres), the t_{2g} d orbitals point between the anions, while the e_g orbitals point directly towards the anions (see Figure 4.25). This raises the energy of the e_g orbitals, and lowers the energy of t_{2g} orbitals relative to an ion in a spherical field. The splitting of these energy levels is designated Δ_o.

we can see that the lobes of the two e_g orbitals point towards the anions, while the lobes of the three t_{2g} orbitals lie between the anions. This results in an energy reduction for electrons in t_{2g} relative to those in the e_g orbitals, as shown in the accompanying energy level diagram in Figure 4.26. The splitting of these energy levels, designated Δ_o, is such that the average energy stays the same. Thus if all five orbitals were each occupied by one electron, the two electrons in e_g would each raise the energy by $\frac{3}{5}\Delta_o$, while the three electrons in t_{2g} would each reduce the energy by $\frac{2}{5}\Delta_o$, giving a net energy change of zero. However if the ion contained only three d-electrons for example, and they were all in t_{2g}, there would be a net reduction in energy of $\frac{6}{5}\Delta_o$. This net reduction is termed the crystal field stabilisation energy, CFSE.

Table 4.2 lists electron configuration and CFSE for high- and low-spin states of transition metal ions in octahedral sites. This table shows that for some ions there is a significant stabilisation of the low-spin state and this tendency opposes the electrostatic interactions which favour a distribution with parallel spins over as many sites as possible (i.e. breaks

Table 4.2. *Electronic configuration and crystal field stabilization energies of transition metal ions in octahedral coordination.* (Burns, 1970).

Number of 3d electrons	Ion	High-spin state Electronic configuration t_{2g}	e_g	Unpaired electrons	CFSE	Low-spin state Electronic configuration t_{2g}	e_g	Unpaired electrons	CFSE
0	Ti^{4+}			0	0			0	0
1	Ti^{3+}	↑		1	$2/5\,\Delta_o$	↑		1	$2/5\,\Delta_o$
2	Ti^{2+}, V^{3+}	↑ ↑		2	$4/5\,\Delta_o$	↑ ↑		2	$4/5\,\Delta_o$
3	Cr^{3+}, V^{2+}, Mn^{4+}	↑ ↑ ↑		3	$6/5\,\Delta_o$	↑ ↑ ↑		3	$6/5\,\Delta_o$
4	Cr^{2+}, Mn^{3+}	↑ ↑ ↑	↑	4	$3/5\,\Delta_o$	↑↓ ↑ ↑		2	$8/5\,\Delta_o$
5	Mn^{2+}, Fe^{3+}	↑ ↑ ↑	↑ ↑	5	0	↑↓ ↑↓ ↑		1	$10/5\,\Delta_o$
6	$Co^{3+}, Fe^{2+}, Ni^{4+}$	↑↓ ↑ ↑	↑ ↑	4	$2/5\,\Delta_o$	↑↓ ↑↓ ↑↓		0	$12/5\,\Delta_o$
7	Co^{2+}, Ni^{3+}	↑↓ ↑↓ ↑	↑ ↑	3	$4/5\,\Delta_o$	↑↓ ↑↓ ↑↓	↑	1	$9/5\,\Delta_o$
8	Ni^{2+}	↑↓ ↑↓ ↑↓	↑ ↑	2	$6/5\,\Delta_o$	↑↓ ↑↓ ↑↓	↑ ↑	2	$6/5\,\Delta_o$
9	Cu^{2+}	↑↓ ↑↓ ↑↓	↑↓ ↑	1	$3/5\,\Delta_o$	↑↓ ↑↓ ↑↓	↑↓ ↑	1	$3/5\,\Delta_o$
10	$Ga^{3+}, Zn^{2+}, Ge^{4+}$	↑↓ ↑↓ ↑↓	↑↓ ↑↓	0	0	↑↓ ↑↓ ↑↓	↑↓ ↑↓	0	0

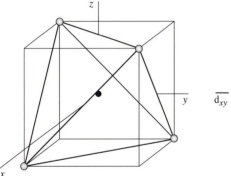
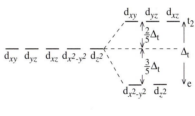

Figure 4.27. Crystal field splitting in tetrahedral coordination. When the transition metal cation (black sphere) is tetrahedrally coordinated by 4 anions (shaded spheres) the t_2 d orbitals point nearer to the anions than the e_g orbitals (see Figure 4.25). This raises the energy of the t_2 orbitals relative to the e_g orbitals, splitting the energy levels by Δ_t. Compare this to the situation in Figure 4.26.

Hund's rule). The result of these opposing tendencies is that while Co^{3+} and Ni^{3+} may exist as low spin states the other transition metals remain in high spin states on the Earth's surface.

Transitions between t_{2g} and e_g orbitals are induced by light passing through the crystal, resulting in absorption at the energy defined by the octahedral crystal field splitting parameter Δ_o. The magnitude of Δ_o depends on the cation, its valence state, and the nature of the anions coordinated to it, as well as the cation–anion distance. The effect of this selective absorption in the visible range is that the transmitted light is coloured.

4.5.2 Effect of coordination around the cation

The analysis above, for an ion in an undistorted octahedral environment can be repeated for other symmetrical coordinations. For example, if the ion is in a tetrahedral site (Figure 4.27) the t_2 orbital lobes are now repelled by the anions to a greater extent than the e orbital lobes and this reverses the energy levels relative to the octahedral case. The crystal field splitting parameter Δ_t is not as great as in the octahedral case, as the orbital lobes are not now directly towards or between the anions. A similar energy level diagram exists for regular cubic coordination, but with a larger relative Δ splitting.

As the symmetry of the coordination is reduced, the energy levels are split further. For example, if an octahedron is distorted to tetragonal symmetry (by extending the c axis for example), the two levels e_g and t_{2g} are split into four. Any further distortion of the octahedron to lower symmetry results in all 5 orbitals having different energies (Figure 4.28). The analysis of the degree of splitting is obviously more

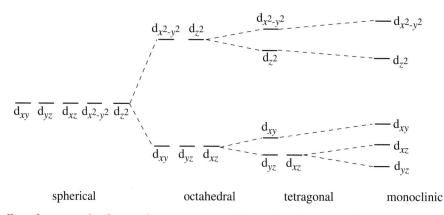

spherical   octahedral   tetragonal   monoclinic

Figure 4.28. The effect of a progressive decrease in symmetry around the cation. In a spherically symmetric field the *d* orbitals are all degenerate, while in octahedral coordination they are split into two groups as described in Figure 4.26. If the octahedron is distorted to tetragonal symmetry by stretching it along the *z* axis the energy levels are split further. (A compression along the *z* axis would reverse the sense of the splitting within each group). A further distortion to monoclinic symmetry completely removes the degeneracy, resulting in five separate energy levels.

complex than in the symmetrical case, and group theoretical methods are needed to describe the symmetry of the site and the resultant energy level diagrams. The net effect of considering crystal field splitting for transition metal ions in different coordinations is to conclude that except for Fe^{3+} and Mn^{2+} in which there is no net CFSE for either octahedral or tetrahedral sites, transition metal ions always prefer octahedral coordination.

4.5.3 Jahn–Teller distortion

For some transition metal ions a distorted coordination always has a lower energy than a regular coordination as a result of the Jahn–Teller effect. The origin of this spontaneous distortion can be seen by considering the case of Cu^{2+} in an octahedral environment. As seen in Table 4.2, Cu^{2+} has electronic configuration $(t_{2g})^6 (e_g)^3$, and the energy level diagram for an undistorted octahedral coordination is shown in Figure 4.29. If the octahedron is distorted however, the energy levels are split into five (Figure 4.29). Although there is no net reduction in the energy associated with the t_{2g} orbitals, the fact that one of the e_g orbitals is half filled results in an overall energy reduction of $\frac{1}{2} \delta_2$ for the distorted state. The distorted state then becomes the ground state of the ion. This will be the case whenever one *d* orbital is completely empty or completely full while another of equal energy is half filled. The largest Jahn–Teller distortions are found when the degeneracy is in the e_g orbitals, which occurs when they are

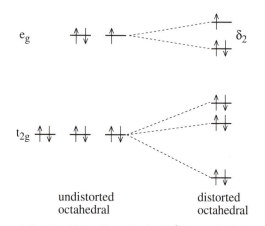

undistorted   distorted
octahedral   octahedral

Figure 4.29. Jahn–Teller distortion for Cu^{2+} in octahedral coordination. Cu^{2+} has filled t_{2g} orbitals, but only three electrons in the e_g orbitals. In an undistorted octahedral site the e_g orbitals are all degenerate, but in a distorted coordination, they are split into two separate orbitals separated by energy δ_2. Since two of the three e_g electrons occupy the lower of these energy levels, the net reduction in energy is $1/2\delta_2$. Thus Cu^{2+} prefers a distorted octahedral environment.

occupied by one or three electrons. Thus Cr^{2+}, Mn^{3+}, Cu^{2+} and Ni^{3+} (low spin) are usually in distorted octahedral environments.

4.5.4 Anisotropy of the absorption – pleochroism

Minerals with symmetry lower than cubic are anisotropic and it is necessary to obtain separate spectra with plane polarized light vibrating along each principal axis of the optical indicatrix (see Chapter 2). For biaxial crystals three spectra, termed the α, β and γ spectra, are taken with the incident vibration

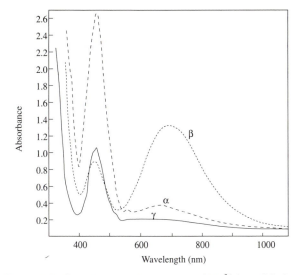

Figure 4.30. Optical spectra due to traces of Mn^{3+} in andalusite, Al_2SiO_5. The three separate spectra are taken with plane polarized incident light, vibrating along the three principal vibration directions, α,β and γ. The differences in absorption with orientation result in a change of colour when a slice of the crystal, viewed in plane polarised light, is rotated on the stage of a transmitted light microscope. This is termed pleochroism. (After Rossman, 1988.)

direction parallel to the axis for each principal refractive index. For uniaxial crystals, two spectra are needed, one with the polarized light vibrating parallel to and one perpendicular to the c axis. Hence they are termed $E \parallel c$ and $E \perp c$ spectra.

A consequence of the optical anisotropy of crystals is that the intensity and wavelength of the absorption due to both crystal field transitions and intervalence charge transitions may be dependent on the orientation of the incident light relative to the crystal structure. This is illustrated by the α-, β- and γ-optical spectra of Mn3+ which can replace some of the Al in the octahedral site of andalusite (Figure 4.30). When the difference in the absorption occurs in the optical range, both the transmittance and the colour of a crystal observed in plane polarized light will vary with orientation. This is termed *pleochroism*.

The strongest pleochroic effects are due to the anisotropy of intervalence charge transitions. Minerals containing both Fe^{2+} and Fe^{3+} ions are often pleochroic, especially when the cations occur in chains of edge-sharing octahedra. Overlapping of the d orbitals of adjacent cations enhances the probability of charge transfer along the chains and hence the absorption is strongest when the incident light is vibrating parallel to such directions.

4.5.5 Intensity and width of absorption bands

The variation in intensity of optical transitions is very pronounced due to the fact that not all transitions between different energy levels occur with equal probability. As with all spectroscopies there are selection rules, derived from quantum mechanics which govern which transitions are allowed and which are forbidden. In practice however there are various factors which weaken these rules so that forbidden transitions still appear but with very reduced intensity. These factors are mainly due to interactions between atoms which produce hybrid electron states. The net effect of these rules is that forbidden transitions are weak, and this accounts for the weak absorption of all d–d crystal field transitions compared to charge transfer transitions which are allowed and hence occur with a high probability.

Another feature of optical spectra is that the absorption bands are generally quite broad. This is due to the thermal vibration of atoms about their mean positions. This changes the interatomic distances and therefore the crystal field splitting parameter Δ also oscillates about some average value. Another cause of broad lines is the overlap of two or more absorption bands when transition metals occupy two or more structurally similar sites in a mineral.

4.5.6 Applications

1. The most obvious application of optical spectroscopy is in the identification of the electronic transitions responsible for colour and pleochroism in minerals. This is not always as straightforward as it may seem since the position of absorption bands for a given transition metal ion depends on the specific structural details of its environment. Thus Cr^{3+} is responsible for both the red colour of rubies and the green colour of emeralds. The dark blue of sapphire is due to a low concentration of both Fe^{2+} and Ti^{4+} in Al_2O_3. The blue colour is due to $Fe^{2+} \Rightarrow Ti^{4+}$ charge transfer. When the Fe^{2+} is absent, or oxidised to Fe^{3+}, strong absorption is lost resulting in pale yellow or greenish-blue colours.

2. When an absorption band has been identified with a particular cation in a crystallographic site, details of valence and coordination and the degree of covalency of cation–anion bonds can be determined. This has provided a lot of information on the crystal chemistry of transition elements in minerals, and their distribution. The effect of increased temperature and pressure on the position of the absorption bands allows the relative distortion of different sites to be evaluated.

3. When a quantitative relationship between the intensity of an absorption peak and the concentration of an element in a site has been established experimentally, it can be used to determine site occupancies. As Fe^{2+} substitutes for Mg in many minerals, this method has been widely applied in the determination of equilibrium distributions of Fe and Mg over different crystallographic sites, and its temperature dependence. In this application it is often compared with Mossbauer spectroscopy (discussed in Section 4.8). Optical spectroscopy requires oriented single crystals for the measurement, but with a microscope mounted spectrometer this can be done on small crystals in thin sections of rocks. It is also more sensitive to low element concentrations.

4. The crystal field site preference energy (CFSE) has a stabilizing effect for some transition elements

in a structure. This can be a significant part of its energy and so influence the relative stability of minerals. For example, crystal field effects stabilize Fe^{2+} in the spinel structure relative to the olivine structure of $(Mg,Fe)SiO_4$. Increasing the amount of Fe therefore expands the stability field of the spinel structure.

5. The electrical and thermal conductivity of minerals is related to processes of charge transfer and absorption. This is of particular interest in the Mantle, and studies of the temperature and pressure effects on the positions and intensities of absorption bands have contributed to predictions of the way in which these properties vary with depth.

6. Spectral remote sensing techniques make use of reflection spectroscopy in which the absorption bands are associated with reflectance minima. The strong absorption of pyroxene in the 5000 cm^{-1} and 10 000 cm^{-1} ranges (Figure 4.31), and its common occurrence in igneous rocks on the surface of terrestrial planets, has enabled different volcanic rock types to be mapped on the moon's surface from Earth-based telescopes equipped with spectrometers.

4.6 X-ray spectroscopy

X-rays are particularly useful probes of the structure of minerals. We have already discussed the diffraction of X-rays and the determination of the average crystal structure from the diffraction pattern (Chapter 3). X-rays are also absorbed by matter, and their energies are such that absorption produces electron transitions from the inner electron shells (see Appendix A). The ejection of an electron from the inner K shell ($1s$ electrons) of an atom involves the highest energy. Less energy is needed to ionize L shell ($2s, 2p$) electrons. The energy of each of these transitions is characteristic of the electron binding energy of each atom.

The ejection of these *photoelectrons* by the incident X-ray beam raises the energy of the ion to an excited state. The relaxation back to the ground state can occur in two ways: either by the *emission* of X-rays as a result of the transfer of electrons from the outer orbitals back to the vacancies in the core shell, or by the secondary ejection of another electron, termed the Auger electron, when part of this X-ray emission is absorbed to eject an electron from a higher shell. The ejection of an Auger electron,

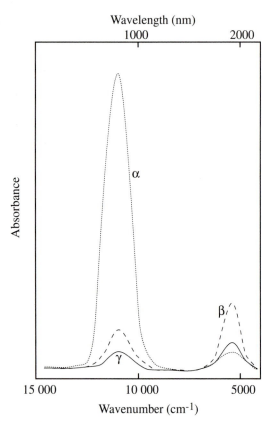

Figure 4.31. Polarized spectra of an iron bearing orthopyroxene showing the strong absorption in the 5000 cm^{-1} and 10 000 cm^{-1} regions.

which leaves the atom doubly ionized, is accompanied by further X-ray emission as the higher level electrons move down to fill the core levels.

This absorption and emission process forms the basis for a number of different spectroscopic methods.

(i) Analysing the energy spectrum of the X-ray absorption.
(ii) Analysing the energy of the electrons emitted from the core shells.
(iii) Analysing the energy of the emitted X-rays.
(iv) Analysing the energy of the Auger electrons.

Each of the spectroscopies associated with these processes have variations, and each is given an acronym. The rapid development of the technology needed to measure the spectra has led to a proliferation of techniques and acronyms, sufficient to confuse all but the practitioners of each Art. Here we will outline the basic principles behind the most important of these methods.

4.6.1 X-ray absorption spectroscopy (XAS) – XANES and EXAFS

Although X-ray absorption spectra have been used for over 50 years, relatively recent developments in the production of extremely intense sources of X-rays from synchrotron radiation have led to the possibility of using fine details of the absorption spectrum to determine local structure.

Synchrotron X-rays are produced by the continuous dissipation of energy when electrons (or positrons) are accelerated in a storage ring at velocities close to that of light. The intensity of such a source is more than 10 orders of magnitude greater than a conventional X-ray tube, and the X-rays have a broad and continuous energy range. This is ideal for probing the range of energies associated with X-ray absorption. Monochromatic X-rays, used for diffraction experiments, may be produced by setting a monochromator crystal at the Bragg angle to the synchrotron X-rays. The diffracted beam, whose wavelength will be determined by the Bragg equation is then used as the incident beam on the sample to be studied. Over two dozen synchrotron radiation sources now exist throughout the world.

When X-rays with a range of energy pass through a sample, absorption will take place at energies characteristic of the ionization energy of the L shell

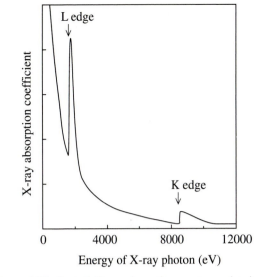

Figure 4.32. Typical X-ray absorption spectrum showing two major features, the L absorption and K absorption superimposed on a smooth curve. The L absorption is due to the ionization of electrons from the L shell; the K absorption at higher energies is due to ionization of electrons from the innermost K shell. The sharp rise of these peaks at the lower energy side is termed the edge.

and K shell electrons. Figure 4.32 shows a typical X-ray absorption spectrum which is a smooth curve increasing very rapidly at low X-ray energies, superimposed on which are the absorption peaks for the K and L transitions. The sharp rise in absorption at the low energy side of the peak is termed the absorption edge.

The absorption edge is not a smooth feature but often has small undulations in intensity both at the low energy and high energy sides. It is convenient to subdivide the absorption spectrum into three regions as illustrated for the K absorption edge in Figure 4.33.

(1) The pre-edge
At energies just below the main absorption edge, small peaks represent the onset of electronic transitions from a core level ($1s$ in this case) to the lowest energy unoccupied or partially occupied level (e.g. $1s \Rightarrow 3d$ in a transition metal). Such transitions have low probabilities and hence produce only small ripples in the absorption spectrum. The exact peak positions depend on details of the oxidation state, site symmetry, and the nature of the bonding.

(2) X-ray absorption near edge structure (XANES)
This part of the spectrum includes the main absorption peak, which is due to the ionization of a core

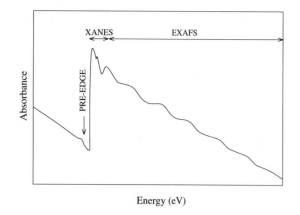

Figure 4.33. A more detailed view of the K absorption shown in Figure 4.32. The peak can be sudivided into three regions, the pre-edge, the main absorption peak (XANES), and the smoothly undulating decreasing intensity (EXAFS).

$(1s)$ electron to a delocalized energy band, or the complete ejection of the electron from the sample. When the energy of the incident X-radiation exactly matches the energy to remove an electron from the $1s$ shell, the probability of absorption is very much greater and strong absorption occurs.

The ejection of an electron from a core level, and hence the absorption of the X-rays is not a simple process when the absorbing atom is surrounded by other atoms. The ejected electron can be multiply scattered by the neighbouring atoms and the interference between the electron waves produces perturbations on the main absorption edge. In the XANES part of the spectrum these perturbations are due to multiple scattering events, and their analysis can provide information about the local structural environment such as site geometry and bond angles to first and second nearest neighbours.

(3) Extended X-ray absorption fine structure (EXAFS)

This region describes the fine intensity variations superimposed on the otherwise smoothly declining intensity of the absorption tail at energies above the absorption edge. These ripples on the absorption spectrum are due to a single scattering process between an emitted electron wave and a neighbouring atom. The interference between the electron wave travelling outwards from an atom, and the same wave scattered by the neighbouring atom produces an interference pattern in much the same way as the electron diffraction process. The ripples are the result of a localized constructive and destruc-

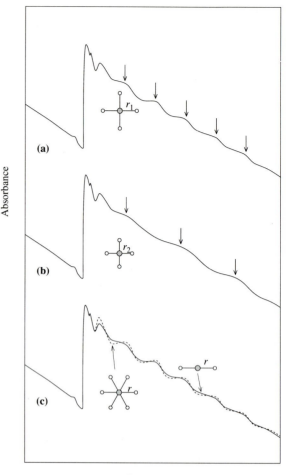

Figure 4.34. The spacing between the ripples in the EXAFS spectrum is reciprocally related to the distance between the absorbing atom and its nearest neighbours. (a) and (b) show the absorbing ion (shaded) in the same coordination but at two different distances r_1 and r_2 from its nearest neighbours. In (c) the nearest neighbour distance r, is the same, but a higher coordination results in a larger amplitude for the undulations (dashed curve). (After Brown *et al.*, 1988.)

tive interference between the electron waves, and the frequency of the ripples is therefore reciprocally related to the interatomic distances between the absorbing and the scattering atoms. The amplitude of the ripples is directly related to the number of nearest neighbours producing the backscattering (Figure 4.34).

The EXAFS ripples on the absorption spectrum can be analysed in much the same way as a diffraction pattern. After separating the ripples from the smoothly declining background, a Fourier transformation produces a real space distribution of the interatomic spacings. This is known as a *radial*

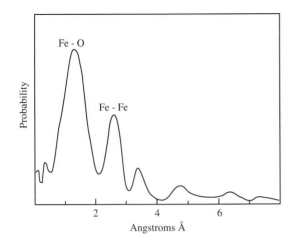

Figure 4.35. A radial distribution function, obtained from the periodicity of the undulations in the EXAFS spectrum, describes the probability of finding a neighbouring atom at a specific distance from the absorber atom (in this case Fe). The first peak is due to the nearest neighbour oxygen atoms, while the second two peaks are related to Fe–Fe pairs. The sample is an iron-bearing clay mineral.

distribution function (RDF) (Figure 4.35) which is a curve giving the probability of finding a second atom as a function of distance from the absorbing atom. The fact that EXAFS is due to a single scattering event makes the theory and interpretation more straight forward that that for XANES.

Being an element specific technique EXAFS can be used to determine a radial distribution function for each element. From this the average distance and coordination number of nearest and next-nearest neighbours, as well as the effect of thermal agitation on interatomic distances can be directly determined. EXAFS is applicable to glasses as well as crystals and has very considerable potential for determining local structural environment.

4.6.2 X-ray emission spectroscopy

One of the ways in which an excited ion can relax to the ground state is by the transfer of an outer orbital electron back to an inner shell, with the emission of X-rays. Each element has a characteristic X-ray emission spectrum composed of a set of sharp peaks, each peak corresponding to the energy of an electron transition from a higher to a lower orbital. Since the difference in the electron energy levels is dependent on the atomic number, the spectrum provides a method for elemental analysis of the sample. Used in this way X-ray emission provides the most com-

mon method for the chemical analysis of minerals and rocks.

(1) Elemental analysis: X-ray fluorescence and electron probe microanalysis

In X-ray fluorescence (XRF) analysis a pellet of a rock or mineral is irradiated by a broad beam of continuous X-radiation and the peak intensities of X-ray emission lines measured. By comparing these to calibrated samples a quantitative elemental analysis can be obtained.

Since the emission depends only on the atom from which it came and not on the nature of the incident radiation which raised it to the excited state, other high energy radiations can also be used. Electron probe microanalysis (EPMA) uses high energy electrons as the incident beam to induce X-ray emission. Although the efficiency of X-ray fluorescence using electrons is not as high, electrons can be focussed onto a small area allowing analysis of sub-micron size spots.

The X-ray wavelengths emitted during bombardment of a mineral surface with electrons consist of all the characteristic lines for each element present. To carry out the compositional analysis of the mineral, these X-ray peaks have to be separated and their intensities measured. This is the role of the detection system. There are two different ways in which this can be done.

1. Wavelength dispersive X-ray analysis. In a wavelength dispersive spectrometer the X-rays are sorted out according to their wavelengths, by using a crystal of known lattice plane spacing (d) to diffract the emitted X-rays. Only the wavelengths which satisfy the Bragg equation (i.e. $\lambda = 2d\sin\theta$) will be diffracted and measured with a suitable detector.

Figure 4.36 illustrates the basic arrangement. To analyse X-rays emitted from a mineral sample, the

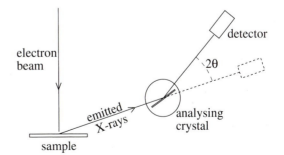

Figure 4.36. Schematic arrangement for wavelength dispersive X-ray analysis of the emission spectrum from a mineral specimen.

reflecting crystal is rotated through a range of angles θ while the detector is rotated simultaneously through an angle 2θ. This preserves the Bragg geometry. Measurement of the angle θ at which output from the detector is maximum enables corresponding values of λ to be calculated. To cover the range of wavelengths excited by a 30kV electron beam (typical of an electron microprobe analyser), several different crystals providing a range of *d* values are required.

The X-ray photons are normally detected and counted using a proportional counter, consisting of a gas-filled cylindrical chamber with an insulated wire held at a high electrical potential. An X-ray photon entering the chamber ionizes the gas and a pulse of current passes between the wire and the body of the chamber. Current pulses are processed to give measurements of count rate, which is proportional to the intensity of X-radiations, and the total counts over the given duration of the analysis (around 60 seconds).

2. Energy dispersive X-ray analysis. This method uses a semiconductor detector to classify the X-radiation according to energy rather than wavelength. The detector is a single crystal disc of silicon doped with some lithium, Si(Li), about 4mm thick. This converts the energy of incident X-ray photons into pulses of current proportional to the photon energy. These pulses are amplified, digitised and fed into a multi-channel analyser, or into the memory of a computer which stores them in a location appropriate to the pulse height, i.e. energy. The energy dispersive spectrometer is thus able to analyse a whole spectrum simultaneously instead of peak by peak as in the wavelength dispersive method. The X-ray emission spectra shown in Figure 3.50 have been collected using an energy dispersive detector.

Each method has its advantages and disadvantages. Wavelength dispersive spectrometers have a better resolution and are more sensitive, and hence are best for precise analyses of trace elements. Energy dispersive spectrometers are generally faster, analysing all elements simultaneously and hence suitable for a fast major element analysis.

Microanalysis attachments are also available as standard accessories to transmission and scanning electron microscopes, by the addition of a detector to collect the emitted X-rays.

New developments in this area are rapid with the possibility of using protons or synchrotron X-rays as

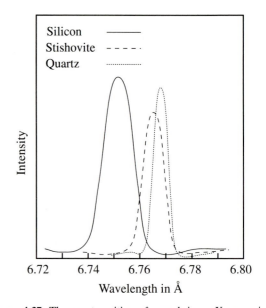

Figure 4.37. The exact position of a peak in an X-ray emission spectrum depends on the local coordination of the emitting atom. In this example the peak is due to the K_β emission of Si in three different structures, silicon (Si atoms tetrahedrally coordinated by Si), stishovite (Si atoms octahedrally coordinated by O) and quartz (Si atoms tetrahedrally coordinated by O). (After White and Gibbs, 1967.)

the incident radiation. Protons are more efficient than electrons for exciting low atomic number elements, while synchrotron radiation with its much higher intensity will allow extremely low concentrations of elements to be measured (in the parts per billion range).

(2) High resolution X-ray emission spectra

In elemental analysis we were only concerned with the intensity of selected peaks in the X-ray emission spectrum, and so the resolution need only be high enough to separate individual peaks from each element present. At higher resolutions than those normally available on instruments designed for chemical analysis, very small shifts in the positions of the peaks as well as intensity changes and additional fine structure can be revealed in the X-ray emission spectrum. These high resolution details are sensitive to the local structure around the emitting atom, and so can potentially provide information on oxidation state, coordination number, and interatomic distances.

Figure 4.37 shows the difference in position of the K_β emission line (due to a $3p \Rightarrow 1s$ transition) of Si in 3 different materials. There is also usually a correlation between peak intensity and position of the spectral lines with oxidation state.

High resolution spectra also enable the construction of molecular orbital energy level diagrams for the various bonding and anti-bonding states in a mineral, and hence comparison with theoretical molecular orbital calculations.

4.7 Photo-electron spectroscopy

The electrons which are ejected from a sample by high energy incident radiation can also be analysed to provide information about electron binding energies. In this case the incident radiation must be monochromatic, so that the energy of the ejected electron can be directly related to the difference between the energy of the incident radiation and the electron binding energy.

Ionization of an electron from an inner shell requires incident X-radiation, and forms the basis of X-ray photoelectron spectroscopy (XPS). Ionizing the valence electrons requires less energy and ultra-violet light can be used, hence UPS (ultra-violet photoelectron spectroscopy). As the electron binding energy is characteristic for each element, measurement of the ejected electron energy provides a means of identifying the element and hence a method of chemical analysis – ESCA (electron spectroscopy for chemical analysis).

The other electron emission stimulated by high energy incident radiation is the ejection of Auger electrons (Figure 4.38) which are often observed at the same time as the direct ionization spectra. Auger electrons may also be produced by an incident electron beam.

The most important application of electron spectroscopies depends on the fact that the energy of the emitted electrons is relatively low (≤ 1.5 keV), and hence they can be very easily stopped by matter. Only electrons emitted from the surface layers of the sample will reach the detector to be analysed. Thus electron spectroscopy is surface sensitive and ideally suited for elemental analysis of near-surface atoms. Oxidation and weathering on surfaces as well as adsorption reactions can be studied in this way. The electron binding energies can potentially also provide information on local structural states although for most elements the energy shifts which can be measured using this technique are too small for positive correlations to be made.

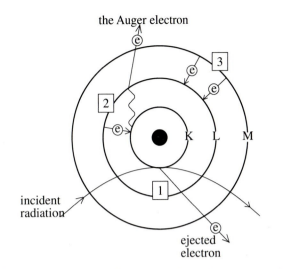

Figure 4.38. Schematic illustration outlining the stages in the production of an Auger electron. [1] the incident radiation ejects an electron from the K shell. [2] an electron from the L shell moves down into the K shell with the emission of an X-ray photon. This X-ray photon ejects an electron from the L shell. This is the Auger electron. [3] electrons from higher levels move into the L shell to replace the two L electrons, with emission of L X-rays.

EELS (electron energy loss spectroscopy) is an analytical technique related to the same processes of absorption and emission discussed above, and is applicable to the analysis of thin electron-transparent materials such as used in transmission electron microscopy. Some of the incident electrons cause ionization of the inner shell electrons, with the consequent emission of X-rays. Those incident electrons which are responsible for the ionizations suffer an energy loss, characteristic of inner shell binding energies. The energy loss spectra have been mainly applied to the chemical analysis of low atomic number elements, for which energy losses are small.

4.8 Mössbauer spectroscopy

We conclude this chapter by returning to a spectroscopic method which relies on transitions which take place in the atomic nucleus. Whereas NMR spectroscopy was only concerned with the small differences in nuclear spin energy levels in a magnetic field, Mössbauer spectroscopy is primarily concerned with the core nuclear energy levels of excited states of specific atomic nuclei in which the energy differences are at the other extreme of the spectrum, in the γ-radiation range. Each energy level will also be associated with a spin quantum number I, but as we have seen in Section 4.1, these spin states are all degenerate (equal in energy) in the absence of a magnetic field, and only have a small energy difference in a strong magnetic field.

The emission of γ-radiation energy from an excited state of a nucleus is associated with its decay to the ground state of the same isotope. If a solid sample containing this particular isotope in the ground state is irradiated with γ-rays of the same energy, γ-rays may be absorbed in a resonance recapture process, and the nucleus raised to the excited state. The exact value of the energy differences between nuclear energy levels depends on the interactions of the nuclear charge with the electric and magnetic fields produced by the immediate environment of the nucleus, especially by its own orbital electrons, and hence by the oxidation state, coordination number etc. By matching the energy of the incident γ-rays to the exact energy level difference for the specific nucleus, absorption will take place and the energy difference be measured.

A potential problem with this process is that because γ-rays have such high energy, their emission and absorption would be expected to produce a recoil in the nuclear mass itself. If the energy associated with the recoil was significant, the energy at which the incident γ-rays were absorbed would not be the same as the energy of the nuclear transition. In a free nucleus this would be the case, but in a solid however, the recoil energy is taken up by the vibrational modes in the entire crystal rather than by the individual nucleus. As the vibrational state of the crystal is quantised, there is a finite probability that there will be some cases where there will be no transfer of energy to the crystal, and hence no recoil. This "recoilless" absorption is the basis of the Mössbauer effect, and hence limits its application to solids.

One feature which reduces the usefulness of Mössbauer spectroscopy is that not all elements have isotopes with suitable nuclear transitions. The nuclear transition energy must be large enough to give γ-radiation, but not so large as to produce a nuclear recoil and disrupt the structure. In this context, the structure itself must have large enough binding energy to fulfill the condition for a high probability of recoilless absorption. There must also be an available source of the γ-rays which has a conveniently long half-life and is not excessively radioactive. Suitable isotopes include ^{57}Fe, ^{119}Sn, ^{121}Sb and

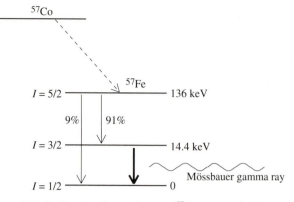

Figure 4.39. Radioactive decay scheme of ^{57}Co to an excited state of ^{57}Fe. The relaxation of ^{57}Fe from this excited state to the ground state involves the emission of various γ-rays. The 3/2 to 1/2 transition emits the 14.4 keV γ-rays which are used in ^{57}Fe Mössbauer spectroscopy.

^{197}Au, but by far the most work has been done with ^{57}Fe which has a natural abundance of 2.17%. Mössbauer spectroscopy has now become a fairly standard technique for characterizing the local environment of Fe in iron-bearing minerals.

For a Mössbauer experiment we need a *source* of γ-rays with the same energy as the nucleus in the sample, as well as a way of slightly modifying this incident energy to match the resonance condition. The source must therefore contain the same isotope as that in the sample, and to emit γ-rays the isotope in the source must itself be in an excited state. However, since the lifetime of these excited states is invariably very short ($\sim 10^{-7}$ s), the excited state must be continuously replenished to ensure a constant supply of γ-rays. To produce ^{57}Fe γ-radiation, the source is ^{57}Co which decays, with a half-life of 270 days, to the excited state of ^{57}Fe. This excited state then decays, emitting the characteristic γ-rays used to probe ^{57}Fe in the sample. The energy of this radiation is 14.4 keV. The decay scheme is shown in Figure 4.39.

The variation in the absorption energy due to the structural environment of ^{57}Fe nuclei in the sample is small, and the energy of the incident γ-rays can be fine-tuned by using the Doppler effect, i.e. moving the emitting source towards or away from the fixed sample. For example, if the source is moved at a constant velocity of $v = 1$ cm/s, the energy change ΔE for 14.4 keV γ-rays would be $\Delta E = E.v/c = \pm 4.8 \times 10^{-7}$ eV (c is the velocity of light), the positive sign for movement towards the sample, and negative for movement away from it. In this way the energy of

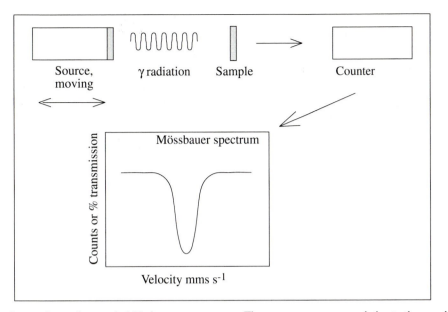

Figure 4.40. Schematic experimental set-up in Mössbauer spectroscopy. The γ-ray source moves relative to the sample, and the counter measures the transmitted γ-ray intensity. The Mössbauer spectrum is a plot of the transmitted γ-ray intensity against the velocity of the source.

the incident γ-rays can be varied and the γ-ray absorption spectrum of the sample determined. The raw data of a Mössbauer spectrum are thus the relative number of γ-rays detected after passing through the sample, as a function of the velocity of the source (Figure 4.40). Modern instrumentation is such that the source is usually moved with a constant acceleration, thus scanning a linear range of velocities, with γ-ray counts recorded by a multi-channel analyses such that the velocity sampled by each channel is a constant.

When a nucleus is placed in a crystal structure, or amorphous solid, the transition energy between ground and excited state will be slightly modified by interactions between the nucleus and the surrounding electric and magnetic fields. These are termed the *hyperfine interactions* and their analysis may reveal the local structural environment of the nucleus. We will consider each of these interactions in turn.

4.8.1 Isomer (or chemical) shift

In the simplest case, if the nucleus in the source and nucleus in the sample have the same environment, the transition energy in each will be the same, and Mössbauer absorption will take place when the source is stationary with respect to the sample (Figure 4.41). If the electron density at the nuclei in

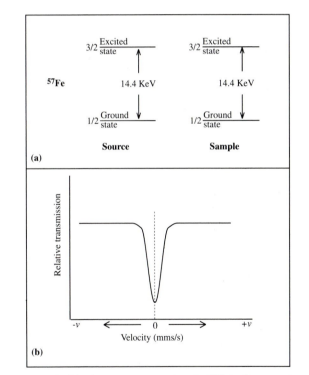

Figure 4.41. Mössbauer resonance when the ^{57}Fe in the source has the same ground state and excited state energy as the ^{57}Fe in the sample being studied. In this case the the Mössbauer spectrum in (b) has an absorption feature at zero velocity of the source.

the source material is different from that in the sample, the Mössbauer transition energy will now be slightly different in the two materials. This will appear as a shift in the position of the absorption

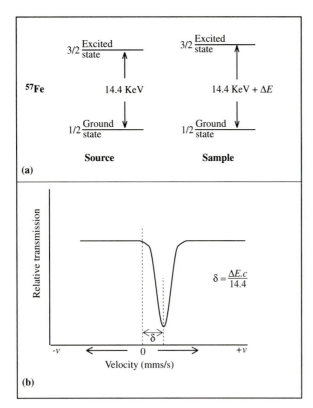

Figure 4.42. Mössbauer resonance when the ^{57}Fe in the sample has a slightly higher energy difference between ground and excited state than the ^{57}Fe in the source. This is due to the difference in structural environment of the ^{57}Fe. In this case the source is moved towards the sample increasing the energy of the emitted γ-rays until they match the transition energy in the sample. The Mössbauer spectrum in (b) has an absorption feature shifted from zero velocity by δ, the isomer shift.

In terms of the spectra in Figure 4.42, the isomer shift δ is the velocity at which maximum absorption occurs, while the difference in relative energy levels between source and sample, ΔE, is equal to the energy shift caused by the Doppler effect at this particular velocity, i.e. $\Delta E = E\delta/c$, where E is energy of the γ-rays from the stationary source, and c is the velocity of light. In practice however, isomer shifts are quoted as δ values relative to some standard material, often metallic iron.

4.8.2 Quadrupole splitting

A nucleus with spin $I \geq 1$ can be considered as an electric quadrupole with a non-spherical distribution of nuclear charge. A quadrupole interacts with electric field gradients around a nucleus, which are generated whenever the nuclear environment has a charge distribution lower than cubic. This interaction between a nuclear quadrupole and an electric field gradient splits the nuclear energy levels into substates with spin I, $(I-1)$, $(I-2)$.. etc, without removing the degeneracy between states having the same value of the spin quantum number I. These substate levels are equivalent to saying that there is a set of restrictions placed on the possible angular orientations of the nucleus with respect to the electric field gradient, and these have different energy.

Quadrupole splitting can be illustrated using the simple case of ^{57}Fe. The ground state has spin $I = \pm 1/2$, while the excited state 14.4 keV higher in energy has spin $I = \pm 3/2$. The presence of an electric field gradient does not affect the ground state, but splits the excited state into $I = \pm 3/2$ and $I = \pm 1/2$ sublevels, as shown in Figure 4.43. Two transitions from the ground state are now possible, resulting in two lines in the Mössbauer spectrum. The degree of splitting of these two lines, which is a measure of the energy difference between the two excited states, depends on the nuclear electric quadrupole moment and the orientation and magnitude of the electric field gradient. For a given nucleus the quadrupole moment is known, but the electric field gradient may arise from the surrounding atoms in the structure as well as from the electronic charge cloud of the atom itself, especially when bonding is predominantly covalent and electrons are delocalized over adjacent atoms. It is often difficult to separate these effects and in complex mineral structures an empirical

peak (Figure 4.42). This shift is known as the *isomer shift*, or *chemical shift* and is proportional to the difference in the electron densities at the nuclear sites in the two materials. The origin of the energy difference lies in the fact that the nucleus is not a point charge and has a finite volume. The nuclear volume is affected by the orbital electron density and any changes in the nuclear volume are reflected in changes in the energy levels.

The electron density is principally dependent on the distribution of the *s* valence electrons, but is also affected by the *p,d* and *f* electron orbitals which influence the spatial distribution of *s* electrons through inter-electronic repulsion. The *p,d* and *f* electrons shield the *s* electrons from the nucleus. The isomer shift is therefore sensitive to any factor that affects the number and/or distribution of the valence electrons, and is thus a probe of oxidation state, coordination and degree of covalency.

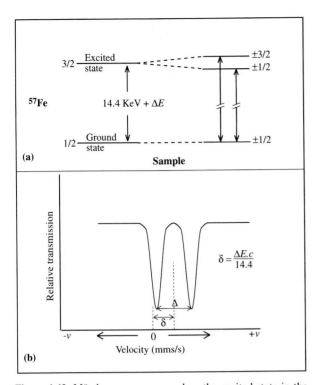

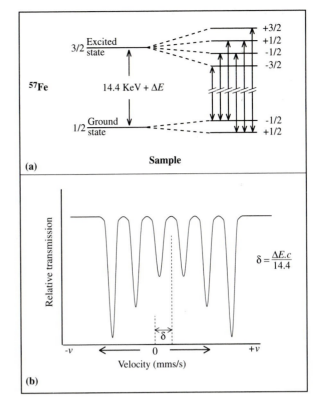

Figure 4.43. Mössbauer resonance when the excited state in the sample is split into two energy levels by the presence of non-cubic symmetry in the distribution of electric charge around the nucleus. This only affects energy levels when $I \geq 1$ and is due to the interaction of the quadrupolar nucleus with electric field gradients. There are therefore two transitions possible, separated by Δ and on either side of the isomer shift position. This is termed the quadrupolar doublet.

Figure 4.44. Mössbauer resonance when the nucleus is in a magnetic field and the energy levels for spin I are split into $2I + 1$ separate levels. In ^{57}Fe the $I = 3/2$ level is split into 4 levels and the $I = 1/2$ level is split into 2. Due to the selection rule that allowed transitions are limited to those for which $\Delta I = 0, \pm 1$, there are six allowed transitions and hence 6 peaks in the Mössbauer spectrum.

approach for the correlation between quadrupole splitting and site distortion is usually employed.

4.8.3 Magnetic splitting

When the nucleus is placed in a magnetic field the energy levels with non-zero spin are split, as we have already seen in Section 4.1. For a spin I, $2I+1$ separate energy levels will arise. In the case of ^{57}Fe, the excited state $I = 3/2$ splits into four levels and the ground state $I = 1/2$ splits into two. Although this might be expected to produce an eight-line spectrum, there is a selection rule for Mössbauer spectroscopy, that allowed transitions cannot change the value of I by more than one, and this restricts the number of lines to six (Figure 4.44).

The splittings between the peaks are directly proportional to the magnitude of the magnetic field at the nucleus. The magnetic field may be externally imposed, as in an NMR experiment, but may also arise when the atom itself has a magnetic moment

due to the presence of unpaired electrons (i.e. a paramagnetic atom). If the spins of the unpaired electrons are all coupled together as in a ferromagnet, the local magnetic field can be quite high producing large magnetic splitting. If however the paramagnetic atoms have random spins, the average magnetic field is zero and no splitting is produced. Applying an external field aligns the spins and the Mössbauer spectrum splits, providing a measure of the degree of magnetic alignment in the sample. Magnetic ordering can also be produced by reducing the temperature to a point where the thermal vibrations are insufficient to counteract the coupling forces between neighbouring paramagnetic atoms. Low temperature Mössbauer spectroscopy can be used to study such magnetic ordering transitions.

4.8.4 Applications

As iron occurs as Fe^{2+} and Fe^{3+} in a large number of minerals, Mössbauer spectroscopy has been a very useful technique in mineralogy for a wide range of problems associated with the local environment of Fe. The fact that it is an element specific technique enables it to be used in a variety of situations and complex structures, but obviously only one aspect of the structure is being investigated and as with most spectroscopic techniques, Mössbauer spectroscopy is generally used in conjunction with other techniques. Here we summarize some of the most common applications.

1. Determination of valence states. The difference between Fe^{2+} with electron configuration $3d^6$, and Fe^{3+} with $3d^5$ electrons involves the addition or removal of d electrons. This indirectly affects the electron charge density around the nucleus by shielding it from the s electron distribution to different extents. (Although the $4s$ electrons are formally lost in Fe^{2+} and Fe^{3+}, there is always some covalent transfer from the neighbouring atoms to this orbital). Fe^{2+} has an appreciably larger isomer shift than Fe^{3+}. This leads to a wide separation of Mössbauer lines and enables a quantitative determination of Fe^{2+}/Fe^{3+} ratios.

2. Oxidation of minerals. This is an obvious extension of the above, but being able to determine the site occupancy of Fe^{2+} and Fe^{3+} ions enables the oxidation to be observed at each individual site. If there are two different crystallographic sites for Fe^{2+}, they will probably oxidise at different rates. This is found to be the case in biotite, for example.

3. Electron delocalization. If a mineral contains both Fe^{2+} and Fe^{3+}, the electrons may be delocalized i.e. change their positions between the different Fe atoms. The rate at which such delocalization takes place increases with temperature until it exceeds the mean life time of the excited Mössbauer state. When this happens the Mössbauer experiment will no longer be able to distinguish between the two and an average valence state of 2.5, with an intermediate value of the isomer shift, will be observed. Such processes can be observed as a function of temperature by Mössbauer spectroscopy. It has been found that for electron delocalization to be effective, the Fe^{2+} and Fe^{3+} sites must be geometrically very similar and linked by common edges to form extended chains or layers.

4. Site occupancy and distortion. The isomer shift increases with increasing coordination number for both Fe^{2+} and Fe^{3+} enabling the detection of even small amounts of Fe in individual sites. X-ray diffraction cannot detect such small amounts of Fe, nor can it normally distinguish between Fe^{2+} and Fe^{3+}. Distortions in coordination, due to bond length and bond angle variations are recorded in the quadrupole splitting. In Fe^{3+} increasing the distortion increases the quadrupole splitting. In Fe^{2+} the situation is more complex as distorting the $3d^6$ electron cloud changes the relative contribution of the electrons and the surrounding structure to the electric field gradient. This leads first to an increase in quadrupole splitting with increasing distortion and then to a decrease. In highly distorted Fe^{2+} coordinations, typical of many silicate structures, increasing the distortion leads to smaller quadrupole splitting.

The most common application of Mössbauer spectroscopy has involved the determination of Fe site occupancy and hence the distribution of Fe and Mg in minerals. As the partitioning of Fe and Mg between different sites is temperature dependent in many minerals, site occupancies as a function of annealing temperature have been used to determine the equilibrium Fe/Mg distributions. Deviations from equilibrium can also be studied in samples which have been cooled at various rates.

5. Magnetic properties. Ordering of magnetic spins in minerals containing paramagnetic ions can be studied as a function of temperature, pressure and applied magnetic field. The nature of the coupling between adjacent ions (i.e. ferromagnetic or antiferromagnetic) can also be distinguished. Application of a magnetic field may also serve to distinguish between two very similar crystallographic sites for which the isomer shift and quadrupole splitting cannot be resolved.

6. The coordination of Fe in silicate melts affects the viscosity, density and crystallisation behaviour. Assuming that quenching the melt to a glass freezes in the structural states in the melt, extensive studies have been carried out on the valence state and coordination of iron in silicate glasses. The variety of small structural differences in the coordination polyhedra in glasses broadens the Mössbauer peaks by producing a distribution of unresolved isomer shifts and quadrupolar splittings. This makes the spectra more difficult to interpret and they are usually supplemented with other spectroscopic methods.

Appendix
Electron energy levels and quantum numbers

Electrons in an atom are described by four quantum numbers.

1. The four electron *shells* surrounding the nucleus are usually designated K,L,M and N, and are assigned the principal quantum number n where n equals 1,2,3 and 4 respectively. These numbers represent the increasing energy levels of the shells, schematically drawn below. (Note that this diagram does not imply the relative locations of the electrons.)

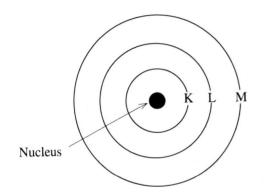

Nucleus

2. These electron shells are split into *sub-shells*, each with its own quantised energy level. These sub-shells are labelled s,p,d and f. These sub-shells are assigned the quantum number l which can have values $l = 0,1,2...(n-1)$ where n is the principal quantum number.

3. Within each sub-shell there are a number of possible *orbitals*. The number of orbitals in a sub-shell equals $2l + 1$, where l is the sub-shell quantum number. Therefore an s sub-shell consists of one s orbital, a p sub-shell consists of up to three p orbitals and a d sub-shell of up to five d orbitals.

The energy values of the orbitals in a sub-shell are normally degenerate. However, in a magnetic field they become split into discrete and quantised *magnetic energy levels*. These energy levels are then described by the quantum number m which can have values from $-l$ to $+l$.

4. Each electron also has a *spin quantum number* s which can have one of two values: $+1/2$ or $-1/2$.

Thus every electron has its own set of quantum numbers describing its energy level. The four quantum numbers of all the electrons in the first three shells and their energy levels are summarised in the Table and Figure.

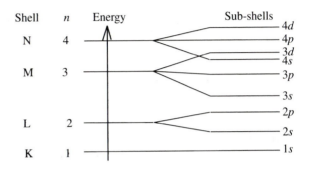

Electronic configuration

The electronic configuration of an element describes how its electrons are arranged in their shells, sub-shells and orbitals. Usually the electronic configuration is given for the element or ion in its ground state, and the following three rules determine the way in which these ground states are built up.

1. Electrons in their ground states occupy orbitals in increasing order of their energy levels, from the lowest energy up. For example the element lithium has 3 electrons. Two of these electrons can occupy the s orbital of the K shell, with quantum number $n = 1$, and the third must go into the s level of the L shell ($n = 2$). The electronic configuration is designated $1s^2 2s^1$.

2. An orbital cannot contain more than two electrons and then only if they have opposite spins. This is the Pauli exclusion principle. In lithium therefore, the two electrons in the K shell have paired spins, while the single electron in the L shell has an unpaired spin.

3. Hund's rule states that the orbitals of a sub-shell must be singly occupied with parallel spins before they can be occupied in pairs.

Table A-1. *Quantum numbers for electrons.*

Shell n		Sub-shell $l = 0,..,n-1$		Orbital $m = -l,..+l$	Spin $\pm 1/2$		Max. no. of electrons	
K	1	s	0	0	+1/2	−1/2	2	
		s	0	0	+1/2	−1/2	2	
L	2			−1	+1/2	−1/2		Total 8
		p	1	0	+1/2	−1/2	6	
				+1	+1/2	−1/2		
		s	0	0	+1/2	−1/2	2	
				−1	+1/2	−1/2		
		p	1	0	+1/2	−1/2	6	
M	3			+1	+1/2	−1/2		
				−2	+1/2	−1/2		Total 18
		d	2	−1	+1/2	−1/2		
				0	+1/2	−1/2	10	
				+1	+1/2	−1/2		
				+2	+1/2	−1/2		

In the transition elements the s orbital in the outer shell is usually filled before the d orbitals in the shell next to the outermost shell. Thus the 'last' electron added is an inner-shell electron rather than an outer-shell electron. This is illustrated by the sequence iron, cobalt, nickel from the first transition series, whose electronic configurations are shown below:

Element		Atomic Number	Electronic configuration			
			K	L	M	N
Iron	Fe	26	$1s^2$	$2s^2 2p^6$	$3s^2 3p^6 3d^6$	$4s^2$
Cobalt	Co	27	$1s^2$	$2s^2 2p^6$	$3s^2 3p^6 3d^7$	$4s^2$
Nickel	Ni	28	$1s^2$	$2s^2 2p^6$	$3s^2 3p^6 3d^8$	$4s^2$

Transition elements thus have unfilled $3d$ orbitals.

Atomic spectra

Each line in an atomic spectrum corresponds to electrons jumping from one energy level to another. For spectra in the visible and near ultra-violet, the transitions predominantly involve changes in the outermost or valence shell electrons. These electrons lie outside the completely filled electron shells which do not participate in transitions in this energy range.

For example, sodium with 11 electrons has the electronic configuration $1s^2\ 2s^2 2p^6\ 3s^1$ and it is only the $3s^1$ electron which is involved in optical transitions. The optical absorption spectrum of sodium vapour is due to excitation of this $3s$ electron into the higher p or d orbitals of the M shell, or to transitions up to the N shell.

When the inner electrons are disturbed the energy involved is much higher, and the best known transitions of this type are due to a single electron moving from the K shell to the L shell i.e., $1s$ up to the $2p$ orbital (absorption) or back again (emission). This transition is known as the K_α transition and involves the absorption or emission of an X-ray photon whose energy is equal to the difference in the two energy levels. An electronic transition from the M shell to the K shell gives rise to the K_β line. Since the L shell has a lower energy than the M shell, the K_α line has a longer wavelength than the K_β line of the same element. K_α lines generally have a higher intensity than K_β lines and are selected for use in X-ray diffraction experiments.

At lower X-ray energies the emissions are due to electron transitions to the L shell, and again characteristic radiations termed L_α, L_β etc. are observed in the X-ray spectrum.

Bibliography

NMR spectroscopy:

ENGELHARDT, G. AND MICHEL, D. (1987). *High-resolution solid-state NMR of silicates and zeolites.* John Wiley and Sons.

KIRKPATRICK, R.J., SMITH, K.A., SCHRAMM,

S., TURNER, G. AND YANG, W-H. (1985). Solid state NMR spectroscopy of minerals. *Ann. Rev. Earth Planet. Sci.* **13**, 29–47.

KIRKPATRICK, R.J. (1988). MAS NMR spectroscopy of minerals and glasses. In: *Reviews in mineralogy* **18**. Ed.: F.C. Hawthorne. Mineral. Soc. America.

STEBBINS, J.F. AND FARNAN, I. (1989). NMR spectroscopy in the earth sciences. *Science*, **245**, 257–63.

ESR spectroscopy:

CALAS, G. (1988). Electron paramagnetic resonance. In: *Reviews in mineralogy* **18**. Ed.: F.C. Hawthorne. Mineral. Soc. America.

McWHINNIE, W.R. (1985). Electron spin resonance and nuclear magnetic resonance applied to minerals. In: *Chemical bonding and spectroscopy in mineral chemistry*. Eds.: F.J. Berry, and D.J. Vaughan, Chapman and Hall.

MARFUNIN, A.S. (1979). *Spectroscopy, luminescence and radiation centres in minerals*. Springer-Verlag.

VASSILIKOU-DOVA, A.B. AND LEHMANN, G. (1987). Investigations of minerals by electron paramagnetic resonance. *Forschr.Mineral.* **65**, 173–202.

Vibrational Spectroscopy:

GHOSE, S. (1988). Inelastic neutron scattering. In: *Reviews in mineralogy* **18**. Ed.: F.C. Hawthorne. Mineral. Soc. America.

HOLLAS, J.M. (1987). *Modern spectroscopy*. John Wiley and Sons.

KITTEL, C. (1976). *Introduction to solid state physics*. John Wiley and Sons.

McMILLAN, P. (1988). Infrared and Raman spectroscopy. In: *Reviews in mineralogy* **18**. Ed.: F.C. Hawthorne. Mineral. Soc. America.

ROSSMAN, G.R. (1988). Vibrational spectroscopy of hydrous components. In: *Reviews in Mineralogy* **18**. Ed.: F.C. Hawthorne. Mineral. Soc. America.

Optical spectroscopy:

BURNS, R.G. (1970). *Mineralogical applications of crystal field theory*. Cambridge University Press.

BURNS, R.G. (1985). Electronic spectra of minerals. In: *Chemical bonding and spectroscopy in mineral chemistry*. Eds.: F.J. Berry and D.J. Vaughan Chapman and Hall.

LANGER, K. (1987). UV to NIR spectra of silicate minerals by microscope spectrometry and their use in mineral thermodynamics and kinetics. In: *Physical properties and thermodynamic behaviour of minerals*. Ed: E. Salje Reidel Publishing.

ROSSMAN, G.R. (1988). Optical spectroscopy. In: *Reviews in mineralogy* **18**. Ed.: F.C. Hawthorne. Mineral. Soc. America.

X-ray spectroscopy:

BROWN, G.E., CALAS, G., WAYCHUNAS, G.A, AND PETIAU, J. (1988). X-ray absorption spectroscopy: applications in mineralogy and geochemistry. In: *Reviews in mineralogy* **18**. Ed.: F.C. Hawthorne. Mineral. Soc. America.

HOCHELLA, M.F. (1988). Auger electron and X-ray photoelectron spectroscopies. In: *Reviews in mineralogy* **18**. Ed.: F.C. Hawthorne. Mineral. Soc. America.

URCH, D.S. (1985). X-ray spectroscopy and chemical bonding in minerals. In: *Chemical bonding and spectroscopy in mineral chemistry*. Eds.: F.J. Berry and D.J. Vaughan Chapman and Hall.

WAYCHUNAS, G.A. (1988). Luminescence, X-ray emission and new spectroscopies. In: *Reviews in mineralogy* **18**. Ed.: F.C. Hawthorne. Mineral. Soc. America.

WOODRUFF, D.P. AND DELCHAR, T.A. (1986). *Modern techniques in surface science*. Cambridge University Press.

Mössbauer spectroscopy:

HAWTHORNE, F.C. (1988). Mössbauer spectroscopy. In: *Reviews in mineralogy* **18**. Ed.: F.C. Hawthorne. Mineral. Soc. America.

MADDOCK, A.G. (1985). Mössbauer spectroscopy in mineral chemistry. In: *Chemical bonding and spectroscopy in mineral chemistry*. Eds.: F.J. Berry and D.J. Vaughan Chapman and Hall.

SEIFERT, F. (1987). Recent advances in the mineralogical applications of ^{57}Fe Mössbauer effect. In: *Physical properties and thermodynamic behaviour of minerals*. Ed: E. Salje Reidel Publishing.

References

BROWN, G.E., CALAS, G., WAYCHUNAS, G.A. AND PETIAU J. (1988). X-ray absorption spectroscopy: applications in mineralogy and geochemistry. In: Spectroscopic methods in mineralogy and geology F.C. Hawthorne (Ed) *Reviews In mineralogy* **18**. Mineralogical Society of America.

BURNS, R.G. (1970). *Mineralogical applications of crystal field theory*. Cambridge University Press.

BURNS, R.G. AND GREAVES, C. (1971). Correlations of infrared and Mössbauer site population measurements of actinolites. *Amer. Mineral.* **56**, 2010–25.

GOLDING, R.M., NEWMAN, R.H., RAE, A.D. AND TENNANT, W.C. (1972). Single crystal EPR study of Mn^{2+} in natural tremolite. *J. Chem. Phys.* **57**, 1912–18.

JEANLOZ, R. (1980). Infrared spectra of olivine polymorphs: α, β phase and spinel. *Phys. Chem. Minerals* **5**, 327–41.

KIEFFER, S.W. (1985). Heat capacity and entropy: systematic relations to lattice vibrations. In: Macroscopic to Microscopic. S.W.Kieffer and A. Navrotsky (Eds.) *Reviews in mineralogy* **14**. Mineralogical Society of America.

KIRKPATRICK, R.J. (1988). MAS NMR spectroscopy of minerals and glasses. In: Spectroscopic methods in Mineralogy and Geology F.C. Hawthorne (Ed) *Reviews in mineralogy* **18**. Mineralogical Society of America.

LANGER, K. (1987). UV to NIR spectra of silicate minerals obtained by microscope spectrometry and their use in mineral thermodynamics and kinetics. In: *Physical properties and thermodynamic behaviour of minerals*. E.K.H. Salje (Ed.) NATO ASI Series C. D. Reidel Publishing Co.

McMILLAN, P. AND HOFMEISTER, A.M. (1988). Infrared and Raman spectroscopy. In: Spectroscopic methods in mineralogy and geology F.C. Hawthorne (Ed) *Reviews in Mineralogy* **18**. Mineralogical Society of America.

PIRIOU, B. AND McMILLAN, P. (1983). The high frequency vibrational spectra of vitreous and crystalline orthosilicates. *Amer. Mineral.* **68**, 426–43.

PUTNIS, A. AND ANGEL, R.A. (1985). Al,Si ordering in cordierite using "magic angle spinning" NMR. II: Models of Al,Si order from NMR data. *Phys. Chem. Minerals* **12**, 217–22.

PUTNIS, A., WINKLER, B. AND FERNANDEZ-DIAZ, L. (1990). In situ IR spectroscopic and thermogravimetric study of the dehydration of gypsum. *Mineral. Mag.* **54**, 123–28.

ROSSMAN, G.R (1988). Optical spectroscopy In: Spectroscopic methods in mineralogy and geology F.C. Hawthorne (Ed) *Reviews in mineralogy* **18**. Mineralogical Society of America.

WHITE, E.W. AND GIBBS, G.V. (1967). Structural and chemical effects on the Si K^{β} line for silicates. *Amer. Mineral.* **52**, 985–93.

5 The crystal structure of minerals I

The ultimate reason for any particular arrangement between the atoms in a mineral structure must lie in the nature of the cohesive forces which hold the structure together, and the equilibrium between these interatomic forces determines their mutual positions. Thus it would be logical to try to develop a classification of crystal structures such that the geometrical description could be related to the symmetries of electron orbits around nuclei. While this is an instructive exercise in a few simple cases, the vast majority of minerals have such complex structures with mixed bonding types that it is not possible at present to predict the structure from the composition. Therefore after a brief general discussion of the relationship between structure and bonding, most of this chapter will deal with the geometrical concepts of space filling, and polyhedral arrangements to describe mineral structures. It is not the aim here to provide a systematic account of all of the important mineral structures, and we will be more concerned with outlining the general principles involved in such a description.

The fundamental question of the extent to which we can predict a mineral structure from theory, and the recent advances that have been made in the computer simulation of structures is discussed in Chapter 8.

5.1 Bonding in crystal structures

The cohesive forces between atoms are determined by the distribution of the outer, or valency, electrons. When atoms participate in bonding these outer electrons interact in ways which determine the symmetry of the atomic structure. At the simplest level of discussion we can recognise three extreme types of bonding, defined in terms of the valence electron distribution.

1. In an atom of a *metallic* element there are usually one, two, or rarely three outer valence elec-trons which are weakly bound to the nucleus and can hence be easily removed to form a positive ion. In a metal structure these valence electrons become deta-ched from individual atoms, and can move freely as part of a 'sea' of electrons which provides a cohesive 'glue' holding the positive ions together. This also results in the high electrical conductivity of metals. There are no directional bonding requirements and the positive ions become regularly ordered, the packing governed by simple geometry based on the close-packing of spheres.

Although we will not be directly concerned with metallic bonding in mineral structures, the close-packed structures of metals provide a geometrical framework which will be useful whenever space-filling arrangements are discussed.

2. On the other hand, in *covalent* bonding, each atom has a tendency to hold on to the electrons in its outer shell, which is incomplete by one or more electrons. A covalent bond is formed by sharing these outer electrons between neighbouring atoms, producing an overlap of adjacent atomic orbitals. The number of covalent bonds which can be formed by an atom depends on the number of electrons in the outer shell, and the overlap of the orbitals defines a direction. This means that the coordination of a covalently bonded atom to its neighbours will be restricted and well-defined.

For example, a carbon atom has four electrons in its outer shell, and hence needs another four to fill it and form a bond. It can therefore be linked to four neighbouring carbon atoms each at the same dis-tance and at the same angle to it. These four neighbours define a tetrahedron: the carbon atom at the centre is said to be *tetrahedrally coordinated* with a *coordination number* of 4 (Figure 5.1(a)). When every carbon atom is tetrahedrally coordinated in this way in a crystal, the result is the *diamond structure* of carbon (Figure 5.1(b)).

Pure covalent bonding however is rare and can

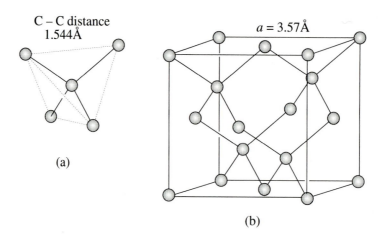

Figure 5.1. (a) The carbon atom at the centre is tetrahedrally coordinated to four neighbouring carbon atoms. These four carbon atoms lie on the corners of a regular tetrahedron (shown in the dashed line). (b) The diamond structure is built up from these tetrahedra, and every carbon atom is coordinated in this way.

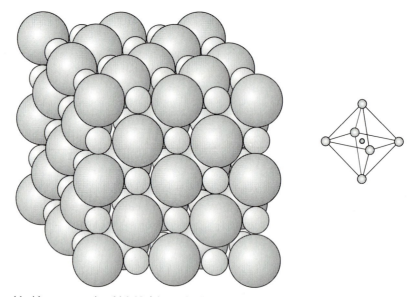

Figure 5.2. The sodium chloride structure, in which Na^+ is octahedrally coordinated by six Cl^- ions. The octahedron, drawn in the same orientation, shows the atoms reduced in size as is usual in 'ball and stick' models.

only occur between atoms of the same element. When the two bonded elements have different electronic structures the sharing arrangement of outer electrons becomes unequal and the bond acquires some polarity (i.e. one end of the bond is not the same as the other). The extreme case of this polarity is in the ionic bond, discussed below.

3. In the *ionic* bond, filling the outer shells of neighbouring atoms is done by transferring electrons from one atom to another so that the resulting positive and negative ions have spherically symmetrical electron distributions. The ionic crystal is made up of alternating cations and anions held

together by electrostatic forces. Each ion surrounds itself with as many ions of the other species as is geometrically possible, and so again the principle of close-packing of essentially spherical atoms applies.

The classic example of an ionic structure is sodium chloride (Figure 5.2) in which the relative sizes of the Na^+ and Cl^- ions dictate that each ion will be surrounded by six neighbours with opposite charge. These six ions form an octahedron around the central ion which is said to be *octahedrally coordinated* with a coordination number of 6 (Figure 5.2). Either Na^+ or Cl^- may be regarded as the central ion: in this respect the ions are interchangeable.

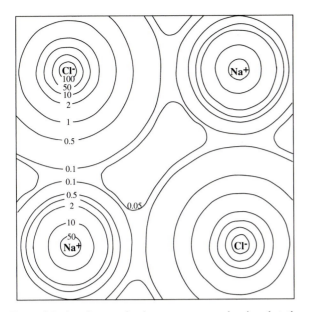

Figure 5.3. An electron density contour map showing that the electrons are concentrated around the ionic centres, with a pronounced minimum in the density between the ions. (After Adams, 1981.)

The concept of *ionic radius* is obviously important in the description of space-filling by dissimilar ions. The definition of ionic radius is not as straightforward as it may at first seem, as it depends on defining the boundary of an ion, a concept which has no precise meaning. The equilibrium separation of two ions can be determined by the diffraction methods discussed in Chapter 3 but a method of apportioning this distance between the two ions has to be found, either from a theoretical bonding model or empirically, by determining a large number of interatomic distances from different structures and devising a self-consistent tabulation of radii.

A more direct approach to the problem of defining ionic radii comes from X-ray diffraction experiments which can produce data on the electron density distribution in a crystal. This is plotted as a contour map, which is a direct 'image' of the ions and their sizes. Figure 5.3 is an electron density map for NaCl which shows the pronounced minimum in the electron density between the ions, as well as its spherical distribution. This method has only been applied to a few substances, mainly as a method of comparing estimates of ionic radii determined in different ways. Experimentally determined radii usually give larger values for cations and smaller values for anions than empirically determined ionic radii. Further discussion of the problems associated with the meaning

and measurement of ionic radii may be found in Adams (1981).

4. Finally, it is necessary to mention a fourth type of bonding which is present in all crystalline solids, but is only important when the other strong bonding mechanisms do not exist. This weak *van der Waals* bond is a universal weak attraction between closely spaced neutral atoms and molecules. The explanation of this general attractive force lies in the fact that even a neutral atom has a charge distribution which fluctuates very rapidly. When two such atoms are brought together the fluctuations in one induce a field around the other and this coupling results in an attractive force. Van der Waals bonding is the dominant cohesive force in crystals of the inert gases, and in organic crystals, but also plays a role in some materials with mixed bonding. A good example is graphite, discussed in Section 2.7, in which there is strong covalent bonding in the layers and weak van der Waals bonding between the layers, resulting in very different C–C distances (Figure 2.15).

The extent to which any of these bonding types applies in a particular crystal is a topic of considerable discussion. Even in a so-called purely ionic crystal such as NaCl there is a small covalent bond contribution due to overlap of adjacent Na and Cl orbitals, and in most minerals of interest both mixed bonding and different types of bonds exist in the same crystal. For our purposes the bonding models serve as a useful introduction to the geometrical aspects of packing and coordination which we will use to describe mineral structures but in most cases we will not make any specific reference to the bond type. Later we shall see that bonding models are also important in the discussion of the energy of a crystal structure, and here the ionic model is very useful, even when discussing structures which have considerable covalency.

5.2 Description of crystal structures

There are various ways of describing a crystal structure. The most precise way is to specify the shape and size of the unit cell (i.e. the crystal system and lattice parameters) and list the coordinates of each type of atom within the unit cell. This gives all the necessary information, but is the least digestible in terms of visualizing the structure and comparing one structure to another. In this section we will look at alternative ways of describing the arrangement of atoms in structures, beginning with the simplest ideas of close-packing and working towards concepts we will use to describe the structures of silicate minerals.

5.2.1 Close-packed structures – hexagonal close-packing and cubic close-packing

Here we are concerned with the most economical way of packing identical spheres in three dimensions. This concept is directly applicable to the structures of metals, but is also useful in discussing many other structures where space-filling is an important consideration.

Figure 5.4 shows the closest-packing of spheres (radius r) in two dimensions, each sphere in contact with six others. Within this close-packed plane there are three close-packed directions at 60°. The unit cell repeat of this two-dimensional layer is hexagonal

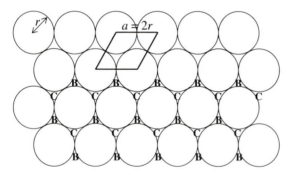

Figure 5.4. A single layer of close-packed spheres with radius r. The hexagonal unit cell of this layer has lattice parameter $a = 2r$. There are two sets of hollows in this layer, labelled B and C respectively.

with lattice parameter $a = 2r$. To build a three-dimensional structure we stack identical layers on top of one another, and the most efficient way of doing this is such that the spheres in one layer rest over the hollows in the layer below. If we refer to the first layer as an A layer, we can see that there are two sets of hollows which the next layer could occupy, labelled B and C. The second layer could occupy either all the B positions (and be referred to as a B layer) or all of the C positions (and become a C layer). In stacking the second layer, the choice of B or C is equivalent to merely rotating the structure through 180° so that there is no difference. Let us choose the B layer (for alphabetical convenience). We now have an AB stacking.

The choice for the third layer however, has very real consequences. Again there are two possibilities; the layer could lie directly above the first A layer, or it could lie above the C positions. The first choice produces an ABA sequence which if repeated in the same way gives a stacking ABABAB...which is called *hexagonal close-packing (h.c.p.)*. The second possibility leads to a stacking ABCABC... which is called *cubic close-packing (c.c.p.)*. We will consider each in turn in a little more detail.

Figure 5.5 illustrates hexagonal close-packing. The unit cell is described by the hexagonal repeat in the A layer ($a_h = 2r$) with the c dimension (c_h) equal to two layer thicknesses. Inspection of this stacking reveals that the atoms in the B layer do not have an environment identical to those in the A layer (the orientation of the interatomic bonds is different), and so the unit cell is primitive with lattice points only at the corners. The close-packed planes are parallel to (001). Many metals have a hexagonal

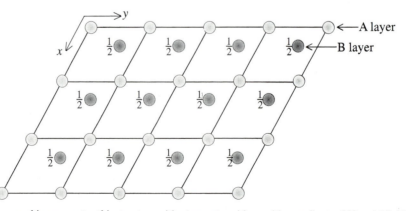

Figure 5.5. Hexagonal close-packing generates this structure with atoms at positions with coordinates 000 and 1/3,2/3,1/2 in each unit cell. The set of atoms at height $z = 1/2$ forms the B layer in the ABAB . . . close-packing, and is shown with darker shading.

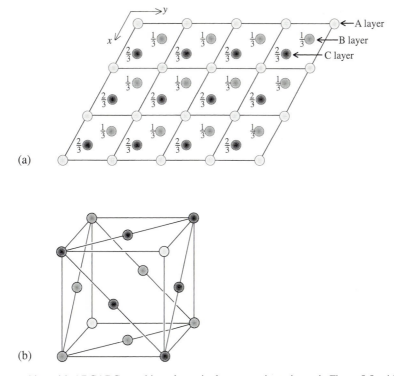

Figure 5.6. (a) Cubic close-packing with ABCABC.. packing, shown in the same orientation as in Figure 5.5, with the A atoms at positions 000, B atoms at 1/3,2/3,1/3 and C atoms at 2/3,1/3,2/3 in this unit cell. In cubic close-packing all of the atoms have identical environments and lie on lattice points. The symmetry generated by this packing is not apparent in this orientation and (b) shows the conventional cubic unit cell. The atoms in each layer are shaded as in (a) to show the relationship between the two orientations. However, the atoms are all identical and all {111} planes in the cubic cell are equivalent and close-packed.

close-packed structure, including magnesium (a = 3.21Å, c = 5.21Å), zinc (a = 2.66Å, c = 4.95Å) and titanium (a = 2.51Å, c = 4.68Å).

Figure 5.6 illustrates the cubic close-packed structure. We *could* describe the unit cell of this structure in the same way as in the h.c.p. structure, i.e. based on the hexagonal unit cell of the base A layer, but now with a *three* layer repeat. This however would completely mask the symmetry which has been created by the ABCABC stacking. In this stacking all the atoms in the A,B and C layers have identical environments, and are hence all at lattice points. If we look at the structure from a different angle, we see that there are now four sets of close-packed planes, and if the unit cell was described as a cube, the close-packed planes would be parallel to the four {111} planes. The cubic unit cell is face-centred, with lattice points at the centre of every face as well as the corners. The close-packed directions are parallel to the diagonals of the cube faces. Copper (a = 3.61Å) silver (a = 4.08Å) , gold (a = 4.07Å) and aluminium (a = 4.05Å) are among the many metals with a cubic close-packed structure.

The relationship between hexagonal and cubic close-packing is important because it also relates many other structures, not necessarily close-packed, where the basic stacking unit has hexagonal symmetry, but where two different stacking modes result in a hexagonal (ABAB) or cubic (ABCABC) structure. In cubic close-packing the lattice parameter a_c = $2\sqrt{2}r$ where r is the radius of the sphere; r is half the repeat in the two-dimensional layer, and so r = $a_h/2$. The body diagonal of the cube ($\sqrt{3}a_c$) is equivalent to 3 layer thicknesses, whereas in hexagonal close-packing the c repeat, c_h = 2 layer thicknesses.

We can thus write that:

$$c_h = \frac{2}{3}\sqrt{3}\,a_c = \frac{2}{\sqrt{3}}\,2\sqrt{2}r = \sqrt{\frac{8}{3}}\,a_h$$

In this way the ratio of the lattice parameters (c_h/a_h) for ideal hexagonal close-packing has been derived simply by considering the cubic close-packed structure, rather than having the more cumbersome geometry of the hexagonal cell. The deviation of a hexagonally close-packed structure from the ideal

√8/3 = 1.63 value is a measure of the deviation from ideal close-packing (e.g. in magnesium c/a = 1.62, in zinc it is 1.86). In ideal close-packing both c.c.p. and h.c.p. structures are equally space-filling, the spheres themselves occupying 74% of the total volume of the unit cell.

As well as the two ideal stacking sequences we could define some other sequence such as ABAC-B.ABACB with a 5 layer repeat. Such sequences are in fact rare in metal structures, but the principle of stacking identical layers to produce different stacking sequences (the resultant structures are called *polytypes*) is an important one, and is discussed in more detail in Chapter 6. A related, and common, phenomenon is that of a stacking fault , in which a mistake is made in an otherwise perfect stacking sequence, e.g. ABCABCACABCABC.. The frequency with which such mistakes are made in a structure depends on the extra energy involved. It is worth noting in this context that in perfect close-packing atoms in both c.c.p. and h.c.p. structures have 12 nearest neighbours with the same bond angles, so that if the energy of the structure depended only on the number of nearest neighbours there would be no difference in the energy of the two structure types, nor would a stacking fault contribute any extra energy. Note however, that deviations from perfect close-packing change the bond angles and also the bond energies.

Some elements adopt different structure types at different temperatures; cobalt is c.c.p. above 500°C and h.c.p. at lower temperatures, when it also contains frequent stacking faults. Evidently, in cobalt the energy balance between the two structures is tipped to favour the c.c.p structure, as that with the minimum energy at higher temperatures. The relationship between a structure and its energy will be a very important theme in many of the topics in this book.

The body-centred cubic structure

Another structure commonly adopted by metals is the body-centred cubic (b.c.c.) structure (Figure 5.7). The atom in the centre of the cube has an identical environment to those at the corners, and hence every atom is at a lattice point. While there are no close-packed layers there are close-packed directions along the cube diagonals and the total volume occupied is 68%, not far from that for closest packing.

Clearly, economy of packing is not the only consideration determining minimum energy in metals adopting this structure, which include sodium (a = 4.22Å), chromium (a = 2.88Å) and iron (a = 2.87Å). Iron is somewhat unusual in having the b.c.c. structure from its melting point down to 1401°C, and then again below 906°C. Between 1401°C and 906°C iron is f.c.c..

5.2.2 Interstitial sites in close-packed structures

Our main interest in discussing close-packing is the fact that from a geometrical point of view many simple mineral structures can be described in terms of approximate close-packing of anions with the generally smaller cations occupying the spaces between them. In such a description, the anions are held apart by the cations but nevertheless the anion distribution is the same as in the close-packed metals described in the previous section.

First we need to look at the shapes of the spaces between such close-packed ions. These interstices lie between any two close-packed layers, and are there-

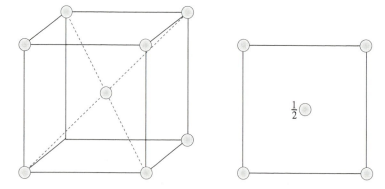

Figure 5.7. The body-centred cubic structure with atoms at 000 and 1/2,1/2,1/2 in each unit cell.

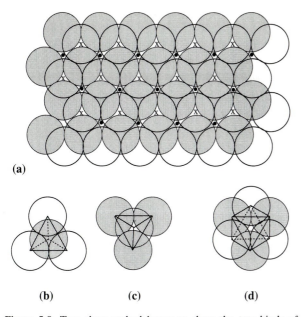

(a)

(b)   **(c)**   **(d)**

Figure 5.8. Two close-packed layers to show the two kinds of interstitial sites between the close-packed atoms. The shaded layer of atoms lies below the layer with no shading. The interstices labelled **x** are enclosed by four close-packed atoms forming a tetrahedron with the apex pointing upwards as shown in (c). There is also another set of interstices enclosed by a tetrahedron pointing down as in (b). The interstices labelled with the black spot are octahedral sites coordinated by six close-packed atoms which form an octahedron as shown in (d).

fore the same in both c.c.p. and h.c.p. structures. In Figure 5.8, two close-packed layers are shown. There are two kinds of interstices between them. The first, shown as a cross (x), lies in the space between four close-packed ions, three from one layer and one above. These four ions form a tetrahedron and the interstitial site is thus a tetrahedral site. Any small ion occupying this site will be tetrahedrally coordinated to its four nearest neighbours. Note also that the tetrahedral sites form two distinct sets, and in the orientation of Figure 5.8 one set has the tetrahedral apex pointing up, the other pointing down (Figure 5.8(b),(c)). Inspection of Figure 5.8 shows that there must be twice as many tetrahedral sites as there are close-packed ions, one above and one below each ion.

The second type of interstitial site lies in the larger space formed by three close-packed ions from one layer and three from the layer above. This is shown as a dot (·) in Figure 5.8(a). The six close-packed ions define an octahedron (Figure 5.8(d)), and the interstitial site is thus termed an octahedral site, with a coordination number of 6. There are half as many octahedral sites as there are tetrahedral sites in a

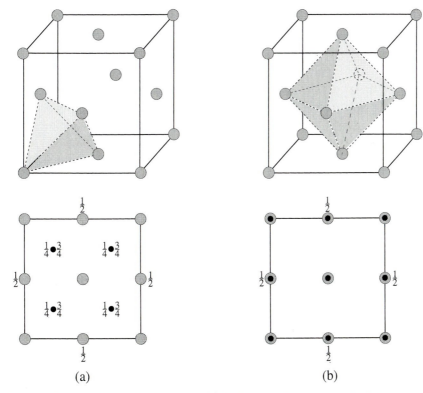

(a)   (b)

Figure 5.9. Interstitial sites in the cubic close-packed structure. (a) In the upper figure one tetrahedral site is shaded, and in the lower figures the positions of all tetrahedral sites are marked by the black circles. (b) In the upper figure one octahedral site is shaded. The octahedral sites lie at the centre, and halfway along the edges of the cube. The lower figure shows the projection of these octahedral sites as black circles, but their coordinates are omitted.

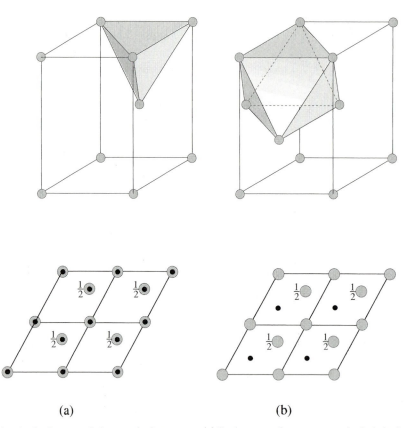

Figure 5.10. Interstitial sites in the hexagonal close-packed structure. (a) In the upper figure one tetrahedral site is shaded, and the lower figure shows the projection of all of the tetrahedral sites as black circles. There are tetrahedral sites above and below each close-packed atom, totalling four sites per unit cell. (b) The position of the octahedral sites with one octahedron shaded. There is another octahedral site immediately below the one shown, giving two sites per unit cell.

close-packed structure; there is therefore one octahedral site associated with each close-packed anion.

Although the interstitial sites are present in the same proportion in both cubic and hexagonal close-packed structures, their orientations in the unit cell appear very different because of the different orientation of the unit cells relative to the close-packing in each case. In Figures 5.9 and 5.10 the positions and orientations of octahedral and tetrahedral sites are shown in each of the structures. For simplicity, only one site is shown in each case. (Note that in the c.c.p. structure there are 4 close-packed ions per unit cell, and therefore 4 octahedral sites and 8 tetrahedral sites; in the h.c.p. structure there are 2 close-packed ions, 2 octahedral sites and 4 tetrahedral sites.)

When considering structures in which close-packed anions are held apart by cations in the interstitial sites, the ratio of the cation size to anion size has long been used as a criterion for deciding

which site a particular cation could occupy. In a stable structure the cation should be in contact with each of its anion neighbours in the coordination polyhedron. The *radius-ratio rules* are a first step in predicting structures based in these simple geometrical considerations. Using such geometry we can calculate the size of a small cation, radius r, which would fit into an interstitial site without distorting the anions (radius R) from exact close-packing. For a tetrahedral site $r/R \geq 0.225$; for an octahedral site $r/R = 0.414$. These are the *minimum* ratios, and we would expect that to keep the anions from touching one another, as expected from the mutual repulsion of like ions, the cations would have to be larger than predicted by these ratios. We would also expect that a cation would occupy a tetrahedral site if $0.414 > r/R \geq 0.225$; when $r/R > 0.414$ the cation would probably occupy the larger octahedral site.

Although radius ratios are useful as a qualitative tool for predicting cation coordination (thus large cations would be expected to have a higher coordination), the criteria must not be taken too literally. Apart from the problem of defining accurate cation radii there is always some degree of covalent bonding in such structures and this will introduce bond overlap in some directions, thus partly invalidating our simple model of spherical ions in contact. A critical discussion on the factors affecting crystal structure can be found in Adams (1981).

5.2.3 Structure types based on close-packing

When crystal structures of compounds are discussed in terms of close-packing, it is not necessarily suggested that the close-packed atoms are touching one another, but merely that they are arranged in an array which is the same as that in close-packed structures. Close-packed anions with cations in interstitial sites is a convenient description, even when the anion array is spread apart due to electrostatic repulsion. If we look at the sodium chloride structure in this way (Figure 5.2 and 5.11(a)), we can describe the anion array as cubic close-packed with the cations occupying all of the octahedral sites. This may seem to be the most logical description. However, apart from the relative cation sizes, the structure could also be described as an array of cations in a c.c.p. arrangement with anions in octahedral sites. From a geometrical point of view, with the atoms represented by points rather than spheres, either description is correct. In the rest of this chapter we will treat structures in a geometrical way, taking care not to draw conclusions about bonding from the occurrence of a particular structure type.

If we consider simple structures based on a close-packed array of ions, which will usually be anions (X), with cations (M) in interstitial sites we can immediately define some important structural types.

1. Cubic close-packing with all octahedral sites occupied. This is the well-known sodium chloride structure, illustrated in Figure 5.2, and again in plan view in Figure 5.11(a).

2. Cubic close-packing with half of the tetrahedral sites occupied. Again the stoichiometry of this structure is MX, and it is named the sphalerite structure, after one of the structures of ZnS. It is illustrated in plan in Figure 5.11(b). Note that the occupied tetrahedral sites are all in one set, with tetrahedral apices all pointing in the same direction.

3. Cubic close-packing with all of the tetrahedral sites occupied. The stoichiometry is now M_2X or

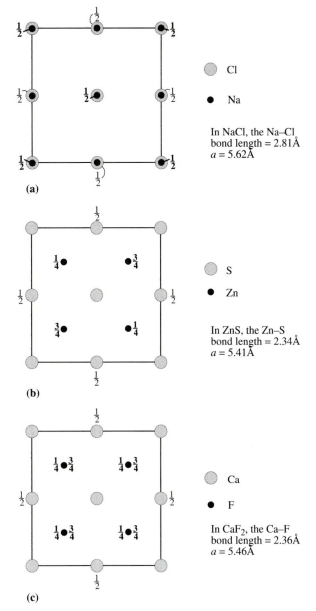

Figure 5.11. (a) The sodium chloride structure (b) the zinc blende structure and (c) the fluorite structure. In each case the z coordinates of the atom shown by the black circle are in bolder print.

MX_2, depending on which ion has the close-packed distribution. The structure is named the fluorite structure, after CaF_2 and is illustrated in Figure 5.11(c). In fluorite it is the the Ca^{2+} ions which are in a close-packed array, with F^- in tetrahedral sites, which given the radius ratio $r_{F-} / r_{Ca}^{2+} = 0.95$ (compared to the minimum of 0.225) makes the close-packing description merely a geometric convenience. The M_2X arrangement, with the anion

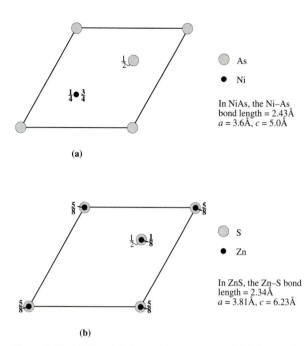

As
Ni

In NiAs, the Ni–As bond length = 2.43Å
$a = 3.6$Å, $c = 5.0$Å

(a)

S
Zn

In ZnS, the Zn–S bond length = 2.34Å
$a = 3.81$Å, $c = 6.23$Å

(b)

Figure 5.12. (a) The nickel arsenide structure and (b) the wurzite structure. In each case the z coordinates of the atom shown by the black circle are in bolder print.

occupying the c.c.p. positions, is known as the 'anti-fluorite' structure.

4. Hexagonal close-packing with all octahedral sites occupied. The type example is the nickel arsenide structure of NiAs, shown in Figure 5.12(a).

5. Hexagonal close-packing with half of the tetrahedral sites occupied. This is termed the wurtzite structure, after another form of ZnS, and is illustrated in Figure 5.12(b). Again one set of tetrahedral sites is occupied. No structures where all of the tetrahedral sites are occupied are known.

5.2.4 Minerals with structures based on close-packing

Many common oxide and sulphide minerals have structures based on one of the simple structure types above, or on some variant of it. We are not interested here in a systematic classification of minerals based on their structure, and will restrict the description to those mineral groups referred to later in the book, particularly with reference to interesting aspects of their compositional variation or their structural behaviour as a function of temperature and pressure. Thus periclase (MgO) and galena (PbS) both with the sodium chloride structure will

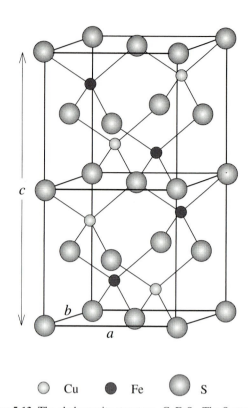

○ Cu   ● Fe   ◉ S

Figure 5.13. The chalcopyrite structure, $CuFeS_2$. The S atoms are cubic close-packed and the Cu and Fe atoms occupy half of the tetrahedral sites in an ordered distribution which doubles the c axis relative to the zinc blende structure.

not be mentioned again. Zinc blende (or sphalerite) and wurtzite, both forms of ZnS with Zn in tetrahedral sites of c.c.p. and h.c.p. arrays of S atoms respectively, will be a topic for further discussion, particularly in relation to polytypes, first mentioned in Section 5.2.1.

The variations on these simple structure types do introduce some important concepts as well as interesting structures and some of these are described below.

The chalcopyrite structure

The composition of the mineral chalcopyrite is $CuFeS_2$ and it has a structure based on the zinc blende, with a c.c.p. array of S atoms. The stoichiometry indicates that half of the tetrahedral sites are occupied, and these all belong to one set, with the apices all pointing in the same direction. If the Cu and Fe atoms were randomly arranged over the tetrahedral sites (i.e. disordered with the probability that a Cu atom could be found at any of the occupied sites was 50%) the formula could be written as

(Cu,Fe)S and the cubic unit cell of the zinc blende structure would be correct. However, the ordering of Cu and Fe atoms in chalcopyrite results in a doubling of the cubic cell in one direction and hence a tetragonal unit cell (Figure 5.13).

This illustrates some concepts which will be important throughout much of this book. First, the concept of disorder in part of a structure. The sulphur array can only be perfectly ordered, but the cation distribution could have any degree of organisation from completely random to completely ordered on the given set of sites. When such disorder exists in a structure, it is confined to relatively high

temperatures, with order setting in as the temperature is lowered. Second is the concept that on ordering the cations, the symmetry is generally reduced and the unit cell size *may* also need to be increased to describe the new periodicity. In chalcopyrite, this is the case, but we will see that it depends on the nature and distribution of the cation sites.

Another question we can ask in relation to the chalcopyrite structure is whether there can be any variation in the Cu:Fe ratio. In the ordered form, each ion is in a specific site and there is little compositional variation possible, but at high temperatures there exists a range of compositions with disordered cations, and hence the cubic zinc blende structure, around the $CuFeS_2$ composition. The shaded area in Figure 5.14 shows the extent of this compositional variation, which includes compositions with excess cations, which must randomly occupy some of the second set of tetrahedral sites. Such a compositional variation is termed a *solid solution* and we will meet many examples of this phenomenon in minerals. In some cases solid solutions exist at ambient temperatures, but it is typically a high temperature phenomenon. As many minerals undergo cooling after crystallising at high temperatures, the behaviour of a solid solution under these conditions, and the change from disorder to order, is one of the principal concerns in mineralogy, and will be the theme of later chapters.

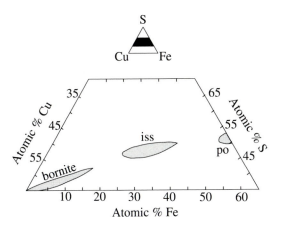

Figure 5.14. The compositional variation possible in the solid solution in the central part of the Cu–Fe–S system. Three solid solutions exist (shaded): the bornite solid solution around Cu_5FeS_4, the intermediate solid solution (iss) around chalcopyrite $CuFeS_2$, and the pyrrhotite solid solution (po). This range of possible compositions exists at high temperatures (600°C) where the Cu,Fe atoms are disordered over the occupied interstitial sites. At low temperature various ordered or partially ordered phases with defined compositions are formed.

The troilite–pyrrhotite ($FeS–Fe_{1-x}S$) solid solution

The structure of troilite, FeS, is based on the NiAs structure with a hexagonal close-packed array of S

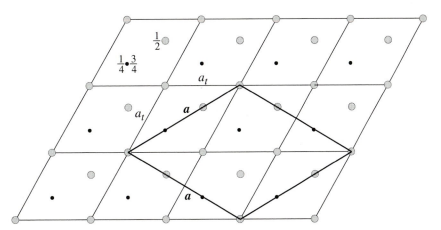

Figure 5.15. Troilite, FeS has the NiAs structure at temperatures above 140°C. Below this temperature the structure distorts so that the new unit cell is hexagonal, with lattice parameters $a = \sqrt{3}a_t$, $c = 2c_t$ where a_t and c_t are the cell parameters of the high temperature cell (see Figure 5.12(a)).

atoms and Fe occupying all of the tetrahedral sites. Below about 140°C the structure undergoes a slight distortion, and can no longer be described by the hexagonal NiAs unit cell. The unit cell of the distorted structure relative to that of the NiAs cell is shown in Figure 5.15 (the displacements of atoms in the distorted cell are too small to be shown). The purpose here is to show that slight distortions result in a unit cell which is related to the parent cell in some simple way. The high temperature form also has higher symmetry, in this case higher translational symmetry (smaller unit cell), than the low temperature form.

At high temperatures the NiAs structure of FeS can exist over a range of compositions $Fe_{1-x}S$, the formula implying that the stoichiometry can vary by removing Fe atoms from the octahedral sites and creating vacancies. Iron sulphides with this general formula are collectively referred to as pyrrhotite. The extent of this omission solid solution is from FeS to around $Fe_{0.875}S$, or Fe_7S_8, so that up to one in every eight octahedral sites could be vacant. In each NiAs unit cell there are two octahedral sites; one in four must contain a vacancy. For the solid solution structure to be described by the NiAs cell the vacant sites must obviously be randomly distributed over the structure, i.e. disordered. Any long-range ordering of vacancies however must necessarily increase the size of the unit cell of the structure.

Ordering of the vacancies in pyrrhotites takes place as the temperature is lowered, and the general principle governing the vacancy distribution is that the electrostatic internal energy of the structure is lowered when the vacancies have a maximum separation. At the composition Fe_7S_8, this can be achieved in quite an efficient way. To illustrate this ordered vacancy distribution it is helpful to look at the NiAs structure in terms of the cation distribution only, as the S array plays no direct role in the ordering process. The cations form layers, parallel to (001) planes and alternating with the hexagonally close-packed S layers. To achieve a maximum vacancy separation only every second cation layer has vacancies and their distribution within such a defective layer is shown in Figure 5.16(a). The layers are then stacked along the c axis, alternating full and defective, and staggered so that an arbitrary position A in one defective layer occupies positions B,C,D,A, in successive defective layers as shown in Figure 5.16(b). This results in a repeat which has a c

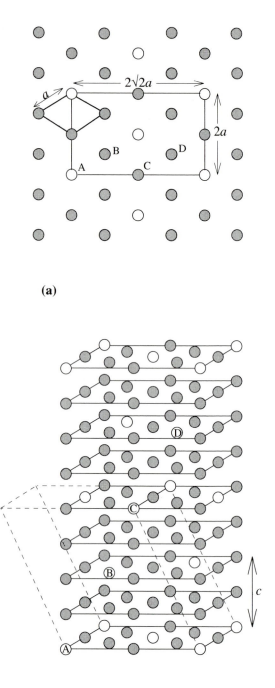

(a)

(b)

Figure 5.16. (a) A layer of Fe atoms in the pyrrhotite, $Fe_{1-x}S$ in which the unfilled circles represent vacant sites. The vacancies are ordered to achieve a maximum vacancy separation. The subcell with cell parameter a is shown in bold, and the supercell due to vacancy ordering, with parameters $2\sqrt{2}a \times 2a$, in a lighter line. In the Fe_7S_8 structure these metal layers are stacked above one another such that a vacancy at site A is placed above sites B,C,D and A in successive defective layers. This stacking is shown in (b). Every alternate metal layer contains vacancies, ordered in the same way, but translated laterally in the stacking sequence. One unit cell, with lattice parameters $2\sqrt{2}a \times 2a \times 4c$ where a,c are the lattice parameters of the subcell, is shown. The dashed line shows an alternative monoclinic unit cell. The sulphur atoms have been omitted for simplicity.

axis four times that of the NiAs cell. The vacancy distribution within the layers also increases the repeats in the layer to $2a \times 2\sqrt{2}a$ where a is the lattice parameter of the NiAs cell.

The ordered unit cell of Fe_7S_8 is called a *superstructure*, *supercell* or *superlattice* of the basic NiAs *subcell*. The structure described above is referred to as the $4c$ superstructure, to distinguish it from other possible superstructures formed by different ordered vacancy distributions. In Figure 5.16(b) the $4c$ unit cell appears to be orthorhombic. In fact, a small additional distortion at low temperatures changes one of the interaxial angles from 90° to 90.5° resulting in a monoclinic structure for which the conventional unit cell is shown in Figure 5.16(b) as a dashed line.

The question that arises next is what happens to a composition somewhere between FeS and Fe_7S_8 as it cools from a high temperature disordered form? We are not yet really equipped to answer such a complex question, but we can consider some possibilities. If a condition for the stability of an ordered $Fe_{1-x}S$ structure is maximum vacancy separation, we can devise ordered structures for compositions more Fe-rich than Fe_7S_8 by introducing extra full cation layers in some regular way. In Fe_7S_8, full (F) and vacancy-bearing layers (V) alternate up the c axis in a sequence VFVFVFVF... (Fig 5.16(b)). If these same layers are stacked VFFVFFVFFVFF... the result is a $6c$ superlattice and a composition $Fe_{11}S_{12}$. In the same way a stacking sequence VFFVFVF-FVFVFFVF... gives a $5c$ repeat and a composition Fe_9S_{10}. How far can we go with this idea? A composition $Fe_{10}S_{11}$ treated this way would have to have an $11c$ repeat which is beginning to sound unreasonable from the point of view of the range over which electrostatic forces can act. And what of other compositions with even less amenable stoichiometries?

As well as considering questions of stability (is ordered $Fe_{10}S_{11}$ more or less stable than a mixture of ordered $Fe_9S_{11} + Fe_{11}S_{12}$?) we also need to consider the *kinetics* of ordering vacancies, both in nature and in any experiment we may carry out to study this problem. Any mineral sample we examine may not yet have achieved its minimum energy state, assuming we know which of the structural possibilities in a complex mineral group is the most stable. Natural pyrrhotites are almost invariably fine-scaled intergrowths, on a submicroscopic scale, of the ordering

possibilities discussed above. The evidence points to a sequential ordering process in which complex superstructures formed in the early stages of ordering gradually transform to simpler, better ordered structures as a function of time. The occurrence, in rocks with a long cooling history, of pyrrhotites which are intergrowths of troilite (FeS) + $4c$ Fe_7S_8 suggests that this may be the most stable state, and that the longer period structures described above may be only stages on the way to achieving this stability.

The corundum structure – hematite and ilmenite

The minerals corundum α-Al_2O_3, hematite α-Fe_2O_3 and ilmenite $FeTiO_3$ have a structure which is a derivative of the NiAs structure in that the oxide ions are approximately hexagonally close-packed and the cations occupy octahedral sites. The cation–anion ratio indicates that only two-thirds of the octahedral sites are filled. The way in which these sites are filled determines the size of the unit cell. Note again that the octahedral sites form layers which lie half-way between the close-packed oxygen layers. In Figure 5.17(a) the pattern of occupancy within the cation layer is shown, and Figure 5.17(b) shows the way the layers are stacked. Each filled octahedron shares a face with another in the layer above or below, forming pairs of Al – Al ions. When octahedra are face-sharing the cations are brought close together and in the corundum structure the mutual repulsion of the Al^{3+} ions moves them off-centre in the octahedron. This distortion puckers the cation layers and leads to a distortion of the oxygen close-packing.

This cation arrangement leads to a tripling of the c axis relative to the ABAB.. repeat of the oxygen stacking, and the unit cell is trigonal, space group $R\bar{3}c$. As is usually the case in the trigonal system, it is described in terms of a hexagonal unit cell.

Hematite Fe_2O_3, also has the corundum structure and ilmenite $FeTiO_3$, is very closely related to it. In ilmenite half of the Fe atoms are replaced by Ti, and the cations are ordered such that each layer contains all Fe or all Ti, alternating so that the close pairs of ions are always Fe – Ti. We have already seen in other examples that cation ordering reduces the symmetry. Hematite has symmetry $R\bar{3}c$; in ilmenite the symmetry is reduced to $R\bar{3}$. As we will see in Chapter 12, this ordering of Fe,Ti in minerals with compositions between ilmenite and hematite

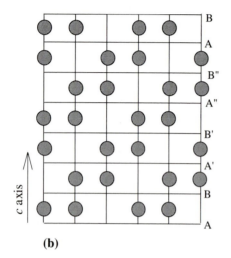

(a)

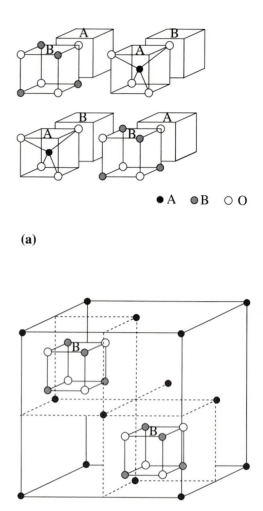

● A   ◉ B   ○ O

(a)

(b)

Figure 5.17. The corundum structure. (a) A single close-packed oxygen layer (unfilled circles) showing the distribution of octahedral site occupancies (filled circles) in the cation layer. (b) The oxygen atoms are hexagonally close-packed, and represented here by the horizontal lines labelled ABAB.. The cation layers lie between the close-packed oxygen layers and are stacked up the c axis, each layer laterally displaced relative to the previous one. This results in a c repeat of three times that for the oxygen packing, indicated by the labelling A B A'B'A''B''.

(b)

Figure 5.18. The spinel structure AB_2O_4. The cubic unit cell contains 32 oxygen atoms in cubic close-packing. To illustrate the distribution of cation sites the structure can be described in terms of the small cubes labelled A and B in (a). A represents an AO_4 tetrahedron, and B a B_4O_4 cube. The arrangement of AO_4 and B_4O_4 cubes shown in (a) puts the B atoms in octahedral sites. To complete the structure this distribution of four AO_4 cubes and four B_4O_4 cubes is placed within the octants of a large cube in which extra A atoms are placed at the corners and face-centres, as shown in (b). Only two B_4O_4 cubes are drawn to show the position of the cubes in the spinel unit cell (after Greenwood, 1968).

(ilmenohematites) plays a very important role in the magnetic properties of rocks.

The spinel structure

The spinel group of minerals has a general composition AB_2O_4, and is named after the mineral spinel, $MgAl_2O_4$. The spinel structure is usually described as a cubic close-packed array of oxygen atoms with the A and B cations occupying one eighth of the tetrahedral sites and one half of the octahedral sites. The occupancy of the interstitial sites results in a face-centred cubic unit cell which is $2\times2\times2$ times that of the basic c.c.p. oxygen array, and therefore there are 32 oxygen atoms in the spinel unit cell. The unit cell contents are therefore $A_8B_{16}O_{32}$. The large number of atoms in the unit cell makes this a difficult

structure to illustrate, and ideally a structure model should be used to locate the sites. Figure 5.18 shows one way of illustrating the regular cation distribution.

One of the characteristics of the spinel structure is its flexibility in the range of cations and cation charge combinations it will accept, making it a structure adopted by over a hundred compounds, many of them important minerals or important

Table 5.1. *Common oxide minerals with the spinel structure*

Mineral Name	Composition	Cell edge a(Å)	u parameter	Structure*
Magnetite	$Fe^{3+}[Fe^{2+}Fe^{3+}]O_4$	8.396	0.2548	Inverse
Magnesioferrite	$Fe^{3+}[Mg^{2+}Fe^{3+}]O_4$	8.383	0.257	Inverse
Chromite	$Fe^{2+}[Cr_2^{3+}]O_4$	8.378		Normal
Magnesiochromite	$Mg^{2+}[Cr_2^{3+}]O_4$	8.334	0.260	Normal
Spinel	$Mg^{2+}[Al_2^{3+}]O_4$	8.103	0.262	7/8 Inverse
Hercynite	$Fe^{2+}[Al_2^{3+}]O_4$	8.135		Normal
Ulvöspinel	$Fe^{2+}[Fe^{2+},Ti^{4+}]O_4$	8.536	0.261	Inverse
Jacobsite	$Fe^{3+}[Mn^{2+}Fe^{3+}]O_4$	8.51		Inverse

* *Many spinels are probably intermediate between normal and inverse.*

commercial magnetic oxides. The reason for this flexibility can be found by looking at the site occupancies in more detail.

Within the space group of spinel ($Fd3m$) the fractional coordinates of the tetrahedral and octahedral sites are fixed at special positions (A atoms at 000 and 7 other symmetry related sites, B atoms at 5/8, 5/8, 5/8 and 15 other related sites). The oxygen atoms are also on special positions, but with fractional coordinates uuu, where the u value can vary. In effect, this means that if the relative sizes of the A and B cations change, their positions remain fixed but the oxygen array expands or contracts around them maintaining the same symmetry throughout. This flexibility is inherent in the spinel structure, by comparison with the perovskite structure discussed in the next section, where variations in relative cation size result in distortions of the structure and a reduction in symmetry. In spinel the u value is an indication of how close-packed the structure is; for ideal close-packing $u = 0.375$.

The cation distribution over the 8 tetrahedral and 16 octahedral sites in the unit cell can vary. In *normal spinels* the 8 A cations in the formula $A_8B_{16}O_{32}$ occupy the eight tetrahedral sites, while the 16 B cations occupy the 16 octahedral sites. However, another extreme case exists where the 8 tetrahedral sites are all occupied by B cations, in which case 8 A and 8B cations randomly fill the 16 octahedral sites. These are *inverse spinels*, and the formula could be written $B_8(A_8B_8)O_{32}$ or B(AB)O_4.

As well as the two extremes of normal and inverse, intermediate (or partly inverse) cation distributions are possible and these may vary with temperature. A completely disordered spinel would have a cation distribution corresponding to $(A_{0.33}B_{0.67})$ $(A_{0.67}B_{1.33})$ O_4. There is no change in symmetry associated with this type of disorder as it only involves a redistribution of cations over sites which are already differentiated i.e. tetrahedral and octahedral. The symmetry changes only when sites which were equivalent in the disordered state (where the average occupancy is the same) become non-equivalent on ordering. If for example, the A and B cations were ordered over the 16 equivalent octahedral sites in an inverse spinel, the symmetry would have to be reduced.

Table 5.1 lists the common oxide minerals with the spinel structure, together with their compositions, u values, cell dimension and structure type. Pure end member compositions of these minerals are rare in nature, and cation substitution, especially between cations of similar size and valence, results in extensive solid solutions between them. An important solid solution is that between magnetite Fe_3O_4 and ulvöspinel Fe_2TiO_4 as it is the main carrier of magnetism in rocks. One of the questions we will look at in greater detail in a later chapter is the fate of spinel solid solutions crystallised at high temperatures. A solid solution implies random substitution of cations for one another i.e. disorder. We have already seen in general terms that disorder at high temperatures tends to give way to order at lower temperatures. We will come back to this topic after discussing the thermodynamic and kinetic factors which control such behaviour.

When magnetite $Fe^{3+}(Fe^{2+}, Fe^{3+})O_4$, becomes progressively oxidised, as is often the case in cooling rocks, the Fe^{2+} will be converted to Fe^{3+} until, when all the Fe is trivalent the charge balance requires a stoichiometry of $Fe_{21.67}O_{32}$ per unit cell. This is a *defect spinel* structure with an average of 2⅓ vacant

Table 5.2. *Composition and cell size of the more common thiospinels*

	Composition	Cell size a (Å)
Linnaeite	Co_3S_4	9.399
Polydymite	Ni_3S_4	9.480
Siegenite	$(Co,Ni)_3S_4$	9.418
Greigite	Fe_3S_4	9.876
Violarite	$FeNi_2S_4$	9.463
Carrollite	$CuCo_2S_4$	9.461
Daubréelite	$FeCr_2S_4$	9.989
Indite	$InFe_2S_4$	10.62

sites per unit cell. Its mineral name is maghemite. Similar defect spinels form along the magnetite–ulvospinel solid solution and are termed titanomaghemites. The ratio of Fe^{2+}/Fe^{3+} in these minerals is controlled by the oxygen partial pressure of their environment, and therefore in any oxidising environment there will always be an excess of Fe^{3+} ions and some extra vacant sites. This affects the magnetic properties of the spinel. Another point of interest is that since the composition of a totally oxidised maghemite can be written Fe_2O_3, which is the composition of hematite, there is a question as to the thermodynamic relationship between the two structures. Ultimately, maghemite should transform to hematite, but yet it is quite common in altered basaltic rocks. The implications of this to magnetism in rocks will be discussed in Chapter 12.

The spinel structure is also adopted by some sulphide minerals, termed thiospinels, whose compositions fall in the range Fe–Co–Ni–S. End-member compositions, as well as some common intermediate compositions are shown in Table 5.2.

Before leaving this section on structures based on close-packed anions it is worth making some final comments on this way of looking at structures. If we apply the concept of radius ratios, discussed in Section 5.2.2, to tetrahedral or octahedral occupancy in spinels, it fails spectacularly. In the type example, $MgAl_2O_4$, which is a normal spinel, the larger Mg^{2+} ion occupies the smaller tetrahedral site, while the smaller Al^{3+} is in the larger octahedral site! Why is $MgAl_2O_4$ normal and not inverse? This has been an endless point of discussion in spinel structures, and many of the arguments are summarized in Greenwood (1968). The factors usually invoked are the relative energetics of mixing cations

on tetrahedral sites (there are not enough tetrahedral sites for all the Al^{3+}), covalency effects, and the stabilisation of transition metal ions in octahedral sites.

Another view, put forward by O'Keeffe and Hyde (1985), which has significant implications for many other structures, is that the concept of anion close-packing with cations 'fitting in' to interstices has outlived its usefulness in many cases. We have already seen in the simple case of the NaCl structure that the cation array and the anion array are equivalent – the structure could be looked at from the point of view that cation–cation repulsions are responsible for the cation distribution and the anions provide charge balance. By looking at the regular cation array in spinels and then adding the anions in such a way that cation–oxygen bond lengths (which are constant from compound to compound for a given coordination) are correct, the u value for the oxygen atoms can be relatively easily calculated. This approach, as well as providing a way out of the inflexibility of radius ratio rules, gains extra credibility when it is pointed out that the $MgAl_2$ array in the spinel structure is the same as that in the structure of the metal alloy $MgCu_2$. The same is true of many other oxide structures with mixed cations. Looking at oxides as stuffed alloys is an alternative geometrical description which sheds new light on many structures.

5.2.5 Structures built from polyhedra

It is often very convenient to describe a structure in terms of the coordination polyhedra around the cations, and the way in which these polyhedra are linked together, by sharing corners, edges or faces. In some structures a description based on close-packing or on polyhedral linking is equally applicable; in other cases, particularly in more complex structures, or covalent structures where close-packing is not evident, the polyhedral description is much more appropriate. The anions become the points at the apices of the polyhedron, so that even in a close-packed structure, a polyhedral arrangement will not completely fill space.

If we consider the NaCl structure in this way, we need to look at the way in which the octahedra around each Na ion are linked. This is not always easy to see from a drawing of the structure, and even more difficult from a plan. Figure 5.19 shows the

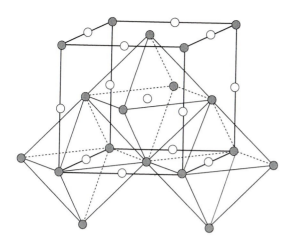

Figure 5.19. The sodium chloride structure, described in terms of edge-shared octahedra.

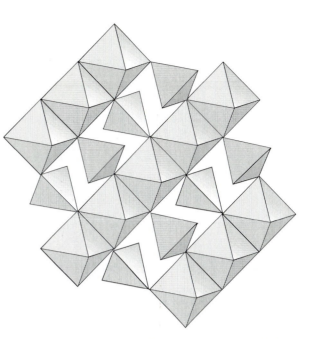

Figure 5.21. Part of the spinel structure showing the linkage between AO_4 tetrahedra and BO_6 octahedra on a {100} face of the cubic unit cell. The unit cell can be described in terms of the arrangement of such layers.

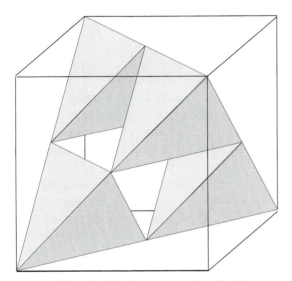

Figure 5.20. The zinc blende structure, described in terms of corner-shared tetrahedra.

octahedral edge-sharing in a unit cell of the NaCl structure. Each octahedral edge is shared by two octahedra and the structure can be described as an infinite array of such edge-sharing octahedra. The spaces between these octahedra are the unoccupied tetrahedral sites in the NaCl structure. In the zinc blende structure of ZnS, the tetrahedra around each Zn ion are linked by corners, each corner being shared by four tetrahedra (Figure 5.20).

In the spinel structure, rows of edge-sharing octahedra run along one diagonal of a cube face, and are cross linked by tetrahedra to form a layer (Figure 5.21) The tetrahedra also link these octahedra to those in the next layer where the rows of octahedra

run along the other diagonal of the cube face. Each corner is shared between three octahedra and one tetrahedron. Complete schemes have been devised, enumerating all the possible linkages between polyhedral networks, but we will not consider them here.

The perovskite structure

The perovskite structure, named after the mineral perovskite $CaTiO_3$, is adopted by many compounds with the general formula ABO_3. The ideal structure is cubic with a BO_6 octahedron at each corner forming an infinite corner-sharing array (Figure 5.22). The A cation is at the centre of the cube and is coordinated to 12 oxygen atoms located at the midpoint of each cube edge. All of the atoms in such an arrangement are on special positions within the unit cell: the A atom has coordinates 1/2,1/2,1/2, the B atom is at 0,0,0 and the oxygen at 0,0,1/2 and symmetrically related positions. In this ideal form the ratio of the A–O to the B–O bond lengths must be equal to $\sqrt{2}$.

When this condition cannot be met, the structure distorts by changing the shape of the octahedron (the cation moving off-centre) or by twisting the linkages between octahedra. These distortions result in a reduction in symmetry. In this respect the

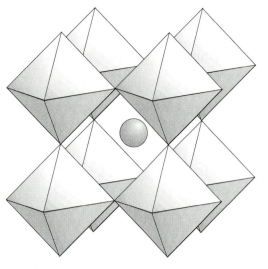

Figure 5.22. The cubic structure of perovskite ABO_3, in which the corner-sharing BO_6 octahedra form a cube with the B atoms at the corners, and with the A atom at the centre.

structure can be contrasted to the spinel structure where the oxygen atoms had the flexibility to accommodate differing cation sizes without distortion because their positions within the structure were not so specifically defined.

The ideal perovskite structure is quite demanding in relation to the relative sizes of the A and B cations. At high temperatures however, when there is more flexibility in the bonds and the atoms themselves have greater vibrational amplitudes, the restriction on the ratio of A–O to B–O bond lengths is relaxed to some extent. This means that even if the cation sizes are not right, the structure can still be cubic at high temperatures, while at low temperatures it will have to distort to accommodate them.

The distortions of the perovskite structure are basically of three types:

(i) If the A cation is too small for the large 12-fold site, the octahedra tilt relative to one another, and so reduce the size of the A site. The type example is $CaTiO_3$.

(ii) If the B cation is too small for the octahedral site, the cation moves off-centre within the octahedron. The type example of this case is $BaTiO_3$, which does not occur as a mineral.

(iii) If the A–O and/or B–O bonds have a pronounced covalency, this results in a distortion of the octahedron itself.

Each kind of distortion may occur independently in a given compound, or in combination, leading to a

great variety of different structural modifications depending on the sizes and the bonding character of the cations. Here we will restrict the discussion to the distortions in $CaTiO_3$ and $BaTiO_3$ and illustrate some of the consequences of this type of behaviour which will also be applicable to other mineral groups.

In $CaTiO_3$ the octahedra tilt relative to one another. In Figure 5.23(a) the tilts of the octahedra are about an axis normal to the page. Adjacent octahedra tilt in opposite senses and are therefore no longer equivalent, as they were in the ideal cubic structure. The size of the unit cell is therefore increased. This tilt pattern also reduces the symmetry to tetragonal. In $CaTiO_3$ the tilts are also slightly out of plane, which further reduces the symmetry to orthorhombic.

In $BaTiO_3$, at temperatures below 120°C, the titanium is displaced very slightly (by ~ 0.1Å relative to the Ti–O bond length of ~ 1.95Å) from the centre of the octahedron towards an oxygen at one of the corners. This also very slightly distorts the octahedron, reducing the symmetry to tetragonal, but retaining the basic unit cell size (Figure 5.23(b)). If the displacements are all in the same direction, there is a net movement of charge resulting in a structure with a dipole moment. This spontaneous electrical polarization is termed *ferroelectricity*. The polarization can be affected by an applied electric field and this leads to considerable applications of perovskites in electronic devices.

The great interest in the perovskite structure in mineralogy is that a perovskite structure of composition around $MgSiO_3$ is generally accepted to be the predominant mineral in the lower mantle (i.e. at depths below around 600km) of the Earth . This means that the most common mineral in the Earth has the perovskite structure. The perovskite structure of $MgSiO_3$ is only stable at very high pressures but has been shown to exist experimentally in apparatus capable of simulating mantle conditions. This will be discussed in greater detail in Chapter 12.

The rutile structure

Rutile, TiO_2 is sometimes described as a distorted hexagonal array of oxygen atoms with titanium in one half of the octahedral sites, but this is taking the close-packing model beyond its usefulness, and the structure is best described in terms of the linkage of octahedra. The structure is shown in Figure 5.24.

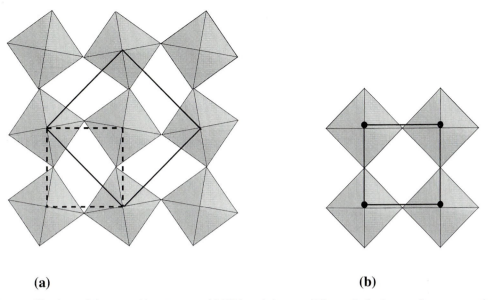

(a)   **(b)**

Figure 5.23. Two modifications of the perovskite structure. (a) Tilting of alternate BO_6 octahedra in opposite senses results in the larger unit cell shown in the bold line. The original perovskite unit cell is shown by the dashed line. (b) Displacement of the B cations from the centre of the octahedra reduces the symmetry to tetragonal while retaining the basic perovskite unit cell.

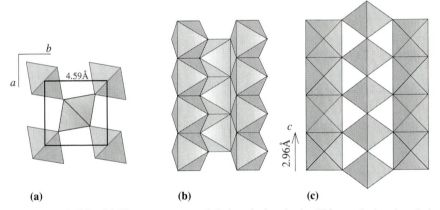

(a)   **(b)**   **(c)**

Figure 5.24. The rutile structure of TiO_2. (a) The arrangement of chains of edge-sharing TiO_6 octahedra viewed along their length. The chains, extending parallel to the c axis are at the corners and the centre of the tetragonal unit cell. (b) A view of the structure along the a axis showing the octahedral chains. (c) A view of the structure along [110] shows how the octahedral chains are joined laterally by sharing corners.

The TiO_6 octahedra form chains parallel to the c axis of the tetragonal unit cell by sharing opposite edges with neighbouring octahedra. The chains of octahedra are connected laterally by sharing corners, thus forming a three-dimensional network. The octahedra are very nearly regular, with two different Ti–O bond lengths of 1.98Å and 1.95Å.

Cassiterite SnO_2 and pyrolusite MnO_2 both have the rutile structure. At high pressures SiO_2 also has the rutile structure and is termed stishovite. It is one of the very few structures where Si occurs in 6-fold coordination with oxygen.

Bibliography

ADAMS, D.M. (1981). *Inorganic solids*. John Wiley and Sons.

GREENWOOD, N.N. (1968) *Ionic crystals, lattice defects and non-stoichiometry*. Butterworths.

O'KEEFFE, M. AND HYDE, B.G. (1985). An alternative approach to non-molecular crystal structures, with emphasis on the arrangements of cations. *Structure and Bonding*, **61**, 77–144.

RUMBLE, D. (Ed). (1976). Oxide minerals. *Reviews in mineralogy*, **3**, Mineralogical Society of America.

SMYTH, J.R. AND BISH, D.L. (1988). *Crystal structures and cation sites of the rock-forming minerals*. Allen and Unwin.

VAUGHAN, D.J. AND CRAIG, J. (1978). *Mineral chemistry of metal sulphides*. Cambridge University Press.

6 The crystal structure of minerals II – silicates

Silicon is the second most abundant element in the Earth's crust and mantle, after oxygen, and since the Si–O bond is considerably stronger than that between any other element and oxygen, it is not surprising that silicate minerals make up the vast majority of rocks. 95% of the Earth's crust is composed of only a handful of mineral groups – feldspars, quartz, amphiboles, pyroxenes and micas, and in this chapter we will be concerned mainly with the structure, composition and behaviour of these groups. Within each group there can be considerable diversity in the composition and the related structural modifications, and in this context, 'behaviour' means the way in which these adaptations of the basic structure are dependent on temperature and pressure in the Earth. We will need to discuss the flexibility of the linkages between the various ions in an attempt to understand the way in which a mineral can respond to changes in its physical and chemical environment.

The description of silicate structures depends to some extent on the model for the Si–O bond. In a purely ionic model composed of Si^{4+} and O^{2-} ions, held together by non-directional electrostatic forces, the oxygen ions would tend to be close-packed with charge balance provided by Si^{4+} and other cations in interstitial sites of the appropriate size. Few silicates however have a density anywhere near that of a close-packed oxygen array, nor are the cation : oxygen radius ratios often consistent with the observed coordinations. In a covalent model a description of the Si–O bond in terms of overlapping orbitals is in good agreement with the observation that in almost all silicates Si is tetrahedrally coordinated to oxygen with the O–Si–O bond angles only slightly deviating from the ideal tetrahedral value of 109.5°. However, neither model alone is satisfactory in every respect and despite the difficulty of defining and measuring 'ionicity', the general consensus is that the Si–O bond is about 50% ionic.

Our description of silicate structures will emphasise the purely geometrical aspects, independently of any particular bonding theory. The use of the terms cation and anion are convenient labels, and similarly, use of relative ionic sizes remains a helpful way of understanding the substitution of one element for another in a structure, without implying a commitment to the ionic model. The observation that $[SiO_4]$ tetrahedra dominate silicate structures has long been the basis for classifying silicate minerals in terms of the way in which the tetrahedra are arranged. $[SiO_4]$ tetrahedra are either isolated from each other (and hence bonded to other cation polyhedra) or connected to other $[SiO_4]$ tetrahedra by corner-sharing. The array of corner-sharing tetrahedra can often be regarded as forming the skeleton of a silicate structure, with other cations occupying suitable interstices. The way in which this skeleton articulates is of prime importance in understanding the adaptations of silicate structures referred to above.

Another feature of many silicate minerals is the substitution of Al for Si in the tetrahedron. Assigning formal charges of Si^{4+}, Al^{3+}, and O^{2-} to the ions indicates that such a substitution must be accompanied by some compensating replacement in the cation content to maintain electrical neutrality.

6.1 The [SiO₄] tetrahedron

Understanding the flexibility of a silicate structure from a geometrical point of view depends on knowing, at least in a qualitative way, the relative rigidity of the various linkages. We begin by considering the individual $[SiO_4]$ tetrahedron (Figure 6.1). Its shape is defined by the Si–O bond lengths and the O–Si–O bond angles, and we need to know the extent to which these are affected by the immediate structural surroundings i.e. does the shape of the tetrahedron depend on the structure it is in.

The dimensions of $[SiO_4]$ tetrahedra have been

determined from a large number of silicate structure determinations by X-ray and neutron diffraction refinements, and the following conclusions can be drawn:

1. The mean Si–O bond length is 1.62 Å. Other cations near the oxygen atom of an Si–O bond also attract the oxygen and tend to lengthen the Si–O bond, and hence there is a positive correlation between the Si–O bond length and the coordination number of the oxygen atom. The strength of the

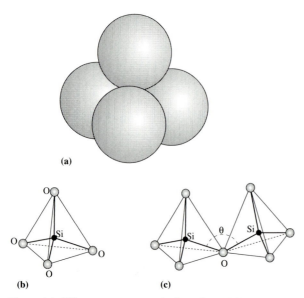

Figure 6.1. Silicate structures are built up from SiO₄ tetrahedra. **(a)** A space-filling model showing four oxygen atoms at the corners of a tetrahedron, obscuring the Si atom at the centre. **(b)** A less realistic, but clearer model of an SiO₄ tetrahedron, showing the Si–O bonds. **(c)** Two SiO₄ tetrahedra sharing a corner to form an Si₂O₇ pair. The angle θ is the Si–O–Si bond angle.

Si–O bond limits the range of bond lengths from about 1.60Å to 1.64Å.

2. When [SiO₄] tetrahedra are linked in a structure, bond lengths between the Si atoms and the bridging oxygen atom O$_{br}$ are on average longer by about 0.025Å compared with the Si–O bond lengths to the non-bridging oxygens. The O$_{br}$–Si–O$_{br}$ bond angle is also slightly smaller than the value when bridging oxygens are not involved. This suggests that the Si atoms are displaced from the centres of the tetrahedra, away from the bridging oxygens, as a result of repulsion between the two Si atoms.

3. When tetrahedra are corner linked, the Si–O–Si bond angle defines the orientation of the tetrahedra relative to one another (Figure 6.1). This bond angle can vary between about 120° and 180°, depending on the local structural environment as well as the temperature and the pressure. The bond angle of a strain free Si–O–Si bond is near 140°.

These experimental data are further reinforced by calculations of the potential energy surface of a pair of linked [SiO₄] tetrahedra as a function of the Si–O$_{br}$ bond length and the Si–O–Si bond angle (Figure 6.2). The energy is represented by contours which show a long narrow valley extending from bond angles of 120° to 180°, surrounded on three sides by steep energy barriers, with an energy minimum at Si–O$_{br}$ = 1.60Å and Si–O–Si = 140°.

These results suggest that the [SiO₄] tetrahedron is relatively rigid over a range of structural environments, and that the expansion and contraction of the silicate skeleton will be related to the ease with

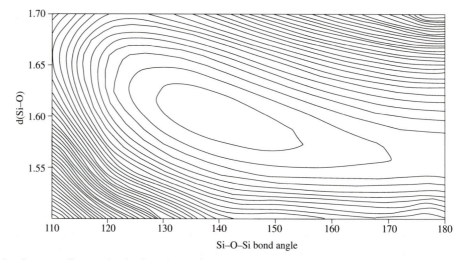

Figure 6.2. Calculated contour diagram showing how the potential energy of an Si₂O₇ pair varies with the Si–O bond length of the bridging oxygen, and the Si–O–Si bond angle. This potential energy surface has a long narrow valley representing low energies for a range of bond angles between 120° and 180°, at an Si–O bond length around 1.6Å. (After Gibbs et al.,1981.)

which the Si–O–Si bond angles can bend to suit any change in the chemical composition, temperature or pressure.

When Al substitutes for Si in a tetrahedron, as it does in a large number of silicates, both Al and Si play a similar structural role and the overall structural description remains the same. However, the [AlO_4] tetrahedron is slightly larger than an [SiO_4] tetrahedron since the Al–O bond (1.75Å) is longer on average than the Si–O bond. When [AlO_4] and [SiO_4] tetrahedra are linked in a structure this size difference is accommodated by a change in the T–O–T bond angle (T = tetrahedral cation).

A further effect of the substitution of Al for Si is that it has long been empirically known, and has more recently been calculated, that the potential energy of a Al–O–Al linkage is greater than that of an Al–O–Si linkage, suggesting that adjacent [AlO_4] tetrahedra are energetically unfavourable. This is termed the 'aluminium avoidance principle' and plays an important role in the ordering of Al and Si among tetrahedral sites in many aluminosilicate minerals. We will discuss this in greater detail when describing the behaviour of specific mineral groups.

6.2 Some generalizations regarding silicate structures and compositions

The compositions of silicate minerals are often rather complex and it is helpful to point out the general relationships between the chemical formula and the way in which the [SiO_4] tetrahedra are linked. We will classify silicates in terms of the increasing degree of polymerization of the tetrahedra, noting again that:

(i)   tetrahedra link by sharing corners
(ii) no more than two tetrahedra can share a common corner i.e. a bridging oxygen
(iii) assigning formal charges of Si^{4+} and O^{2-} is a convenient way of ensuring charge balance in the formula, and gives the tetrahedron a net negative charge i.e. $[SiO_4]^{4-}$.

1. At one extreme the [SiO_4] tetrahedra are isolated from each other and there are no bridging oxygens. The silicon to oxygen ratio in the chemical formula is therefore 1:4, e.g. olivine Mg_2SiO_4.

2. When two tetrahedra form linked pairs (dimers, as in Figure 6.1) in a structure the bridging oxygen is common to both, and hence in determin-

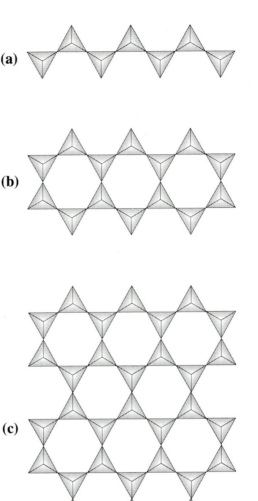

Figure 6.3. (a) Part of an infinite single chain of SiO_4 tetrahedra with each tetrahedron sharing two corners, as in the pyroxene structure. (b) Part of an infinite double chain of SiO_4 tetrahedra with half the tetrahedra sharing two corners, and the other half sharing three corners, as in the amphibole structure. (c) Part of an infinite sheet of SiO_4 tetrahedra, each tetrahedron sharing three corners, as in the sheet silicate structures.

ing the Si:O ratio each bridging oxygen counts as 1/2. In a structure made up of dimers each tetrahedron has one bridging and three non-bridging oxygens. The Si:O ratio is thus 1:3.5 and the net charge on the dimer is $[Si_2O_7]^{6-}$ e.g. rankinite $Ca_3Si_2O_7$.

3. In single-chain silicates (Figure 6.3(a)) every tetrahedron has two bridging and two non-bridging oxygens giving an Si:O ratio of 1:3. The infinite chains have a net charge of $[SiO_3]_n^{2n-}$ e.g. in the pyroxene group as in enstatite $MgSiO_3$.

4. In double-chain silicates (Figure 6.3(b)) two single chains are joined such that half of the tetrahedra have two bridging and two non-bridging oxygens (Si:O = 1:3) , while the other half have three bridging and one non-bridging oxygen (Si:O = 1:2.5), giving a net Si:O ratio of 2:5.5. The chains have a net charge of $[Si_4O_{11}]_n^{6n-}$ e.g. in the amphibole group as in anthophyllite $Mg_7Si_8O_{22}(OH)_2$. The OH groups in amphiboles are independent of the tetrahedra.

5. When tetrahedra form infinite sheets, each tetrahedron has three bridging and one non-bridging oxygen, i.e. Si:O = 1:2.5. The sheets have a net charge of $[Si_2O_5]_n^{2n-}$ e.g. in the mica, talc $Mg_6Si_8O_{20}(OH)_4$. Again the OH groups are independent of the tetrahedra.

6. Finally when all tetrahedra share corners with other tetrahedra, a framework silicate is formed, with four bridging oxygens per tetrahedron and an Si:O ratio of 1:2, as in quartz SiO_2.

Although this scheme of relating a formula to the structure type has its merits, there are complications:

(i)   when Al substitutes for Si in the tetrahedron the overall (Al+Si):O ratio should be considered. For example in the framework minerals of the plagioclase feldspars, albite $NaAlSi_3O_8$ and anorthite $CaAl_2Si_2O_8$ both have an (Al+Si):O ratio of 1:2.

(ii) in many cases Al may substitute for Si in tetrahedra, but may also exist elsewhere in the structure in sites independent of the tetrahedral array, often in octahedral sites. This is generally distinguished in the way the formula is written, by bracketing the tetrahedral group, e.g. the mica, muscovite has a formula $K_2Al_4[Si_6Al_2O_{20}](OH)_4$.

(iii) in Al_2SiO_5 (kyanite, andalusite or sillimanite) the Al are in octahedral sites, some of whose oxygen atoms are not connected to the isolated tetrahedra, thus giving an Si:O ratio less than 1:4.

6.3 Silicates with isolated [SiO₄] tetrahedra

In this group isolated [SiO₄] tetrahedra are linked together by the other cations which lie between them. The minerals in this group include the olivines, the garnets, and the aluminium silicate

minerals kyanite, andalusite and sillimanite, as well as zircon, sphene, topaz, staurolite and chloritoid. Only the first three groups will be discussed here.

6.3.1 The olivine minerals

The olivine minerals have a general formula M_2SiO_4 where M is Mg, Fe^{2+} or Ca. The structure can be described from a number of points of view. Figure 6.4(a) shows the arrangement of [SiO₄] tetrahedra in a projection down the *a* axis of the orthorhombic unit cell. The isolated tetrahedra point alternately up and down along rows parallel to the *c* axis. In this

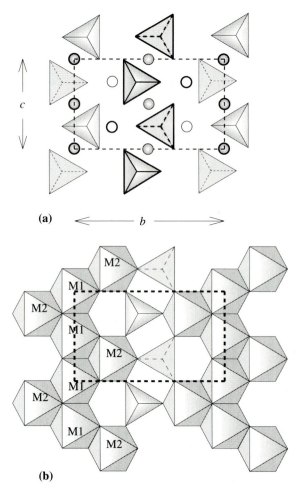

Figure 6.4. The structure of olivine, M_2SiO_4. **(a)** A projection of the structure down the *a* axis of the orthorhombic unit cell (dashed line), showing isolated SiO₄ tetrahedra alternately pointing up and down and forming rows along the *c* axis. The tetrahedra are joined by the M cations (circles) also forming rows along *c*. These rows are at two levels within the unit cell: at height 0, indicated by the heavier line, and at height *a* = 1/2, in the lighter line. M1 cations are shown as filled circles, M2 cations as open circles. **(b)** The lower level of the unit cell shows the linkage between the M1 and M2 octahedra and the row of SiO₄ tetrahedra.

projection there are rows at two levels in the unit cell, the lower level (at $a = 0$) drawn in a heavier line and an upper level (at $a = 1/2$); in each level therefore the rows are separated by the b lattice parameter, and the top half of the cell is related to the lower level by a b glide plane (i.e. $b/2$ translation and a mirror reflection.)

Within each level the tetrahedra are linked via octahedra which contain the M cations. Figure 6.4(b) shows these linkages forming a layer in the lower level. There are two kinds of octahedra, labelled M1 and M2. The M1 octahedra share edges to form ribbons parallel to the c axis. Ribbons in one layer are connected to those in the layer above by the M2 octahedra, also by edge-sharing. The M1 and M2 octahedra taken together form a three-dimensional network. In this idealized description the polyhedra are regular; in the real structure both M1 and M2 are somewhat distorted, M2 being slightly larger and more distorted than M1.

The olivine structure can also be described as an approximately hexagonally close-packed oxygen array with M and Si cations in one half of the octahedral and one eighth of the tetrahedral sites respectively. The close-packed layers lie in the plane of the projection in Figure 6.4 i.e. parallel to the (100) plane. If the oxygen atoms were perfectly close-packed, the M1 and M2 octahedra would be regular.

The response of the olivine structure to an increase in temperature is a significant increase in the sizes of the M1 and M2 sites, while the [SiO₄] tetrahedra remain virtually unchanged. This is because the Si–O bond is stronger than the M–O bond. An increase in pressure has the reverse effect on the M sites, and this has led to the generalization that in many polyhedral structures an increase in pressure has the same structural effect as a decrease in temperature. Changing the size of the octahedral cation expands the M site in much the same way as an increase in temperature – the effect on the M–O bond length of an increase in cation size from 0.69Å (Ni^{2+}) to 1.00Å (Ca^{2+}) is the same as that of an increase in temperature from 20°C to 1000°C. At high temperatures therefore we can expect that larger cations will be able to substitute in the octahedral sites i.e. the degree of solid solution possible increases.

Most natural olivines have compositions in the range Mg_2SiO_4 (forsterite) to Fe_2SiO_4 (fayalite)

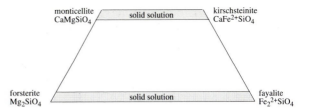

Figure 6.5. Compositions of natural olivines. The mineral olivine refers to the solid solution between forsterite and fayalite. Less common is monticellite, which also has solid solution due to Mg ⇔ Fe substitution.

(Figure 6.5). There is a complete solid solution between the end members and virtually no ordering of Fe^{2+} and Mg between the M1 and M2 sites at room temperature, although there is a slight preference for the Fe^{2+} to occupy the M1 site. Despite this slight non-ideality, the Mg_2SiO_4–Fe_2SiO_4 solid solution has often been taken as an example of an 'ideal' solid solution, in which the structure shows no energetic preference for either Mg or Fe^{2+}. This is reflected in linear changes in physical properties as a function of Fe,Mg content, and a measure of the birefringence for example, or the lattice parameters, can be used to determine the chemical composition.

In the Ca-bearing olivines, Ca^{2+} occupies the M2 sites while the solid solution between monticellite $CaMgSiO_4$ and $CaFeSiO_4$ involves random mixing of Mg and Fe^{2+} over M1 sites. There is virtually no solid solution between the Ca-bearing olivines and Fe,Mg olivines under normal geological conditions, as the structure cannot tolerate the strain which would be produced by random substitution on M2 sites of cations of such different sizes (see Table 5.1). At higher temperatures we would expect the degree of substitution to increase, and this is confirmed by experiments which show that at 1000°C there is about 5% Ca ↔ (Fe,Mg) substitution, increasing to around 20% at 1450°C.

Ca_2SiO_4 exists as an olivine structure over a limited temperature range but the distortions exerted on the [SiO₄] tetrahedra by the large octahedra make this rather unstable and it readily transforms to different structures. The mineral chrysoberyl, Al_2BeO_4 also has the olivine structure.

At the high pressures which exist in the Earth's mantle Fe,Mg olivine transforms to a spinel structure with a density increase of around 10%. This is a very important structural transformation related to the discontinuity in seismic wave velocities at a depth near 400 km, and will be discussed in some detail in Chapter 12.

6.3.2 The garnet minerals

The silicate garnets consist of a multicomponent solid solution, with general formula $A_3^{2+}B_2^{3+}Si_3O_{12}$, where A is Ca^{2+}, Mg^{2+}, Fe^{2+} or Mn^{2+} and B is Al^{3+}, Fe^{3+} or Cr^{3+}. The body-centred cubic unit cell is large (containing 8 formula units) and complex, and our main interest is the extent of solid solution and cation ordering within the different garnet minerals. The structural arrangement is shown in Figure 6.6. The [SiO_4] tetrahedra alternate with BO_6 octahedra by corner sharing along $x, y,$ and z directions of the cubic unit cell forming continuous linkages in three dimensions, with the larger A cations in a distorted cubic 8-fold coordination.

The [SiO_4] tetrahedra are distorted by an the amount which depends on the size of the cation in the AO_8 distorted cubes with which the tetrahedra share two opposite edges. However, the relatively rigid tetrahedra can accommodate to varying A-cation sizes by a rotation which increases the size of

the A sites and therefore the shared BO_6 octahedral edge as well. The effect of an increase in temperature is similar, in that the rotation of the [SiO_4] tetrahedron can accommodate the expansion of the less rigid polyhedra. However, the amount of possible rotation depends on the relative A and B cation size – for example, if A is Ca^{2+} and B is Al^{3+} the tetrahedron is already fully rotated to accommodate this large cation size difference, whereas with A = Mg^{2+} and B = Al^{3+} more rotation is possible. It follows that in the former case the thermal expansion of the structure will be lower than in the latter.

Natural garnets are commonly divided into two groups – those in which the A cation is Ca^{2+}, and those in which A is not Ca^{2+} and B is Al^{3+}. The end members and their cell dimensions are listed below.

Pyrope	$Mg_3Al_2(SiO_4)_3$	11.46Å
Almandine	$Fe_3^{2+}Al_2(SiO_4)_3$	11.53
Spessartine	$Mn_3Al_2(SiO_4)_3$	11.62Å
Uvarovite	$Ca_3Cr_2(SiO_4)_3$	12.02Å
Grossular	$Ca_3Al_2(SiO_4)_3$	11.85Å
Andradite	$Ca_3Fe_2^{3+}(SiO_4)_3$	12.05Å

The two groups are sometimes known as 'pyralspite'

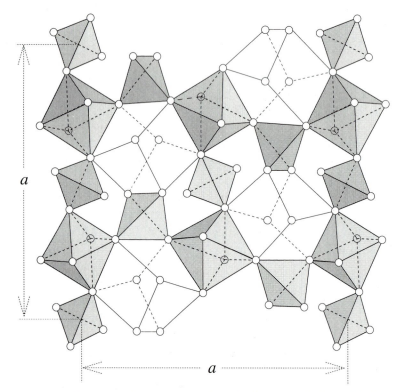

Figure 6.6. Part of the garnet structure showing the linkage between the isolated SiO_4 tetrahedra and the BO_6 octahedra in a slab of the structure parallel to the face of the cubic unit cell. The large eight-coordinated sites are left unshaded. The rest of the unit cell with side a is made up by arranging these slabs of structure along the third axis. (After Novak and Gibbs, 1971.)

and 'ugrandite', the terms being derived from the names of the three end-member minerals in each group.

Despite the complexity of the structure and the diversity of the cations, the way in which the polyhedra articulate results in a linear correlation between the radius of the A and B cations and the cell dimensions. A comparison of the cell dimensions of the end members therefore gives some indication of the possibility of solid solutions between them. As a first approximation therefore, it is reasonable to expect extensive solid solution within each series, but limited solid solution between them. Natural garnets never occur as pure end-member compositions, and the degree of solid solution possible is therefore quite important.

The temperature–composition response of the garnet structure again suggests that at higher temperatures solid solutions should be more extensive. The A site, which is somewhat too large for Mg^{2+} (which accounts for pyrope being a high pressure phase stable in the deep crust) can be expanded by [SiO₄] rotation to be just large enough for Ca^{2+}. An increase in temperature would promote solid solution, but as the temperature falls, we would expect the degree of Mg and Ca solid solution to be restricted. The resultant unmixing (or exsolution) would result in the separation of Mg-rich and Ca-rich regions within the crystal.

Natural garnets with the general composition $(Fe,Mg,Ca)_3Al_2(SiO_4)_3$, which form in metamorphic rocks at temperatures in excess of about 700°C, are quite common and do appear to be homogeneous solid solutions on an optical microscopy scale. While there is a tendency for Ca and Mg to unmix, we will also need to consider the scale on which such unmixing might occur, relative to the time scale available and the rate at which Mg and Ca can diffuse within a garnet structure. We will return to this topic when we discuss solid solutions and unmixing processes in more detail in Chapter 11.

Another phenomenon which we have mentioned in Chapter 5 is the tendency for cations which are disordered at high temperatures to become more ordered at lower temperatures. In the $Ia3d$ space group of garnet all of the octahedral sites are symmetrically equivalent and therefore the B cations must be disordered. Any ordering would result in a reduction in symmetry: equivalent sites would become non-equivalent. Such cation ordering has

been observed in garnet of composition around $Ca_2Al_{1.3}Fe_{0.7}(SiO_4)_3$ where the octahedral sites become differentiated into those which contain more Fe^{3+} and those with less, and hence it becomes necessary to define two types of octahedral sites M1 and M2. M1 has about 40% Fe, 60% Al, while M2 has 25% Fe, 75% Al. A statistically disordered distribution at this composition would have all sites with 35% Fe, 65% Al average occupancy. Thus the ordering is rather slight, but does lead to a reduction in symmetry from cubic to orthorhombic.

At high pressure in the Earth's mantle $(Mg,Fe)SiO_3$ is thought to exist as a garnet structure, with Si in octahedral as well as tetrahedral sites. This suggestion is based on high pressure experiments and the presence of such a mineral, termed majorite, in shock veins of meteorites. This will be discussed in more detail in the section on high pressure minerals in Chapter 12.

6.3.3 The Al₂SiO₅ minerals: kyanite, andalusite and sillimanite

These three minerals have the same composition, but exist as different structures depending on the temperature and pressure conditions. The general term for the existence of a number of different structures of the same chemical compound is *polymorphism*, and the polymorphs of Al_2SiO_5 have been widely studied in petrology as they are used as indicators of pressure and temperature of metamorphism in crustal rocks.

Their structures have a number of features in common. In all three minerals straight chains of edge-sharing AlO_6 octahedra extend along the c axis (Figure 6.7). These octahedra contain half of the Al in the structural formula. The remaining Al atoms are in coordination which is different in each mineral: 6-fold sites in kyanite, 5-fold sites in andalusite and 4-fold sites in sillimanite. These Al-polyhedra alternate with [SiO₄] tetrahedra, also along the c axis, linking together the AlO_6 chains (Figure 6.8). The cell dimensions of these three structures are controlled by the relative orientation of the AlO_6 octahedra. In each case the length of the c-axis repeat is 2 octahedra.

Our main interest in these minerals will be in their relative stability fields as a function of pressure and temperature. Kyanite is 14% denser than andalusite and 11.5% denser than sillimanite, so that on this

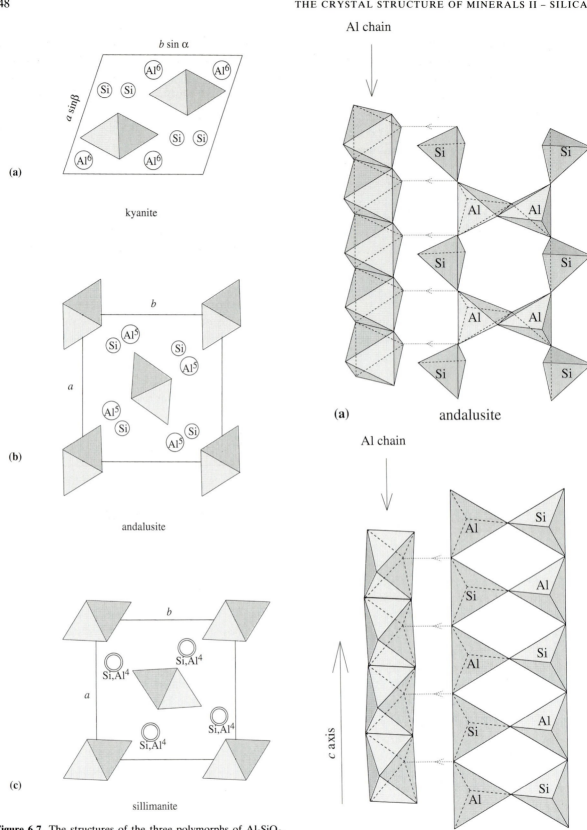

Figure 6.7. The structures of the three polymorphs of Al_2SiO_5, showing the relative positions of the AlO_6 octahedra (shaded) which form chains along the c axis. The projections of the positions of the other Al polyhedra and SiO_4 tetrahedra are shown as circles. The coordination of the Al is given by the superscript.

Figure 6.8. The linkage between the AlO_6 chains and the other Al polyhedra and SiO_4 tetrahedra in **(a)** andalusite, and **(b)** sillimanite.

basis we would expect kyanite to be stable at the highest pressures and lowest temperatures. Andalusite is the low pressure phase and sillimanite is stable at high temperatures and moderate pressures. The determination of the exact form of the equilibrium phase diagram is one of the sagas of metamorphic petrology, and we will discuss the difficulties with its determination and the reasons that one polymorph may persist indefinitely within the stability field of another, in Section 12.1.1.

Another interest in this mineral group concerns the mineral mullite, which is not common in nature, but is one of the most important ceramic refractory materials. It has a structure related to that of sillimanite. In sillimanite, the Al in the 4-fold tetrahedral sites and the [SiO$_4$] tetrahedra are linked by corners to form double chains of tetrahedra parallel to the c axis. These chains of alternating Al and Si tetrahedra, provide the lateral linkage between the AlO$_6$ octahedral chains (Figure 6.8(b)). The [AlO$_4$] tetrahedra play a similar structural role to the [SiO$_4$] tetrahedra and the sillimanite structure can be described in terms of these ordered double chains (Figure 6.9(a)).

In mullite, extra Al substitutes for Si in these tetrahedral chains. For electrostatic charge balance,

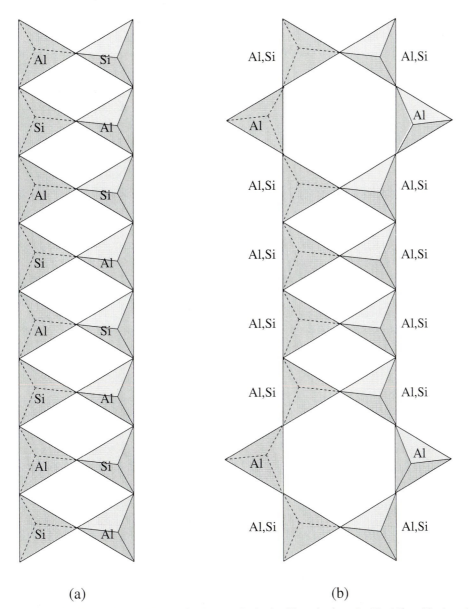

(a)   (b)

Figure 6.9. A comparison of the double chain of SiO$_4$ and AlO$_4$ tetrahedra in sillimanite (see also Fig 6.8) and in the related structure of mullite. In mullite the ordering of Al,Si along the chain is lost and extra Al atoms in the chain result in oxygen vacancies.

such a substitution requires removing some O atoms, according to the scheme $2Al^{3+} + \square = 2Si^{4+} + O^{2-}$, where $\square$ represents a vacant oxygen site. The general formula for mullite is then written $Al_{4+2x}Si_{2-2x}O_{10-x}$ where x is the number of missing oxygen atoms per formula unit. The x value can vary between about 0.17 and 0.6, and so mullite can exist with a wide range of Al:Si ratios. The absence of these O atoms makes some of the tetrahedral sites effectively disappear and the extra Al atoms introduced by the substitution go into new tetrahedral positions as shown in Figure 6.9(b). Furthermore, the Al and Si atoms in the original tetrahedral double chain become disordered, and as the Al,Si alternation is lost the c axis repeat is reduced by half relative to that of sillimanite.

When sillimanite is heated to above 1545°C it is converted to mullite + melt while the mullite remains stable to ~1850°C. The stability of mullite, in view of its apparently defective structure, will be a topic for further discussion in Section 12.5.1.

6.4 Single chain silicates

In this section we will describe two mineral groups in which the [SiO$_4$] tetrahedra form linear single chains with two bridging oxygen atoms per tetrahedron as shown in Figure 6.10. In the *pyroxenes*, the periodicity of the chain is 2, i.e. the chain repeats after every two tetrahedra. In a related mineral group, the *pyroxenoids*, longer chain periodicities occur, such as in wollastonite CaSiO$_3$, where the the periodicity is 3. The Si:O ratio in these single chain silicates is 1:3 with a net charge of $(SiO_3)_n^{2n-}$. The chains are

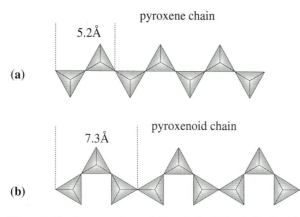

Figure 6.10. A comparison of the single SiO$_4$ chains in **(a)** pyroxenes with periodicity of 2 tetrahedra, and **(b)** pyroxenoids with periodicity 3 tetrahedra.

cross linked by cations, generally in octahedral sites. The flexibility of the chain is able to provide stable geometrical configurations with a very extensive range of cations and over a wide range of temperature and pressure conditions.

6.4.1 The pyroxenes

The various modifications of the pyroxene structure are best understood by beginning with an idealized model which shows the basic topology, and then considering the effect of cation size and temperature on this topology. There are many analogies between single chain and double chain silicates in this respect, so that an understanding of the behaviour of pyroxenes will be directly applicable to other chain silicates. The pyroxene structure will be therefore be described in some detail.

Figure 6.11(a) shows a straight single chain, which in all pyroxene structures extends along the c axis, and defines the c parameter of the unit cell (~5.2Å). The view of the structure down the c axis (Figure 6.11(b)) shows the chains end-on, and the way they are stacked back to back forming layers parallel to (100) planes. The b axis repeat of pyroxenes (~8.9Å) is defined by this stacking arrangement. This projection of the structure is common to all pyroxenes but it does not show how the chains are arranged parallel to their length, nor does it show any differences between the chains. These two important features define the different pyroxene structural groups.

The cation positions in pyroxenes are of two types. The M1 sites lie between the apices of opposing tetrahedra; the M2 sites lie between their bases (Fig 6.11(b)). The M1 sites are smaller, and are almost regular octahedra; M2 sites are larger, more distorted and may be octahedra when containing a smaller cation, or 8-fold sites when occupied by a larger cation. The M sites form edge-sharing chains which run parallel to the silicate chains.

The clinopyroxene structure

Figure 6.12 shows the arrangement of the silicate chains along their length in the clinopyroxenes and defines the orientation and repeat of the a axis (~9.7Å). The unit cell is monoclinic with a β angle of around 106°, hence the term *clinopyroxene*. In this idealized structure the chains are all the same and are symmetrically related to one another. The

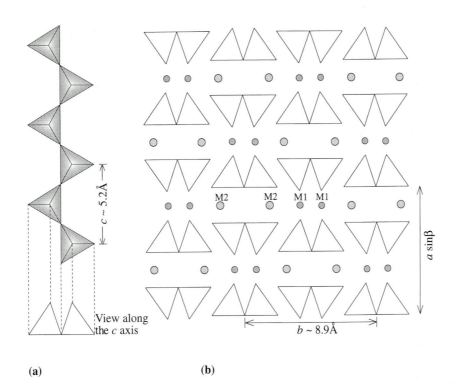

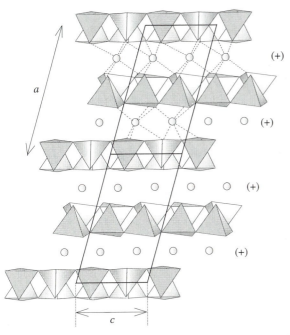

Figure 6.11. (a) A single pyroxene chain which extends along the c axis and below, a schematic representation of this chain viewed end-on. **(b)** The arrangement of SiO_4 chains in the pyroxene structures, viewed along the c axis. The M1 cations form chains of edge-sharing octahedra between the apices of the tetrahedra, while the larger M2 octahedra form similar chains between the bases of the tetrahedra.

Figure 6.12. A view of the clinopyroxene structure along the b axis showing the tetrahedral chains along their length. In this view the chains form layers parallel to (100) planes. These layers are staggered along the c axis giving the monoclinic unit cell (two unit cells are outlined). The displacement of the chains defines the orientation of the M1 octahedra, here labelled +. The definition of this octahedral orientation is described in Figure 6.21. (After Burnham, 1966.)

space group is C2/c; the C-centring of the (100) face of the unit cell can be seen in the projection in Figure 6.11(b) and the diad axis, which passes through the M1 and M2 sites relating the chains, can be seen in Figure 6.12. The different pyroxene structures will be referred to by their space groups as convenient (and informative) labels.

The way in which the silicate chain links to the cation polyhedra is important in understanding its response to changes in composition, temperature and pressure. As we have previously noted, the size and shape of the $[SiO_4]$ tetrahedron remains the same over a wide temperature and pressure range. The M1 and M2 sites do however expand and contract significantly, and tetrahedral chain must be able to change its length to accommodate it. Figure 6.13(a) shows the way in which a straight silicate chain is linked, by its apices, to an edge-shared octahedral chain. This is the fully extended situation – if the octahedra were any larger, the pyroxene chain could not accommodate them in this way. A contraction of the octahedra is accommodated by a rotation of the individual $[SiO_4]$ tetrahedra which shortens the chain length. Alternate tetrahedra along the chain always rotate in opposite senses, but there are two different rotation possibilities (cf. Figures 6.13(b) and 6.13(c)) leading to two types of chains, termed the O-chain and the S-chain. If adjacent chains are not of the same type they lose their equivalence and the symmetry of the space group is reduced.

We have now described enough of the structural detail to begin to discuss pyroxene minerals. Because the structure is controlled by the size of the M sites, i.e. by composition, temperature and pressure, it is more convenient to describe the various structures in terms of their specific chemistry. The most important pyroxene minerals, represented by

the general formula $ABSi_2O_6$, have compositions in which A may be Ca^{2+}, Mg^{2+}, Fe^{2+}, or Na^+ in the M2 site and B is Mg^{2+}, Fe^{2+}, Fe^{3+}, or Al in the M1 site. The compositions of the Ca,Fe,Mg pyroxenes are generally shown on the pyroxene quadrilateral (Figure 6.14).

Clinopyroxene minerals

The diopside $CaMgSi_2O_6$ – hedenbergite $CaFeSi_2O_6$ solid solution has a structure close to the ideal pyroxene topology with space group C2/c and a chain angle ϕ (see Figure 6.13(b)) between 166° (diopside) and 164.5° (hedenbergite). The Ca^{2+} in the M2 site is in a distorted 8-fold site, with the (Mg^{2+}, Fe^{2+}) in M1. As the temperature increases, both M1 and M2 sites expand linearly, with more of the expansion taken up by the M1 octahedron. The increase in the chain angle to accommodate this expansion is about 3° over a 1000°C temperature range.

Augite (C2/c) has a very similar structure and behaviour. It is essentially part of the same solid solution, but separately named because of its characteristic composition as a major constituent of basaltic rocks. There is some substitution of Al^{3+} for Si^{4+} in the tetrahedra, charge compensated by the presence of higher valence cations (eg Fe^{3+}, Al^{3+}, Ti^{4+}) cations on the M sites. The substitution of (Fe^{2+},Mg) for Ca has little effect on the structure in this composition range.

At the Ca-poor side of the pyroxene quadrilateral this situation changes. In pigeonite, the C2/c structure is only possible at high temperatures (~900°C, but very compositionally dependent). The chain angle ϕ reaches about 174°, and in keeping with the symmetry, all the chains are equivalent. As the temperature falls however, the small M cations cannot support the contraction of the structure and

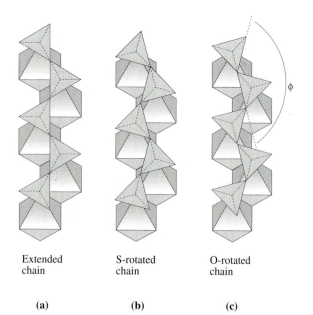

Extended chain	S-rotated chain	O-rotated chain
(a)	**(b)**	**(c)**

Figure 6.13. The linkage between a single SiO_4 chain and a chain of octahedra. **(a)** A fully extended chain. **(b)** and **(c)** When the octahedra become smaller, due to lower temperature or occupancy by a smaller cation, the tetrahedral chains have to shorten by rotation of the individual tetrahedra to achieve a successful linkage. There are two possible senses of tetrahedral rotation, termed S-rotation or O-rotation. In this figure the octahedral chains in (b) and (c) have been reduced in length by 4% relative to (a). (After Cameron and Papike, 1982.)

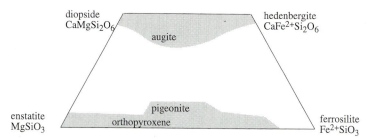

Figure 6.14. The pyroxene quadrilateral showing the approximate compositions of common pyroxene minerals. The shaded areas represent the extent of solid solution in naturally occurring pyroxenes. At high temperature there is complete solid solution between augite and pigeonite.

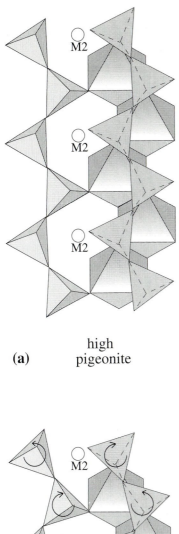

(a) high pigeonite

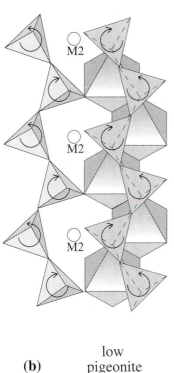

(b) low pigeonite

Figure 6.15. A comparison of the structures of high and low pigeonite. In high pigeonite the tetrahedral chains are almost straight and are symmetrically related. In low pigeonite adjacent chains are rotated in opposite senses (cf. Figure 6.13), losing their equivalence and resulting in the C2/c → P2₁/c symmetry change.

the chains undergo a sudden distortion. Chains in alternate (100) layers (see Figure 6.12) rotate in opposite senses (Figure 6.15) and lose their equivalence. The diad axes through the M sites are also lost and the space group symmetry is reduced to P2₁/c. At room temperature the chain angles for the two non-equivalent chains are ~149° and ~170° respectively. The coordination of the M2 site is reduced from 8-fold to a rather irregular 7-fold site. This structural collapse in high temperature pigeonite is referred to as the high – low pigeonite transition. This transition is discussed in greater detail in Section 7.3.4.

On the basis of these structural considerations, we can expect a complete solid solution between high pigeonite and augite at high temperatures, but as the temperature falls, and the pigeonite structure distorts, the structural differences will limit the amount of solid solution possible between low-Ca and high-Ca pyroxenes.

Orthopyroxenes

In pyroxenes which contain virtually no Ca^{2+}, and hence small cations in both the M1 and M2 sites, a more fundamental structural reorganization is preferred at low temperatures. A new structure, with the silicate chains rearranged along their length, becomes stable. Figure 6.16 shows a projection of the structure along the b axis. The unit cell is orthorhombic with an a lattice parameter ~2asinβ relative to clinopyroxenes (Figure 6.17). The b and c parameters remain the same. The space group is Pbca, and the silicate chains are non-equivalent. The chain rotations are however in the same sense, but with chain angles of ~167° and ~145° in alternate (100) layers. In this structure, the coordination of the M2 site is further reduced to 6-fold, in keeping with the smaller cation content. The M2 site in orthopyroxenes is too small for Ca^{2+}, and hence there is virtually no solid solution possible between orthopyroxene and the clinopyroxenes.

The solid solution between Mg and Fe orthopyroxenes is not ideal and at lower temperatures there is considerable ordering of Fe and Mg between M1 and M2 sites, Fe preferring M2. This type of ordering, discussed in Section 8.9.3, does not reduce the symmetry as the two sites are already symmetrically distinct. As the temperature increases above ~500°C, the Fe,Mg distribution becomes increasingly disordered.

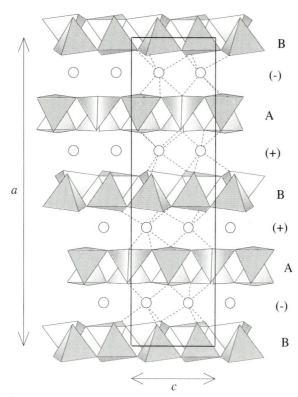

Figure 6.16. A view of the orthopyroxene structure along the *b* axis showing the tetrahedral chains along their length (cf. Figure 6.12). The chain arrangement along the *c* axis results in non-equivalent chains, labelled A and B, and an alternation of octahedra in the + and − orientation (defined in Figure 6.21). One unit cell is outlined. (After Burnham, 1966.)

Orthopyroxenes (P*bca*) transform to clino-pyroxene (C2/*c*) at high temperatures, consistent with our general view that the more expanded M sites are better accommodated by the C2/*c* structure. When the temperature is reduced again it is often the case that the C2/*c* structure transforms to the P2$_1$/*c* structure, as in the high – low pigeonite transition, rather than back to the more favourable P*bca* orthopyroxene structure. The problem is basic-ally one of relative kinetics since the C2/*c*⇒P2$_1$/*c* transition involves a distortion which is very fast, and 'easy' in terms of structural reorganization. The C2/*c*⇒P*bca* transition involves a major structural change which involves bond breaking and is rela-tively much more 'difficult', and hence sluggish. The problem is that in experiments, as well as in nature, the structure we see may not be the energetically most favoured one due to the slow kinetics of transforming to a more stable state. This is a compli-cation which we will discuss in Sections 7.2 and 9.6, where we return to the discussion of polymorphism in these pyroxenes.

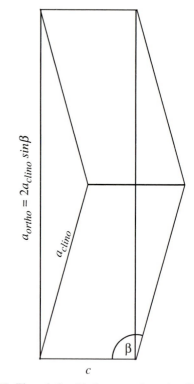

Figure 6.17. The relationship between the unit cells of clinopy-roxene and orthopyroxene.

Protoenstatite

At the enstatite (MgSiO$_3$) composition there is another structural modification we need to consider. Above about 1000°C a structure termed *protoensta-tite* is formed. Protoenstatite is orthorhombic, P*bcn*, and has almost straight silicate chains which are equivalent to one another. The chain arrangement is such that the *a* parameter of the unit cell is halved relative to P*bca* orthopyroxene. As enstatite is cooled to lower temperatures, all the previously mentioned pyroxene structures (P*bca*, C2/*c*, P2$_1$/*c*) appear, but their relative stabilities at the enstatite composition are still uncertain. This is due to similar kinetic problems to those described above, but the favoured idea is that the orthopyroxene structure P*bca* is the stable at low temperature, and trans-forms on heating to the P*bcn* structure. The two clinopyroxene structures play a kinetically- con-trolled role during cooling at this composition.

Proto-, ortho-, clino-: stacking sequences in pyroxene structures

Another convenient way of looking at these three pyroxene structure types is to describe them in terms of the stacking arrangement of the silicate chains.

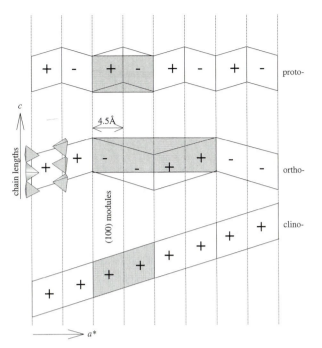

Figure 6.18. The relationship between the structures of proto-, ortho-, and clino-pyroxenes in terms of the stacking of layers of chains which form modules of the structure parallel to (100) planes. The modules are 4.5Å thick. There are two possible positions for each module, corresponding to relative displacements of adjacent chains along their length. These positions are labelled + and −. The unit cell for each structure is shaded. (After Buseck *et al.*, 1982.)

The *b* and *c* parameters of all three structures are similar; the *a* lattice parameter depends on the chain arrangement. Figure 6.18 shows the chain arrangements schematically in *b* axis projections. As noted before, the chains are arranged in layers parallel to (100) with a layer thickness ~4.5Å. These layers, or slabs of structure, are then stacked relative to one another in different ways in the three pyroxene structure types. The relationship of one slab to the next along the *a* axis is by a *b*-glide operation and can be in two different directions, designated + or − in Figures 6.12 and 6.16. The stacking sequence in clinopyroxenes is then +++++..., in orthopyroxenes is ++−−++−−++..., and in the protoenstatite structure is +−+−+−+−..... (Note that the stacking sequence −−−−− is twin-related to +++++).

Looking at structures in terms of arrangements of modular units is often a very helpful way of understanding the relationship between structures and the possible mechanism by which one can transform to the other if the physical conditions change. Structures related to one another by different stacking arrangements of the same modules are termed polytypes, and occur in many mineral groups. Some more detailed aspects of polytypism in minerals are discussed in Chapter 12.

Jadeite – omphacite – diopside

In high pressure crustal rocks pyroxenes are enriched in Al in octahedral coordination, as in jadeite $NaAlSi_2O_6$ which has the $C2/c$ diopside structure with Al in M1 sites and Na in M2 sites. Above about 700°C there is complete solid solution between jadeite and diopside involving a coupled substitution: $Na^+ + Al^{3+} \Leftrightarrow Ca^{2+} + Mg^{2+}$, thus maintaining overall charge balance. There is also some replacement of Al^{3+} by Fe^{3+}, and Mg^{2+} by Fe^{2+} but this does not affect the discussion below.

At lower temperatures there is a tendency for the cations to order within the M1 and M2 sites. As the $C2/c$ structure requires disorder within the M1 and M2 sites, cation ordering must result in a lower symmetry. Omphacite, at compositions around $(Na_{0.5}Ca_{0.5})(Al_{0.5}Mg_{0.5})Si_2O_6$, is a cation ordered pyroxene with a structure essentially the same as diopside but with two distinct M1 and two M2 sites. This reduces the symmetry to $P2/n$. However it is not possible to achieve a structure in which both Mg,Al and Na,Ca are independently fully ordered in alternate sites, as local charge balance requires the Na^+ to be as near as possible to the Al^{3+}. Thus the degree of Mg,Al order on M1 sites is coupled with the Ca,Na ordering on M2 sites. The best ordered structure appears to be a compromise solution in which alternate M1 sites are occupied by Mg^{2+} and Al^{3+} while alternate M2 sites are statistically occupied by 1/4Na, 3/4Ca and 3/4Na, 1/4Ca (Figure

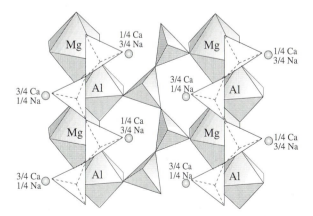

Figure 6.19. The structure of omphacite $(Na,Ca)(Al,Mg)Si_2O_6$. Alternate M1 sites are occupied by Mg and Al, and alternate M2 sites are statistically occupied by 1/4Na, 3/4 Ca and 3/4Na, 1/4Ca.

6.19). The coupling between ordering on M1 sites and M2 sites is linear – any reduction in the degree of order on M1 must be balanced by an equivalent reduction in the ordering on M2.

Model pyroxene structures as I-beams

Having come to the end of the description of the pyroxene structures we return, briefly, to the beginning. Part of understanding any complex structure and its relationship with other structures is in the way we describe it. How we describe it depends on the feature we are emphasising, for in any description which attempts to simplify a structure some details are inevitably lost. We have described pyroxene structures with some emphasis on the flexibility of the silicate chain. It would not therefore be appropriate to depict the chain as a rigid unit. In describing the polytypic relationships between proto-, ortho- and clino-pyroxenes however, a modular description makes the relationships logical and easily understood.

The reason for introducing yet another model description of the pyroxene structures at the end is that it will make double chain silicates easier to

understand, as well as helping to summarize many aspects of pyroxene structures.

In Figure 6.20, a pair of opposing chains and the M1 sites between the apices is shaded, forming an I-shaped beam which extends along the c axis (compare with Figure 6.11(b)). In this projection the I-beams are all arranged in the same relative position in all pyroxene structures. An I-beam is designated + or − depending on the orientation of the octahedra between the chain apices. Viewed down the c axis, as in Figure 6.11(b), the octahedra can have a triangular face pointing downwards (+) or upwards (−) (Figure 6.21) depending on the relative positions of the chains along their length. These octahedral orientations are also shown in Figures 6.12 and 6.16. The chains themselves are designated A and B if they are non-equivalent, and the chain rotations are shown as O and S.

In summary, Figure 6.22 shows the pyroxene structures in terms of the I-beam model.

6.4.2 The pyroxenoids

At the beginning of the description of the pyroxene structure, Figure 6.13(a) showed the linkage between a fully extended single silicate chain and an octahedral column, with the observation that if the cation octahedra were any larger, the pyroxene chain could no longer accommodate them in this way. In wollastonite $CaSiO_3$, which is also a single chain silicate, a different solution therefore has to be found. Figure 6.23 shows such a wollastonite chain, typical of the pyroxenoid group of minerals, and its

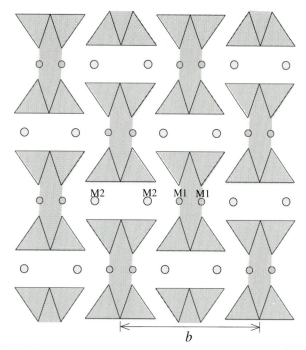

Figure 6.20. The I-beam representation of the pyroxene structure. Each pair of opposing tetrahedral chains, together with the M1 sites between them is shaded, and represents an I-beam, extending along the c axis. The various pyroxene structures can be conveniently represented in terms of such I-beams (see Figure 6.22).

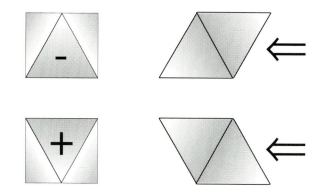

Figure 6.21. The definition of the '+' and '−' orientations of the M1 octahedra in pyroxenes. When the octahedron is viewed along the direction of the arrow (the c axis), as in the figures on the right, the triangular face of the octahedron facing the viewer of the octahedron may either point up towards the viewer (the '−' orientation) or down towards the viewer (the '+' orientation). The two front views are shown on the left.

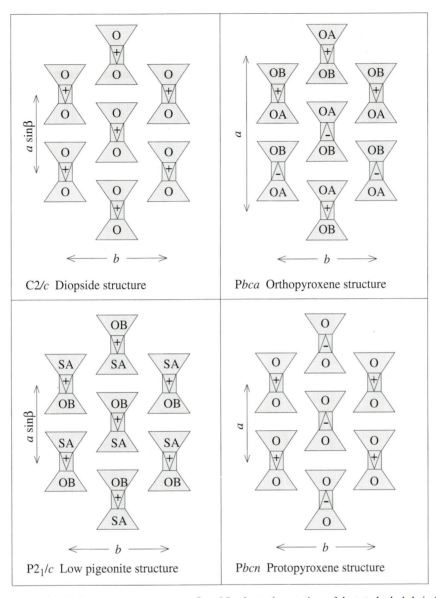

Figure 6.22. I-beam diagrams for the four pyroxene structures. O and S refer to the rotations of the tetrahedral chain (see Figure 6.13); M1 octahedra are labelled + and − according to the definition in Figure 6.21, and A and B refer to non-equivalent chains. (After Papike and Cameron, 1976.)

linkage to an octahedral chain. Two edges of [CaO$_6$] octahedra have about the same length as 3 [SiO$_4$] tetrahedra, thus forming a chain with a periodicity of 3, running parallel to the octahedral column.

Our interest in this mineral group is to point out how the chain periodicity is related to the cation size, the temperature and the pressure in the (Ca,Mn)SiO$_3$ pyroxenoids (Figure 6.24). There is virtually no solid solution between diopside CaMgSiO$_3$ and wollastonite CaSiO$_3$ and no minerals of intermediate composition exist. However, as Ca in wollastonite is replaced by Mn^{2+}, we again have the situation of a silicate chain having to respond to smaller octahedra. The way this is achieved in the pyroxenoids is shown in Figure 6.25. If a three-tetrahedra slice of the wollastonite chain, denoted W, is taken as one end-member, and a two-tetrahedra slice of straight clinopyroxene-type chain, denoted P, is taken as the other end-member, then all of the intermediate structures can be described in

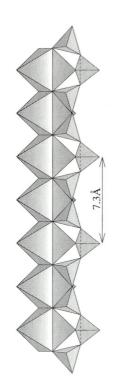

Figure 6.23. The linkage between a chain of tetrahedra and an octahedral chain in wollastonite CaSiO₃. The relative size of the SiO₄ tetrahedra and CaO₆ octahedra require twisting of the tetrahedral chain to be able to achieve a satisfactory linkage. The result is a chain repeat of 3 tetrahedra corresponding in length to two octahedral edges.

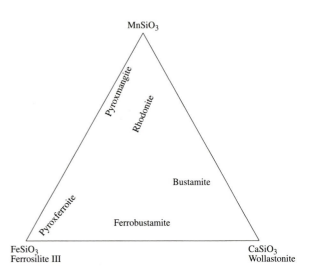

Figure 6.24. Compositional fields of the common pyroxenoid minerals in the CaSiO₃ – MnSiO₃ – FeSiO₃ system. Bustamite has a structure similar to wollastonite with a chain repeat of 3.

terms of interleaving these two types of modules to form chains with different periodicities. Thus:

rhodonite with a chain periodicity of 5 has a stacking sequence <WPWP..>;

pyroxmangite with a periodicity of 7 has a stacking sequence <WPPWPP...>, and

ferrosilite III with a periodicity of of 9 has a stacking sequence <WPPPWPPP...>.

The cation sites associated with the P modules are better able to accomodate smaller cations than those in the W modules, and so the stability fields of the different stacking sequences will be strongly dependent on those factors which influence the size of the octahedra. We would expect therefore that decreasing the mean cation size and the temperature but increasing the pressure will stabilize pyroxenoid structures with a smaller proportion of W modules, i.e. those with the longer chain periodicities. The pressure–temperature stability fields of a number of different (Ca,Mn)SiO₃ compositions have been studied. Figure 6.26 shows the general form of these relationships for the MnSiO₃ composition, the exact

positions of the phase boundaries being very dependent on the exact composition i.e. mean cation size.

Triclinic and monoclinic wollastonites

Another feature of the wollastonite structure is that, just as in proto-, ortho- and clinopyroxene, different stacking sequences of the silicate chain along its length gives rise to different structures. The basic wollastonite structure is triclinic, with the chains extending along the *b* axis. Figure 6.27 shows the way the chains are arranged in (100) layers in the triclinic structure. These (100) layers of structure, one triclinic unit cell thick are the modules which are used to build up the other wollastonite structures. In triclinic wollastonite they are merely repeated continuously with respect to the previous cell. One of the features of this structure is that the oxygen atoms are separated by a distance *b*/2 along the edge of the module, so that a displacement of *b*/2 of one module relative to its neighbour introduces no appreciable strain. Thus in building the structures there are always two possible positions of one module relative to the previous one.

Schematic diagrams of different stacking sequences of (100) modules are shown in Figure 6.28. A module in a continuous position with respect to the previous one is denoted T, and G is one that is displaced by *b*/2. The basic triclinic structure is then TTTTT. It is termed 1T wollastonite (a 1-module repeat, and a triclinic cell). There are obviously

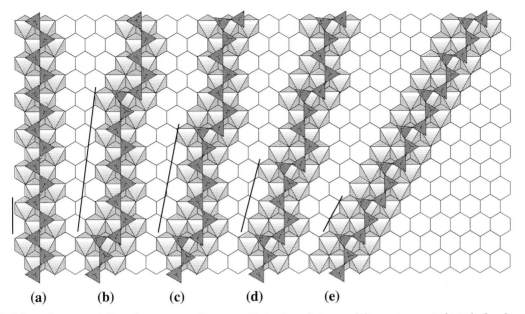

Figure 6.25. Schematic representation of pyroxene and pyroxenoid structures in terms of the arrangement of tetrahedra along the chain and the coordination of the chain to the underlying octahedra. **(a)** The straight chain of clinopyroxene, with a 2 tetrahedra repeat, denoted P. **(e)** The wollastonite chain with a 3 tetrahedra repeat, denoted W. **(b)** Ferrosilite III has a chain denoted PPPW with a repeat of 9 tetrahedra along the chain. **(c)** Pyroxmangite has a chain denoted PPW with a repeat of 7 tetrahedra along the chain. **(d)** Rhodonite has a PW chain with a repeat of 5. The full line against each chain shows the repeat length and the changing orientation of the chain. (After Liebau, 1985.)

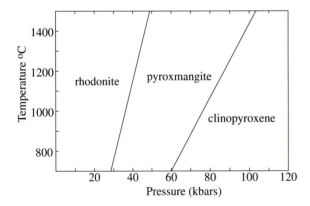

Figure 6.26. Pressure – Temperature diagram showing the stability regions of rhodonite, pyroxmangite and pyroxene structures for the composition $MnSiO_3$. Increasing the pressure and/or decreasing the temperature is equivalent to the effect of decreasing the cation size, and hence $MnSiO_3$ forms the pyroxmangite and then the pyroxene structure as the pressure is increased. (After Akimoto and Syono, 1972.)

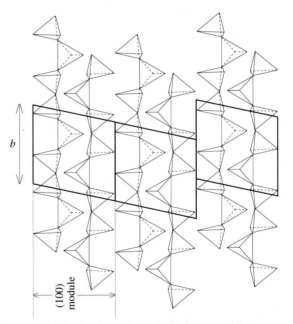

Figure 6.27. The structure of wollastonite in terms of the arrangement of chains along the b axis. A (100) module is a layer of the structure whose thickness defines the triclinic unit cell of 1T wollastonite. There are two possible positions for subsequent modules: either translated along the a axis of this triclinic cell as in the two unit cells outlined on the left, or displaced by b/2 as is the case for the module on the right. (After Hutchison and McLaren, 1976.)

limitless possibilities for stacking sequences, but the most common wollastonite structure has a stacking sequence TGTGTG... a 2-module repeat and a monoclinic cell (hence 2M wollastonite). Other stacking sequences which have been described are TTGTTG (3T wollastonite), TTTGTTTG (4T wollastonite) and TTTTG... (5T wollastonite). Of the many possible polytypic stacking combinations a few

are found regularly, indicating that there must be energy minima associated with particular sequences.

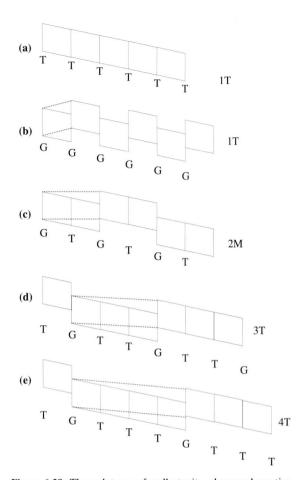

Figure 6.28. The polytypes of wollastonite, shown schematically in terms of the stacking of the (100) modules described in Fig 6.27. **(a)** Continuous translation of the modules results in 1T wollastonite. **(b)** If every module is related by a $b/2$ glide relative to the previous one, the structure is still 1T, but is twin related to (a). The twin related unit cell is shown by the dotted line. **(c)** 2M wollastonite **(d)** 3T wollastonite **(e)** 4T wollastonite. The unit cell is shown by the dotted line in each case. (After Henmi *et al.*, 1983.)

6.5 Double chain silicates – the amphiboles

The structure and the behaviour of the amphibole minerals are similar in many ways to the pyroxenes and we will make comparisons between the two mineral groups throughout this section. We will not therefore treat this group in as much detail, although the complexities, particularly due to the wider range of chemical compositions of the amphiboles are greater.

The essential feature of the amphibole minerals is the double chain of [SiO$_4$] tetrahedra, which can be thought of as two single chains joined by corner-sharing, with a mirror plane along the join (Figure 6.29). The mirror plane is preserved in all amphi-

boles, even when the chains are not straight. Half of the tetrahedra have two bridging and two non-bridging oxygens; the other half have three bridging oxygens and one non-bridging. The general formula of the chain is therefore $(Si_4O_{11})_n^{6n-}$. The chains extend along the c axis defining the c axis repeat (~5.2Å) common to all amphibole structure types. The stacking of the chains is analogous to that in pyroxenes, forming (100) layers of chains with apices alternately pointing up and down, defining the b axis repeat (Figure 6.29(b)).

Again by analogy with the pyroxenes the cation sites in the structure are defined by their position relative to the apices and bases of the [SiO$_4$] tetrahedra in the double chains. The sites between tetrahedral bases of adjacent chains are termed the M4 sites (cf. the M2 sites in pyroxenes), and the smaller sites between the opposed tetrahedral apices are the M1, M2 and M3 sites (cf. the M1 sites in pyroxenes). The coordination of the M4 sites is 8 when occupied by a larger cation such as Ca^{2+}, but reduces to 6 when occupied by a smaller cation such as Mg^{2+} or Fe^{2+}. The M1, M2 and M3 sites are octahedral. The double chains lead to a third type of cation site which lies between the rings formed by opposed tetrahedral bases in the chains. This large A site as it is called, may be vacant, partially filled, or fully occupied by Na and/or Ca in some amphiboles. Finally, (OH)$^-$ or F$^-$ lies in the centre of the hexagonal rings, at the level of the tetrahedral apices. All of these sites are shown in Figure 6.29(b). The space group of this ideal amphibole stucture is C2/m. An I-beam, analogous to that in pyroxenes, is shaded.

The basic relationships between the structure type, its composition and the temperature and pressure are analogous to the pyroxenes. Again the silica tetrahedra remain essentially inert during expansion or contraction of the cation sites, the linkage between the tetrahedral chain and the cation polyhedra being maintained by rotation of the individual tetrahedra which reduces the dimensions of the chain (Figure 6.30). Apart from the different mineral names, and their more complex chemistry, the structural modifications in the amphibole quadrilateral (Figure 6.31) can be interpreted in the same way as in pyroxenes.

1. In Ca-rich amphiboles the C2/m structure exists over the whole temperature range and over a wide range of compositions. It also exists in Ca-poor amphiboles at high temperatures.

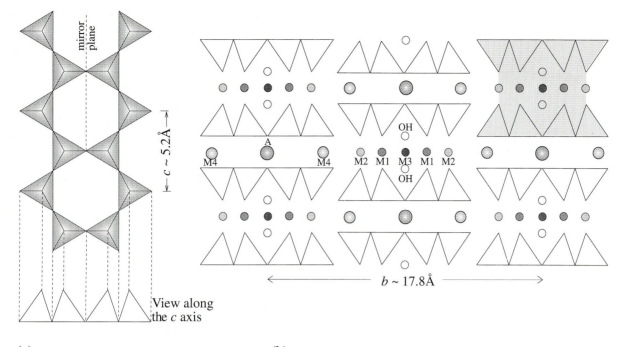

(a)   **(b)**

Figure 6.29. (a) A double chain of SiO_4 tetrahedra in amphiboles, extending along the c axis, and below, a schematic representation of this double chain viewed end-on. Here the chains are straight – in practice they are always slightly rotated as in Figure 6.30. **(b)** The arrangement of double chains in the amphibole structures, viewed along the c axis. The M1, M2 and M3 cations form chains of edge-sharing octahedra between the apices of the tetrahedra, while the larger M4 octahedra form similar chains between the bases of the tetrahedra. The A sites and the OH sites lie in the rings formed along the double chain. One I-beam, analogous to that in pyroxenes, has been shaded.

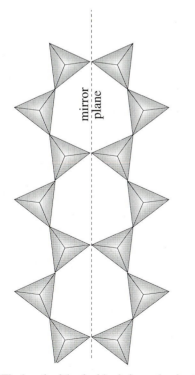

Figure 6.30. The length of the double chain can be shortened by a rotation of the tetrahedra. Tetrahedra related by the mirror plane rotate in opposite senses, preserving the mirror plane.

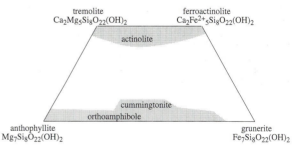

Figure 6.31. The amphibole quadrilateral showing the approximate compositions of simple amphibole minerals. The shaded areas represent the extent of solid solution in naturally occurring minerals. At high temperature there is complete solid solution between cummingtonite and actinolite (cf. Figure 6.14).

2. In Ca-poor amphiboles, just as in the pyroxenes, a collapse of the chain takes place on cooling, reducing the symmetry to $P2_1/m$. The loss of the C-centring is due to alternate (100) layers of chains kinking by different amounts, although the sense of rotation of the tetrahedra is the same, unlike that in the analogous $C2/c \Rightarrow P2_1/c$ transition in pyroxenes. Hence there are only O-rotated chains in amphiboles (compared with O and S chains in pyroxenes).

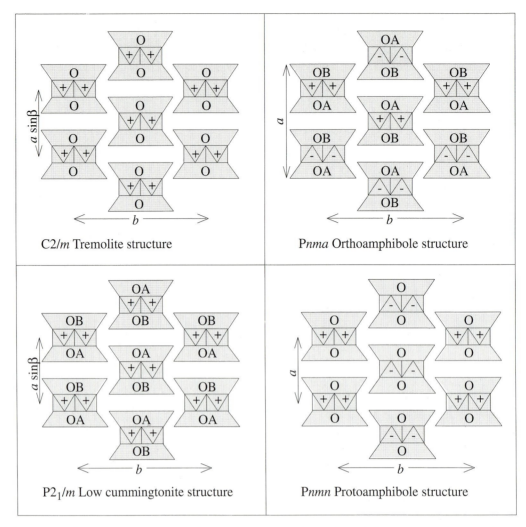

Figure 6.32. The amphibole structures represented by I-beam diagrams analogous to those in the pyroxene structures in Figure 6.22. (After Papike and Cameron, 1976.)

When the opposing chains become non-equivalent they are termed A and B.

3. In the Ca-free amphiboles an orthorhombic structure (P*nma*) is formed at low temperatures by changing the relative positions of the chains along their length. The *a* lattice parameter of this cell compared to that in clinoamphibole is $a_{ortho} = 2a_{clino} \sin\beta$.

4. A high temperature protoamphibole (P*nmn*) also exists at the pure Mg end member at high temperatures.

The amphibole structures are summarized as I-beam diagrams in Figure 6.32, which should be compared with the analogous diagram for pyroxenes in Figure 6.22. The relationships between proto-ortho- and clinoamphiboles can also be expressed in

terms of (100) stacking sequences analogous to those in the pyroxenes (Figure 6.18).

The general composition of amphibole minerals may be described by the formula

$$A_{0-1}B_2C_5T_8O_{22}(OH,F)_2.$$

A = the large A site which may be vacant or contain varying amounts of Na/Ca.

B = the content of the M4 site which in the most common amphiboles may be Ca,Na,Fe^{2+} or Mg;

C = Fe^{2+},Mg, Fe^{3+} or Al in the M1,M2 and M3 sites,

T = Si and Al in the tetrahedra. The limit of this substitution appears to be (Al$_2$Si$_6$).

This profusion of cation sites makes the amphiboles the mineral group with the widest chemistry, and so able to crystallise in rocks over a broad range of bulk composition. The range of chemical substitutions possible make amphibole nomenclature a difficult problem, even after the recent rationalization. A typical complex solid solution is that of "hornblende" a very common constituent of many igneous and metamorphic rocks. The substitution scheme can be described with tremolite $Ca_2Mg_5Si_8O_{22}(OH)_2$ as a starting point. First, the entry of Na into the A sites is balanced by the substitution of Al^{3+} for Si^{4+} in the tetrahedra giving a formula $NaCa_2Mg_5(AlSi_7)O_{22}(OH)_2$. Second, the substitution of a trivalent ion such as Al for Mg in the M1,2 and 3 sites balanced again by substitution of Al^{3+} for Si^{4+} in the tetrahedra leads to a formula $Ca_2(Mg_4Al)(AlSi_7)O_{22}(OH)_2$. A combination of both substitutions, i.e. $NaAl_3$ for $MgSi_2$ gives a composition $NaCa_2(Mg_4Al)(Al_2Si_6)O_{22}(OH)_2$ which may be regarded as one end member of the hornblende series. The composition of most hornblendes lies between this and tremolite, with a further substitution of Fe^{2+} for Mg.

The general principles of the behaviour of such complex solid solutions will be similar to that in the pyroxenes. At high temperatures, solid solutions are possible over most of the compositional range in amphiboles. At lower temperatures however structural changes in the chains, controlled by the size of the cations in the M4 site and the A site, lead to a reduction in the degree of solid solution possible. In general we would expect immiscibility between amphiboles where (i) the M4 site is Ca-rich in one amphibole and Ca-poor in the other, or (ii) where the M4 site is occupied by Na in one amphibole and Ca in the other, or (iii) where the A site is occupied in one amphibole and vacant in the other.

6.6 The layer silicates

The layer silicates are made up from two basic building blocks: a sheet of $[SiO_4]$ tetrahedra (the *tetrahedral sheet*) and a sheet of edge-sharing octahedra (the *octahedral sheet*). In the tetrahedral sheet each $[SiO_4]$ tetrahedron shares three corners, forming a continuous sheet with general formula $(Si_2O_5)_n^{2n-}$. The unbonded tetrahedral apices of the sheet all point in the same direction as shown in Figure 6.33. These apices connect the tetrahedral sheet to the octahedral sheet, the apical O atoms of the tetrahedra shared with an O atom from the octahedral sheet. The octahedral sheet is made up of

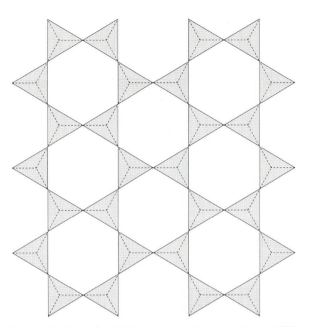

Figure 6.33. Part of an infinite tetrahedral sheet, made up of SiO_4 tetrahedra all pointing in the same direction, and sharing three corners. This sheet is one of the basic building units in the layer silicates.

an array of edge sharing octahedra with either (OH) groups or O atoms at the corners.

When the octahedra contain a divalent ion such as Mg^{2+} or Fe^{2+}, charge balance between the sheets is achieved when all sites are occupied. As there are three octahedra within each hexagonal ring of tetrahedra (see Figure 6.35), the sheet is termed a *trioctahedral* sheet (Figure 6.34(a)). When a trivalent ion such as Al^{3+} or Fe^{3+} occupies the octahedra, only two thirds of the octahedra are occupied, and the sheet is termed a *dioctahedral* sheet (Figure 6.34(b)). The occupied octahedra in a dioctahedral sheet are invariably more distorted, the vacant site becoming significantly larger than the occupied sites around it.

The layer silicate structures are built up from different stacking combinations of these two basic sheets, one sheet connected to the other as shown in Figure 6.35. From the outset we can see that this type of linkage between the tetrahedral and the octahedral sheets imposes restrictions on their relative sizes. An ideal fit between a flat $[Si_2O_5]$ sheet and an octahedral sheet would require the octahedral cation radius to be around 0.70Å, i.e. slightly smaller than Mg^{2+}. In many layer silicates there is considerable substitution of Al^{3+} for Si in the tetrahedral sites as well as of M cations in the

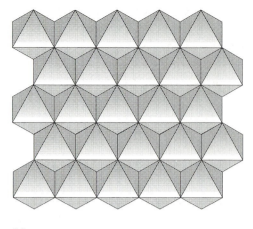

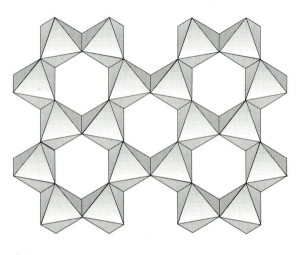

(a)

(b)

Figure 6.34. (a) Part of an infinite sheet of edge-sharing octehedra. When all the octahedra are occupied the sheet is termed *trioctahedral*. **(b)** In a *dioctahedral* sheet one third of the octahedra are unfilled, and these vacant octahedra are ordered as shown. The distortion of the filled octahedra enlarges the vacant site.

octahedra, leading to a variation in the mesh sizes. One consequence of the misfit between the sheets is that at some compositions only very small crystals can grow, as the strain imposed by any misfit will increase with the area of the layer. The clay minerals, in which this is the case, form a subgroup of the layer silicates.

The layer silicates are classified by the way in which the layers are built up of tetrahedral and octahedral sheets, and by the way in which the layers are stacked relative to each other. The repeat of the layers defines the (001) basal spacing of the unit cell, and this spacing is used to identify the type of stacking present. In finely crystalline material the

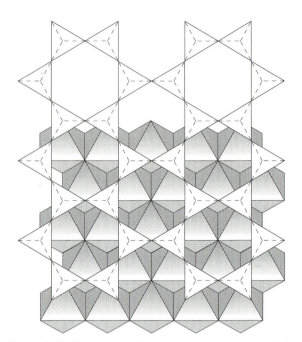

Figure 6.35. The way in which a sheet of downward-pointing tetrahedra links to an octahedral sheet below.

basal spacing is often the main diagnostic feature.

There are three main groups of layer silicate minerals. In the 1 : 1 *layer silicates*, the layers are made up from one sheet of [SiO$_4$] tetrahedra, combined with one octahedral sheet. These layers are charge balanced and only weak bonds hold successive layers together in the stack (Figure 6.36(a)). In the 2 : 1 *layer silicates*, an octahedral sheet is sandwiched between the apices of two tetrahedral sheets (Figure 6.36(b)). These Tet–Oct–Tet layers are either held together by weak van der Waals forces if they are neutral (e.g. as in talc), or may have cations between them for charge balance, if substitutions in either sheet result in a residual layer charge (e.g. as in the micas). In the 2 : 1 : 1 *layer silicates* an additional octahedral sheet is sandwiched between each T–O–T layer (Figure 6.36(c)).

Two features of layer silicates are of particular interest from a structural point of view. The first is how they cope with any mismatch between the tetrahedral and octahedral sheets, and second is the occurrence of polytypism due to different modes of stacking of the layers.

6.6.1 Tetrahedral – octahedral mismatch in the layer silicates

The simplest 1:1 layer silicate is *kaolinite* Al$_2$Si$_2$O$_5$(OH)$_4$, which is dioctahedral with Al in two thirds

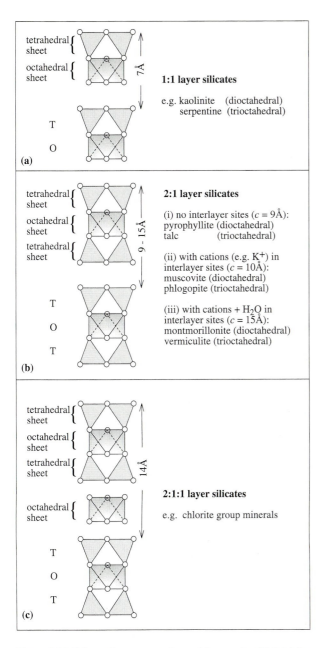

Figure 6.36. Schematic representations of the way in which **(a)** the 1:1 layer silicates **(b)** the 2:1 layer silicates and **(c)** the 2:1:1 layer silicates are built up from tetrahedral and octahedral sheets. The minerals within each group are named according to whether the octahedral sheet is dioctahedral or trioctahedral, and the nature of the interlayer cations.

of the octahedral sites. The mesh size of the Al octahedral sheet is smaller than that of the tetrahedral sheet, so a better fit is achieved when the [SiO_4] tetrahedra rotate about an axis normal to the sheet, with adjacent tetrahedra rotating in opposite senses in much the same way as in the amphiboles

(Figure 6.30). This reduces the tetrahedral mesh size. In kaolinite, the tetrahedra have to rotate by about $9°$ to achieve a good fit.

In other 1:1 layer silicates, the degree of mismatch depends on the cations occupying the tetrahedral and octahedral layers. The rotation of the tetrahedra needed to satisfactorily link the two sheets decreases by replacing the Al^{3+} ion by the larger Mg^{2+} ion in the octahedral layer, and increases by a replacement of Si^{4+} by the smaller Al^{3+} ion in the tetrahedral layer.

Mg^{2+} occupies the octahedral sites in the trioctahedral *serpentine* minerals, $Mg_3Si_2O_5(OH)_4$. The mismatch is smaller in serpentines than in kaolinite and there are three forms of serpentine, each reflecting a different attempt at coping with this tetrahedral – octahedral mismatch.

1. In *lizardite* (Figure 6.37(a)) the layers remain flat as in kaolinite, and the tetrahedral rotation is around $3.5°$.

2. In *chrysotile*, the tetrahedra are tilted slightly so that their apices no longer lie in a flat plane. This increases the distance between the apices to achieve a fit with the octahedral mesh. The effect of this is to give a curvature to the whole layer (Figure 6.37(b)), and continued growth of the curved layer results in a crystal in which the layers are rolled up as in a carpet. The larger the misfit, the smaller the the radius of curvature and hence the tighter the roll.

3. In *antigorite*, the tilting of the tetrahedra which leads to the curved layers is accompanied by a periodic switching of the direction of the tetrahedral layer, producing a structure which has been compared with a stack of corrugated iron (Figure 6.37(c)).

Figure 6.38 shows a high resolution transmission electron micrograph of amphibole being replaced by the layer minerals, talc, lizardite and chrysotile.

In the 2:1 layer minerals, with tetrahedral sheets on either side of the octahedral sheet, curvature of the layers on any macroscopic scale is not a possibility. In talc, $Mg_3(Si_2O_5)_2(OH)_2$ the layers are flat (Figure 6.39(a)) and the fit is achieved by a tetrahedral rotation of about $3.5°$. If the smaller Al^{3+} ion is substituted for Mg in the octahedra, this rotation has to be increased: in pyrophyllite $Al_2(Si_2O_5)_2(OH)_2$ the rotation angle is $10°$. In cases where the octahedral sheet is too large for the tetrahedral sheet, it is much more difficult to stretch the tetrahedral sheet to fit. In compositions where this is the case the octahedral

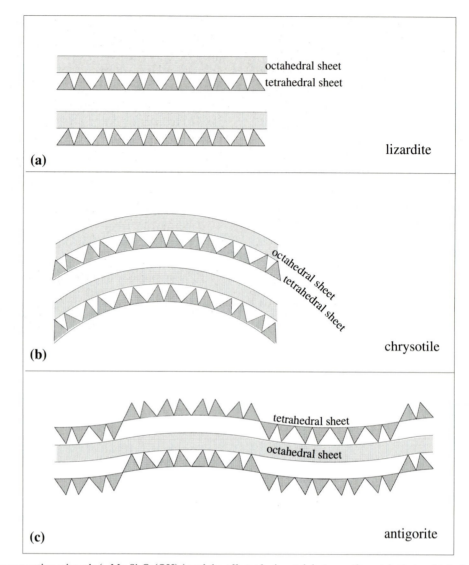

Figure 6.37. The serpentine minerals (~Mg$_3$Si$_2$O$_5$(OH)$_4$) and the effect of mismatch between the octahedral and tetrahedral sheet. **(a)** In lizardite the sheets remain planar, although the mismatch prevents the crystals from growing to any appreciable size. **(b)** In chrysotile, the sheets are curved to achieve a better fit, and the crystal rolls up like a carpet (the curvature is exaggerated in this figure). **(c)** In antigorite the sheets are also curved, but the tetrahedral sheet switches its orientation, producing a corrugated effect.

sheet has to split into ribbons to reduce the strain (Figure 6.39(b)). The minerals in which this happens (e.g. sepiolite and palygorskite) are not very common.

6.6.2 Polytypism in the micas

The mica minerals belong to the 2:1 layer silicates in which cation substitution in either the tetrahedral (e.g. Al^{3+} for Si^{4+}) or octahedral sheets (e.g. M$^+$ for M^{2+}) results in a overall negative charge on the layers. The charge is compensated and the layers are bonded together by large, positively charged *inter-layer cations*, most commonly K$^+$, Na$^+$ or Ca^{2+}. Muscovite, KAl$_2$(Si$_3$Al)O$_{10}$(OH)$_2$ (Figure 6.40) is a typical example.

Two features of this structure lead to the polytypism observed in the micas. First, in the tet – oct – tet sandwich the two opposing tetrahedral sheets are not directly opposite one another. There is an offset (the 'stagger vector' shown in Figure 6.40) necessary to produce the octahedral coordination in the octahedral sheet. Secondly, the hexagonal symmetry of the tetrahedral sheet means that this offset can be in any one of 6 directions. For a given direction of the offset in one layer, the important point is how

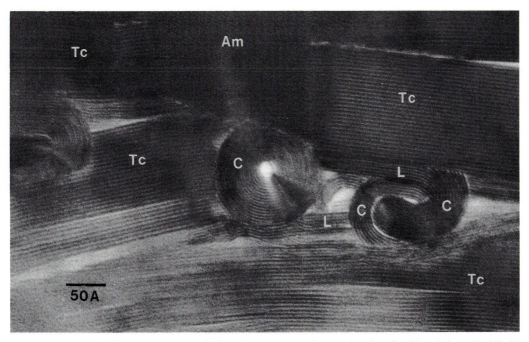

Figure 6.38. Amphibole (Am) partially replaced by talc (Tc) and several types of serpentine, lizardite (L) and chrysotile (C). (From Buseck and Veblen, 1981.)

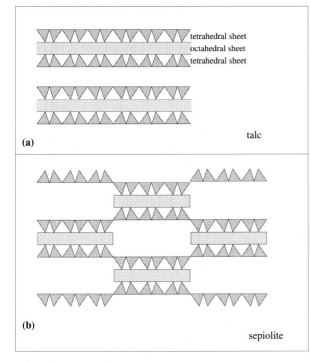

(a) tetrahedral sheet / octahedral sheet / tetrahedral sheet

talc

(b)

sepiolite

Figure 6.39. The effect of a mismatch between the tetrahedral and octahedral sheet in the 2:1 layer structure. **(a)** In talc the mismatch can be accommodated by rotation of the tetrahedra. **(b)** When the mismatch becomes too great the sheets can no longer remain continuous and the octahedral sheet is split into ribbons. The width of the ribbons depends on the degree of mismatch.

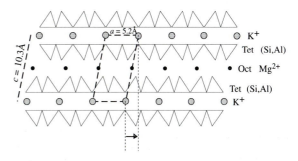

Figure 6.40. The structure of muscovite mica, viewed parallel to the layers. The 2:1 layer is made up of two opposing tetrahedral sheets containing (Si,Al) with the Mg^{2+} octahedral sheet between. The layers are stacked with an offset (the stagger vector) shown by the length of the arrow. This defines a monoclinic cell (dashed line). K^+ occupies the interlayer positions.

the offsets in consecutive layers are related to one another. There are 6 possible interlayer stacking angles: 0°, 60°, 120°, 180°, 240° and 300°. Figure 6.41 shows how different sequences of these stacking angles leads to the 6 standard mica polytypes. The polytypes are labelled according to the number of layers in the repeat unit (before the stagger directions coincide again), and the symmetry of the unit cell (M – monoclinic, T – trigonal, O – orthorhombic and H – hexagonal).

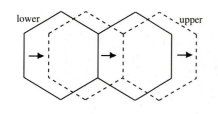

(a)

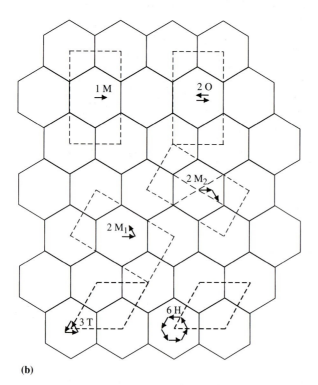

(b)

Figure 6.41. (a) The stagger vector (arrowed) relating two mica sheets. The open hexagonal mesh represents the hexagonal mesh of a tetrahedral sheet in the mica structure. **(b)** Polytypism in the micas. Six different polytypes can be produced by different combinations of the orientation of the stagger vector. In each case the repeat is shown as a multiple of the 10.3Å *c* repeat of muscovite (Figure 6.40), and the crystal system of the polytype indicated (M: monoclinic; O: orthorhombic; T: trigonal; H: hexagonal). The unit cell is outlined (dashed) in each case.

The basic layer repeat in micas (001) is ~10Å. The most common muscovite polytype in nature is 2M with a 20Å (001) layer spacing.

6.6.3 A note on clay minerals

Clay minerals are layer silicates with a grain size < 2μm. Any of the layer silicates described above could qualify as a clay mineral, but additionally there are more hydrated forms in which water molecules occupy the interlayer position. *Montmo-*

rillonite for example is a clay mineral based on pyrophyllite but with the capacity to gain and lose interlayer water – hence *swelling clays*. *Illite* is essentially fine-grained muscovite, with some of the K^+ leached out and replaced by weakly bound water or H_3O^+ ions. Similarly, *vermiculite* is a leached biotite with some interlayer water.

6.7 Biopyriboles

The similarities in the structures of the pyroxenes, amphiboles and micas, with opposing chains of width 1, 2 and ∞ respectively, and with cations between the apices, have led to a description of these structures in terms of modular slabs. This approach has gained added significance by the discovery of chain silicate minerals with triple chains as well as periodic combinations of different chain widths. The term *biopyribole* refers to the whole range of possible mineral structures from the micas (e.g. *biot*ite) to amphi*bole*s and *pyr*oxenes.

The modular approach is most conveniently visualized in terms of I-beams. Figure 6.42 shows the structure of an amphibole which has been sliced up into (010) slabs with thickness *b*/4 (~4.5Å). There are two types of slabs: the central part of the I-beam, if repeated indefinitely would produce a mica structure, hence this slab is termed M; the slabs on either side produce the pyroxene structure when repeated, and are labelled P. On this basis pyroxenes and micas consist of only one type of slab (PPPPP... and MMMMM...), while in amphiboles these slabs alternate (MPMPMP...). M and P slabs can fit together

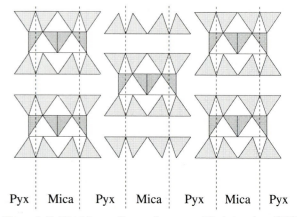

Pyx | Mica | Pyx | Mica | Pyx | Mica | Pyx

Figure 6.42. The I-beam diagram for an amphibole, cut into (010) slabs by the dotted line. The slabs are of two types: Mica (M), because a repetition of such slabs generates a mica structure, and Pyx (P), because their repetition generates a pyroxene structure. (After Thompson, 1981.)

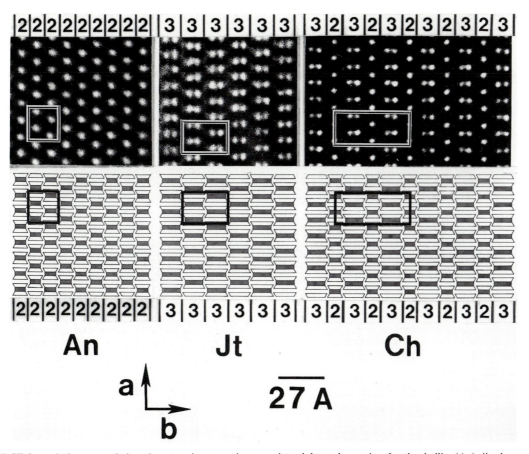

Figure 6.43. High resolution transmission electron microscope images, viewed down the c axis, of anthophyllite (An), jimthompsonite (Jt) and chesterite (Ch). The white spots are the projected positions of the A-sites, which are located between the I-beams. The structural interpretation is shown in terms of I-beam diagrams, and unit cells and chain widths are indicated. (From Veblen and Buseck, 1979.)

in any order, producing a whole family of structures collectively known as the biopyriboles. This modular approach is reminiscent of polytypism, although the M and P slabs each have a different structure and composition : the analogous term used to describe structures related by stacking two different modules is *polysomatism*.

The alteration of amphiboles at low temperatures and hydrothermal conditions leads to the formation of layer silicates, and in some cases this change from double chains to layers proceeds via well-defined intermediate structures. The prediction and discovery of single crystals of such minerals in the mid 1970s revitalized this modular approach to silicate chain mineralogy. A triple chain mineral (and hence MMPMMP... or more simply, 33... etc.) and one in which double and triple chains alternate (MPM-MPMPMMP... or 2323... etc.) have been termed jimthompsonite and chesterite respectively, (the first after Prof Jim Thompson who predicted their occur-

rence, the second after the locality). Figure 6.43 shows high resolution transmission electron micrographs of anthophyllite, and these two new minerals, together with their I-beam diagrams for comparison. Other sequences in lesser amounts, as well as intergrowths of different chain widths with various degrees of organization have also been found by electron microscopy. Figure 6.44 is an example of the way in which double and triple chain material is intergrown with well-ordered chesterite.

Just as there are ortho- and clino- forms of pyroxene and amphibole, depending on the relative displacements of the chains along their length, an analogous situation exists in the wider chain biopyriboles. Hence ortho- and clinojimthompsonite can be represented by stacking sequences of $+ + - - + +$ $- -..$ and $+ + + + + +...$ (cf. Figure 6.32 for the amphiboles).

The general composition of a P module is $B_4T_4O_{12}$, while an M module is $AB_3T_4O_{10}(OH)_2$. (A: A site as

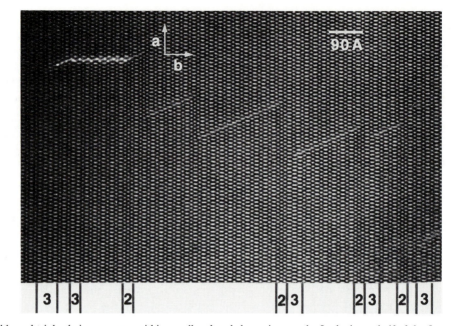

Figure 6.44. Double and triple chain sequences within a well-ordered chesterite sample. In the lower half of the figure blocks of structure where double and triple chains do not alternate rigorously are indicated by "2" and "3". These strips of double and triple chain material terminate at faults in the crystal, above which the chesterite is perfectly ordered. (From Veblen and Buseck, 1980.)

in the amphiboles, B: other cations, T: tetrahedral site). An amphibole unit (PM) is therefore $B_4T_4O_{12}$ + $AB_3T_4O_{10}(OH)_2$ = $AB_7T_8O_{22}$ $(OH)_2$, and pure Mg-jimthompsonite (MMP) with a vacant A site has an ideal composition $Mg_{10}Si_{12}O_{32}(OH)_4$.

Apart from its crystal–chemical interest the importance of this approach to chain silicate minerals is in understanding the mechanisms by which the alteration sequence pyroxene ⇒ amphibole ⇒ mica takes place, and also in interpreting the origin of structural defects. This will be discussed further in the next chapter.

6.8 The framework silicates

In the framework silicates all $[SiO_4]$ tetrahedra share their corners with others, generally forming rather open three-dimensional networks. If no ion substitutes for Si, the entire framework has the composition SiO_2 and all valence bonds are satisfied. When Al substitutes for Si in the tetrahedra, interstitial cations are required to maintain charge balance. The openness of these framework structures results in rather large interstitial cation sites compared to those in the chain silicate minerals. Thus Na^+ and Ca^{2+} will be considered as 'small' cations compared to K^+, while ions such as Mg^{2+} are too small to play

a role in these structures. The aluminosilicate framework minerals are by far the most abundant minerals in the Earth's crust, the feldspars making up about 65% by volume. Mineralogically they are also the most interesting and challenging from the point of view of the response of the structure to temperature, pressure and composition changes.

A general feature of framework structures is that at high temperatures, they have more open expanded structures with the maximum symmetry allowed by the tetrahedral linkage pattern. At lower temperatures they tend to crumple slightly, reducing the size of the interstitial cavities where any cations would be sited. The crumpling is therefore constrained to some extent by the size of these interstitial cations. The crumpling of the structures is achieved by a rotation of the $[SiO_4]$ tetrahedra, and in many cases when the tetrahedral tilting is not constrained by symmetry to be in a particular plane, this can be achieved without very significant departures from the ideal Si–O–Si bond angles.

6.8.1 The silica minerals

Silica, SiO_2, occurs in a number of different forms in the Earth. Quartz, the most common crystalline polymorph is stable up to 857°C; tridymite is the stable form from 857°C to 1470°C, and then cristobalite from 1470°C up to the melting point at 1713°C. The high pressure forms of silica are coesite, stable in the deep crust of the Earth, and stishovite which is thought to be stable in the Earth's mantle. Stishovite has the rutile structure and is one of the very few known materials in which Si occurs in octahedral coordination with oxygen. The stability relationships of the SiO_2 polymorphs are shown in Figure 6.45. In low temperature environments at the Earth's surface and on the ocean floor, silica also occurs in amorphous and partly crystalline forms.

The crystal structures of quartz, tridymite and cristobalite are not merely simple modifications of each other, but represent quite different ways of linking tetrahedra by all corners. To change one polymorph to another requires breaking bonds and creating a new structure, i.e. it is a *reconstructive* transformation. As such it is inherently difficult, in the sense that on cooling tridymite to below 857°C for example, it will only transform to quartz very slowly, or not at all if the temperature is reduced too quickly to a point where there is insufficient thermal energy in the structure to break and remake Si–O bonds. We will discuss such behaviour in more detail in a later chapter, although it is necessary to be aware of the kinetic constraints in transformations to be able to appreciate the relationships between the silica structures.

Quartz, tridymite and cristobalite all have *high* and *low* forms. The high form has the more symmetrical structure, while the low form can be considered to be a slightly 'crumpled' version, with bonds bent and the symmetry reduced. These high–low transformations do not require bond-breaking, merely bond-bending, generally a very rapid process that occurs instantaneously when the temperature is reduced below a particular value. This is termed a *displacive* transformation, and we have already mentioned similar transformations in perovskites, and in pigeonite. Note that the low forms of tridymite and cristobalite do not have a stability field in Figure 6.45. This is because the quartz structure is the more stable, although low tridymite and low cristobalite appear during cooling of the high forms when the cooling rate is too fast for the reconstructive structural transformation to take place.

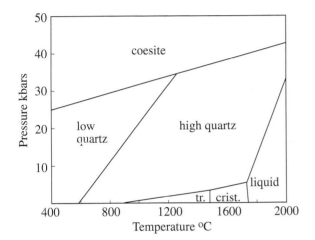

Figure 6.45. The fields of stability, in pressure–temperature space, of the polymorphs of silica.

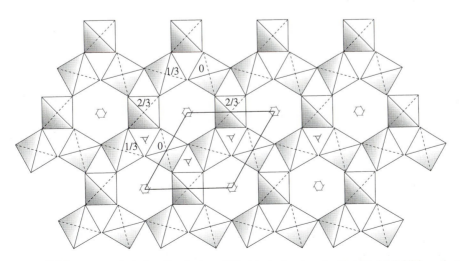

Figure 6.46. The structure of high quartz projected on the *ab* plane. SiO4 tetrahedra at vertical heights 0,1/3,2/3 form three-fold spirals, and these spirals are arranged in an hexagonal array to give a framework structure, each tetrahedron sharing four corners. A unit cell is outlined.

Quartz

High quartz has a structure based on chains of tight three-fold spirals parallel to the *c* axis, the vertical heights of the tetrahedra labelled as 0, 1/3, and 2/3 around the screw triad axis marked in Figure 6.46. These spirals are joined by their corners to form a larger double spiral, with two non-intersecting threads, around the hexad axis. The fact that these spirals could be right-handed or left-handed leads to the growth of crystals which are either right-handed or left-handed and hence are mirror images of each other. This is termed *enantiomorphism* and may be evident from the crystal morphology or from optical properties.

These general features of the high-quartz structure are retained in low quartz, the form stable below 573°C. The relationship between high and low quartz is more easily seen by considering Figure 6.47 in which the structure is simplified to show the Si atoms only. In high quartz the hexad axis is evident (Figure 6.47(a)); in low quartz rotation of the tetrahedra around the *c* axis reduces this hexad axis to a triad axis, and hence low quartz is trigonal (Figure 6.47(b)).

Tridymite and cristobalite

Tridymite and cristobalite have some similarities in the tetrahedral linkage pattern and so it is convenient to discuss them together. Both have a structure which can be described in terms of layers formed from [SiO$_4$] tetrahedra linked, with alternate tetrahedra pointing up and down, to form hexagonal rings (Figure 6.48(a)). These layers are then stacked on top of each other, joined by the tetrahedral apices. The hexagonal symmetry of the individual layer allows the possibility of linking them in different stacking sequences. Tridymite has a two-layer repeat: ABABAB and a hexagonal structure with the layers parallel to (001) planes. Cristobalite has a three layer repeat ABCABC... and a face-centred cubic structure with the layers parallel to (111) planes.

This relationship between tridymite and cristobalite is reminiscent of that between hexagonal close-packing and cubic close-packing although the structures here are far from being close-packed. Furthermore, in tridymite the B layer is not simply a laterally translated A layer, but involves an inversion (turning the layer upside down) before linking it onto the A layer. The stacking sequence in tridymite

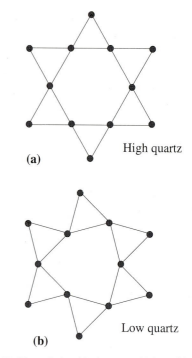

Figure 6.47. The relationship between high and low quartz. **(a)** High quartz, showing only the projection of the Si atoms onto the *ab* plane. **(b)** The distortion of the structure to give low quartz involves rotation of the tetrahedra and a reduction in symmetry from hexagonal to trigonal.

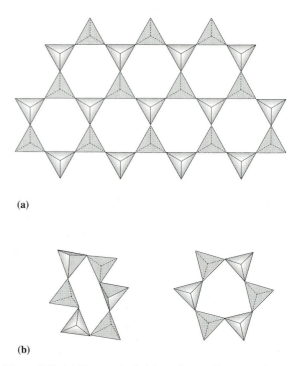

Figure 6.48. (a) The tetrahedral layer from which both tridymite and cristobalite structures are built. The tetrahedra are linked in rings, with alternate tetrahedra pointing up and down. **(b)** Two types of distortions of these rings in the low tridymite structure: the oval distortion (left), and the trigonal distortion (right).

should more properly be denoted A ∀ A ∀ A ∀ ... In cristobalite no such inversion of the layers is involved and the ABCABC... stacking is staggered in the same way as in cubic close-packing. Nevertheless, the basic geometry of the dimensional relationships between tridymite and cristobalite remains the same as that between hexagonal and cubic close-packed structures as described in Section 5.2.1.

Despite these similarities in the structure of tridymite and cristobalite, transformation from one to the other is a reconstructive process. The fact that both structures are made up of identical layers leads to the structural possibility of mixing different stacking sequences leading to a mechanism for very fine-scale intergrowths between the two polymorphs.

Low tridymite and low cristobalite are both distorted forms of the high structures forming at temperatures below about 150°C and 230°C respectively. These displacive transformations reduce the symmetry of tridymite to orthorhombic, and cristobalite to tetragonal. The detailed nature of these transformations is not well understood because they are quite complex, but in general, the distortions involve a reduction in the symmetry of the hexagonal rings of tetrahedra. In low tridymite for example, two types of distortions appear to take place simultaneously, one resulting in ditrigonal rings the other in more oval-shaped rings (Figure 6.48(b)).

As noted above, at these temperatures the quartz structure is the most stable, and the formation of low tridymite for example, indicates that high tridymite has failed to transform to quartz on cooling, as required by the stability diagram (Figure 6.45).

Silica in marine environments
Silica deposits on the ocean floor originate from accumulations of the siliceous remains of diatoms, radiolarians and other organisms which secrete shells of amorphous silica. This form of silica is often termed Opal-A (siliceous ooze). The recrystallisation of this deposit after burial on the ocean floor ultimately results in the formation of chert, a rock composed of very finely crystalline quartz. The sequence in this post-burial recrystallisation (or *diagenesis*) illustrates a general principle which will be relevant in other mineral groups. The recrystallisation of Opal-A is not directly to quartz which is the most stable state, but via another crystalline form termed Opal-CT. Opal-CT has a structure which can

be loosely described as a disordered intergrowth of tridymite and cristobalite-type stacking sequences. The fact that the most stable state (quartz) does not form directly indicates that kinetic factors must control the crystallisation, i.e. that Opal-CT must be kinetically favoured over quartz. This is borne out by the observation in deep sea deposits that the rate of transformation from Opal A – Opal CT – chert is very dependent on the chemistry of the surrounding sediments and hence that of the fluid phase.

6.8.2 Kalsilite – nepheline : stuffed tridymites

When part of the Si^{4+} is replaced by Al^{3+} in a tetrahedral framework structure, interstitial cations are required to provide charge balance. These cations, which occupy the large cavities in the framework, also control the distortion of the structure by imposing their own requirements on the most favourable cation–oxygen bond lengths. The structures of kalsilite $KAlSiO_4$, and nepheline $(K,Na)AlSiO_4$, are stuffed derivatives of high tridymite in which half of the Si is replaced by Al and the compensating cations K or Na occupy the channels between the rings of tetrahedra. The difference between the structures is in the way these rings of tetrahedra are distorted. In these structures we do not appear to have a situation where the rings can adapt continuously to a changing interstitial cation size: rather, there are three different-sized rings available and these are combined in different arrangements depending on the composition. These three types of rings have already been mentioned in the tridymite structure (Figure 6.48): hexagonal, ditrigonal and oval.

In the structure of kalsilite, $KAlSiO_4$ (Figure 6.49(a)), the K atoms all occupy the centres of ditrigonal rings of tetrahedra. The layers of rings are stacked as in tridymite, i.e. ABAB... with the alternate layers rotated through 180° relative to one another so that the ditrigonal rings in alternate layers point in opposite directions, as indicated by the arrow in Figure 6.49(a). The Al,Si atoms are ordered in alternating tetrehedra according to the aluminium avoidance principle (no adjacent Al tetrahedra).

In nepheline, which has an ideal composition $K_{0.25}Na_{0.75}AlSiO_4$, there are two kinds of tetrahedral rings (Figure 6.49(b)): hexagonal rings containing the larger K atoms, and oval rings with Na

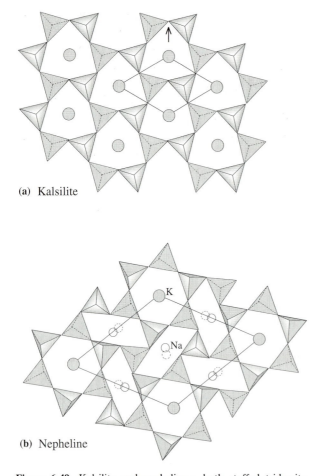

(a) Kalsilite

(b) Nepheline

Figure 6.49. Kalsilite and nepheline – both stuffed tridymite structures, with half of the Si replaced by Al in the tetrahedra, and charge compensating cations in the distorted rings. **(a)** The structure of kalsilite, $KAlSiO_4$, with K^+ in the centre of the trigonally distorted rings. The structure is built up by an ABAB . . . stacking of the layer shown here. Alternate layers are rotated through $180°$ so that the direction indicated by the arrow is reversed in adjacent layers. **(b)** The structure of nepheline, $(K_{0.25},Na_{0.75})AlSiO_4$ in which ovally distorted and trigonally distorted rings contain Na and K respectively. The uncertainty in the position of the Na atom within the oval rings is indicated by the full and dashed circles.

proportions to achieve the most favourable overall cation environment. The fact that these structures are rare suggests that this is not a particularly satisfactory arrangement, but the structures do demonstrate the fact that in these stuffed tridymites the framework appears to be unable to adapt to different cation sizes by continuously changing the shape of the cavity to suit.

Finally, one further complication in the nepheline structure is that the Na atom has difficulty in achieving a proper coordination on all sides with the oxygen atoms in the oval site, and each of the three Na atoms in the unit cell pulling at their neighbouring oxygens causes a statistical distribution of these oxygen atoms over three slightly displaced positions. The oxygen atoms may be statistically distributed in space as well as time. Natural nephelines also have Ca^{2+} substitution in the Na^+ sites, causing K^+ vacancies, which may affect up to 1/3 of these sites. The distribution of the vacant sites affects the distortion of the structure and at low temperatures the real structure of nepheline involves a combination of K-vacancy ordering and coupled atomic displacements superimposed on the ideal structure described above.

6.8.3 The feldspars

The stuffed framework structure of the feldspars is a very successful solution to the structural problem of accommodating K,Na and Ca in an aluminosilicate mineral. The feldspars have a general formula MT_4O_8 with between 25% and 50% of the Si replaced by Al in the T sites, and the M sites occupied by Na^+, K^+, Rb^+, Ca^{2+}, Sr^{2+}, or Ba^{2+}. The compositions of most natural feldspars lie in the $KAlSi_3O_8$–$NaAlSi_3O_8$–$CaAl_2Si_2O_8$ triangle (Figure 6.51), in which the shaded region represents the extent of high temperature solid solution.

All feldspars have a similar tetrahedral linkage pattern which may undergo different kinds of distortions and hence lower the symmetry. The origin of these structural modifications, and the interplay between the various factors which play a role in tending to lower the symmetry as the temperature is reduced, has been a major preoccupation in mineralogy for decades. Not only are feldspars by far the most common minerals, but the whole spectrum of their possible structural states can be found in rocks with different cooling histories, providing a unique

atoms. The alternating arrangement of these two types of rings within the layers doubles the *a* axis of nepheline relative to kalsilite and also requires that succeeding layers are stacked virtually directly on top of one another as in tridymite; relative rotation of the layers as in kalsilite is not possible. The Si,Al ordering scheme is the same as that in kalsilite.

Minerals with compositions lying between kalsilite and nepheline are rare, but the structural principles involved in this mineral group are well illustrated by a brief look at two intermediate structures (Figure 6.50). All three ring types are utilized in different

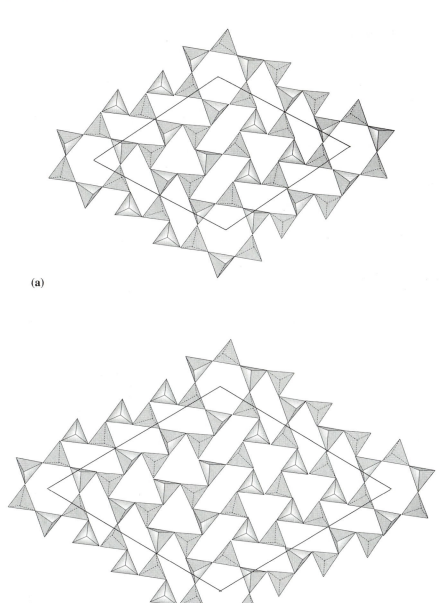

(a)

(b)

Figure 6.50. Structures such as **(a)** trikalsilite and **(b)** tetrakalsilite, are built up by alternating oval and trigonal rings in the tetrahedral layers. With K^+ in the hexagonal and trigonal rings and Na^+ in the oval rings, this scheme provides a mechanism for building structures with compositions between that of kalsilite and nepheline.

opportunity for studying the way a structure responds to a slowly changing geological environment.

We begin with the ideal high temperature feldspar structure of *sanidine* $KAlSi_3O_8$. The Al and Si are distributed at random so that the average occupancy of each tetrahedron is 25%Al,50%Si. The structure is quite complicated and we will build it up from somewhat idealized units to describe its essential features. The basic construction of the framework is of rings of four tetrahedra with alternate pairs of

vertices pointing in opposite directions (Figure 6.52a). These rings are then joined in layers as shown in Figure 6.52b, in which the layers are viewed down the *x* axis.

In the third dimension, the rings are joined to one another by the apices, forming crankshaft-like chains parallel to the *x* axis (Figure 6.53). In Figure 6.53(b) the crankshafts are represented by lines joining the centres of the tetrahedra, and the way these crankshafts are related to each other by mirror

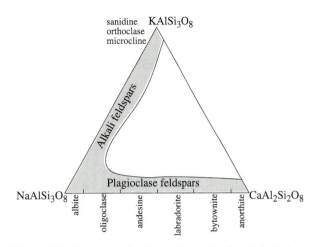

Figure 6.51. The extent of solid solution in alkali and plagioclase feldspars at high temperature. The plagioclase feldspars are subdivided according to composition, as indicated.

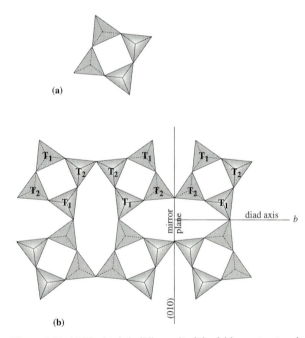

Figure 6.52. (a) The basic building unit of the feldspar structure is the four-membered ring of tetrahedra with a pair of tetrahedra pointing up and a pair pointing down. **(b)** The four-fold rings are joined to form a layer in which the rings are related by mirror planes parallel to (010) and diads parallel to the b axis. Two sets of individual tetrahedra are distinguishable in this layer, and are labelled T_1 and T_2. The T_1 tetrahedra are all related to one another by symmetry, as are the T_2 tetrahedra. Cations occupy the large oval-shaped cavities between the rings.

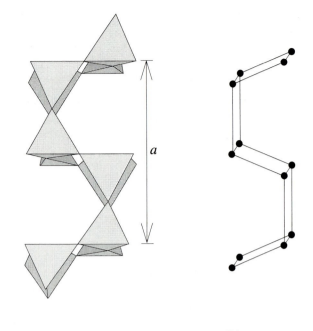

(a)   **(b)**

Figure 6.53. The linkage between the four-fold rings, shown at right angles to the view in Figure 6.52. The rings in one layer are linked by their apices to rings in the layers above and below, forming a crankshaft-like chain. In **(b)** this chain is abstracted further showing only the Si atoms at the centres of the tetrahedra.

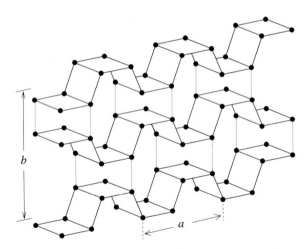

Figure 6.54. A perspective view of the feldspar structure showing the way the crankshafts described in Figure 6.53 are reflected by (010) mirror planes. The dotted lines are an aid to show some of the mirror-related atoms. The cations occupy the large open cavities between the crankshafts. (after Megaw, 1973)

planes is shown in Figure 6.54. The K atoms lie in the mirror planes between the crankshafts, occupying the largest cavities in the framework.

Although this is an idealized description, and in the real structure of high sanidine there is some rotation of the tetrahedra, the symmetry remains the same as in the figures above. The structure is monoclinic, space group C2/m and has the highest symmetry possible in the feldspars. It will be treated as the parent structure from which all other feldspar structures can be derived. It is important to note that this C2/m space group not only requires that the shape of the unit cell be monoclinic, but that there are only two distinguishable sets of tetrahedra, labelled T_1 and T_2 in Figure 6.52(b). The significance of this is that the symmetry of the C2/m structure can be reduced by any distortion which changes the shape of the unit cell, but also by any Al,Si ordering which requires the specification of more than two types of sites.

Structural distortion and Al,Si ordering in the feldspar structure

Before describing the various modifications of the feldspar structure it is worth making a few general observations on the factors that promote a change from the high sanidine C2/m structure to lower symmetries. We will describe each separately, although it is important to realize from the outset that these different modes of symmetry reduction do not operate independently of one another.

1. The expanded high temperature framework tends to collapse around the interstitial cation as the temperature falls. A sufficiently large cation such as K or Ba is able to prevent the collapse, and the framework remains extended, while a smaller atom such as Na or Ca allows the framework to distort, reducing the symmetry to triclinic, space group $C\bar{1}$. Figure 6.55 shows how the distortion around the cation site changes the cation–oxygen bond lengths and changes the interaxial angle α from 90° in the monoclinic C2/c structure to 93.4° in the triclinic $C\bar{1}$ structure.

In Na-feldspar the C2/m structure is stable above 980°C and the $C\bar{1}$ structure forms below this temperature. As the structural collapse is controlled by the cation site size, the temperature of the $C2/m \Rightarrow C\bar{1}$ transition is very dependent on the cation content of the M site. The temperature at which the structure collapses increases sharply as the composition moves from Na-feldspar towards Ca-feldspar, and decreases sharply from Na-feldspar towards K-feldspar, shown schematically in Figure 6.56. In the Na,Ca feldspars the structure has the collapsed structure right up to the melting point, while in the K-rich feldspars the structure retains C2/m symmetry to well below room temperature. The $C2/m \Rightarrow C\bar{1}$ distortion is displacive and instantaneous.

2. Another tendency in aluminosilicate structures is to order the tetrahedral Al,Si at lower temperatures. By comparison with the structural distortion,

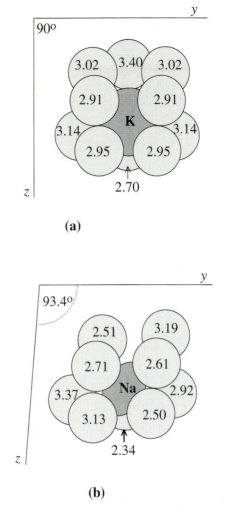

(a)

(b)

Figure 6.55. (a) In sanidine, $KAlSi_3O_8$, the oxygen coordination around K^+ is symmetrical and the unit cell is monoclinic (α = 90°). (b) In high albite, $NaAlSi_3O_8$, the oxygen coordination around Na^+ is distorted and the unit cell is triclinic (α = 93.4°). The position of the Na^+ ion is smeared over a number of positions and hence represented by an ellipsoid. The number on each oxygen atom is the cation – oxygen distance in Å. (After Liebau, 1985.)

Al,Si ordering is an extremely slow process, involving breaking very strong Si–O and Al–O bonds.

From the point of view of the structural change produced by ordering, we must note that the C2/m structure has only two independent sets of tetrahedral atoms, T_1 and T_2 both present in equal numbers. Therefore when the Al:Si ratio is 1:3 as in the alkali feldspars, it is not possible to distribute the atoms between these T sites in an ordered way. Any tendency to order Al into a particular T_1 site for example, with Si on the remaining T sites can only be accomplished by a reduction in symmetry which allows four independent T sites. This is shown in

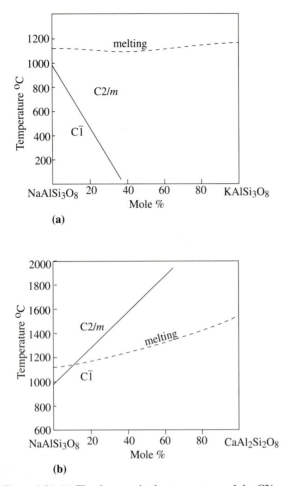

Figure 6.56. (a) The decrease in the temperature of the C2/m ⇒ C$\bar{1}$ distortion as more K$^+$ ions are present in the structure. **(b)** The increase in the temperature of the C2/m ⇒ C$\bar{1}$ distortion as more Ca^{2+} ions are present in the structure. In this latter case the transition is truncated by the melting curve and hence the C2/m structure only exists over a narrow temperature interval in albite-rich compositions.

Figure 6.57(a), which is a slightly different projection of a pair of crankshafts, allowing Al,Si distributions to be more clearly seen. Both the mirror planes and the diad axes are lost and the symmetry is reduced to triclinic, space group C$\bar{1}$, coincidentally the same as that for the structural distortion. Complete Al,Si order can be achieved within the C$\bar{1}$ structure with an Al:Si ratio of 1:3. In Figure 6.57(a) the T sites are relabeled T_{1o} , T_{1m}, T_{2o}, T_{2m} to indicate that pairs of sites which are now different were related by a mirror plane in the high symmetry form. The Al orders into the T_{1o} site resulting in a slight distortion of the structure since the [AlO$_4$] tetrahedron is slightly larger than the [SiO$_4$] tetrahedron.

Al,Si ordering in anorthite, CaAl$_2$Si$_2$O$_8$ with an Al:Si ratio of 1:1, is quite different from that in the alkali feldspars. The same symmetry problem does not exist, for it would be possible to put all the Al into T_1 sites and all the Si into T_2 sites. The fact that this does not happen is due to the unfavourable energetic effect of Al occupying neighbouring tetrahedral sites in aluminosilicates, as mentioned in Section 6.1. A different ordering scheme, in which there is a strict alternation of Al and Si in neighbouring tetrahedra is preferred, and is illustrated in Figure 6.57(b). This alternation of Al and Si results in a doubling of the *c* axis dimension of the unit cell.

3. Although we have described the structural distortion separately from the Al,Si ordering both processes are strongly dependent on one another. In the alkali feldspars for example, the way in which Al,Si ordering can take place in a C$\bar{1}$ collapsed structure in which four different T sites already exist, is very different from that in the expanded C2/m structure with only two distinguishable T sites. Conversely, albite with the C$\bar{1}$ structure should revert to the C2/m structure on heating above 980°C. However, if there is any degree of Al,Si ordering, the C2/m structure cannot form until all the order is also lost.

Thus although the displacive mechanism in itself is instantaneous, the symmetry change is controlled by the very much slower process of Al,Si ordering or disordering. The two processes are said to be *coupled* and the way the structure actually behaves on cooling and heating depends on the way in which these two component parts of the behaviour interact. A formal theory exists to deal with such behaviour and we will discuss this in a later chapter.

The alkali feldspars

Albite, NaAlSi$_3$O$_8$ is monoclinic C2/m above 980°C (*monalbite*) but collapses around the small Na atom to the triclinic C$\bar{1}$ structure below this temperature. At this stage while there is very little Al,Si order it is termed *high albite*, but below 700°C Al,Si ordering begins and can proceed without any further symmetry change until at low temperatures fully ordered *low albite* is formed. As noted above however, the structural collapse and the Al,Si ordering are strongly coupled, so that as the degree of order changes further structural collapse can take place, and vice versa. Al,Si ordering is promoted by the displacive distortion of the structure and so occurs at a higher

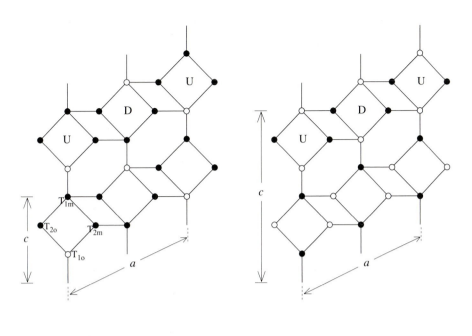

(a)   (b)

Figure 6.57. A perspective view of the crankshafts (shown from a different angle in Figure 6.54) in the feldspar structure showing the different Al,Si ordering schemes in **(a)** the alkali feldspars in which Al:Si = 1:3 and **(b)** anorthite, with Al:Si = 2:2. The anorthite ordering scheme results in a doubling of the c lattice parameter of the unit cell. Filled circles: Si; open circles: Al. The labels U and D in each figure refer to upward and downward facing squares formed by the centres of linked tetrahedra. (After Megaw, 1973.)

temperature than it would if there was no structural collapse possible.

Sanidine is the high temperature, monoclinic $C2/m$, form of $KAlSi_3O_8$. The large K atom prevents the displacive transformation and the structure remains monoclinic until Al,Si ordering begins at about 450°C. This is a very low temperature for a process which involves breaking strong Si–O and Al–O bonds, and hence ordering proceeds at a very slow rate. In natural K-feldspars, various degrees of Al,Si order can be found, depending on the thermal history of the rock. To fully appreciate the nature of these intermediate states of order we need to understand some concepts which will be covered in Section 7.3.5 when K-feldspars will be discussed again. However, at this stage we can make some general observations.

1. Ordering within the $C2/m$ structure involves a discrimination of T sites. The extreme slowness of the ordering process, and the fact that any one part of a crystal is not in direct communication with another, i.e. the forces which order Al and Si are local, leads to the possibility that some order can be achieved in a crystal on a local scale while the average symmetry of whole crystal remains mono-

clinic. This is the case in *orthoclase*, a form of K-feldspar from rocks which cooled at a moderate rate.

2. Al,Si ordering on a local level eventually results in a macroscopic symmetry change to the triclinic $C\bar{1}$ *microcline* structure. A sequence of increasing order from intermediate microcline to maximum microcline is recognised in terms of a gradual change in the shape of the unit cell from monoclinic (with interaxial angles α and γ at 90°) to triclinic (maximum microcline has $\alpha = 90.7°$, $\gamma = 87.7°$). The relationship between orthoclase and microcline is essentially due to the scale over which Al,Si order is correlated in a crystal.

3. Although the symmetry change in K-feldspar is driven by Al,Si ordering rather than a displacive collapse as in Na-feldspar, a coupling between Al,Si ordering and the tendency for the framework to collapse needs to be invoked to explain the detailed behaviour of K-feldspar. In other words, increasing the degree of Al,Si order allows the structure to distort.

There is complete solid solution with $C2/m$ structure between $NaAlSi_3O_8$ and $KAlSi_3O_8$ at high temperatures, but as the temperature is decreased, the difference between the size of Na and K, as well as the different structural behaviour of the two end members, leads to reduced miscibility. At temperatures below about 300°C there is virtually no Na – K substitution possible in alkali feldspars. Alkali feldspars in which the crystals show intergrowths of Na-rich and K-rich regions which form by unmixing from a cooling solid solution are termed *perthites*. These will be discussed in Section 11.3.4

The plagioclase feldspars

In the plagioclase feldspars, between albite $NaAlSi_3O_8$ and anorthite $CaAl_2Si_2O_8$ the structure must combine the complexities of the structural changes of the end members, with a solid solution in which the Si:Al ratio varies. In this section we are only concerned with the principal structure types and a general appreciation of the structural problems encountered by these minerals on cooling, and a more detailed account will be given in Chapter 12.

In pure anorthite $CaAl_2Si_2O_8$, the small Ca atom is unable to support the expanded monoclinic C2/m structure at any temperature below the melting point. Similarly, the tendency to order Si and Al in the tetrahedra is greater than in the alkali feldspars, since the Si:Al ratio of 1:1 means that in a framework structure any amount of disorder results in the formation of Al–O–Al linkages. As Al occupancy in adjacent tetrahedra is energetically unfavourable, framework structures with Si:Al ~1 have to be heated to very high temperatures before total disorder occurs. In pure anorthite, this disordering temperature is estimated at above 2000°C, well above the melting point. Therefore, anorthite has an essentially Al,Si ordered structure, which as described above, involves the doubling of the *c* axis of the unit cell relative to that of albite. Ordered anorthite is triclinic with space group $I\bar{1}$. Partial disordering is allowed in anorthite while still retaining the $I\bar{1}$ symmetry, and up to 20% disorder may be induced by annealing anorthite just below its melting point. A completely disordered anorthite structure would have the same symmetry and unit cell as high albite, i.e. $C\bar{1}$.

Below ~240°C there is a further structural collapse in anorthite, distinct from the high temperature collapse described for the albite end member. This low temperature, displacive transition reduces the symmetry still further, to $P\bar{1}$, and involves a restriction in the size of the Ca^{2+} site.

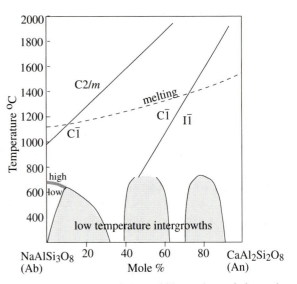

Figure 6.58. Summary of the stability regions of the various feldspar structures in the plagioclases. At high temperature the phase boundaries between C2/m, $C\bar{1}$ and $I\bar{1}$ structures are truncated by the melting curve. At low temperature the shaded areas represent regions in which the plagioclase consists of intergrowths barely visible by optical microscopy. (After Carpenter, 1987.)

There is complete solid solution between albite (An_0) and anorthite (An_{100}) above about 700°C, the result of the substitution $Na^+ + Si^{4+} = Ca^{2+} + Al^{3+}$. Over most of the compositional range (Figure 6.58), the solid solution has the collapsed $C\bar{1}$ structure with no Al,Si order i.e. the high albite structure, while on the anorthite rich side the solid solution has the $I\bar{1}$ structure due to the increasing tendency to order Al,Si as the Al:Si ratio approaches 1. Only in pure anorthite however, is it possible to achieve complete order with strictly alternating Al,Si.

At low temperatures the plagioclases appear, on an optical microscope scale, to be homogeneous across the whole compositional range. On an electron microscope scale however, a crystal of plagioclase is made up of intergrowths between regions with different composition and different Al,Si ordering schemes. As in all mineral behaviour, the structures formed in the intermediate plagioclases represent an attempt by the system to minimise the energy by a reorganisation of the framework and by ordering the atomic positions. Attempting to find an ordering scheme compatible with a changing Al:Si ratio as well as keeping the strain energy in the framework at a minimum involves the formation of a new kind of structure, termed an *incommensurate structure* which will be discussed in Section 12.3.4. Here we will simply outline the nature of the prob-

lem and some of the observations in natural plagioclases.

The behaviour of the plagioclase solid solution on
cooling is complicated by two factors.

1. The distribution of Al and Si must be linked to
that of Ca and Na by the need to maintain local
charge balance. Any Ca,Na ordering would have to
be compatible with the Al,Si ordering scheme.

3. The Al,Si ordering schemes in $NaAlSi_3O_8$ and
$CaAl_2Si_2O_8$ are fundamentally different and there is
no simple ordering scheme in the feldspar structure
for Al:Si ratios between 1:3 and 2:2. CaAl substitution for NaSi into an ordered albite structure or the
reverse substitution into ordered anorthite necessarily introduces disorder. In the case of albite it also
results in the formation of Al–O–Al linkages, which
may explain why the albite ordering scheme tends to
be restricted to albite-rich compositions (Figure
6.58), while the $I\bar{1}$ anorthite ordering scheme can
exist over a wider compositional range.

If pure albite and pure anorthite are the only two
fully ordered structures possible, it could be argued
that the most stable state for a plagioclase at any
intermediate composition might be to unmix into an
intergrowth of albite + anorthite. This is not observed to be the case in natural plagioclases. There are
two possible explanations: it may be due to the
extreme sluggishness, even on a geological time
scale, of such a complete reconstruction of the Al,Si
framework and the diffusion of Al and Si at temperatures below 700°C. Alternatively, there may be an
energetically more favourable structural solution.

The observations on natural plagioclase crystals
indicate that they are not homogeneous but are
made up of three kinds of intergrowths, all dominated by the existence of ordered intermediate structures, referred to above as incommensurate. These
complex structures have the basic $C\bar{1}$ structure on
which is superimposed a periodicity which is not an
integral multiple of the unit cell. This periodicity,
with a wavelength of ~50Å, describes a modulation
in the degree of Al,Si and Na,Ca ordering through
the crystal, and represents a compromise ordering
scheme, satisfying local charge balance requirements
as well as having the flexibility to adjust to changing
a Al:Si ratio by altering the wavelength of the
modulation. There are two such incommensurate
structures recognized in feldspars, here labelled e_1
and e_2.

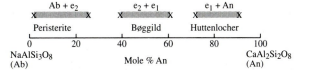

Figure 6.59. The three types of intergrowths in low temperature
plagioclase feldspars. The shaded regions indicate the bulk compositions in which the intergrowths occur, and the crosses indicate
the compositions of the two coexisting phases in the intergrowth.

Figure 6.59 summarizes the three types of intergrowths present at different bulk compositions in
natural plagioclases:

1. The central region at bulk compositions
between An_{40} and An_{60} is made up of a lamellar
intergrowth, on a scale of about 0.1μm, of two
incommensurate structures. These are the e_1 and e_2
structures and they differ in composition by up to
20% An. This is termed the Bøggild intergrowth.

2. The albite-rich plagioclases between about An_2
and An_{20} occur as lamellar intergrowths of pure
ordered albite and the e_2 structure plagioclase. This
is termed the peristerite intergrowth.

3. The anorthite-rich plagioclases between about
An_{65} and An_{90} occur as lamellar intergrowths of
ordered anorthite and the e_1 structure plagioclase.
This is termed the Huttenlocher intergrowth.

The slow kinetics of any process involving the migration of Al and Si at these temperatures makes the
interpretation of these intergrowths in natural intermediate plagioclases difficult. In particular, to what
extent do these intergrowths represent free energy
minima? The interplay between the need to reduce
the energy of the system (thermodynamics) and the
rate at which this can occur (kinetics) becomes very
important in the interpretation. Only an understanding of these factors at quite a detailed level will
enable us to unravel the complexities of the plagioclases.

6.8.4 Cordierite

The final framework silicate we discuss in this chapter is cordierite, $(Fe,Mg)_2Si_4Al_5O_{18}$. As well as
being an important mineral in metamorphic rocks, it
provides a good example of the ways in which Al,Si
ordering can be studied in a structure, and the effect
of this ordering on the thermodynamics. This in turn
leads to an evaluation of the importance of the state

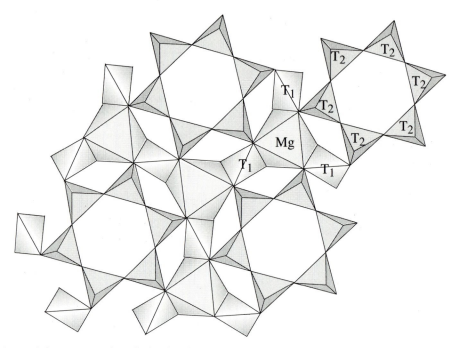

Figure 6.60. One layer of the structure of cordierite showing the six-fold rings of tetrahedra, labelled T_2 connected via the tetrahedra labelled T_1. The Mg^{2+} cations lie in octahedral sites.

of Al,Si order in cordierite involved in metamorphic reactions where the appearance of cordierite is an important indicator of metamorphic grade. In Section 12.4.1, cordierite is discussed in some depth, as part of a case study. Here we introduce its structure and polymorphism.

The structure of cordierite is unusual for a framework since topologically it has the features of a ring silicate and has been classified as such in the past. It is based on six-fold rings of Al,Si tetrahedra (termed the T_2 tetrahedra) joined laterally and vertically by T_1 tetrahedra, which also may contain Al or Si. The Mg (or Fe) cations occupy octahedral sites between the rings. Figure 6.60 shows one layer of this structure and the connectivity of the tetrahedra. Layers are stacked vertically above one another so that the rings form infinitely long channels parallel to the c axis. Figure 6.61 shows how the T_1 tetrahedra (shaded) connect the rings to form these channels. (See Figures 3.35 and 3.36 for high resolution transmission electron micrographs of the cordierite structure.)

In each unit cell there are 9 tetrahedra: $3T_1$ and $6T_2$ tetrahedra over which the 4 Si and 5 Al atoms have to be distributed. If we disorder them randomly over all 9 sites, the structure is hexagonal and each

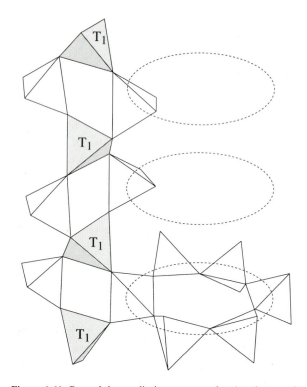

Figure 6.61. Part of the cordierite structure showing the way the T_1 tetrahedra (shaded) connect the T_2 rings to form channels. The unshaded tetrahedra belong to the T_2 rings. Each T_2 tetrahedron is joined to 2 T_1 and 2 T_2 nearest neighbour tetrahedra in the framework, while each T_1 tetrahedron is connected to 4 T_2 tetrahedra.

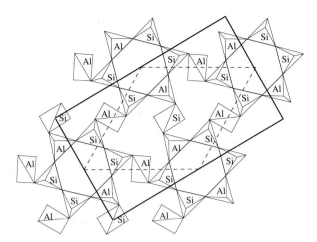

Figure 6.62. The distribution of Si,Al in ordered cordierite. The hexagonal cell of the disordered structure is dashed, and the orthorhombic cell of the ordered structure is shown in full line.

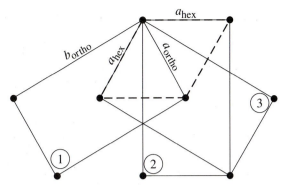

Figure 6.63. Three equivalent orientations of the orthorhombic cell derived from the hexagonal cell (dashed line). The black spots are lattice points which are common to both cells. The three orientations of the orthorhombic cell can be related to the three different pairs of T_2 tetrahedra which are occupied by Al (see Figure 6.62).

site has occupancy $4/9$ Si, $5/9$ Al. However there is another virtually random distribution we could consider, and that is that the occupancy of T_1 and T_2 sites is different, but that within these sites the cations are disordered. This is found to be the case in hexagonal high temperature cordierite. The 3 T_1 sites contain 2Al and 1Si, randomly distributed, while the other 2 Al and 4Si are randomly distributed over the 6 T_2 sites. In this disordered state cordierite is hexagonal with space group P6/*mcc*. Natural hexagonal cordierite is termed indialite.

Hexagonal cordierite is only stable above about 1450°C, indicating the very strong tendency for Al,Si ordering. During Al,Si ordering an opposite pair of T_2 tetrahedra are occupied by the 2 Al atoms, destroying the six-fold axis and reducing the symmetry to orthorhombic, space group C*ccm*. The distribution of Al,Si in the T_1 sites is also ordered such that no Al–O–Al bonds exist in the fully ordered structure (Figure 6.61).

The relationship between the hexagonal and orthorhombic unit cells is shown in Figure 6.62. Since the orthorhombic unit cell has a C-face-centred lattice, the ordering transformation does not change the volume of the unit cell *per lattice point*, and hence does not involve the formation of a superlattice. The orthorhombic structure is also slightly distorted relative to the hexagonal cell, so that the angle between the original *a* axes is no longer exactly 60°.

There are three equivalent orientations of the orthorhombic unit cell which could form from the

hexagonal structure, and in terms of Al,Si distribution this corresponds to the 2 Al atoms in the T_2 ring sites occupying one of the three pairs of opposite tetrahedra (Figure 6.63). Since these are all equivalent possibilities, we would expect to find twinned crystals as a result of ordering.

Although natural cordierites crystallise well within the stability field of the ordered orthorhombic form, there is evidence that metastable Al,Si disorder may be an important factor in controlling the early nucleation and growth of natural cordierite. From an experimental point of view, cordierite is an ideal mineral in which to study the effects of Al,Si ordering, since the high ordering temperature allows ordering experiments to be carried out at temperatures where the kinetics of the process are relatively fast. The synthesis of samples with well-defined degrees of Al,Si order allows the study of the thermodynamics, kinetics and mechanism of ordering to be carried out. Many of the results of such studies can be applied to other mineral groups where low order–disorder temperatures make experiments kinetically impossible.

Bibliography

LIEBAU, F. (1985). *Structural chemistry of silicates*. Springer-Verlag.

MEGAW, H. (1973). *Crystal structures: a working approach*. W.B. Saunders Co.

Reviews in mineralogy published by the Mineralogical Society of America provides concise review articles on most of the mineral groups discussed in this chapter.

Volume 2: *Feldspar mineralogy*, 2nd Edition 1983. P.H.Ribbe (Ed)

Volume 5: *Orthosilicates*, 2nd Edition 1982. P.H.Ribbe (Ed)

Volume 6: *Marine minerals*, 1979. R.G.Burns (Ed) Includes chapters on silica polymorphs and clay minerals.

Volume 7: *Pyroxenes*, 1980. C.T.Prewitt (Ed)

Volume 9A: *Amphiboles and other hydrous biopyriboles – mineralogy.* 1981. D.R.Veblen (Ed)

Volume 13: *Micas*, 1984. S.W.Bailey (Ed)

Volume 19: *Hydrous phyllosilicates (exclusive of micas)*, 1988. S.W.Bailey (Ed)

References

AKIMOTO, S. AND SYONO, Y. (1972). High pressure transformation in MnSiO₃. *Amer. Mineral.* **57**, 76–84.

BURNHAM, C.W. (1966). Ferrosilite. *Carnegie Inst. Washington Year Book*, **65**, 285–90.

BUSECK, P.R., NORD, G.L. AND VEBLEN, D.R. (1982). Subsolidus phenomena in pyroxenes. In: *Pyroxenes* C.T.Prewitt (Ed) *Reviews in Mineralogy* Vol. 7. Mineralogical Society of America.

BUSECK, P.R. AND VEBLEN, D.R. (1981). Defects in minerals as observed with high resolution transmission electron microscopy. *Bull. Mineral.* **104**, 249–60.

CAMERON, M. AND PAPIKE, J.J. (1982). Crystal chemistry of silicate pyroxenes. In: *Pyroxenes* C.T. Prewitt (Ed) *Reviews in Mineralogy* Vol. 7. Mineralogical Society of America.

CARPENTER, M.A. (1987). Thermochemistry of Al/Si ordering in feldspar minerals. In: *Physical properties and thermodynamic behaviour of minerals*. E.K.H. Salje (Ed.) NATO ASI Series C. D.Reidel Publishing Co.

GIBBS, G.V., MEAGHER, E.P., NEWTON, M.D. AND SWANSON, D.K. (1981). A comparison of experimental and theoretical bond length and angle variations for minerals, inorganic solids and molecules. In: *Structure and Bonding I.* M. O'Keeffe and A. Navrotsky (Eds) Academic Press.

HENMI, C., KAWAHARA, A., HENMI, K., KUSACHI, I. AND TAKEUCHI, Y. (1983). The 3T, 4T and 5T polytypes of wollastonite from Kushiro, Japan. *Amer. Mineral.* **68**, 156–63.

HUTCHISON, J.L. AND MCLAREN, A.C. (1976). Two dimensional images of stacking disorder in wollastonite. *Contribs. Mineral. Petrol.* **55**, 303–9.

LIEBAU, F.(1985). *Structural chemistry of silicates*. Springer-Verlag.

MEGAW, H. (1973). *Crystal Structures: a working approach*. W.B. Saunders Co.

NOVAK, G.A. AND GIBBS, G.V. (1971). The crystal chemistry of the silicate garnets. *Amer. Mineral.* **56**, 791–825.

PAPIKE, J.J. AND CAMERON, M. (1976). Crystal chemistry of silicate minerals of geophysical interest. *Rev. Geophysics and Space Physics*, **14**, 37–80.

THOMPSON, J.B. Jr. (1981). An introduction to the mineralogy and petrology of the biopyriboles. *Reviews in Mineralogy* Vol 9A Mineralogical Society of America.

VEBLEN, D.R. AND BUSECK, P.R. (1979). Chain-width order and disorder in biopyriboles. *Amer. Mineral.* **64**, 687–700.

VEBLEN, D.R. AND BUSECK, P.R. (1980). Microstructures and reaction mechanisms in biopyriboles. *Amer. Mineral.* **65**, 599–623.

7 Defects in Minerals

In an ideal crystal structure every unit cell is identical and each has a specified shape, size and cell content. The structure represents the chemical composition and atomic arrangement which defines its stability in a particular environment in the Earth. However, in a real mineral structure there are always some local violations or defects in this perfect arrangement (e.g. Figure 7.1). We have mentioned some of these in previous chapters, e.g. a structure will contain some point defects such as vacant sites, or one atom may be replaced by an impurity atom. There may also be some disorder among cation sites, such as Al,Si disorder in many aluminosilicates. Another class of defect we have encountered includes stacking mistakes in silicates, e.g. pyroxenes in Section 6.4.1. In all of these cases the presence of the defects will affect the properties of the mineral.

The importance of defects however, is that their effect on many mineral properties by far outweighs their concentration. The reactivity, and the way that atoms can diffuse through a solid mineral is controlled by the presence of vacant sites in a structure. Hence the response of a mineral structure to changes in its physical and chemical environment is controlled by these *point defects* which affect perhaps only one unit cell in every 10 000. Point defects are also responsible, in many cases, for the colour of gemstones when impurity atoms are present in an otherwise perfect structure (e.g. red rubies, green emeralds and blue sapphires are all due to "impurity" atoms). The deformation of minerals (i.e. the response to an external stress) is entirely controlled by the way in which *line defects*, termed dislocations, are generated and move through a crystal. On a geological scale the deformation of rocks builds mountains and keeps continents moving.

Planar defects are two-dimensional surfaces within a crystal structure, where the perfect structure on one side of the defect is related to that on the other

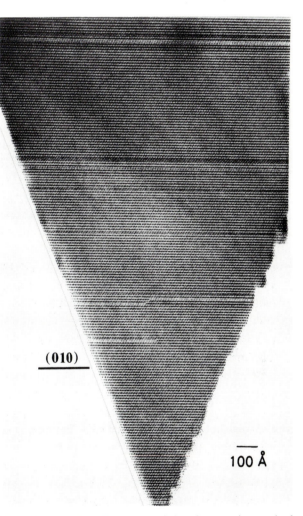

Figure 7.1. A high resolution transmission electron micrograph of a pyroxenoid mineral (babbingtonite) which has a structure similar to rhodonite with periodicity along the silicate chain of 5 (see Section 6.4.2). The planar defects on (010) planes are mistakes in the periodicity along the chain. (Micrograph courtesy of M. Czank)

by a displacement or a rotation element of symmetry. There are many different types of planar defects, which are not always strictly planar but may be curved and quite complex internal surfaces. The movement of such defects may control the mechanism by which a crystal structure undergoes a solid state transformation to a different structure, and during the actual transformation such defects may be very numerous. Those left in a mineral represent the frozen-in remnants of this transformation and record aspects of the transformation mechanism.

The surface of a mineral itself is the most obvious yet commonly overlooked defect. The boundary between a crystal and the surrounding medium, which may be a fluid or gas phase, or a different solid, is the termination of the regular atomic bonding pattern and produces an imbalance of forces that results in a surface tension. Since a grain boundary relates crystals with different orientation (and possibly different structure and composition) it falls into a different category from the defects discussed here, and is not easily describable by traditional crystallographic techniques. Nevertheless, the importance of grain boundaries as sites for chemical interchange and reaction as well as pathways for diffusion makes an understanding of their structure and chemistry an important aspect of mineralogy. With our present knowledge of the nature of surfaces in complex structures, this topic is beyond the scope of this book.

Defects are generally classified in terms of their dimensionality – hence point defects, line defects and planar defects. In this chapter we will discuss the various types of defects in turn, pointing out some aspects of their significance in relation to processes in minerals.

7.1 Point defects

All crystals at temperatures above absolute zero contain some point defects which are generally vacant atomic sites or substituted impurity atoms. Up to a certain concentration, the presence of point defects actually reduces the free energy G of a crystal and is therefore favourable. To understand why this is so we need to discuss the concept of free energy in a little more detail, and this is an opportunity to introduce some basic aspects of thermodynamics which will be discussed more fully in the following chapter.

The Gibbs free energy, G, of a crystal is that thermodynamic quantity which is a minimum when a crystal is in equilibrium with its surroundings. It is defined by the expression

$$G = H - TS \qquad (7.1)$$

where the enthalpy, H, is the internal energy at constant pressure. The internal energy is the sum of all the electrostatic energy terms due to interatomic forces and the kinetic energy terms due to vibrational motions. The entropy, S, is a measure of the state of disorder in the crystal and T is the temperature in kelvins.

The creation of a vacancy or other point defect requires energy and hence increases the enthalpy of the crystal. The structure is locally distorted and the bonding requirements are less well satisfied. This increase in the enthalpy, ΔH, however, is accompanied by an increase in the entropy ΔS, since the defects increase the disorder in an otherwise perfect crystal. The value of the entropy change is associated with the number of ways of distributing defects within the structure, and for small defect concentrations the entropy increase is greater than the enthalpy increase. The change in free energy, ΔG, due to the defects is thus:

$$\Delta G = \Delta H - T\Delta S \qquad (7.2)$$

For small defect concentrations, the $-T\Delta S$ term dominates at all temperatures above 0K, and the change in free energy ΔG is negative. In Figure 7.2, the minimum in the G-curve represents the equilibrium defect concentration at temperature T.

The enthalpy and entropy terms are not in themselves very temperature dependent, and so at higher temperatures the $T\Delta S$ term becomes greater and the free energy minimum occurs at a higher defect concentration. This is in accord with our general expectation that crystal structures will have more

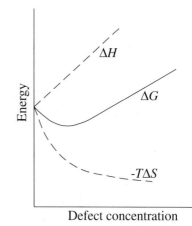

Figure 7.2. The change in free energy ΔG with defect concentration. Both the enthalpy ΔH and the entropy ΔS increase, but the resultant free energy has a minimum corresponding to the equilibrium defect concentration at the particular value of T.

defects and tend to more disordered states at higher temperatures. Thus at higher temperatures there are more vacancies, a greater accommodation of impurity atoms, and a greater degree of cation disorder (e.g. Al,Si disorder in aluminosilicates).

This very simplified analysis of the situation is at this stage sufficient to enable us to see the general form of the free energy versus defect concentration curve and to appreciate why point defects occur. We will return to the theme of the energy of minerals in more detail in the next chapter.

7.1.1 Schottky defects and Frenkel defects

The nature of point defects in complex mineral structures is not very well understood, and most of our basic ideas come from studies in simpler compounds such as halides, e.g. NaCl, AgCl. It is worth outlining the main types of point defects as an introduction. The simplest case is one in which a vacant cation site in a structure is balanced by a vacant anion site to maintain electrical neutrality. Such a pair of vacancies is called a *Schottky defect*, although the two vacancies are not necessarily directly associated with each other in any way. In a crystal of NaCl for example a pair of vacancies such as shown in Figure 7.3 is equivalent to one Schottky defect. The situation shown in Figure 7.3 is schematic; in practice the vibrating atoms will accommo-

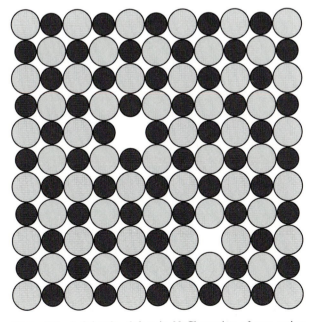

Figure 7.3. A Schottky defect in NaCl consists of one cation vacancy and one anion vacancy.

date the defect to some extent by moving towards the vacant site, or away from it, depending on the type of bonding. This *relaxation* of the structure minimizes the enthalpy associated with the vacancy.

We can calculate the equilibrium number of Schottky defects in a crystal of MX type (e.g NaCl) at any temperature if we can evaluate the enthalpy change ΔH and the entropy change ΔS associated with defect formation. The derivation is outlined in Appendix A to this chapter, and results in the expression

$$n_s = N e^{-\Delta H/2RT} \qquad (7.3)$$

where n_s is the equilibrium number of Schottky defects at temperature T, N is the total number of cation and anion sites and R is the gas constant (8.3 JK^{-1} mol^{-1}). In NaCl, the value of ΔH is about 220 kJ mol^{-1} and hence the fraction of sites vacant due to Schottky disorder at 300K is 1.6×10^{-19}. At 1000K this fraction increases to 2.3×10^{-6} or about 1 in 430 000 atoms, i.e. in any one direction in the crystal about 1 in 75 atomic positions are unoccupied. These figures are given to indicate the magnitude of the equilibrium defect concentrations in simple MX compounds. (N.B. If we were considering single vacant sites, rather than the pairs which define a Schottky defect, the expression above would be modified to $n_v = N e^{-\Delta H/RT}$, where n_v is the equilibrium number of vacancies.)

It is worth noting that if a material is heated to high temperature and allowed to achieve its equilibrium defect concentration, and then rapidly cooled (i.e. *quenched*) to room temperature the defect concentration will not, in general, have time to reequilibrate. A high defect concentration can be frozen-in in this way and subsequently affect any material properties measured at room temperature.

Another possible type of point defect occurs when an atom moves from a site, leaving a vacancy, and is placed in an alternative *interstitial* site which is normally unoccupied. This is a *Frenkel defect* and again leaves the overall charge balance and stoichiometry unaffected. Figure 7.4 shows a schematic diagram of a Frenkel defect in AgCl which also has the NaCl structure, but in which Frenkel defects have a lower enthalpy of formation than Schottky defects. The calculation of the equilibrium number of Frenkel defects is analogous to that for Schottky defects, except that the total number of lattice sites and available interstitial sites need not be the same.

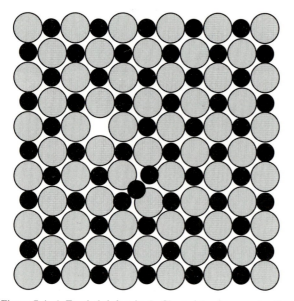

Figure 7.4. A Frenkel defect in AgCl consists of a vacant cation site, with the Ag$^+$ ions occupying an interstitial position, not normally occupied in the ideal structure.

The general form of the equations and the overall temperature dependence of the defect population are the same.

Schottky and Frenkel defects are referred to as *intrinsic defects* since their numbers are controlled by intrinsic properties of the structure, related to the size of the interatomic forces. The presence of impurities, variations in oxidation state etc. also constitute defects which are *extrinsic* and can vary from one crystal to another.

7.1.2 Impurity atoms and atomic substitutions

Minerals are never pure compounds and may contain a wide variety of substituted atoms. We can treat these substitutions in much the same way as defects, in that there will be an equilibrium concentration of impurity atoms which will depend on the enthalpy and entropy associated with the substitution. This is usually called the *enthalpy and entropy of mixing*. Some minerals only allow small deviations from their pure end-member composition, e.g. quartz SiO_2, while in others there is a complete solid solution between two extreme compositions, e.g. in the olivines the composition can range from Mg_2SiO_4 (forsterite) to Fe_2SiO_4 (fayalite). As long as the increase in the $T\Delta S$ term outweighs any increase in the enthalpy due to the substitution, a solid solution will be thermodynamically stable. The temperature

dependence of the impurity concentration is the same as for other forms of disorder: the higher the temperature, the greater the substitution possible. We will return to the very important topic of solid solutions in minerals in the next chapter.

7.1.3 Point defects as colour centres in crystals

Anion vacancies, such as chlorine vacancies in NaCl, are regions in which there is locally more positive than negative charge. An electron in the vicinity of the anion vacancy can feel the influence of this positive charge and may become bound to it. An electron trapped in this way will have a series of energy levels available to it, and transitions between energy levels may be in the visible part of the electromagnetic spectrum. The defect then acts as a colour centre, or F-centre (*Farbenzentrum*, in German) and imparts a characteristic colour to the crystal.

In NaCl, F-centres can be produced by heating a crystal in sodium vapour. This causes the diffusion of sodium into the crystal and increases the number of anion vacancies. The NaCl crystal becomes a greenish yellow. KCl treated in this way becomes blue. Various kinds of defect centres have been recognized in alkali halides, and provide an interesting source of phenomena which have been extensively studied by ESR spectroscopy (Section 4.3). The bound electron has unpaired spin and therefore an electron paramagnetic moment which can be detected by ESR.

Electrons can be generated in other minerals by high energy radiation, and these electrons then migrate to defects where they are trapped, and may produce colour. The nature of the defects is not easily determined from the often complex ESR and optical spectra which have been used to study them. In quartz, the most widely studied mineral, the point defects are associated with oxygen vacancies, or with the trace substitution of trivalent cations for Si^{4+} (e.g. Al^{3+}, Fe^{3+} or Ti^{3+}). Artificial colours can be induced in slightly impure quartz by irradiation. Blue feldspars, green diamonds and blue calcite are other examples of this process. In nature, pleochroic haloes are produced in some minerals when they are irradiated by tiny inclusions of radioactive minerals such as zircon or allanite.

The presence of trace concentrations of transition metal cations in otherwise colourless minerals can also impart colour if the electronic energy levels of the impurity cation are to provide transitions in the appropriate visible spectral range (Section 4.5).

7.1.4 Point defects and diffusion

The existence of point defects was first proposed to attempt to explain the bulk diffusion of atoms and ions through simple crystal structures, notably metals and alkali halides. In such structures it is clear that for an atom to move it must have a vacant site to move into, or it must occupy an interstitial site before it can move on into a neighbouring one. Figure 7.5 shows some of the atomic mechanisms of diffusion which have been proposed in a simple idealized structure, and it is possible to calculate their relative probabilities from estimates of the interatomic forces. Vacancy mechanisms are always the most likely, and despite the low concentration of point defects, they play a major role in controlling diffusion.

The ultimate aim of theoretical diffusion studies is to be able to relate the measured bulk transport properties such as diffusion and conductivity to an atomic mechanism of defect motion. In complex mineral structures this is much more complicated than in alkali halides, mainly since the number of possible defect types is greater, and more than one

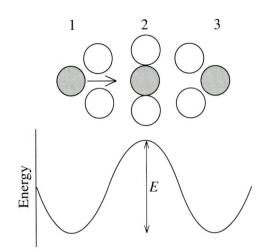

Figure 7.6. Three stages in the migration of the shaded atom from one site to another. The starting position (1) and the final position (3) are equilibrium positions with free energy minima. The intermediate position involves lattice distortion and a maximum in energy, giving rise to an energy barrier.

type may be operating in the crystal at any one time. Whatever the specific mechanism, each time an atom moves from one site to another it has to overcome an energy barrier, which represents the less stable position which it has to occupy between the two more stable sites. Figure 7.6 shows this in terms of an atom squeezing through a narrow gap to move into a neighbouring vacant site.

The energy barrier, E, which has to be overcome for an atom to move from one site to another is termed the activation energy for diffusion. The energy required to jump over this barrier comes from the thermal motion of the atoms. The probability, p, that an atom has energy E, relative to the mean thermal energy kT (k is Boltzmann's constant), is solved by statistical thermodynamics, from which the standard Boltzmann distribution expression is

$$p = \exp\left(-E/kT\right) \qquad (7.4)$$

More often we deal with mass transfer of atoms, in which case the equivalent expression becomes

$$p = \exp\left(-E/RT\right)$$

where E is the energy per mole and R is the gas constant.

The diffusion rate, D, will be proportional to this probability, so that we may write

$$D = c \cdot \exp(-E/RT) \qquad (7.5)$$

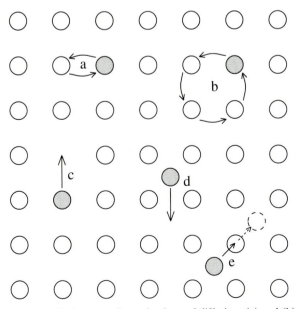

Figure 7.5. Various atomic mechanisms of diffusion. (a) and (b) are exchange mechanisms without involving vacancies, (c) is a vacancy migration mechanism, (d) and (e) are interstitial migration mechanisms.

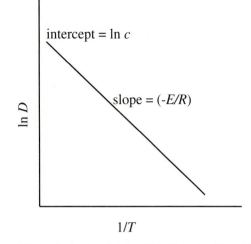

Figure 7.7. An Arrhenius plot, in which the natural logarithm of the rate, in this case ln D where D is the diffusion rate, is plotted against $1/T$, where T is the temperature in kelvins. In thermally activated processes this gives a straight line, the slope determined by the activation energy E for the process.

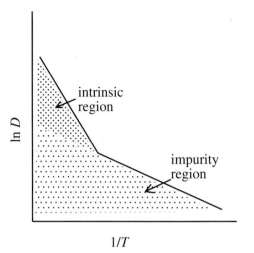

Figure 7.8. An Arrhenius plot of the diffusion rate as a function of temperature may have two straight-line segments. At low temperatures, diffusion is dominated by impurities and other extrinsic defects which migrate with a low activation energy. At higher temperatures the vacancy concentration increases, and the migration of these intrinsic defects, associated with a higher activation energy, is the dominant mechanism of bulk diffusion.

where the pre-exponential constant, c, will contain terms such as v, the atomic vibrational frequency (the atom makes v attempts per second at surmounting the energy barrier, each time with a probability p of success), the number of defects in the crystal, and the distance between atomic sites. Each atomic mechanism will have a different activation energy, and so by measuring the diffusion rate it may be possible to identify the predominant mechanism.

The activation energy for any thermally activated process can be determined from experimental rate data at a number of temperatures. A graph, known as an *Arrhenius plot*, of the natural logarithm of the rate (ln Rate) is plotted against the reciprocal of the temperature in kelvins ($1/T$). In the case of diffusion in equation (7.5) above,

$$\ln D = \ln c - (E/RT). \qquad (7.6)$$

Thus the slope of the Arrhenius plot is $(-E/R)$ as shown in Figure 7.7.

Arrhenius plots of the bulk diffusion rate through different crystals generally fall into two straight line segments (Figure 7.8). This shows that there are two different mechanisms operating in the different temperature regimes. The low temperature part has a lower activation energy, while at higher temperatures a different process, with a higher activation energy is predominantly responsible for diffusion.

The explanation for this is that at low temperatures, the number of intrinsic defects (such as Schottky or Frenkel defects) is low, and diffusion is due to extrinsic impurities etc. in the crystal. The number of these extrinsic defects is fixed, and independent of temperature. As the temperature increases the number of intrinsic defects increases exponentially (eqn.7.1) until at some temperature they take over as the main diffusion mechanism. In this high temperature regime the measured activation energy is now the sum of the activation energy required to produce the intrinsic defect plus the activation energy for diffusion.

In olivine, $(Fe,Mg)_2SiO_4$, for example this break in the slope of the Arrhenius plot occurs at around 1125°C, above which the activation energy for Fe,Mg diffusion is about 260 kJ mole^{-1}, and below which it is around 125 kJ mole^{-1}. From our simple analysis we could estimate that the activation energy for the formation of intrinsic defects responsible for Fe,Mg diffusion in olivine is therefore about 135 kJ mole^{-1}.

7.1.5 Defects and non-stoichiometry

Schottky defects and Frenkel defects are both stoichiometric, i.e. they do not alter the cation: anion ratio. Some compounds however, can accommodate relatively large numbers of vacant sites, especially at high temperatures, and this has the effect of significantly changing the nominal cation: anion ratio. For example, the stoichiometry of a compound such as FeS can be changed very easily at high temperature by exposure to a vacuum or to a high sulphur vapour pressure. The composition, written $Fe_{1-x}S$, is that of the mineral pyrrhotite (see Section 5.2.4) which at high temperatures can exist with $0 < x < 0.125$, and a random distribution of vacant cation sites. At such large concentrations, the defects interact with each other, and there is a tendency for them to order in some way. At high temperatures this tendency may be overcome by the entropy gain due to disorder and hence the vacancy disordered state may be stabilized. In pyrrhotite at low temperatures the defects order in a regular way, driven by the electrostatic repulsion between them (Section 5.2.4).

Another classic example is that of wustite, $Fe_{1-x}O$ which has the NaCl structure with vacancies in the Fe positions. When this non-stoichiometric oxide is quenched rapidly from high temperatures in an attempt to freeze in the defect distribution, analysis of diffraction data shows that the vacancies aggregate, or form clusters, rather than being randomly distributed. These clusters bear a strong resemblance to fragments of the spinel structure of Fe_3O_4, the next higher oxide, and hence non-stoichiometric $Fe_{1-x}O$ could be interpreted as a partly ordered structure with 'microdomains' of Fe_3O_4. Calculations of the lattice energies for various defect distributions also confirm that the defect clusters are more stable than random defects.

Both of these examples suggest that while large departures from stoichiometry can be described by random defects at high temperature, as the temperature is decreased there will be a tendency to assimilate the defects into some type of ordered structure. There are many examples of such behaviour in minerals, and we will meet some of these in later chapters.

Another very important aspect of the non-stoichiometry in the two examples above is that the defect concentration will depend on the partial vapour pressure of the volatile constituent, in this case sulphur or oxygen. For example, we may write:

$$(1-x)\, FeO + x/2\, O_2 \Leftrightarrow Fe_{1-x}O$$

The extra oxygen atoms bond to the surface of the original FeO and Fe atoms diffuse from the bulk of the crystal to combine with these surface oxygens, leaving behind Fe vacancies. The extent to which non-stoichiometry can be accommodated depends on the energy of formation of these defects. To maintain charge neutrality the formation of each defect requires the removal of two electrons, and the process will occur only if the cations are able to lose and gain electrons easily. Transition elements can adopt variable valence with relatively small ionisation energies, and in this case the electron loss is achieved by converting two Fe^{2+} ions into Fe^{3+} ions for every vacancy formed.

The oxidation of magnetite, Fe_3O_4, which has the spinel structure (see also Section 5.2.4) is another example where defect formation accommodates a change in oxygen fugacity. The formation of Fe vacancies can extend as far as the composition $Fe_{21.67}O_{32}$, the mineral termed maghemite, which has the defect spinel structure. This composition is the same as that of the more stable mineral hematite, Fe_2O_3. However the transformation of a spinel structure to the hematite structure involves a change in the oxygen close-packing from cubic to hexagonal. At high temperatures ($> \sim 500°C$) this transformation is readily achieved, and hematite is formed during the oxidation of magnetite. At lower temperatures such a major structural change is difficult, and the defect spinel structure may persist indefinitely, despite the fact that it is thermodynamically less stable than hematite.

In maghemite at low temperature a reduction in the free energy is achieved by vacancy ordering: at the composition $Fe_{21.67}O_{32}$ there are on average $2\frac{1}{3}$ vacancies per spinel unit cell, which means that any ordered array of vacancies will need a unit cell at least three times as large to produce a long-range ordered structure. Natural vacancy-ordered maghemite has a tetragonal unit cell with lattice parameters ($a \times a \times 3a$) where a is the size of the spinel unit cell.

In oxides of non-transition elements a change in the oxygen partial pressure results in a more limited deviation from stoichiometry since the defects have a considerably greater energy of formation.

Many of the most important mineral groups contain iron, and their defect populations will vary as some power of the oxygen partial pressure (usually referred to as the oxygen fugacity, $f(O_2)$, assuming an ideal gas). Any physical property which depends on defect density will therefore be highly dependent on the oxygen fugacity. For example, olivine,

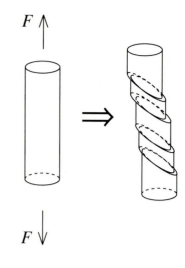

Figure 7.9. The dependence of the Fe–Mg interdiffusion coefficient on the oxygen fugacity $f(O_2)$, at different Fe contents. The diffusion rate increases with Fe concentration and with oxygen fugacity, confirming that vacancies which are formed by the oxidation of Fe^{2+} to Fe^{3+} control diffusion. (After Buening and Buseck, 1973.)

$(Fe,Mg)_2SiO_4$, will only have a stoichiometric cation to anion ratio at a specific value of $f(O_2)$ depending on its composition, and the temperature. At a higher $f(O_2)$ there will be a predominance of oxidation defects such as metal vacancies and Fe^{3+} centres; under reducing conditions the predominant defects are associated with excess metal, e.g. interstitial cations and oxygen vacancies. Experimental measurements of Fe,Mg diffusion in olivines demonstrate the dependence of the diffusion rate on the oxygen fugacity, and on the iron content (Figure 7.9), confirming that diffusion is by a vacancy mechanism. Diffusion coefficients are discussed in more detail in Section 11.2.

It is known that olivine from several natural environments can contain substantial numbers of cation vacancies, up to the composition $Fe_3 \square Si_2O_8$, where $\square$ denotes a vacancy. At low temperatures the vacancies tend to order, and naturally occurring ordered structures found around this composition have been given the name laihunite.

7.2 Line defects

The dynamic processes which are continually taking place at high temperatures deep in the Earth exert

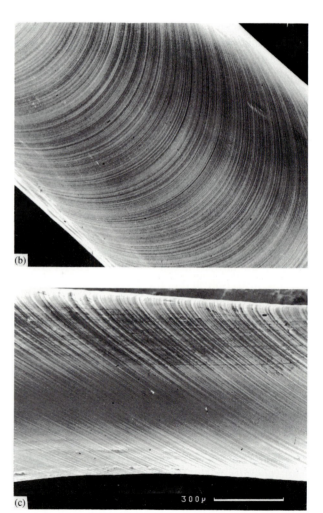

Figure 7.10. (a) An exaggerated illustration showing how the application of a tensile force on a single crystal results in deformation by slip on sets of parallel planes in the structure. (b),(c) The individual slip steps are very small and appear as fine lines on the surface of the crystal, in this case in two views of single crystal wire of cadmium metal, imaged by scanning electron microscopy. (Micrograph courtesy of T. Page.)

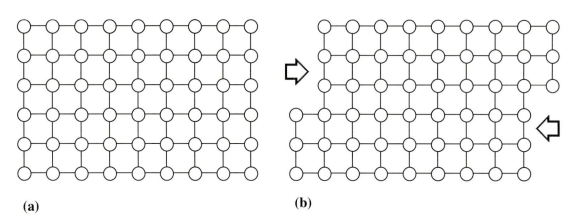

(a) **(b)**

Figure 7.11. (a) A single crystal before deformation, and (b) after deformation. The applied shear has caused the crystal to slip by one lattice repeat, producing small slip steps on the surface.

stresses on the mineral grains in the rocks. If a mineral is stressed beyond its elastic limit it becomes permanently deformed. Under slow strain rates and relatively high temperatures crystalline materials can deform, without brittle fracture, by a process of *slip* where one part of the crystal is translated relative to an adjacent part, as shown schematically in Figure 7.10a. The surface steps produced by deformation are large enough to be observed by microscope (Figure 7.10b,c). Ultimately the deformation of individual minerals, as well as the movement of one mineral grain over another is responsible for the large-scale translation of material in the Earth's crust and mantle.

In this section we will briefly describe the deformation of individual crystals. The theory of single crystal deformation was developed in metals, where it was established that slip takes place on specific crystallographic planes, generally close-packed planes, with slip directions parallel to the shortest lattice vectors. The slip plane and slip direction together define the *slip system*. For example, in a cubic close-packed metal the close-packed planes are {111} planes, and within these planes the shortest lattice vectors are close-packed directions which have the general form <1$\bar{1}$0>. The slip system is written {111}<1$\bar{1}$0>. In olivine a common slip system is (100)[001], meaning that slip occurs on (100) planes in the [001] direction.

The concepts needed to describe mineral deformation are most easily understood with reference to simple cubic lattice such as that in Figure 7.11a. The early ideas (before about 1930) on crystal slip considered ways in which one part of the crystal could

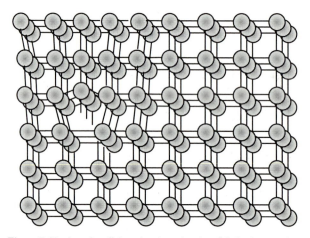

Figure 7.12. An edge dislocation in a simple cubic lattice consists of an extra half-plane of atoms terminating within the crystal. The termination of this extra half-plane (the core) defines the dislocation line along which there is local distortion due to the disturbed bonding.

rigidly pass over the lower part, with all atoms moving simultaneously (Figure 7.11b). This would produce the required slip step and successive movements on parallel planes would result in a macroscopic deformation. However, the theoretical stresses required to produce such simultaneous movement of atoms over a slip plane are about a thousand times higher than those found experimentally, and to account for this discrepancy the concept of *dislocations* was introduced.

In its simplest form a dislocation is a narrow line defect in which a plane of atoms stops within the crystal producing a line of dangling bonds (Figure 7.12). The *dislocation line* is defined by the edge of this 'extra' plane of atoms, and in the immediate

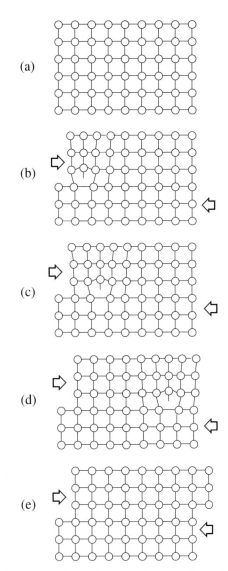

Figure 7.13. Successive steps in the migration of a single dislocation during deformation. The applied stress generates the dislocation which migrates from left to right until it passes out of the crystal leaving a slip step on the surface. The strain at any instant is confined to the dislocation core.

vicinity of the dislocation line the structure is distorted. When a stress is applied to a crystal, dislocations are generated within it (Figure 7.13a) and these move through the crystal such that the dangling bond is moved from one atomic plane to the next (Figure 7.13(b)–(d)). At any instant only those atoms around the dislocation line are disturbed. The deformation is propagated along the slip plane by the movement of the dislocation line until it reaches the surface of the crystal and a small slip step is produced (Figure 7.13(e)). The stress required to move a dislocation line is very much less than that

needed to move a whole plane of atoms simultaneously over one another, and the theory of deformation was developed on the assumption that such dislocations existed in crystals, long before they were able to be observed experimentally.

7.2.1 The Burgers vector of a perfect dislocation

The displacement and direction of the deformation associated with a dislocation is defined by the *Burgers vector* **b**, as described in Figure 7.14. One conventional way of describing the Burgers vector is as follows: In a perfect crystal we can describe a path starting from any point S, tracing out a circuit with equal numbers of lattice steps in opposite directions and returning to the original point. If a dislocation is present, as in Figure 7.14, the same circuit fails to close and will finish at point F. The closure failure SF defines the magnitude and direction of the Burgers vector **b**.

In a perfect (or unit) dislocation the Burgers vector is always equal to a translation vector of the lattice, and the dislocations with the least energy are those with the smallest Burgers vectors. The energy of a dislocation is proportional to $|\mathbf{b}|^2$. Thus as a general guide, the most common dislocations in a material will have Burgers vectors equal to the shortest lattice vectors. However, as we shall see below, in many complex mineral structures dislocations with Burgers vectors which are a fraction of a lattice vector may be favoured. These *partial dislocations* introduce yet another type of defect in a structure as they move through it, and they will be discussed later in this section.

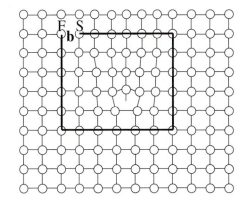

Figure 7.14. Determination of the Burgers vector **b** of a dislocation by carrying out a Burgers circuit. Starting from a point *S* a circuit with equal atomic steps in opposite directions (in this case 6 × 5 atomic steps) around the dislocation core, terminates at point F. The failure to close the circuit indicates the presence of a dislocation, and the Burgers vector **b** is defined as the vector **FS**.

To determine whether a dislocation will move when a stress is applied, it is necessary first to resolve the applied stress onto the slip plane and then determine the component of this shear stress which is parallel to the Burgers vector. If the value of this resolved shear stress in the direction of **b** exceeds some critical value characteristic of the interatomic bond strength of the material, the dislocation will glide along the slip plane. Metals are ductile even at relatively low temperatures due to the non-directionality of the bonding and the ease of dislocation movement. At the other extreme, covalent materials such as diamond are very hard because the displacement around the dislocation line is very localized and the force required to move a dislocation is very high.

7.2.2 Edge dislocations and screw dislocations. Dislocation loops

A dislocation is described by specifying both the Burgers vector and the direction of the dislocation line. In Figures 7.12–7.14 the dislocation line is perpendicular to the page and the Burgers vector is at right angles to it. Both of these directions lie in the slip plane over which the atomic displacements take place. Such a dislocation is termed an *edge dislocation*.

In a *screw dislocation* the Burgers vector and the dislocation line are parallel. This is more difficult to visualize and requires a three dimensional illustration (Fig. 7.15). For a screw dislocation to move through the crystal the applied stress must again have a component parallel to the Burgers vector, and the progressive migration of the dislocation line and the resulting deformation of the crystal is illustrated in the sequence in Figure 7.15.

Since a dislocation line defines the boundary between one region of the crystal in which the slip has already taken place and another which has not yet slipped, it follows that it cannot terminate within the crystal. A dislocation line can form a closed dislocation loop, or its two ends can terminate at the surface of the crystal. Dislocation loops are quite common in many materials, and the generation of new dislocations during the progressive deformation of a crystal usually involves the formation of such loops. Within a dislocation loop the Burgers vector remains the same but the dislocation line changes direction. Thus in one part of the loop the disloca-

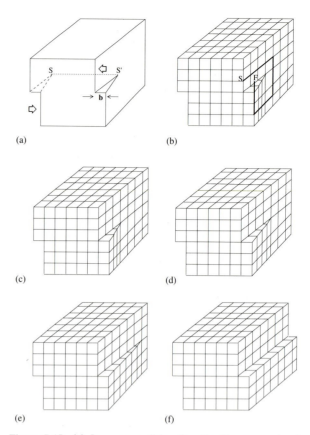

Figure 7.15. (a) In a screw dislocation the Burgers vector is parallel to the dislocation line S'S, and during deformation the line moves at right angles to the applied stress as shown in the sequence (c) to (f). As the dislocation moves out of the crystal slip steps are left on the surface. A Burgers circuit in (b) confirms the Burgers vector as **FS**.

tion will be a pure edge dislocation, while in that part at right angles it will have pure screw character. In between, the dislocation will have intermediate properties. To easily appreciate the atomic displacements around a dislocation loop requires a three dimensional model, but Figure 7.16 shows a sketch in which two layers of atoms parallel to the slip plane contain a quarter dislocation loop which changes from pure edge to pure screw at its ends.

To achieve a macroscopic deformation large numbers of dislocations need to move through a crystal. Dislocations may be generated at the surface, at defects, or at intergrowths within the crystal. During deformation, dislocations move like strings through the structure. When they cross they may interact with one another, the resulting local atomic displacements giving rise to a new dislocation with a Burgers vector which is the vector sum of the original dislocations. There are many possible interactions of this

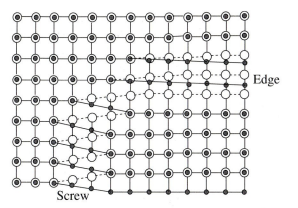

Figure 7.16. Part of a dislocation loop in which an edge dislocation becomes a screw dislocation as the dislocation line turns through 90°. The diagram represents two layers of atoms on a simple cubic lattice, the lower level unshaded, the upper level shaded. At the edge dislocation the Burgers vector, due to the termination of the extra half-plane of unshaded atoms, is at right angles to the dislocation line. As the dislocation line moves around the loop the same Burgers vector becomes parallel to the dislocation line.

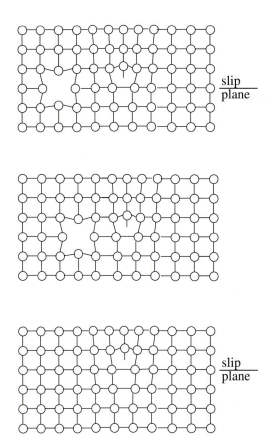

Figure 7.17. The interaction of a line of vacancies with an edge dislocation results in the edge dislocation moving from one slip plane to another, a process termed dislocation climb.

type, but one common phenomenon which is the result of dislocations interacting is that they can become tangled in such a way that their movement is impeded. This effectively hardens the material against further deformation. These dislocation pile-ups are responsible for the *work-hardening* of metals when they are beaten or cold-rolled. To remove the work-hardening the metal must be heated to a high enough temperature to allow atomic diffusion to occur so that the defects can migrate out of the crystal. The process of untangling the dislocations and reorganising them into stable networks is termed *recovery*.

7.2.3 The interaction of point defects and line defects – dislocation climb

When deformation takes place at a temperature high enough for atomic diffusion to take place on a significant scale, point defects such as vacant sites can migrate through a structure and hence interact with dislocation lines. This allows the possibility for an edge dislocation to move from one slip plane to another. When dislocations move along a single slip plane the process is termed *dislocation glide* while the movement of an edge dislocation out of its slip plane is termed *dislocation climb*. Figure 7.17 shows the way in which the interaction of a vacancy with an edge dislocation results in climb at the previously vacant site.

Screw dislocations on the other hand involve pure shear and do not normally interact with vacancies. Screw dislocations do not climb in the same way as edge dislocations. With the Burgers vector parallel to the dislocation line, screw dislocations are not tied to specific slip planes, and at higher temperatures may *cross-slip* from one slip plane to another intersecting plane which contains the same Burgers vector.

Dislocation climb and cross slip are important mechanisms of deformation at high temperatures. They are also involved in the way in which recovery takes place. Tangled dislocations store strain energy in the crystal which will be released if the dislocations can be organised into more stable arrays. One such recovery process is the formation of *low-angle grain boundaries*, in which dislocations within a single crystal line themselves up as shown in Figure 7.18. The dislocation network produces a small angular difference in the orientation in the crystal on either side. The result is that the original deformed

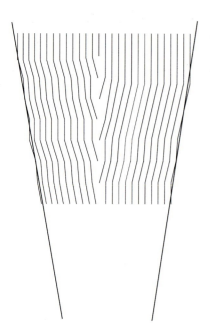

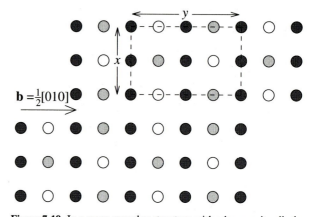

Figure 7.19. In a more complex structure with a large unit cell, the movement of a dislocation with a Burgers vector which is a fraction of the lattice repeat (a partial dislocation) generates a stacking fault across which the periodicity of the structure is broken. In this case the stacking fault involves a mistake in the order of the lighter-shaded atoms.

Figure 7.18. When dislocation lines arrange themselves in an array the result is a small angular misorientation (here exaggerated) between the parts of the crystal on either side of the dislocation array. Such rearrangement of dislocations into arrays takes place during recovery and reduces the overall strain energy. The dislocation array defines a low-angle sub-grain boundary between two parts of the same crystal.

single crystal becomes a mosaic of dislocation-free regions with a small (1°–2°) angular mismatch. When such crystals are observed in thin-section with a polarizing optical microscope, this angular mismatch results in the extinction position varying slightly throughout the crystal, a phenomenon known as undulatory extinction.

Another mechanism of ridding the crystal of tangled dislocations is the nucleation and growth of new dislocation-free crystals which grow through the deformed or partially recovered structure. This process is *recrystallization*, and results in a new undeformed polycrystalline state with high-angle grain boundaries (i.e. no particular orientational relationship between the grains). Such recrystallisation is generally a higher temperature process than recovery.

7.2.4 Partial dislocations and stacking faults

When a perfect dislocation, in which the Burgers vector is a translation vector of the lattice, passes through a crystal, it leaves the structure intact merely translating one part of the crystal by a whole unit cell. In the simple cubic lattice we have used to illustrate dislocations, the Burgers vector is always a lattice vector, but in more complex and realistic structures, this need not be the case, and Burgers vectors which are some fraction of a lattice repeat are allowed, and may be more favourable. These *partial dislocations* do not replace the translated part of the structure in the correct position to reestablish perfect long-range order. Figure 7.19 is a simple illustration in which a partial dislocation with Burgers vector $\mathbf{b} = \frac{1}{2}[010]$ has moved through the crystal. The slip plane remains a two-dimensional defect in the structure. Such a planar defect is termed a *stacking fault* and will be discussed in more detail in the following section.

In order to return the crystal structure to its correct state, a second partial dislocation travelling in the same direction would have to pass over this slip plane (i.e. the plane of the stacking fault) This would complete the translation which would have taken place had the original dislocation been perfect. Partial dislocations often travel in pairs, and they may be quite closely spaced.

The fact that partial dislocations exist at all must mean that there is some energetic advantage in carrying out the translation in two stages. The overall energy balance depends on the energy of the two partial dislocations relative to that of a perfect dislocation, and on the energy of the stacking fault between the two partials. If the stacking fault energy is relatively high, the partial dislocations will travel close together, assuming of course that the sum of

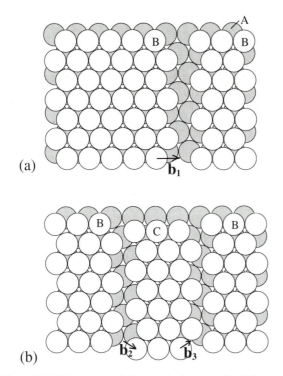

(a)

(b)

Figure 7.20. Two layers of close-packed atoms, the A layer and the B layer, with a vacant row of atoms in the B layer, corresponding to the termination of an edge dislocation. In (a) the movement of this edge dislocation involves a translation $\mathbf{b_1}$ the Burgers vector. (b) This same net translation can be achieved by two partial dislocations $\mathbf{b_2}$ and $\mathbf{b_3}$. Between these two partial dislocation the stacking sequence of the close-packed layers is changed, the B layer occupying the C position (in the conventional close-packing nomenclature).

their energies is less than that of a perfect dislocation. Since the energy of a dislocation is proportional to $|\mathbf{b}|^2$, a two-stage process is often favourable if the displacement due to a partial dislocation can match the structure in some satisfactory way (which is the same as saying that the stacking fault energy is not too high).

The best known examples of partial dislocations are in cubic close-packed metals such as copper and the principle involved is directly relevant to similar phenomena in minerals. Figure 7.20 shows two close-packed layers A and B of the structure with a vacant row of atoms in the B layer, corresponding to the dislocation line of an edge dislocation. A perfect dislocation would require a translation equal to the Burgers vector $\mathbf{b_1}$ in Figure 7.20a. It is easy to see how, if we imagine spherical atoms, it may be easier to take the route through the troughs in the close-packed A layer, and hence the two successive translations $\mathbf{b_2}$ and $\mathbf{b_3}$ (Fig.7.20b). If these two partial

dislocations are separated, the stacking sequence between them is incorrect. In terms of the cubic close-packed structure ABCABCABC... , these intervening atoms are occupying C positions in the stacking sequence (see Section 5.2.1). The B layer has been converted to a C layer by the partial dislocation, and all the other translated layers above are similarly renamed relative to the undisplaced part of the crystal (i.e. B → C, C → A, A → B). The stacking sequence between the two partial dislocations is now ABCA$\underline{C}$ABCA... etc.

Notice that the stacking fault has changed the stacking sequence from cubic close-packing to hexagonal close-packing, CACA, in the immediate vicinity of the slip plane. If a partial dislocation moved on every second close-packed layer the entire stacking sequence could be changed from cubic close-packing to hexagonal close-packing.

There are many instances in minerals where, under one set of physical conditions a structure with a particular stacking sequence is stable, while under a different set of conditions a structure with the same composition but a different stacking sequence becomes more stable. For example, the olivine (hcp oxygen packing) and spinel (ccp oxygen packing) structures of $(Mg,Fe)SiO_4$ are related in this way, although the cation positions must also be taken into account. In the pyroxenes and pyroxenoids (Section 6.4) the structure of the layers may be more complex than the simple close-packing example, but a partial dislocation can still be invoked to describe the geometric relationship between adjacent layers. Where changing physical conditions reverse the relative stability of such structures, a structural phase transformation has to take place if equilibrium is to be maintained. It is possible that in such cases the movement of partial dislocations through the structure, as well as being part of a deformation process, may also be involved in the transformation mechanism.

In the example of triclinic and monoclinic wollastonites, described in Section 6.4.2, (100) layers of the structures are related by displacement of $b/2$ where b is the lattice parameter. In terms of partial dislocations, this could be achieved by a partial dislocation with Burgers vector ½[010] moving on (100) planes.

In most minerals the role of partial dislocations in structural phase transformations is not well understood. The fact that structures can be related geo-

metrically by shear-related modules in this way does not necessarily imply that the presence of stacking faults are due to the movement of partial dislocations. In wollastonite for example, the observed stacking faults may also be growth features. This type of example however does highlight the need to understand the way in which deformation, partial dislocations, and structural changes are related. This is discussed in more detail in Chapter 12.

7.2.5 Observation of dislocations

The theory of dislocations was highly developed long before individual dislocations could be observed directly by transmission electron microscopy (TEM). The observation of dislocations by TEM is a good illustration of the principles involved in the various types of electron microscope imaging which were described in Sections 3.13–3.15. Diffraction contrast imaging is the most widely applicable way of studying dislocations, although high resolution microscopy produces more direct and spectacular images.

High resolution images of dislocations
In modern high resolution electron microscopes, which can produce images of the crystal structure with a resolution of better than 2Å, dislocations can be imaged directly. When the electron beam is parallel to the dislocation line of an edge dislocation the image looks similar to the schematic diagrams we used at the beginning of this section. The Burgers vector can be measured directly from the image, although the resolution is still often insufficient to be able to see the local structure at the dislocation core.

Figure 7.21a is a high resolution micrograph of an edge dislocation in silicon. The electron beam is along <011>. The dislocation line is parallel to the electron beam direction (i.e. normal to the page) and the defect is more easily seen when the micrograph is viewed at a low angle to the page. To interpret the defect we need to know the relationship between the high resolution image and the crystal structure. Figure 7.21b shows a projection of the diamond structure on a plane normal to the electron beam. In this projection, pairs of atoms (circled), image as the white spots in the electron micrograph. A Burgers circuit around the dislocation core will now confirm that the Burgers vector is of the type $\mathbf{b} = \frac{1}{2}<011>$.

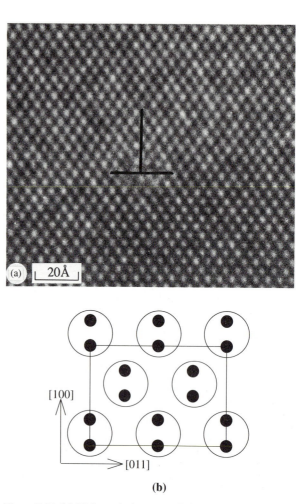

Figure 7.21. (a) High resolution transmission electron micrograph of silicon, with the electron beam incident along an <011> zone axis. The edge dislocation with Burgers vector $\mathbf{b} = \frac{1}{2}[011]$ is marked ⊥. The horizontal line marks the slip plane, and the vertical line shows the extra half-plane of atoms. The dislocation line is therefore normal to the page. The image is best viewed at a low angle. Try a Burgers loop around the dislocation core to confirm the Burgers vector! (from Bourret *et al.*, 1982) (b) A projection of the diamond structure looking along an <011> direction shows that the white spots in the electron micrograph correspond to pairs of atoms (shown circled).

High resolution imaging is not a simple and routine method of studying dislocations. Clearly, to form an image such as that in Figure 7.21a requires that the dislocation is exactly aligned relative to the electron beam, and that the specimen is thin enough to allow a high quality structure image to be produced. This is experimentally very demanding, although it is the only way to image the core of the dislocation directly. There have been very few studies of dislocations in minerals using direct imaging owing mainly to the difficulties in interpretation of very high resolution images of defects in complex structures.

Lattice fringe images of dislocations

If only one set of lattice planes are imaged, the extra half plane of atoms associated with an edge dislocation can be easily seen. Again, the image in Figure 7.22 is taken with the electron beam parallel to the dislocation line, and hence the Burgers vector lies in the plane of the page. The dislocation in Figure 7.22a is narrow, the disturbance being limited to a small region around the dislocation core. In some materials the dislocation core can be much wider, as seen in Figure 7.22b where it is difficult to define the exact position of the dislocation line. By counting the number of fringes on either side of the defect the existence of a pure edge dislocation can be confirmed.

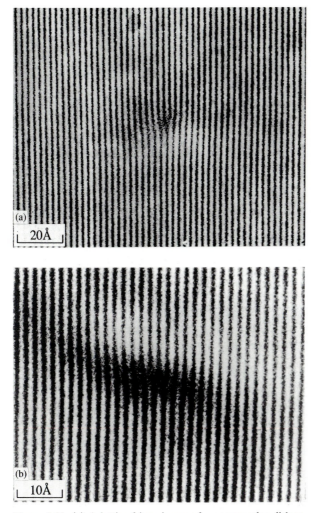

Figure 7.22. (a) A lattice fringe image of a narrow edge dislocation with the electron beam parallel to the dislocation line (i.e. normal to the page). (b) A lattice fringe image of a wider edge dislocation where the exact position of the dislocation core is more difficult to define. The darker region around the core is due to diffraction contrast.

Diffraction contrast TEM of dislocations

The principle of diffraction contrast TEM (Section 3.15) is that any defect or intergrowth which modifies the way in which the electron beam is locally diffracted will appear in a bright field or dark field image. Consider Figure 7.23 in which an edge dislocation is schematically illustrated in a thin crystal flake. The electron beam direction is now approximately at right angles to both the dislocation line (which is normal to the page) and the Burgers vector. In the immediate vicinity of the dislocation core the orientation of the lattice planes is slightly different to that in the rest of the crystal. This means that the way in which these lattice planes diffract electrons will be different in the distorted and undistorted parts of the crystal.

By slightly tilting the crystal relative to the beam it is possible to set the undistorted planes in the exact Bragg reflection (note that the Bragg angle θ for short wavelength electrons is very small), in which case the planes around the dislocation core will not diffract so strongly. Alternatively, as shown in Figure 7.23 a slight adjustment of the crystal will set the distorted planes into the exact diffracting condition, while the other lattice planes will only weakly diffract. A bright field image taken in this latter position will show depleted intensity in those parts of the crystal strongly diffracting i.e. around the dislocation line. The dislocation line will appear as a dark string against a paler background, as shown in the micrograph in Figure 7.24a. In a dark field image, using the diffracted beam, the dislocation line will appear brighter against a dark background (Figure 7.24b).

In Figure 7.24, most of the dislocations image as long curved lines because they lie in the plane of the thin specimen. When a dislocation line is inclined to the specimen surface, it intersects the top and bottom of the crystal and its image appears as a short straight segment (Figure 7.25). A dislocation cannot terminate within a crystal: the ends are the intersections with the surfaces.

7.2.6 Determination of the Burgers vector of a dislocation by diffraction contrast TEM

Consider Figure 7.26a, which is similar to Figure 7.23 except that the third dimension is now shown. Let us assume also that the thin crystal flake is oriented with the electron beam parallel to [001] and

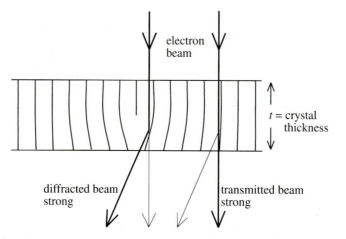

Figure 7.23. Schematic illustration of diffraction around a dislocation core. In this case the electron beam is diffracted more strongly by the tilted lattice planes to one side of the dislocation core than in the undistorted parts of the crystal. The transmitted beam is depleted in intensity around the dislocation line, and in a bright-field image the dislocation line will appear darker than the rest of the crystal. On the other hand, the diffracted intensity is greater around the dislocation line and in a dark field image using this diffracted beam, the dislocation line will appear lighter than the rest of the crystal.

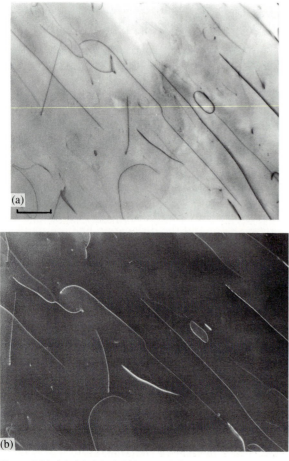

Figure 7.24. (a) A bright field image of dislocations in experimentally deformed diopside. The dislocation lines lie virtually in the plane of the foil and therefore image as long thin lines. (b) A dark field image of the same area, with the contrast now reversed. The length of the scale bar is 1μm. (Micrographs courtesy of J.C. Doukhan).

the Burgers vector of the edge dislocation is $\mathbf{b} = [010]$. Not all of the lattice planes are distorted by the presence of the dislocation. Those planes whose normals are parallel to the Burgers vector (shown dashed in Figure 7.26a) will be distorted most, and hence diffraction from these planes will be most affected. These dashed planes are (010) planes in our figure. On the other hand, planes such as (100), with normals at right angles to the Burgers vector will not be affected at all by a perfect edge dislocation. The determination of the Burgers vector therefore depends on the fact that some spots in the electron diffraction pattern will be affected by the presence of the dislocation, while others will be unaffected. This is the same as saying that the information about the dislocation will be only contained in those electron beams diffracted from the distorted planes. Dark field images using these electron beams will reveal the presence of the dislocation, while dark field images using reflections from unaffected planes will not show the dislocation.

The a^*b^* electron diffraction pattern appropriate for the situation in Figure 7.26(a) is shown in Figure 7.26(b). In the context of defect analysis the reciprocal lattice vector (normal to the lattice plane) is denoted by $\mathbf{g}$. Hence $\mathbf{g} = 100$ refers to the 100 spot in the diffraction pattern, which defines the orientation and spacing of (100) planes in the structure. The discussion in the previous paragraph can now be formulated as follows: if the reciprocal lattice vector $\mathbf{g}$ is parallel to the Burgers vector $\mathbf{b}$ the dislocation

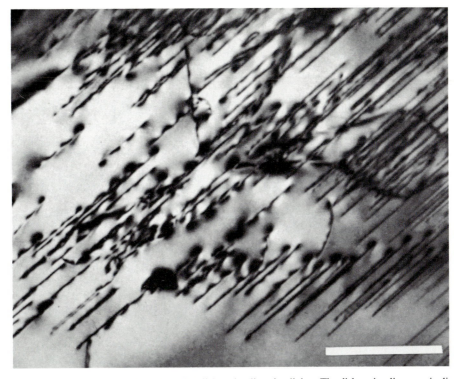

Figure 7.25. A bright field image of a large number of parallel dislocation lines in olivine. The dislocation lines are inclined to the surface of the thin specimen and hence appear as short segments, the ends of the dislocation lines marking their intersection with the top and the bottom of the specimen. The length of the scale bar is 1μm.

will be visible in a dark field image using the diffraction spot defined by **g**. At the other extreme, when **g** is at right angles to **b**, the dislocation will be invisible.

This contrast criterion for dislocations is usually defined in terms of the scalar product of the vectors **g** and **b** as: If **g·b** = 0, the dislocation will be invisible (or out of contrast), and the procedure for determining the Burgers vector from successive dark field images is referred to as **g.b** analysis. For example, if a dislocation is invisible for **g** = $02\bar{2}$ and also for **g** = 111, the Burgers vector must be in the direction $[2\bar{1}1]$ since both **g·b** = $02\bar{2}$. $2\bar{1}1$ = 0 and 111. $2\bar{1}1$ = 0.

Although the general principle of diffraction contrast at dislocations is straightforward, there are a number of complications:

1. The **g·b** criterion only works in theory for elastically isotropic materials, and so in most minerals it will not be strictly applicable. This means that instead of dislocations being totally invisible, their image contrast will be weak rather than strong.

2. Figure 7.26a shows essentially flat planes, and further suggests that planes such as (001) for which **g** is perpendicular to **b** will be unaffected by the edge

dislocation. This is not strictly true. Although the (001) planes in this example will not be distorted as strongly as the (010) planes there will be some displacement of these planes around the dislocation line in an edge dislocation. For a pure screw dislocation the same problem does not exist since the Burgers vector and the dislocation line are parallel. So therefore while the **g·b** = 0 criterion is sufficient for a pure screw dislocation, for an edge dislocation we need to specify an additional criterion to eliminate planes such as (001) in the example above. The correct criterion for invisibility for an edge dislocation is **g·b×u** = 0, where **u** is the unit vector parallel to the dislocation line. The invisibility criterion for edge dislocations is therefore restricted to **g** vectors for those planes parallel to the Burgers vector and normal to the dislocation line.

3. Finally, for the quantitative analysis of dislocations, ambiguities in interpretation arise if there are many planes diffracting simultaneously, i.e. if there are a large number of bright spots in the diffraction pattern. This may be avoided by tilting the specimen so that there is only one strongly diffracted beam. Bright field and dark field images are then comple-

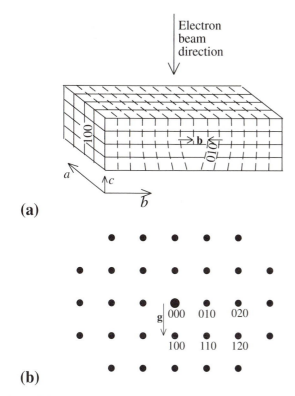

(a)

(b)

Figure 7.26. (a) A three-dimensional view of an edge dislocation with Burgers vector **b** on an (001) slip plane, showing that the planes most distorted by the presence of the dislocation are those normal to the Burgers vector, i.e. the (010) planes. Planes at right angles to **b**, such as (100) are virtually unaffected. (b) The electron diffraction pattern for this beam direction. A dark field image using a 100 reflection (i.e. **g** = 100) will not detect the presence of the dislocation. When **g·b** = 0 the dislocation is invisible.

mentary and there is no ambiguity about which planes are responsible for the diffraction contrast. By tilting the specimen, different diffracted beams can be excited in turn and **g·b** analysis carried out for each one.

When the diffraction pattern contains the transmitted beam and *one* strongly diffracted beam *two-beam conditions* are said to exist. The theory of diffraction contrast which leads to the **g·b** = 0 criterion assumes two-beam conditions, although in practice this is not always easy to achieve in minerals. This is because, from the Ewald sphere construction (Section 3.7), two-beam conditions require that only the origin and one other reciprocal lattice point lie on the Ewald sphere. This can be achieved more easily if the reciprocal lattice spacings are large (as in metals). Most minerals have large unit cells, and hence small reciprocal lattices.

Despite these reservations and difficulties, the diffraction contrast analysis of dislocations is the

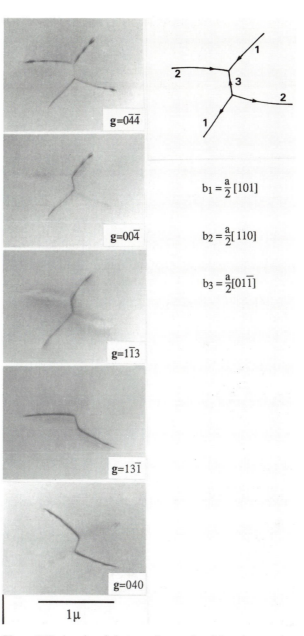

Figure 7.27. A series of electron micrographs of three intersecting dislocations (labelled 1,2,3) in spinel, illustrating the visibility/invisibility of the different dislocations when viewed under various diffracting conditions. The operative strongly diffracting beam, defined by the reciprocal lattice vector **g** is marked on each micrograph. Using the **g·b** criteria the Burgers vector for each dislocation can be determined. (Micrographs courtesy of C. Willaime.)

major tool for quantitatively analysing the processes of deformation in minerals.

Example: Analysis of dislocations in MgAl₂O₄ spinel
Figure 7.27 shows a series of bright field images of three intersecting dislocations, labelled 1, 2 and 3.

Table 7.1 *Possible values of* **g·b** *for the three dislocations under the reflections in Figure 7.27.*

g	½[110]	½[101]	½[011]	½[1$\bar{1}$0]	½[10$\bar{1}$]	½[01$\bar{1}$]	No. 1	No. 2	No. 3
13$\bar{1}$	2	0	1	−1	1	2	Inv.	Vis.	Vis.
040	2	0	2	−2	0	2	Inv.	Vis.	Vis.
11$\bar{3}$	0	2	1	1	−1	−2	Vis.	Inv.	Vis.
00$\bar{4}$	0	−2	−2	0	2	2	Vis.	Inv.	Vis.
044	2	2	4	−2	−2	0	Vis.	Vis.	Inv.

Note: From inspection of this table the only consistent values for the Burgers vectors for the dislocations are those given in Figure 7.27.

By changing the orientation of the crystal, different planes can be put into the diffraction condition. The strongly diffracting planes in each case are specified by the reciprocal lattice vector **g** on each micrograph. In each case one of the dislocations is invisible and hence **g·b** = 0. Since the shortest lattice vector in a face-centred cubic structure has the general form ½<110>, the analysis involves determining all possible **g·b** values and comparing these with the contrast observed for each dislocation. This is summarized in Table 7.1.

7.2.7 Other methods of observing dislocations

X-ray topography

One of the disadvantages of transmission electron microscopy for the study of dislocations is that, as a result of the preparation method, only a small volume of the crystal can be observed, and therefore the dislocation density has to be high if there is to be a good chance of actually seeing a dislocation in the field of view. The dislocation density is generally measured in terms of the length of dislocation line per unit volume of material i.e. cm/cm^3 ≡ cm^{-2}. In deformed minerals, the dislocation density is high (10^8–10^{10} cm^{-2}) and relatively large numbers of dislocations will be present within the thin flake of material. If the dislocation density is very low ($<10^3$ cm^{-2}), as is the case in synthetic quartz crystals grown for technological applications, transmission electron microscopy would not be an appropriate technique to determine the dislocation density. Most thin flakes would show no dislocations at all.

A technique which gives a large-scale view of the dislocation density in a crystal is X-ray topography. The method is non-destructive and the crystal can be a slice up to 10mm thick, depending on the absorp-

tion of X-rays by the material. The principle of X-ray topography is similar to that of diffraction contrast TEM in some respects, namely that the slight differences in the spacing and orientation of the lattice planes around a dislocation line change the way that the X-rays are diffracted.

There are a number of different geometrical arrangements possible, the most common being the Lang Projection Topograph, illustrated schematically in Figure 7.28. A beam of X-rays passes through a suitable filter and collimator and the monochromatic radiation is incident onto the crystal slice. The orientation of the specimen is arranged so that the incident beam is diffracted from one set of crystal planes in the crystal. The diffracted beam passes through a narrow slit and onto a photographic plate which records the intensity of the diffracted beam. The film and the crystal are linked together and move across the path of the X-ray beam, so that as the X-ray beam scans the crystal the photographic plate maps out the way in which the diffracted intensity varies. In this sense the principle is similar to that of dark-field imaging in a TEM.

Dislocations in the crystal locally change the diffracted intensity from the region around the dislocation line, and an image of the line appears on the photographic plate. Figure 7.29 shows an X-ray topograph of a quartz crystal with the dislocation lines appearing dark against a paler background. No magnification of the image is possible and the X-ray topographs are the same size as the crystal specimen. A low dislocation density can be measured in this way, and by producing X-ray topographic images using different diffracted beams, the Burgers vectors can be determined in a manner directly analogous to the **g·b** method in dark field TEM. However, isolating different diffracted beams is not experimentally

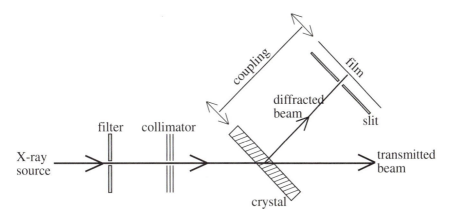

Figure 7.28. Schematic illustration of the principle of Lang X-ray topography. A monochromatic X-ray beam is directed onto a crystal which is set up so that a diffracted beam passes through a slit onto a photographic plate. As the crystal and film are translated any variation in the diffraction in the crystal, due to dislocations or other defects, is mapped onto the film as a variation in diffracted intensity.

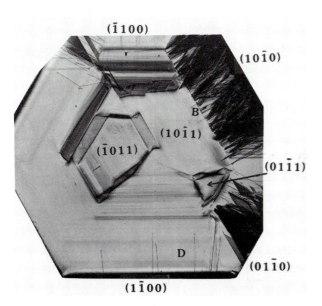

Figure 7.29. X-ray topograph of a quartz plate, 1.5mm thick, normal to the *c* axis. The $10\bar{1}0$ X-ray reflection was used to form the image. The major growth sectors are indicated on the topograph, namely the $\{10\bar{1}0\}$ prism faces and the $\{10\bar{1}1\}$ rhombohedral faces. The dark contrast at B is due to dislocation bundles which originate at the boundaries between growth sectors. Isolated dislocation lines can be seen at D. Planar defects can also be observed as dark contrast in the lower right-hand part of the crystal. (From Ser *et al.*, 1980. See also Authier and Zarka, 1977.)

straightforward and relatively few studies have used X-ray topography in this way.

X-ray topography is sensitive to any change in the lattice spacings and so is also commonly used to map out the distribution of impurities in a crystal, especially when they are present in a regular distribution, such as in the growth zones commonly present in

most minerals. In the X-ray topograph (Figure 7.29) the growth zones image as fine banding parallel to the crystal faces.

Dislocation decoration

When minerals containing ferrous ions are oxidised, magnetite and/or hematite may be formed. Since dislocations are high energy sites within the structure, the oxidation process will occur preferentially along the dislocation line. Small particles of iron oxides decorate the dislocation line making it visible by optical microscopy. This technique of dislocation decoration to study deformation mechanisms was first developed for iron-bearing olivines and has become a standard method of determining dislocation density and distribution in experimental deformation studies. It has the advantage that large volumes of crystal can be observed, and it is also more useful at dislocation densities around 10^6–10^7 cm^{-2}, which are too high for X-ray topography, yet a little low for a convenient TEM study of the distribution of dislocations. To determine Burgers vectors the decoration technique must be supplemented by TEM diffraction contrast observations of individual dislocations.

The technique has been extended to include iron free minerals such as pure Mg-olivine (forsterite), by diffusing Fe^{2+} into the structure prior to the oxidation procedure. The ferrous ions are diffused into the forsterite by placing it in contact with natural iron-bearing olivine at a temperature which will have a minimal effect on the dislocations which are to be studied.

Etch pits

Sites on the surface of a crystal where dislocation lines emerge are high energy sites and are therefore more reactive. When a fresh surface of a crystal is treated with a suitable reagent for a short time the material is etched more rapidly in the distorted regions around the dislocation line. The *etch pits* which form on the surface can be seen under an optical microscope or a scanning electron microscope (Figure 7.30). The most common application of this technique is in the evaluation of materials for technological applications.

Another feature of etch pits is that they commonly demonstrate the symmetry control of the dissolution rate. In such a case they have a shape and orientation which corresponds to the projection of the point group symmetry of the crystal onto the particular face concerned. Thus an etch pit on a (111) face of a cubic crystal with point group m3m should show 3-fold symmetry, while on a (100) face the etch pit should have 4-fold symmetry. This method has been used to obtain information about the possible point group of an unknown crystal.

7.2.8 Experimental studies on the deformation of minerals

Since the late 1960s experimental apparatus has been available which can plastically deform silicate minerals by compressing a specimen under conditions of high temperature and confining pressure. Experiments can be carried out at a constant strain-rate in which a piston compresses the mineral sample at a constant speed, and the *stress–strain curve* determined (e.g. Figure 7.31(a)). Alternatively, the stress can be kept constant and the strain measured as a function of time, producing a *creep curve* (Figure 7.31(b)). There are variants of both types of experiments, but the aim of all such experimental studies is to determine the deformation mechanisms by relating the macroscopic deformation, under a wide variety of experimental conditions, to the nature of the dislocations and their distribution as observed by transmission electron microscopy. The microscopic deformation mechanism involves a description of the generation of dislocations, how they glide, climb and interact with each other at the

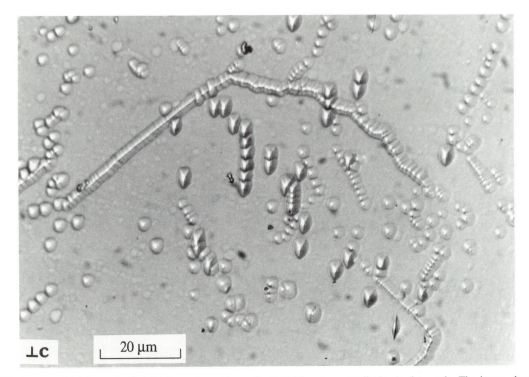

Figure 7.30. An optical micrograph of etch pits on the surface of a diopside cut perpendicular to the *c* axis. The image shows isolated dislocations as well as arrays of dislocations defining low angle grain boundaries. (from Wegner and Christie, 1985)

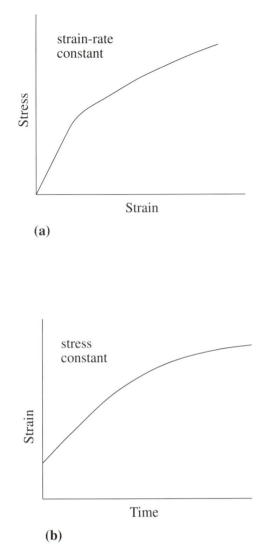

(a)

(b)

Figure 7.31. (a) An example of a stress–strain curve during the deformation of a sample at a constant strain-rate. (b) If the stress is kept constant the result is a creep curve which measures the strain as a function of time.

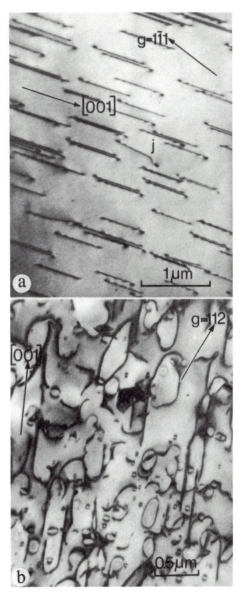

Figure 7.32. (a) Straight screw dislocations in olivine deformed experimentally at 600°C. (b) Dislocation loops in olivine deformed at 1000°C. (From Phakey *et al.*, 1972.)

various stages of deformation described in the stress–strain or creep curves.

In experimentally deformed olivine, for example, the dislocation microstructure varies with temperature and less sensitively, with strain rate. When deformed at relatively low temperatures (~600°C) olivine contains straight screw dislocations with Burgers vector **b** = [001] (Figure 7.32(a)), even though this is not the shortest lattice vector. As such dislocations are generated as loops, it follows that the edge segments of such loops must have moved much faster through the crystal than the screw segments, thus leaving behind the straight screw dislocations

observed in the thin TEM specimens. When edge segments are observed, the slip plane and hence slip system can be determined. The two main slip systems in olivine are (100)[001] and {110}[001]. (100) is the close-packed plane of oxygen atoms. When deformed at ~1000°C, dislocation loops (Figure 7.32(b)) become apparent, suggesting that the velocity of edge and screw components is now approximately the same. Cross slip of the screw segments and climb of the edge segments produces complex tangles. At higher temperatures recovery and recrystallization begins to take place.

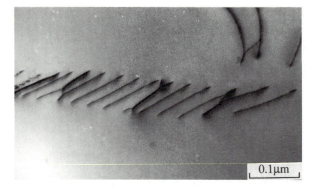

Figure 7.33. Part of a sub-grain boundary in olivine (cf. Figure 7.18). (Micrograph courtesy of M. Drury.)

Recovery involves the migration of dislocations to form walls which separate low-angle grain boundaries as described in Figure 7.18. Figure 7.33 shows part of such a sub-grain in a natural olivine.

Two typical examples of the microstructures associated with high temperature recovery and recrystallisation processes are shown in the electron micrographs in Figure 7.34. Figure 7.34(a) is an olivine crystal made up of a large number of sub-grains. The boundaries are made up of dislocation arrays similar to that in Figure 7.33. Figure 7.34(b) shows the growth of a clear, strain-free crystal eliminating dislocations as it grows into a strongly deformed and sheared olivine crystal during high temperature recrystallisation.

Another feature of high temperature deformation in olivine is the appearance of dislocations with Burgers vectors [100] and [010]. Despite the fact that **b** for an [010] dislocation is 10.2Å, which is a large dislocation with high energy, no dissociation into partial dislocations has been observed, presumably because of the high energy of the stacking fault which would be produced. Note however that in the presence of a hydrous phase, stacking faults terminated by partial dislocations have been observed (see Section 7.3.3).

In chain silicates, dissociation of dislocations is a commonplace phenomenon and the passage of partial dislocations through the structure can transform one polymorph into another . In pyroxenes the tetrahedral chains extend along [001] and their stacking forms layers parallel to (100) planes (Figure 6.11). The dominant slip system in both orthorhombic and monoclinic pyroxenes is (100)[001] which indicates that the deformation moves the chains

relative to one another along their length. The relationship between ortho- and clinopyroxenes is in this arrangement of the chains (Section 6.4.1). When orthopyroxenes are deformed, the perfect dislocations on (100) planes are dissociated into two partial dislocations according to the reaction:

$$[001] \Rightarrow 0.83\,[001] + 0.17\,[001]$$

The first of these partial dislocations changes the stacking arrangement of the (100) layers from that in orthopyroxene to that in clinopyroxene. These stacking faults are effectively thin lamellae, one or two unit cells thick, of clinopyroxene within the orthopyroxene structure. The movement of the partial dislocation converts one structure locally into the other by a process which is essentially a shear transformation (Figure 7.35). These lamellae may terminate within the crystal at partial dislocations with **b** = 0.17 [001], as expected from the above dissociation.

The role of partial dislocations in pyroxenes provides another example of the way polymorphic structures can be transformed from one to the other by shearing on parallel planes. We have already mentioned wollastonite (Section 7.2.4) in a similar context. Deformation and the glide of partial dislocations producing stacking faults can enhance such polymorphic transformations between structures. For example, if orthopyroxene is deformed within the stability field of clinopyroxene, the stacking faults (which locally have the clinopyroxene structure) have a negative stacking fault energy. This favours their extension through the structure and provides a transformation mechanism from ortho- to clinopyroxene.

Deformation microstructures have been experimentally studied in most of the important rock-forming minerals and further details can be obtained from the references at the end of this chapter. Since similar dislocation features have been seen in experimentally deformed and naturally deformed minerals, it should be possible in principle to determine the mechanisms and conditions of deformation in naturally deformed rocks. There have been many studies of the dislocation microstructure in natural minerals, together with interpretations of the deformation mechanisms. The main problem however is that rocks generally have a complex deformational history and the observed dislocations in individual mineral grains will probably record the stresses in

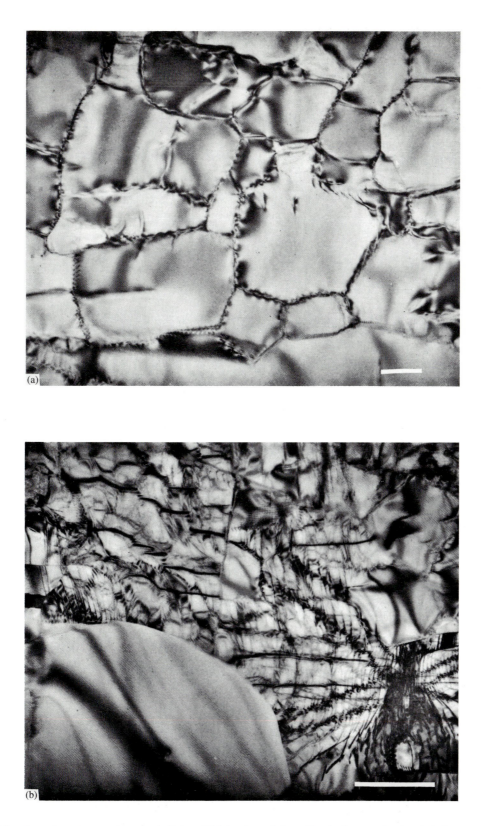

Figure 7.34. Dislocation microstructures in natural olivine which has been deformed and subsequently annealed. The strain energy of the dislocations is reduced by two possible mechanisms: (a) The formation of sub-grain boundaries in a recovered olivine. The boundaries are made up from dislocations which have migrated to form walls separating sub-grains which are misoriented by several degrees. (b) A partially recrystallised olivine in which a new, dislocation-free grain is growing into a heavily deformed olivine. The length of the scale bar in each case is 1mm. (From Lally *et al.*, 1976.)

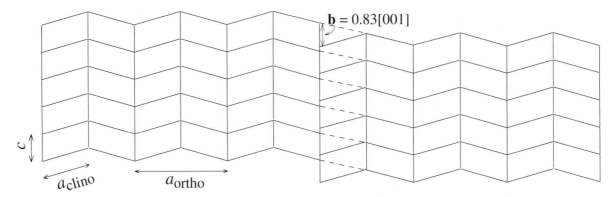

Figure 7.35. A partial dislocation in orthopyroxene, moving on a (100) plane, with Burgers vector **b** = 0.83 [001], creates a stacking fault which is equivalent to generating an extra unit cell of the clinopyroxene structure, shown by the dashed lines.

only the last stages of the deformation. Dislocation microstructures may be easily 'reset' by later heating and deformation, and in any interpretation of deformational history reference must always be made to the large scale deformational features of rocks which are more likely to preserve structures from a sequence of deformational events.

It should be noted that the glide of dislocations represents only one of the possible mechanisms of deformation of rocks. Other mechanisms which may operate are (i) diffusive mass transfer in which material diffuses away from regions under high stress to those under lower stresses (ii) the movement of crystals by sliding along grain boundaries and (iii) deformation twinning (Section 7.3.5) and (iv) brittle fracturing. Discussion of the general topic of deformation is beyond the scope of this book.

7.3 Planar defects

A planar defect involves a violation of the long-range order and crystal symmetry along a two-dimensional surface. Not all two-dimensional defects are strictly planar, and some may be curved in quite complex ways; for convenience we will treat them all under the same heading. Planar defects include stacking faults, Wadsley defects (also termed crystallographic shear planes), twin planes, and antiphase domain boundaries. We will discuss each in turn, emphasising how the defects are formed and how they can be observed, giving examples from mineral systems. The relevance of the defects in mineral transformations will be discussed in more detail in Chapter 12.

7.3.1. Stacking faults

The simplest type of stacking fault is a planar defect which involves a displacement of the structure by a displacement vector **R** which lies in the fault plane. The glide of a partial dislocation through a structure, as discussed in the previous section, produces such a stacking fault. The Burgers vector **b** of the partial dislocation becomes the displacement vector **R** of the resulting stacking fault. Stacking faults are most common in structures which can be described in terms of layers. These layers may be close-packed layers as for example in the zinc blende and wurtzite structures of ZnS, or they may be layers as in the layered mica structures, or also layers such as the (100) layers of single chains in the pyroxene structure. Many structures can be loosely described in terms of layers across which a displacement can occur without involving any major structural misfit and thus produce a fault with a low stacking fault energy. Generally the displacement vector may leave some aspect of the structure unfaulted, as in Figure 7.19, such that the stacking fault refers to only the cation distribution, for example, rather than the whole structure.

Stacking faults can occur during the growth of the crystal, or during a structural transformation from one layer structure to a different structure made up of similar layers (as in the transformation of one polytype to another), or during deformation by the glide of partial dislocations. Such faults are particularly common in chain and sheet silicates.

When stacking faults are isolated they are clearly describable as defects (Figure 7.36). When they are

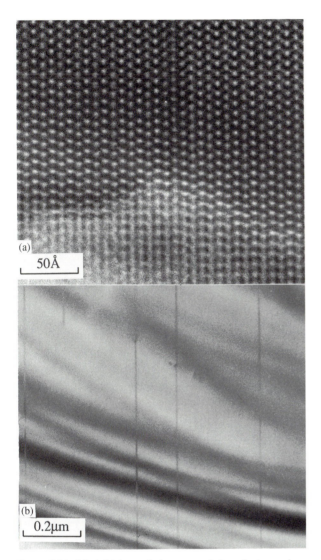

Figure 7.36. (a) A high resolution transmission electron micrograph of a single stacking fault in zoisite. By viewing the image at a low angle the relative displacement of the crystal on either side of the fault can be determined. (b) A diffraction contrast electron micrograph of a number of stacking faults, imaged as thin lines across the otherwise uniform diffraction contours.

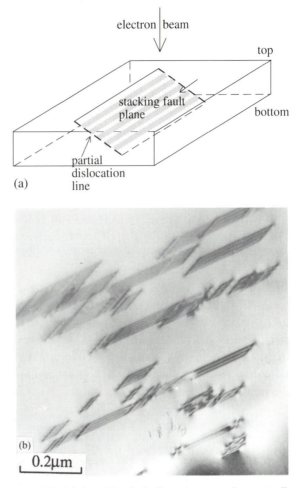

Figure 7.37. (a) A stacking fault, shown here as a plane extending from the top to the bottom of a thin crystal flake terminates at a partial dislocation line (heavy dashed line) within the crystal. The crystal above and below the projected area of the stacking fault plane is the superposition of two wedges of material in which the structure is displaced across the fault plane. When viewed by transmission electron microscopy, diffraction contrast thickness fringes (shaded) appear to lie along the stacking fault plane parallel to its intersection with the crystal surface. (b) Transmission electron micrograph of inclined segments of stacking faults in experimentally deformed sillimanite. The stacking faults are formed when partial dislocations move apart. Thus each stacking fault terminates at a partial dislocation at each end. (From Doukhan et al., 1985.)

periodic they become part of the structural principle describing some layering sequence, as in polytypism. In wollastonite for example the various stacking sequences (see Figure 6.28) could be described in terms of periodic stacking faults within the basic 1T triclinic structure, different periodicities producing different polytypes. In such cases however it is not particularly helpful to refer to them as stacking faults since they are responsible for generating a new long-range ordered structure.

Stacking faults cannot be seen by optical microscopy, but may be imaged by transmission electron microscopy in both high resolution and diffraction contrast modes. In high resolution images the electron beam must be parallel to the stacking fault plane to be able to interpret the nature of the defect. Figure 7.36(a) shows a high resolution image of a single stacking fault in zoisite. The displacement vector $\mathbf{R}$ relating the two parts of the crystal on either side can be determined directly from the micrograph – in this case $\mathbf{R} = 1/4 \,[001]$.

In diffraction contrast mode the same stacking fault is visible as a thin line with different contrast to the surrounding crystal (Figure 7.36(b)). The contrast on both sides of the stacking fault is the same since the structure is in the same orientation and has only been displaced along the stacking fault plane. The local structure in the plane is however different and hence will diffract differently.

When a stacking fault is inclined to the surface of the crystal as shown in Figure 7.37(a), the diffraction contrast image will show a series of alternate bright and dark fringes running parallel to the intersection of the stacking fault plane with the top and bottom of the thin flake (Figure 7.37(b)). These fringes arise due to the superposition of two wedges of material displaced relative to one another by the displacement vector **R**. The direction of the displacement vector **R** can be determined from the fact that the fault is invisible when $\mathbf{g}\cdot\mathbf{R} = 0$, by analogy with the invisibility criterion for dislocations. If two different diffracted beams $\mathbf{g_1}$ and $\mathbf{g_2}$ can be found to satisfy the invisibility criterion then **R** will be the common direction to the two diffracting planes.

7.3.2 Wadsley defects

When the displacement vector associated with a planar stacking fault is at an angle to the fault plane the local chemical composition around the fault is different from that in the bulk structure (Figure 7.38). The presence of such defects therefore changes the stoichiometry of the crystal. We noted in an earlier section (7.1.5) that in a non-stoichiometric crystal a high concentration of point defects, such as vacant sites, is not tolerated by the structure and there is a tendency, particularly at lower temperatures, to order vacancies in some way (e.g. in $Fe_{1-x}S$ by ordering with maximum separation, in $Fe_{1-x}O$ by clustering). This could be described as the assimilation of the defects to incorporate them as a new structural element. Another way of solving the structural problem is to eliminate the point defects by the formation of Wadsley defects.

The way in which Wadsley defects arise is best illustrated by a non-mineral example in which the general principles were first studied. The tungsten oxide structure of WO_3 is made up of corner-sharing WO_6 octahedra that link up via shared oxygen atoms to form a cubic three-dimensional array (Figure 7.39(a)). If this compound is placed in a reducing

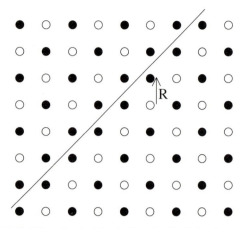

Figure 7.38. When the stacking fault vector **R** relating two parts of a crystal is not parallel to the fault plane, the composition of the crystal is not conserved. In this case the stacking fault is richer in black atoms than the bulk.

atmosphere, vacant oxygen sites are formed. The oxygen vacancies migrate to certain planes within the crystal (Figure 7.39(b)) where they are eliminated by a condensation process in which the structure 'closes up' to form a plane containing edge-sharing octahedra (Figure 7.39(c)). This change in the anion coordination along the plane, termed a Wadsley defect, eliminates the oxygen defects. In Figure 7.39(c), there are blocks of four edge-sharing octahedra along the planar defect which lies on a {102} plane of the structure.

Very small changes in the stoichiometry of WO_3 are accommodated by this process of vacancy elimination by changing the spacing and orientation of the Wadsley defects. A composition $WO_{2.998}$ already causes the introduction of irregularly spaced Wadsley defects on {102} planes as shown in Figure 7.39(c). At a composition $WO_{2.95}$ the defects become ordered into a parallel set with constant spacing, and the spacing between them is determined by the exact stoichiometry. On further reduction to a composition below $WO_{2.93}$ the {102} Wadsley defects would need to be too closely spaced, and a different orientation for the defect plane is adopted. Along a {103} plane of the WO_3 structure six edge-shared octahedra form (Figure 7.39(d)) and further changes in stoichiometry can be accommodated by changes in the spacing of these ordered Wadsley defects. The ordered array of defects produces a superlattice in which the periodicity of the basic WO_3 cell is modified by the superimposed periodicity of the planar defects. The

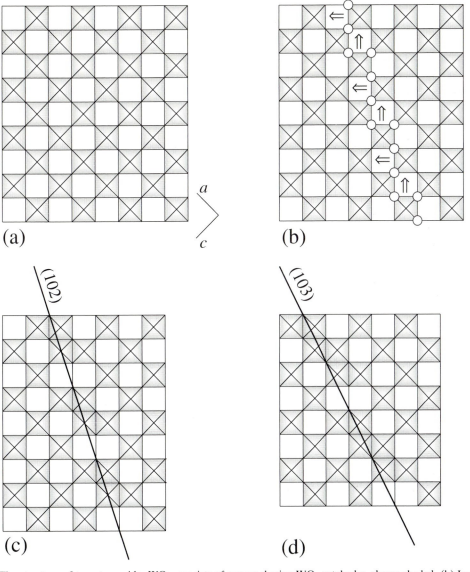

Figure 7.39. (a) The structure of tungsten oxide, WO_3, consists of corner-sharing WO_6 octahedra, shown shaded. (b) In slightly reduced tungsten oxide, the oxygen vacancies (open circles) migrate onto certain planes in the structure. These vacancies can be eliminated if the octahedra condense into the vacant sites as shown by the arrows. This is equivalent to a shear on (102) planes and the result (c) is a planar defect on (102) along which the octahedra share edges. In (d) the planar defect is on (103) and corresponds to a greater degree of reduction.

new unit cell will therefore be some multiple of the cubic subcell.

Ordered Wadsley defects are also termed *crystall-ographic shear (CS) planes* and the structures which use this principle to change their stoichiometry are called CS structures. These include oxides of tungsten, molybdenum, vanadium and titanium. In niobium oxide, Nb_2O_5, the CS principle is further extended since the CS planes occur in *two* sets intersecting at right angles. With one set of CS planes the unreduced material between the planes

forms thin slabs, while in the two-dimensional case the unreduced material forms infinite columns. The dimensions of the column can be changed, and columns of two or three different sizes can be arranged in an ordered way to produce a remarkably complex series of structures.

Crystallographic shear structures have been very widely studied in solid state chemistry, both by X-ray diffraction methods and high resolution transmission electron microscopy which can image the individual octahedra and hence their connectivities.

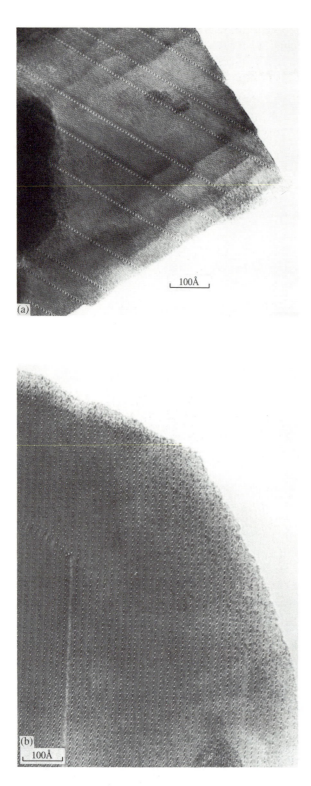

Figure 7.40. In high resolution transmission electron microscopy, the rectangular columns of vacant sites (2×1 octahedra in width) along the CS plane image as white spots, and so define the CS plane boundary. (a) In slightly reduced WO_3, CS planes form on {102} planes, almost regular in this micrograph and at around 100Å spacing. (b) In more strongly reduced tungsten oxide ($WO_{2.88}$) the CS planes form a quite well ordered array on {103} planes. (From Tilley, 1979.)

TEM is particularly applicable when the CS planes are disordered. Figure 7.40(a, b) shows two typical high resolution images of widely spaced {102} CS planes in slightly reduced WO_3, and well ordered {103} CS planes in $WO_{2.88}$ ($W_{18}O_{52}$).

Although Wadsley defects and CS structures are not common in mineral structures, the principle is an important one and may be applicable when we consider the way in which minerals accommodate changes in stoichiometry. When rutile, TiO_2, is experimentally reduced at high temperatures, the formation of oxygen vacancies is accompanied by the reduction of some of the Ti^{4+} ions to Ti^{3+} or Ti^{2+}. The oxygen vacancies are eliminated by forming CS planes along which the edge-sharing octahedra of the stoichiometric rutile structure (Figure 5.24) become face-sharing. In the early stages the CS planes are parallel to {132} planes, while greater degrees of reduction are accommodated on {121} planes. The changeover occurs at a composition between $TiO_{1.93}$ and $TiO_{1.90}$. The substitution of Fe^{3+} for Ti^{4+}, which commonly occurs to some extent in natural rutile, should have a similar effect in that oxygen vacancies are produced. CS planes have been observed in synthetic Fe and Cr-doped rutiles.

7.3.3 Planar defects and solid-state chemical reactions

In a number of mineral systems the mechanism of solid state reaction involving a change in the chemical composition takes place under certain conditions by the propagation of planar defects from the surface into the bulk of the crystal. The best studied of these are the hydration reactions in chain silicates, from the anhydrous single chain pyroxenes $\Rightarrow$ amphibole $\Rightarrow$ sheet silicate. The structural principles involved in this sequence have been discussed in Section 6.7 on biopyriboles.

The replacement of anhydrous pyroxenes by amphiboles or sheet silicates is a common phenomenon whenever such minerals are in contact with hydrous fluids at high temperatures. Often the replacement is *topotactic* which means that the phases share certain crystallographic orientations, for example, the chain orientation in an amphibole replacing a pyroxene may be unchanged. The early stages of such a replacement reaction may take place by the migration of a pair of amphibole double

Figure 7.41. An I-beam diagram of a pair of double chains of amphibole replacing pyroxene, thereby generating a chain multiplicity defect in the pyroxene structure. (after Veblen, 1991)

chains into the single chain pyroxene (Figure 7.41). This produces a *chain-width or chain multiplicity defect* within the pyroxene. Frequently the amphibole lamellae are wider as shown in the high resolution electron micrographs (Figure 7.42) where amphibole lamellae two, six and eight chains wide have grown into the pyroxene. Such replacement involves negligible net shear in the pyroxene host structure.

Chain-width planar defects are common in many amphiboles undergoing retrograde metamorphism which eventually results in their replacement by sheet silicates. Figure 7.43 shows a single triple-chain defect within amphibole. With more extensive replacement, quadruple and wider chains may be intercalated into the amphibole structure as part of this process (Figure 7.44). Under certain conditions the triple-chain mineral jimthompsonite and the ordered double-triple chain phase chesterite (see Section 6.7) may form.

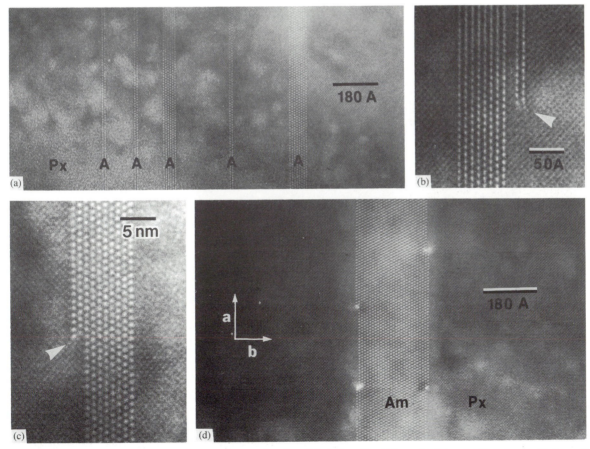

Figure 7.42. High resolution transmission electron micrographs of amphibole lamellae (A) growing into an orthopyroxene (Px), parallel to (010) in the pyroxene matrix. (a) The thin lamellae of amphibole are from left to right 2,2,6,2, and 8 amphibole chains wide. (b) The termination (arrowed) of an ampibole lamella 2 chains wide. (c) A ledge two chains wide on an amphibole lamella. (d) The lamellae thicken by the growth along ledges, which here are one amphibole chain wide. (From Veblen and Buseck, 1981.)

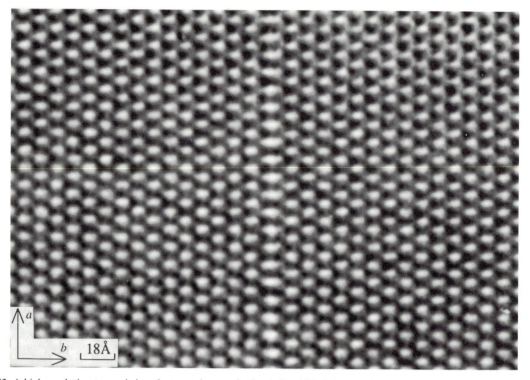

Figure 7.43. A high resolution transmission electron micrograph of a chain-width defect consisting of a thin lamella one triple-chain wide, intercalated into double chain material (anthophyllite). For an interpretation of such images see Figure 6.43.

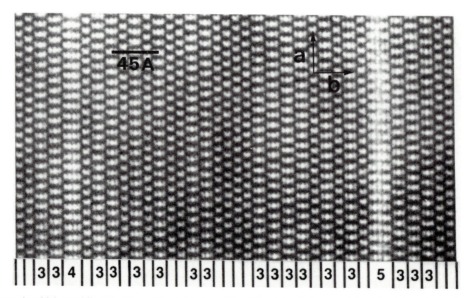

Figure 7.44. An area in which amphibole has been almost totally replaced by wider chain material, with triple and occasionally wider chains disordered through the host. The double-chain material is not labelled. (from Veblen and Buseck, 1979).

A different type of planar defect involving silicate chains is found in the pyroxenoid mineral group (Section 6.4.2). In the $(Ca,Mn)SiO_3$ pyroxenoids the structural principle involved is the stacking of modules which have either the wollastonite (W – periodicity 3) or the clinopyroxene (P – periodicity 2) chain arrangement. The chain periodicity of any pyroxenoid structure is determined by the stacking sequence of these W and P modules. Each module has a distinct composition and so the stacking

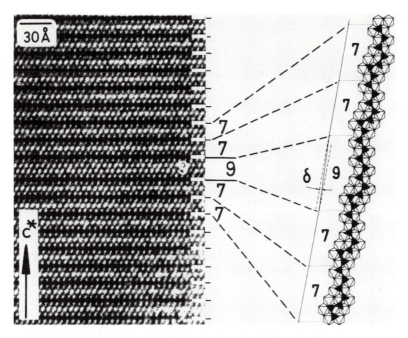

Figure 7.45. High resolution micrograph of pyroxmangite (Ca,Mn)SiO$_3$, with chain periodicity 7, containing a chain periodicity fault, involving the intercalation of a 9-repeat slab. The diagram on the right shows that the defect is due to an extra P module added to the chain. (From Czank and Liebau, 1980.)

sequence provides a sensitive means for producing small changes in the composition. Mistakes in a well ordered stacking sequence constitute a *chain length or chain periodicity fault* in the structure. An example is shown in Fig 7.45 in which an ordered stacking sequence in pyroxmangite in which the chain periodicity is 7 (WPP) is broken by a single intercalated slab with a chain periodicity of 9 (WPPP). Written in terms of a module sequence this defect involves the intercalation of an extra P module thus:

WPPWPPWPP**P**WPPWPP…

A final example is the rather unusual occurrence of planar defects associated with the hydration of olivine at high temperature and pressure. Under most crustal conditions olivine hydrates fairly readily to produce serpentine minerals (Section 6.6.1) but olivine crystals recovered from upper mantle rocks may contain planar defects on (001) and (021) planes with a displacement vector **R** = 1/4<011> (Figure 7.46). The displacement was determined using **g·R** criteria and can be seen directly in the high resolution electron micrograph in Figure 7.47. Since the displacement vector is not parallel to the defect plane, the olivine stoichiometry must change. IR spectroscopy confirms that the defects are associated

with OH groups and Figure 7.48 is an illustration of the structure of the defect.

Stacking faults in chain silicates

In this section we have we have described three different types of stacking faults in chain silicates and here we briefly summarize these.

1. The first involved a mistake in the *chain arrangement*. Examples include the defects on wollastonite (100) planes, characterized by a displacement vector **R** = 1/2[010], which lead to the differences in the wollastonite polytypes. A similar type of fault relates ortho- and clinopyroxenes and amphiboles.

2. The second type of fault is a mistake in the *chain multiplicity*, such as the intercalation of a triple chain into a double-chain amphibole structure.

3. The third type is the mistake in the *chain periodicity* as occurs in pyroxenoids. In each case the planar defect is a product of the structural mechanism which relates families of silicate structures.

There are very many examples in other mineral systems of similar structural relationships between minerals related by stacking modules. When the modules have the same chemical composition the

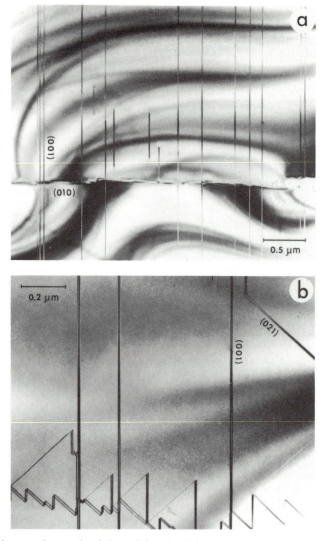

Figure 7.46. Low magnification electron micrographs of planar defects in olivine from a kimberlite rock. (a) Planar defects on (001) planes, some terminating at partial dislocations. The horizontal feature along (010) is a thin band of serpentine, a common hydration product of olivine. The olivine on either side of this band is slightly misoriented as shown by the displacement of the diffraction contours. (b) Planar defects on {021} forming zig-zag lines. (From Kitamura *et al.*, 1987.)

various stacking arrangements are referred to as polytypes; when the modules are different and different stacking arrangements result in changes in chemical composition the various structures are termed polysomes.

7.3.4 Antiphase domain boundaries

An antiphase boundary is an interface within a crystal across which there is a mistake in the translational symmetry. The interface, which is generally curved, arises when a crystal transforms from a higher to a lower symmetry retaining a similar structure but with the loss of some translational symmetry element. Three examples will illustrate the most common origin of antiphase boundaries (APBs) in minerals. All three examples describe structural transformations as the origin of APBs, although there are some instances where they could be produced during the crystallization.

(i) The high–low pigeonite transformation

As explained in Section 6.4.1, pigeonite at high temperatures has space group C2/c with all of the pyroxene chains equivalent. At lower temperatures the chains undergo a sudden distortion as the individual tetrahedra rotate. The sense of rotation in the

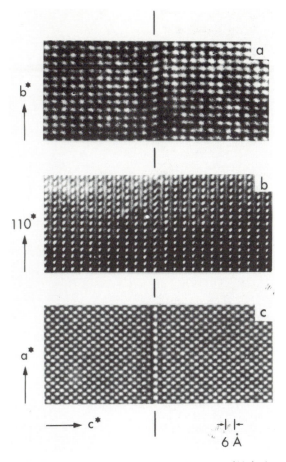

Figure 7.47. Lattice images of the planar defects on {001} planes taken in three different orientations. From these images the displacement vector can be determined as **R** = ¼<011>. (From Kitamura *et al.*, 1987.)

chains at the corners of the original unit cell is opposite to that in the chains through the centre of the cell (see also Figures 6.15 and 6.22). The chains thus lose their equivalence, and the space group symmetry is reduced to P2$_1$/c. The change from a C-centred to a Primitive unit cell is illustrated schematically in Figure 7.49.

If the distortion of the chains in one part of the crystal is not correlated with that in another, there is the possibility of the formation of an interface across which the sequence of chains is incorrect. The displacement vector relating the two regions of the crystal across this APB is 1/2[110] which is the lattice translation lost in the C ⇒ P transformation.

Another way of describing the origin of the APB is that the low temperature primitive cell can have two possible origins – 000 or 1/2,1/2,0 leading to the formation of two types of regions in the crystal related by the vector 1/2[110]. These domains are sometimes referred to as translational variants produced as a result of the loss of translational symmetry. The domains are called *antiphase domains*.

The local crystal structure at an antiphase boundary depends on its orientation. In some orientations the boundary has a structure similar to the C-centred cell, with a relatively large M2 cation site which can accommodate a calcium ion. Note that the C ⇒ P transformation occurs in clinopyroxenes where the Ca content in the M2 sites is too low to maintain extended silicate chains. These shorten to reduce the size of the M2 site. The response of the Ca ions in the P2$_1$/c structure is to diffuse to the

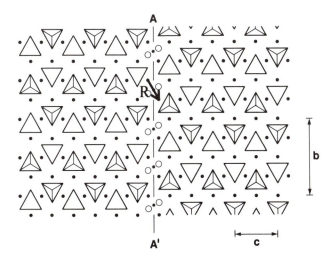

Figure 7.48. The structure of olivine projected onto the (100) plane, showing the displacement associated with the planar defect (A-A′). The displacement vector **R** is shown as an arrow. The open circles are oxygen atoms not bonded to tetrahedra. (From Kitamura *et al.*, 1987.)

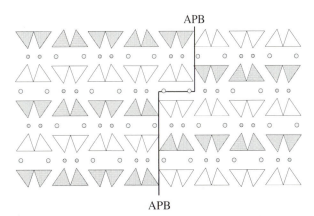

Figure 7.49. In the transformation from C2/c to P2$_1$/c, the pyroxene chains lose their equivalence and form two sets (shaded and unshaded). There are two equally probable ways of achieving this (equivalent to interchanging shaded and unshaded chains), resulting in domains in which the chain distortions are not in register. Antiphase boundaries (APBs) separate such domains.

antiphase boundaries where the cation sites are more favourable. One effect of this is that the coarsening of antiphase domains, which would be expected as the state of order of the mineral increases with time, is controlled by the diffusion of Ca (as well as trace amounts of other large cations which may be preferentially sited on the boundary). The presence of Ca reduces the mobility of the boundaries.

The migration of incompatible cations to defects is a general phenomenon in all materials, and is part of the process of minimizing the overall free energy of the structure. The ultimate distribution of domains is always the result of a compromise between this attempt by the structure to reach a free energy minimum and the kinetics of the processes involved.

(ii) Cation ordering in omphacite

At high temperatures the pyroxene omphacite $(Na,Ca)(Al,Mg)Si_2O_6$ has a disordered cation distribution within the M1 and M2 sites and a space group $C2/c$ (Section 6.4.1). Cation ordering reduces the symmetry to $P2/n$, so during the ordering process the loss in translational symmetry in transforming from a C to a P cell is the same as that in pigeonite, although the structural changes involved in the phase transformations are very different. The antiphase domains formed will be related across the antiphase boundaries by a vector which is the lattice translation lost in the symmetry change, i.e. 1/2[110].

The domain boundary defines a mistake in the cation ordering pattern, and locally the structure resembles that of the disordered state. At low temperature, where the ordered state is more stable, the ordered domains will tend to grow, reducing the overall area of the boundaries, and hence the free energy.

(iii) Al,Si ordering in Ca-rich plagioclases

In the previous examples the symmetry change involved a change in the lattice type, but no change in the size of the unit cell. In the case of transformations in the anorthite rich plagioclases (Section 6.8.3) the situation is more complicated and involves changes in the lattice type as well as the unit cell size. This distinction is however, trivial since in both cases the size of the *primitive* unit cell changes and the principle is exactly the same: antiphase domains are likely to be formed whenever the translational symmetry is reduced, and they will be related by the lattice translation lost in the transformation.

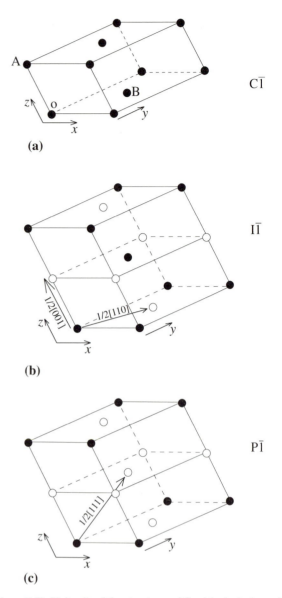

Figure 7.50. Unit cells of the structures of Ca-rich plagioclases. In each case lattice points are filled circles. (a) The high temperature $C\bar{1}$ cell, in which Al,Si atoms are disordered. (b) The $I\bar{1}$ cell of the Al,Si ordered structure. The c axis of the unit cell is doubled and the open circles are lattice points in the high temperature $C\bar{1}$ form which are lost in the transformation to the $I\bar{1}$ structure. (c) In low temperature anorthite, a displacive transition reduces the symmetry to $P\bar{1}$, involving the loss of the translation vector ½[111].

The anorthite structure of $CaAl_2Si_2O_8$ with a disordered Al,Si distribution is triclinic with space group $C\bar{1}$. Ordering the Al,Si atoms results in a doubling of the c axis of the unit cell and a change of space group to $I\bar{1}$. Figure 7.50(a, b) illustrate the loss of translational symmetry associated with the transformation. In the doubled $I\bar{1}$ cell the vectors 1/2[001] and 1/2[110] are no longer lattice translations, whereas they were lattice vectors in the $C\bar{1}$ cell. Both of these vectors therefore are possible candidates for displacements across antiphase boundaries. This is equivalent to saying that the points O, A, and B in Figure 7.50(a), are all possible origins for the doubled $I\bar{1}$ cell. However, points A and B are equivalent to one another within the $I\bar{1}$ cell, being related by the lattice vector of the type $1/2<111>$ (Figure 7.50(b)) i.e. the vectors $1/2[001]$ and $1/2[110]$ are equivalent in the I-centred cell. There will therefore be two types of antiphase domains associated with the $C\bar{1} \Rightarrow I\bar{1}$ ordering transformation, related across the antiphase boundaries by the displacement vector $1/2[110]$.

At a lower temperature there is another phase transformation in anorthite, involving a distortion of the structure around the Ca^{2+} site. This further reduces the symmetry from $I\bar{1}$ to $P\bar{1}$ but there is no change in the unit cell size. The loss of translational symmetry is again associated with the formation of antiphase domains. As shown in Fig.7.50(c), the $I\bar{1} \Rightarrow P\bar{1}$ transition involves a loss of the $1/2[111]$ lattice vector and hence the antiphase domains formed will be related by this displacement across the boundary.

Imaging antiphase domains

Since they involve only translational rather than orientational differences, antiphase domains and boundaries cannot be seen by optical microscopy. With electron microscopy however the antiphase boundaries can be imaged since the structural difference at the boundary affects the diffraction of electrons. Although high resolution structure imaging can be used, the fact that APBs are generally curved makes exact orientation of the defect relative to the electron beam difficult, and most imaging is done in the diffraction contrast mode.

One feature of the phase transformations in the above examples is that they affect only one aspect of the structure, i.e. the basic structure remains the same but is modified in some way that reduces the translational symmetry (increases the volume associated with each lattice point). This is equivalent to saying that the reciprocal lattice remains essentially the same but its size is reduced by the addition of extra reflections. In the case of the transformation from a C lattice to a P lattice as in pigeonite and omphacite, reflections of the type $h + k$ even, which are systematic absences for the C cell, appear in the diffraction pattern of the Primitive structure. The structural information regarding the C $\Rightarrow$ P tran-

sition, as well as any mistakes associated with domain formation, will be contained in these extra reflections. Antiphase boundaries are therefore imaged by TEM in dark field mode using the extra reflections, commonly termed superlattice reflections, which arise during the structural transformation.

Figure 7.51(a) is a typical TEM image of antiphase boundaries, shown in pigeonite. The contrast of the domains on either side of the boundaries is the same, and the boundary itself is imaged as a thin ribbon intersecting the top and bottom surfaces of the thin crystal flake. When the boundaries are inclined to the surface they may image as alternate dark and light fringes, as in the case of stacking faults described above. If the antiphase domains are small and the boundaries strongly curved, the overlapping of adjacent boundaries can produce an image where it is difficult to see the domains themselves (Figure 7.51(b)). This effect becomes more pronounced in thicker crystal flakes.

It is instructive to consider the case of anorthite once again from the point of view of imaging using the extra reflections produced by the structural transformations. The primary reciprocal lattice spots due to the high temperature, disordered $C\bar{1}$ cell are termed *a* reflections, and these are shown in a part of the reciprocal lattice drawn out in Figure 7.52(a). Described in the $C\bar{1}$ cell the indices of the *a* reflections would have the condition that $h + k$ must be even, with no restriction on the l reflection. However, since the unit cell of low temperature anorthite has a doubled c parameter, it is convenient to describe all reflections in terms of this doubled cell. This effectively doubles the l index of all reflections so that the *a* reflections now have the form $h + k$ even, l even.

The transformation from the $C\bar{1}$ cell to the $I\bar{1}$ cell removes some of the restrictions on allowed reflections since only the condition $h + k + l$ is even must be satisfied for a reflection to occur. This allows the possibility of reflections with $h + k$ odd, l odd, and extra reflections of this type will appear in the diffraction pattern. These are termed the *b* reflections, and carry the information regarding Al,Si ordering and the distribution of domains associated with this transformation. Dark field images using the *b* reflections thus show the domain boundaries associated with translational mistakes in the Al,Si ordering pattern.

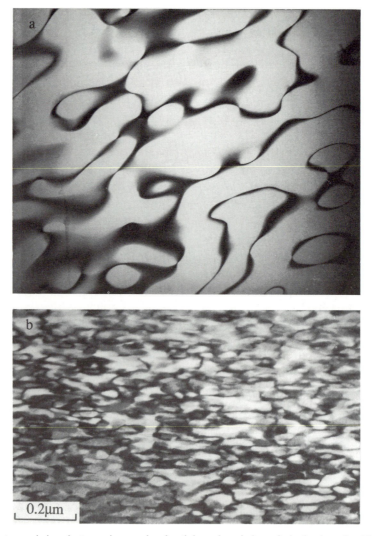

Figure 7.51. Low resolution transmission electron micrographs of antiphase domain boundaries in pigeonite. The boundaries are imaged as dark ribbons whose projected width depends on their orientation in the thin specimen foil. In (b) very small domains result in overlapping domain boundaries in the image. (Micrographs courtesy of G. Nord and M.A. Carpenter).

The further transformation from the $I\bar{1}$ cell to the $P\bar{1}$ cell removes all restrictions on the allowed reflections and so more reflections will appear in the diffraction pattern. These have been traditionally defined as c reflections when $h + k$ is even and l is odd, and d reflections when $h + k$ is odd and l is even. Imaging with either of these reflections shows the antiphase boundaries associated with mistakes in the pattern of displacements associated with this transition.

Figure 7.52(b, c) shows dark field electron micrographs of antiphase domain boundaries imaged with b and c reflections respectively (circled in the electron diffraction pattern in Figure 7.52(a)). The domains are usually referred to as b

domains and c domains. (In this example, the material is $CaAl_2Ge_2O_4$ which has the anorthite structure with Ge substituted for Si. This analogue material has the same sequence of transitions as natural anorthite).

It should be noted that antiphase boundaries can also be seen in bright field images if the appropriate extra reflection is strongly operating. In this case the contrast would be reversed relative to that in the dark field image. However, it is always preferable to image the boundaries in dark field, when the operative **g** vector for the reflection can be determined. As in the case of other planar defects, the antiphase boundary will be invisible if $\mathbf{g} \cdot \mathbf{R} = 0$, where **R** is the displacement vector across the boundary.

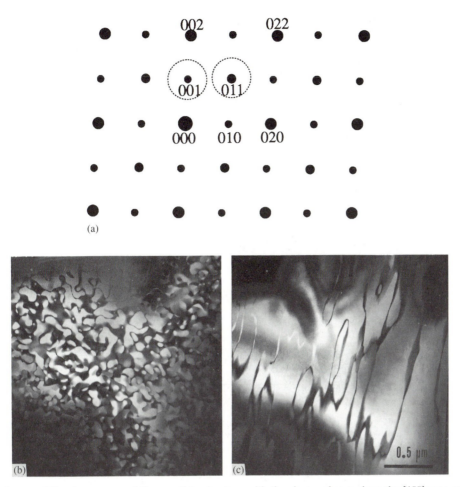

Figure 7.52. An electron diffraction pattern of the anorthite structure with the electron beam along the [100] zone axis, indicating the various types of reflections present. The strong *a* reflections such as 020,002 and 022 define the sub-lattice due to the basic feldspar structure, the *b* reflections such as 011 are due to Al,Si ordering (the C$\bar{1}$ to I$\bar{1}$ transformation), and the *c* and *d* reflections (001 and 010 respectively) are due to the low temperature displacive transition. The circled reflections were used to form the dark field transmission electron micrographs in (b),(c) which are of the anorthite-structured synthetic material CaAl$_2$Ge$_2$O$_8$. (a) Using a *b* reflection, e.g. (011), the fine antiphase domain structure is associated with mistakes in the Al,Ge ordering. (b) The same area of the crystal imaged with a *c* reflection e.g. (001), showing the coarser antiphase domains associated with the displacive transition. (From Müller *et al.*, 1987.)

7.3.5 Twin boundaries

Twin boundaries are analogous in a sense to antiphase boundaries. While APBs relate two domains by a displacement vector and hence involve a mistake in the translational symmetry, twin boundaries separate domains which are related by a point symmetry element i.e. a mirror, a rotation axis or an inversion axis. Antiphase domains represent the *translational variants* associated with a reduction in translational symmetry, and the displacement vector across the antiphase boundary is a lattice vector lost during the transformation. By analogy, twin domains represent the *orientational variants* assoc-

iated with a reduction in the point group symmetry, and the geometric relationship between the domains is defined by the symmetry element lost during the transformation.

Twin boundaries may be formed during phase transformations but also form quite commonly during growth, and in some minerals, during deformation. We will illustrate the origin of twin boundaries by describing a number of typical examples from different mineral systems.

(i) The cubic–tetragonal transformation in leucite

Leucite KAlSi$_2$O$_6$ has a framework structure of (Si,Al)O$_4$ tetrahedra which are arranged in four-

Figure 7.53. Optical micrograph of twinning associated with the cubic – tetragonal transition in leucite.

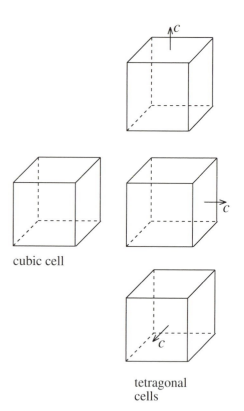

Figure 7.54. The transition from a cubic to a tetragonal cell allows three equally probable orientations of the tetragonal c axis (six if $+c$ is not equivalent to $-c$). Domains of these different orientations within a crystal are twin related.

membered and six-membered interconnected rings. It is a rather open structure with the large K ions in the structural cavities. One of the characteristic features of leucite thin sections under a polarizing microscope is the presence of repeated lamellar twinning, often in several orientations (Figure 7.53).

Above about 665°C leucite has cubic symmetry with point group $m3m$. Below this temperature a displacive transformation, involving a distortion of the structure, reduces the symmetry to tetragonal point group $4/mmm$. On cooling below about 630°C, a second transformation reduces the symmetry yet further, to the tetragonal point group $4/m$. Both transformations are very rapid and the high temperature cubic structure cannot be retained even during fast cooling. In each reduction of symmetry there is the possibility of forming twin domains, although there are some important differences between the two cases.

In the first transformation from cubic $m3m$ to tetragonal $4/mmm$, the orientational variants are associated with the three different possible orientations of the c axis of the tetragonal unit cell. Any of the three cubic crystallographic axes becomes an equally likely candidate for the tetragonal c axis. This is shown in a general way in Figure 7.54. Given that domains of these three orientations are likely to form in a crystal on cooling through the transformation, the next question is what controls the orienta-

tion and spacing of the twin boundaries, i.e. how are these domains arranged in the crystal. Here it is important to recognize that the cubic ⇒ tetragonal transformation involves a change in the lattice parameters i.e. the shape of the unit cell (Figure 7.55). Fitting together domains in different orientations necessarily involves local strains at the twin boundary, while on the other hand, having many domains in all three orientations averages out these local strains over the whole crystal. If only one domain of the tetragonal structure was formed, the whole crystal would have to change its dimensions. The distribution and orientation of transformation twin boundaries is such that the overall *macroscopic* strain is minimized.

For a transformation from $m3m$ to $4/mmm$ symmetry it can be shown theoretically, by considering the strain orientation for each domain and making the strain components equal at the boundary, that the domain walls must be parallel to {101} planes, and each pair of domains is related by a reflection of the lattice across these {101} boundaries. In the

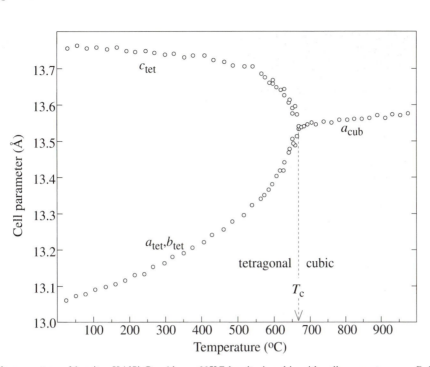

Figure 7.55. The cell parameters of leucite, $KAlSi_2O_6$. Above 665°C leucite is cubic with cell parameter a_{cub}. Below this temperature leucite transforms to tetragonal symmetry with lattice parameters a_{tet} and c_{tet} which diverge as the temperature decreases. (From Palmer *et al.*, 1989.)

tetragonal point group 4/*mmm*, there are four different orientations of planes corresponding to the set {101} i.e. (101), (10$\bar{1}$), (011) and (01$\bar{1}$), which accounts for the various lamellar twin orientations observed in Figure 7.53. It is important to note however that the angle between these four twin plane orientations is defined in the cubic structure i.e. they are 90° apart, even though on a unit cell level the change in lattice parameters means that there will be an angular deviation between the tetragonal (101) planes and their cubic counterparts. This angular deviation must be taken up by strain at the twin boundaries.

There are a number of consequences of this local strain which deserve some mention. As the *c* and *a* lattice parameters of the tetragonal phase diverge on cooling, as shown in Figure 7.55, the strain at twin boundaries increases. The response of the structure depends on the elastic coefficients. Up to a certain point the strain is taken up by elastic distortion. If the strain is too great for the local strains to be taken up on twin boundaries, the lamellar twins may become internally twinned by a second generation of lamellar twins inside the first. This is more likely to occur if the initial lamellar twinning is relatively coarse. The extent to which further twin generation

takes place will depend on the energy of the twin boundaries. Alternatively, the misfit at the boundary may be taken up by dislocations, which means that the lattice planes will no longer match up across the boundary. The strain is released by the presence of dislocations along the boundary. This is referred to as a *loss of coherence*, where coherent boundaries refer to perfect lattice plane matching across the interface, the strains being taken up by elastic distortion.

Another outcome of the local strain generated at the twin boundaries is that twin junctions become regions of maximum strain, and the elastic distortions are greatest at points of maximum curvature. Two phenomena result. First, right-angled domain walls become curved near the intersection, and second, two adjacent right-angled domain walls combine to form a 'needle twin' to reduce the surface area of distorted domain wall. Once the junctions have merged to form a needle twin, the surface energy is markedly reduced when the needle migrates back into the matrix and finally disappears completely. These twin microstructures and their development are shown in Figure 7.56.

The general features of this type of twinning are common to many minerals in which a transformation

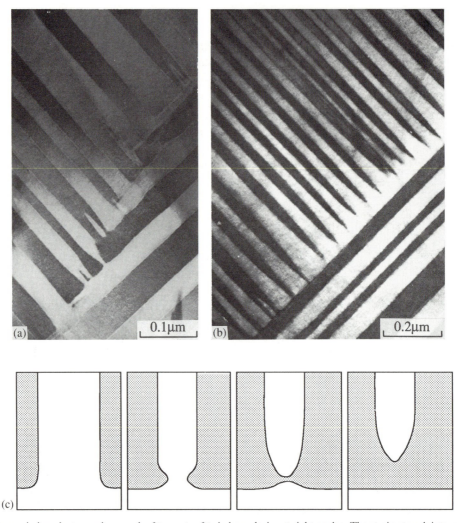

Figure 7.56. (a) Transmission electron micrograph of two sets of twin boundaries at right angles. The strain at such intersections causes the twin walls to become curved. The strain is further reduced with the formation of needle-shaped twins, eventually migrating away from the intersections, as shown in (b). (c) Schematic diagram showing, from left to right, the development of a needle twin from a pair of intersecting twin walls. (After Palmer *et al.*, 1988.)

involves both a loss of of point group symmetry and a change in the shape of the unit cell. If the symmetry change in a structural transformation allows the formation of twins, then strain is always the key factor in determining their distribution.

The second transformation in leucite, from point group 4/mmm to 4/m has been shown to be associated with the delocalization of K ions in the cavities of the structure. The symmetry change also results in a second twinning event. The structure remains tetragonal, and there is no change in the shape of the unit cell, although there may be an overall change in volume. In the absence of a shape change there are no oriented strains in the crystal as a result of the transformation, and therefore any twinning will not

be constrained in the same way as the lamellar twinning associated with the first transformation. The twinning which arises in the 4/mmm to 4/m transition is due to the loss of the {110} mirror planes. This means that adjacent twin domains will be related by this symmetry element, i.e. adjacent domains will have common c axes, but their a and b axes are interchanged. Thus a plane (hkl) is no longer symmetrically equivalent to a plane (khl). However, the a and b axes still have equal dimensions and so there is no change in the orientation of lattice planes across the twin boundary. The boundary orientations are therefore not constrained, and curve through the crystal in an apparently random way.

Such twins where there is an exact lattice coincidence across the twin boundary are termed *merohedral*, while the lamellar twins with an approximate lattice coincidence are termed *pseudomerohedral*. Merohedral twins are not particularly common in minerals, but in natural leucite both lamellar and merohedral twins are always present. Both the lamellar twins and the merohedral twins are imaged in the transmission electron micrographs shown in Figure 7.57.

Imaging of twins

From the above discussion it is clear that the lamellar twins can be imaged by optical microscopy as long as they are coarse enough to be resolved. The difference in the angular orientation of the adjacent domains in lamellar twins means a difference in optical orientation as well. Under crossed polars therefore, the two twin-related domains will go to extinction at slightly different angles, corresponding to the mismatch in the optic orientations caused by the lattice parameter changes.

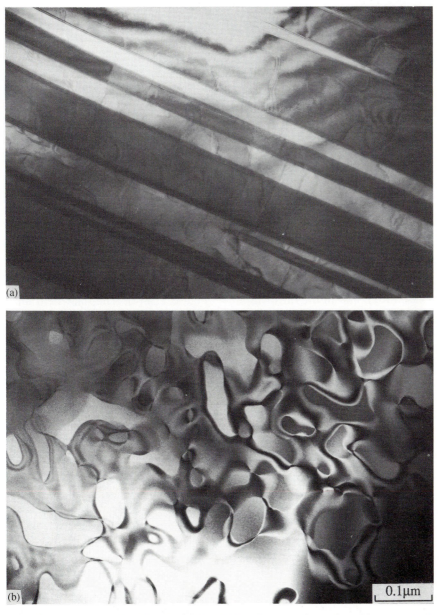

Figure 7.57. Transmission electron micrographs of leucite. (a) Parallel lamellar twins due to the cubic – tetragonal transition. Within the twins weak contrast due to another set of curved twin domain walls can be seen. (b) The curved domains within one of the parallel twins is imaged at higher magnification and slightly different orientation. (See also Palmer *et al.* 1988.)

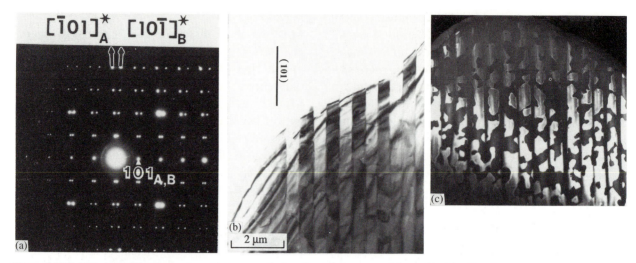

Figure 7.58. (a) Electron diffraction pattern taken across the lamellar twin intergrowth imaged in (b). The diffraction pattern consists of the superposition of two identical diffraction patterns, one from each twin, by rotating one pattern through 180° about the axis normal to the twin plane. The twin plane is (101) and the corresponding row of spots in the diffraction pattern is shared by both twins. The image (b) and the diffraction pattern (a) are in the correct relative orientation to illustrate the relationship between the orientations in real and reciprocal space. See also Fig.3.42. (c) A dark field image (from a different area) using a reflection from one of the lamellar twins. The merohedral twins appear within the imaged set of lamellae as irregularly shaped black–white domains. (From Heaney and Veblen, 1990.)

In transmission electron microscopy using diffraction contrast mode, lamellar twins have different contrast also due to the fact that they have a different orientation and hence diffract differently. An electron diffraction pattern taken across lamellar twins (Figure 7.58(a)) shows the presence of two diffraction patterns superimposed on one another by the twin operation. Both domains share a common twin plane, in this case (101), and in the diffraction pattern the spots normal to the (101) planes are common to both domains. In the rows parallel to this common row the diffraction spots are paired due to the superposition of the diffraction patterns from each twin domain.

Analysis of the electron diffraction pattern is very important in identifying a lamellar microstructure such as that in Figure 7.58(b) as due to twinning. As we shall see in Chapter 11 very similar microstructures result from solid state exsolution (precipitation) reactions in minerals where the lamellae have a compositional as well as structural difference. Electron diffraction patterns from exsolution lamellae also show the superposition of two diffraction patterns, but without the strict geometric relations imposed by twinning.

Merohedral twins cannot be seen optically, nor can they be easily inferred from diffraction patterns. There is no orientational difference between domains, and the lattice matching means that diffraction patterns from twin-related domains are exactly superimposed. However, in the case of leucite, the non-equivalence of (hkl) and (khl) planes in the merohedral twins results in differences in the structure factors for diffraction. TEM imaging in diffraction contrast mode shows this difference in structure factor as each domain diffracts to a different extent, as shown in the dark field micrograph in Figure 7.58(c). The dark field image was made the 420 diffracted beam from one set of lamellae. Therefore only one set of lamellae is imaged, and within this set the merohedral twins image as black/white irregularly shaped domains. The 420 beam was used because the structure factor for 420 and 240 are appreciably different, increasing the contrast between the merohedral twin domains.

Curved twin boundaries can look very similar to antiphase domain boundaries, especially in an image such as Figure 7.57(b), although by tilting the crystal, differences in the contrast from adjacent domains become apparent. In antiphase domains, the contrast on either side of the boundary is always the same.

Straight twin boundaries are not necessarily lamellar, as shown in the electron micrograph of twin domains in natural perovskite $CaTiO_3$ (Figure 7.59). This micrograph also illustrates the fringes observed when the twin domain boundary is inclined to the surface of the thin specimen.

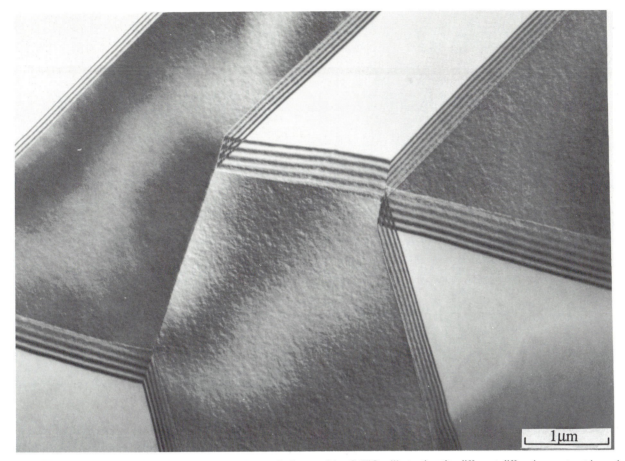

Figure 7.59. Straight domain boundaries in a natural crystal of perovskite $CaTiO_3$, illustrating the different diffraction contrast in each domain, as well as the fringes present on inclined domain walls. When the domain wall is seen edge-on, no fringes are present. (From Doukhan and Doukhan, 1986.)

(ii) The monoclinic – triclinic transformation in K-feldspar

The origin and consequences of the monoclinic to triclinic transformation in alkali feldspars is a major topic in itself, and in this section we will consider only one aspect of it, namely, the twinning which arises as a result of the symmetry change in K-feldspar. In the potassium end member $KAlSi_3O_8$, the symmetry change from monoclinic (space group $C2/m$) to triclinic (space group $C\bar{1}$) is due to the ordering of Al,Si atoms within the structure. The disordered monoclinic form, sanidine, is stable at high temperature, while below about 450°C the triclinic form, microcline, with an ordered Al,Si distribution is thermodynamically more stable. On slow cooling, the ordering transformation accompanied by the symmetry change, takes place.

Note however that, as discussed in Section 6.8.3, the same change in symmetry takes place in Na-feldspar at about 980°C, but for very different reasons. In Na-feldspar a spontaneous displacive collapse of the structure is responsible for the transformation, while in K-feldspar the Al,Si ordering is a very slow process. In geometrical terms the twinning in both cases is the same although the differences in the kinetics of the two processes has a marked effect on the scale and distribution of the twins. In this section we will be mainly concerned with a description of the twinning in the low temperature K-feldspar, microcline.

Optically, microcline is characterized by cross-hatched twinning of varying scale and complexity (Figure 7.60) which arises during the slow Al,Si ordering process and the associated monoclinic to triclinic symmetry change. The only microcline crystals without twinning are those which have grown directly in the ordered triclinic state at low temperatures in aqueous environments. The transformation

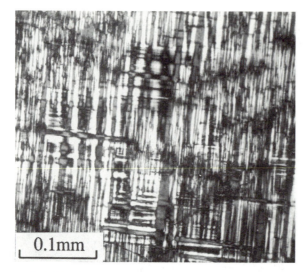

Figure 7.60. Optical micrograph of the typical cross-hatched twinning in thin sections of microcline.

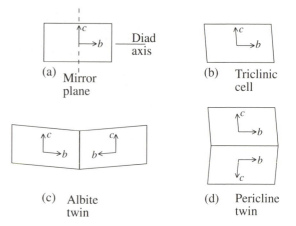

Figure 7.61. The distortion of the monoclinic cell (a) to the triclinic cell (b) involves a loss of the mirror plane and the diad axis. The two equivalent orientations of the triclinic cell can form twin domains related by either of the symmetry elements lost in the transition. Thus albite twins (c) are related by a mirror plane, and pericline twins (d) are related by a diad axis.

from $C2/m \Rightarrow C\bar{1}$ involves a loss of both the diad axis and the mirror plane at right angles to it. Either of these symmetry elements can relate twin domains in the triclinic structure. The change in shape of the unit cell in the monoclinic–triclinic transformation in K-feldspar is shown in Figure 7.61. Triclinic domains related by the (010) mirror plane are termed *albite twins*, while those related by the diad axis parallel to the *b* axis are *pericline twins*.

The distortion of the monoclinic cell can take place in two equally likely ways, i.e. to the 'left' or to the 'right' as shown in Figure 7.61. These two orientational variants can be related in two different ways, either by the (010) mirror planes (albite twins), or the diad axis of the monoclinic cell (pericline twins). Figure 7.61 shows how a combination of albite and pericline twinning leads to the existence of four different orientations of triclinic domains within the crystal. The albite twin plane and pericline twin planes are always perpendicular, as defined in the monoclinic point group, and the fact that the *b* axis is no longer perpendicular to the (010) planes in the triclinic structure results in a local strain at twin boundary intersections, similar in principle to that described in leucite. The cross-hatched twinning in microcline is due to the presence of both albite and pericline twins in the crystal, although the cross-hatching is most likely to be due to a superposition of some parts of the crystal which are albite twinned and other parts that are pericline twinned, rather than the intersection of twin planes.

The effect of the twinning is to minimize the change in the shape of the crystal undergoing the monoclinic–triclinic transformation. If, as in orthoclase, the twinning is on a submicroscopic scale and all four domain orientations are equally represented, the crystal will be optically monoclinic. On an electron microscope scale the microstructure of orthoclase (Figure 7.62) is similar to that in the optical micrograph of microcline, except on a very much finer scale. This suggests that in orthoclase there are domains of triclinic feldspar related to one another in the same way as in microcline, and that therefore orthoclase has a high degree of short-range Al,Si order despite being macroscopically monoclinic. The very early stages of such transformations present some interesting problems common to many mineral groups, and the issues will be discussed in more detail in Section 7.3.6.

Coarsening of orthoclase leads to an imbalance in the distribution of triclinic domains as some grow at the expense of others. On a macroscopic scale this leads to a gradual change in the average deviation from monoclinic symmetry, measured by the *obliquity*, which is the angle between the b^* and the b axes. In monoclinic feldspar the obliquity is zero, while in *maximum microcline* it can reach 0.7°. During the coarsening of the twinning, so-called *intermediate microclines* can have values between zero and 0.7° as a result of this imbalance in the distribution and volume of the domains.

0.1μm

Figure 7.62. Transmission electron micrograph of the tweed microstructure in orthoclase. The broad dark bands across the image are diffraction contours.

In Na-rich feldspars where the monoclinic transformation is a spontaneous, high temperature event, the distribution of albite and pericline twins is always coarser and more regular. The change in shape of the unit cell is much greater than in K-feldspar, the obliquity being ~4° in pure albite (Figure 6.53(b)), and as a consequence twin domains of *either* albite twinning *or* pericline twinning occur in separate parts of the crystal. Most of this obliquity in the lattice shape occurs in the first 50–100° below the transformation temperature. Variations in the spacing and distribution of the twin planes often occur within a single crystal, and are likely to be due to the inhomogeneous stresses that develop during the transformation. Substitution of potassium into Na-feldspar reduces the obliquity of the triclinic structure, as well as the temperature of this displacive transformation (see Section 6.8.3 and Figure 6.54).

(iii) Growth twinning

Twin boundaries within crystals can also arise by mistakes during growth. In the simplest cases, where a crystal consists of two twin-related individuals in contact, as for example in Figure 7.63(a), the twinning must have occurred either at the nucleation stage or very soon after. There are many examples of mineral specimens which exhibit this type of twinning, with the twin individuals either in contact along a plane, or interpenetrating. In such cases twinning at the nucleation stage must be a thermodynamically or a kinetically favoured process, but is not well understood.

When the growth mistakes occur many times at the growing face the twinning is repetitive, and often regular. Most plagioclases show repeated twinning on the albite law (Figure 7.63(b)). Since plagioclases grow with triclinic symmetry, the twinning cannot be

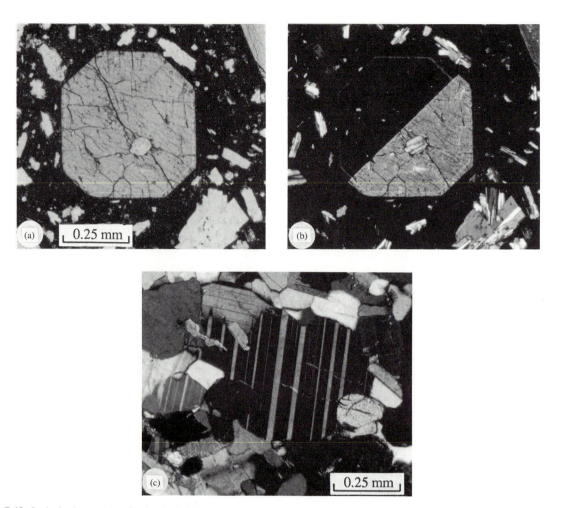

Figure 7.63. Optical micrographs of a simple {100} twin in a pyroxene crystal from a volcanic rock (a) with uncrossed polars and (b) with crossed polars. (c) Optical micrograph of repeated albite twinning in plagioclase.

attributed to a transformation and it may be a growth feature. In this case growth must have proceeded by successive surface nucleation events, with an equal probability of either twin orientation. The periodicity of the twinning is not easily explained without invoking some external influence to provide a stress. Equal numbers of twin domains of opposite sign may be more compatible with an isotropic stress environment. Pericline twins also occur in plagioclases, but with a lower frequency. Both types of twins can also be produced by glide processes, and it is not easy to determine the origin of the twinning without a detailed examination of the twin boundary by TEM (see below).

(iv) Deformation (or glide) twinning
Twinning can be induced in many minerals by the application of a shear stress. The mechanism of such

mechanical twinning is related to the glide of dislocations on slip planes, as discussed in Section 7.2. Instead of deforming inhomogeneously by slip on relatively few slip planes, a crystal may deform homogeneously in shear by the formation of twins (Figure 7.64). Each atomic plane parallel to the twin plane moves over the one below it by a fraction of the lattice spacing, and the whole process is virtually instantaneous. Deformation twinning is generally thought to take place by a coordinated movement of partial dislocations moving through successive planes. Most explanations of the origin of such a coordinated movement invoke the existence of screw dislocations at right angles to the partial dislocations, with the screw dislocation acting as a 'spiral staircase' around which the partial dislocations move.

Deformation twinning is very easily induced in

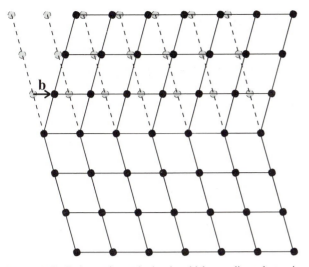

Figure 7.64. Deformation twinning in which coordinated atomic displacements on successive planes result in a macroscopic shear. The original crystal before twinning is shown by the dashed lines and shaded circles. After deformation the two parts of the crystal are related by a mirror plane.

calcite crystals and can be experimentally demonstrated by applying pressure with a knife edge. Calcite in deformed rocks always shows characteristic lamellar twinning produced by glide. High temperature deformation experiments have been carried out to relate the applied stress to the twin density as a function of the grain size of the crystals. These relationships can then be used to estimate the values of stress in naturally deformed calcite rocks. Application is limited to regions with a simple deformational history, without later overprinting, such as a fault plane in a rock.

Both albite and pericline twins can also be induced in plagioclase by deformation and may occur alone or as apparently intersecting sets, producing a type of cross-hatching reminiscent of microcline. Albite twins are more frequent than pericline twins in deformed plagioclases and appear to form at lower shear stresses. It is not always a simple matter to distinguish growth twins from glide twins on the basis of their morphology. In general, glide twins are often lenticular, tapering to a point, and also occur in clusters of variable thickness. They may be related to internal deformation features such as cracks etc. or to the grain boundaries. Growth twins are less frequent, often have stepped contact planes and may thicken or thin independently of one another. In metamorphic rocks where growth takes place under the influence of an external stress, it may be impos-

sible to assign a specific origin to a twinning microstructure.

Another issue which arises in the deformation twinning of ordered albite, for example, is that the glide movement transforms T_{1o} sites into T_{1m} sites and vice versa, and the same with T_{2o} and T_{2m} sites. If the feldspar has an ordered Al,Si distribution, then T_{1o} (the Al sites) and T_{1m} sites are not equivalent, and hence not related by the twin plane. Strictly speaking therefore, such a structure is not a true twin, although it will be indistinguishable optically from a true twin which would form in the case where the T_1 sites were all equivalent, i.e. if the feldspar was disordered.

It has been observed that there is a relation between the ease of formation of glide twins and the degree of Al,Si order in albite – ordered albite being very difficult to twin by mechanical stress. This is presumably due to the unfavourable Al,Si distribution caused by the glide twin boundary, as noted above. No such effect is observed in ordered anorthite where the Al and Si atoms occupy alternate tetrahedra and their distribution is not affected by glide twinning.

7.3.6 The early stages of transformation twinning – tweed structures

In the discussion above, on the origin of crosshatched twinning in microcline, we noted that in the early stages of the transformation, as in the K-feldspar termed orthoclase, the microstructure observed by TEM appears to resemble very finely twinned microcline, although on average the crystal retains the high temperature monoclinic symmetry. This is often termed a 'tweed' microstructure and is found to be quite a common phenomenon as the precursor to a macroscopic symmetry change in many materials, both metallic and non-metallic. Since there is a general principle involved, it is worth considering the nature of this tweed microstructure in a little more detail.

In our discussion of the strain involved in changing the shape of the unit cell of a structure undergoing a symmetry change, we pointed out that if the symmetry change allowed twin-related orientations to occur, then the macroscopic strain could be reduced by producing a domain structure. When the transformation involves a local diffusion controlled event, such as Al,Si ordering, there is no correlation

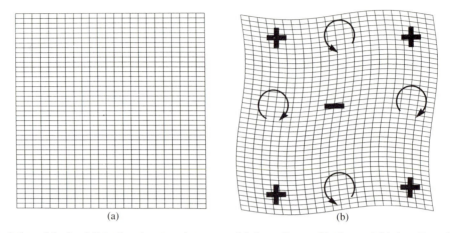

Figure 7.65. A simulation of the local distortions in a tweed structure. (a) An undistorted lattice, and (b) the effect of superimposing two orthogonal transverse sinusoidal distortions on this lattice. The resulting structure contains distorted regions in a twin related relationship (labelled + and −), separated by regions in which the undistorted lattice is rotated alternately clockwise and anticlockwise.

between the atomic rearrangement in one part of the crystal and another, except via the strain which results from this local ordering – (N.B. an AlO_4 tetrahedron is larger than an SiO_4 tetrahedron). These local strain fields interact throughout the whole crystal to reduce the overall macroscopic strain.

The evidence from transmission electron microscopy is that the early stages of this process should be described as two planar transverse distortion waves lying at right angles to one another through the crystal – one parallel and one perpendicular to the b axis. If the waves are described as sinusoidal, there are no defined boundaries and hence no twin planes. Figure 7.65 illustrates the effect of two such sinusoidal waves on the local distortions of a lattice. The lattice is similar to the cb section of the monoclinic cell in Figure 7.61, although the following argument is quite general. Two opposite distortions of the monoclinic cell are generated by the distortion waves. These are labelled '+' and '−' in Figure 7.65(b), and can be equated with the two equivalent distortions from the monoclinic to the triclinic cell. The important observation is that between the '+' and '−' regions, the lattice is rotated either clockwise or anticlockwise, while *retaining the undistorted monoclinic symmetry*. These rotated regions between each sheared region are necessary to balance the strain produced by coexisting domains distorted in opposite senses.

Since the change in shape of the monoclinic cell is due to the ordering of Al and Si atoms, the rotated

regions could be considered as transitional between the two ordered alternatives in different orientations, or alternatively, we could describe the distorted regions as embryonic domains of order within a disordered matrix. The distribution of the domains is controlled by the way the strain is propagated through the structure via the distortion waves. The fact that these distortion waves are parallel and perpendicular to the original monoclinic b axis suggests that these are elastically softer directions in the structure involving the least energy strain energy.

Coarsening of the tweed structure involves atomic diffusion and a reversal of the state of Al,Si order in adjacent domains i.e. converting a '+' region to a '−' region or vice versa. The minimum energy of coexistence of '+' and '−' domains corresponds to either the albite twin-law or the pericline twin-law, and both are in competition as coarsening proceeds. However, as long as '+' and '−' regions coexist in a checkerboard array, there must be undistorted regions in between. This is why ultimately, the twinned ordered structure can have domains related by either albite or pericline twinning, but not both.

Coarsening and continued ordering therefore involves unravelling the tweed structure into its two separate components – separating the warp from the weft, in the weavers' terminology. The needle twins shown in Figure 7.56(b) are part of this process of withdrawing one set of twins from the other. In K-feldspar the coarsening process is controlled by Al,Si ordering which only begins at temperatures below about 450°C. At these temperatures the diffusion of Al and Si is so slow that even in rocks cooled extremely slowly over geological time the cross-hatching has still not sorted itself out on a macroscopic scale, and microclines appear to have coexisting albite + pericline twinning. Only by electron microsopy has it been shown that albite-related and pericline-related domains are spatially separated.

Another example of a very similar phenomenon is found in cordierite $Mg_2Al_4Si_5O_{18}$, where again Al,Si ordering results in a symmetry change with a tweed microstructure produced as a precursor to twinning. In cordierite Al,Si ordering begins below ~1450°C, a temperature where Al,Si diffusion is fast enough for ordering, and the associated microstructural development, to go to completion. Cordierite is discussed in a case study in Chapter 12.

Appendix
How to determine the equilibrium number of Schottky defects in a crystal of composition MX (M: cation, X: anion)

To determine the equilibrium number of Schottky defects, we need to evaluate the change in entropy ΔS and enthalpy ΔH as a function of defect concentration, and then minimize the free energy ΔG, where

$$\Delta G = \Delta H - T\Delta S$$

Let us assume that we have n_s Schottky defects per cm^3 (i.e. n_s vacant cation sites and n_s vacant anion sites) in a crystal which contains a total of N cation sites and N anion sites. The entropy associated with these defects is termed a *configurational entropy* because it arises from the number of different ways, W, of randomly distributing the n_s vacancies over the N available sites in the crystal.

In general, entropy is defined by the Boltzmann equation:

$$S = k \ln W \tag{1}$$

where k is Boltzmann's constant (1.38×10^{-23} joules deg^{-1}).

W is derived from probability theory and is given by the formula:

$$W = \frac{N!}{(N-n)!\, n!} \tag{2}$$

For the case of Schottky defects, we have n_s cation vacancies to distribute over N cation sites, and n_s anion vacancies to distribute over N anion sites.

For the cations

$$W_c = \frac{N!}{(N-n_s)!\, n_s!} \tag{3}$$

For the anions

$$W_a = \frac{N!}{(N-n_s)!\, n_s!} \tag{4}$$

If the defects are independent of one another, the total number of configurations is $W_c.\,W_a = W^2$.

The change in entropy is therefore
$$\Delta S = k \ln(W)^2$$
$$= 2k \ln W$$

$$\Delta S = 2k \ln\left[\frac{N!}{(N-n_s)!\, n_s!}\right] \tag{5}$$

We can simplify this expression by applying the standard approximation to Stirling's theorem:

$$\ln N! = N \ln N - N$$

and substituting into equation (5) above:

$$\Delta S = 2k\{N\ln N - (N-n_s)\ln(N-n_s) - n_s \ln n_s) \tag{6}$$

If the enthalpy change associated with forming a single Schottky defect is ΔH_s, the enthalpy change for n_s defects is $n_s\Delta H_s$.

Therefore the overall free energy change is given by

$$\Delta G = \Delta H - T \Delta S$$
$$= n_s \Delta H_s - 2kT\{ N \ln N - (N-n_s) \ln (N-n_s) - n_s \ln n_s \} \tag{7}$$

At equilibrium, ΔG must be a minimum at the given temperature T

Therefore

$$\left(\frac{\partial G}{\partial n_s} \right)_T = 0 \tag{8}$$

Differentiating equation (7) and simplifying gives

$$\Delta H_s - 2kT[\ln (N-n_s) - \ln n_s] = 0$$

$$\Delta H_s = 2kT \ln \left[\frac{N-n_s}{n_s} \right]$$

Therefore

$$n_s = (N-n_s) e^{-\Delta Hs/2kT}$$

If $N \gg n_s$

$$n_s \sim N e^{-\Delta Hs/2kT} \tag{9}$$

In this form ΔH_s is in joules per defect and k is in joules deg^{-1}. It is more usual to express everything in terms of moles and equation (9) becomes:

$$n_s \sim N e^{-\Delta Hs/2RT} \tag{10}$$

where R is the gas constant ($= kN_A$, N_A is Avogadro's number, 6.02×10^{23}). $R = 8.314$ joules K^{-1} mole^{-1}.

Two important assumptions which have been made in this calculation are:

(a) that the concentration of defects is sufficiently small so that they do not interact with one another and hence are truly randomly distributed throughout the crystal, and

(b) that the energy to form defects is independent of temperature. This would only be true if the volume of the crystal did not change on heating.

Neither of these assumptions is strictly correct, and a more sophisticated treatment would produce a refined equation. However the general form of the equation, and the conclusion that all crystals in equilibrium will contain point defects whose concentration depends exponentially on temperature and on their energy of formation, remains unchanged.

Bibliography

BARBER, D.J. AND MEREDITH, P.G. (Eds.) (1990). *Deformation processes in minerals, ceramics and rocks*. Unwin Hyman.

HAYES, W. AND STONEHAM, A.M. (1985). *Defects and defect processes in non-metallic solids*. Wiley.

HENDERSON, B. (1972). *Defects in crystalline solids*. Edward Arnold.

HULL, D. AND BACON, D.J. (1984). *Introduction to dislocations*. Pergamon.

SALJE, E. (1990). *Phase transitions in ferroelastic and coelastic crystals*. Cambridge University Press.

TILLEY, R.J.D. (1987). *Defect crystal chemistry and its applications*. Blackie.

References

AUTHIER, A. AND ZARKA, A. (1977). Observation of growth defects in spodumene crystals by X-ray topography. *Phys. Chem. Minerals* **1**, 15–26.

BOURRET, A., DESSEAUX, J. AND RENAULT, A. (1982). Core structure of the Lomer dislocation in Ge and Si. *Phil. Mag.* **A45**, 1–21.

BUENING, D.K. AND BUSECK, P.R. (1973). Interdiffusion of the cations Mg and Fe in olivines. *J. Geophys. Res.* **78**, 6852–62.

CZANK, M. AND LIEBAU, F. (1980). Periodicity faults in chain silicates: A new type of planar lattice fault observed with high resolution electron microscopy. *Phys. Chem. Minerals* **6**, 85–93.

DOUKHAN, J-C., DOUKHAN, N., KOCH, P.S. AND CHRISTIE, J.M. (1985). Transmission electron microscopy investigation of lattice defects in Al_2SiO_5 polymorphs and plasticity induced polymorphic transformations. *Bull. Mineral.* **108**, 81–96.

DOUKHAN, N. AND DOUKHAN, J-C. (1986). Dislocations in perovskites $BaTiO_3$ and $CaTiO_3$. *Phys. Chem. Minerals* **13**, 403–10.

HEANEY, P.J. AND VEBLEN, D.R. (1990). A high temperature study of the low-high leucite phase transition using the transmission electron microscope. *Amer. Mineral.* **75**, 464–76.

KITAMURA, M., KONDOH, S., MORIMOTO, N., MILLER, G., ROSSMAN, G.R. AND PUTNIS, A. (1987). Planar OH-bearing defects in mantle olivine. *Nature* **328**, 143–5.

LALLY, J.S., CHRISTIE, J.M., NORD, G.L. AND HEUER, A.H. (1976). Deformation, recovery and recrystallisation of lunar dunite 72417. *Proc. Lunar Sci. Conf.* **7th**, 1845–63.

MÜLLER, W.F., VOJDAN-SHEMSHADI, Y. AND PENTINGHAUS, H. (1987). Transmission electron microscopy study of antiphase domains in $CaAl_2Ge_2O_8$ feldspar. *Phys. Chem. Minerals.* **14**, 235–7.

PALMER, D.C., PUTNIS, A. AND SALJE, E. (1988). Twinning in tetragonal leucite. *Phys. Chem. Minerals* **16**, 298–303.

PALMER, D.C., SALJE, E. AND SCHMAHL, W.W. (1989). Phase transitions in leucite: X-ray diffraction studies. *Phys. Chem Minerals.* **16**, 714–719.

PHAKEY, P., DOLLINGER, G. AND CHRISTIE, J.M. (1972). Transmission electron microscopy of experimentally deformed olivine crystals. *Geophys. Monogr.* **16**, 117–38.

SER, A., BIDEAU, J.P., CLASTRE, J. AND ZARKA, A. (1980). Etude des défauts de croissance dans des monocristaux naturels de quartz. *J. Appl. Cryst.* **13**, 50–57.

TILLEY, R.J.D. (1979). The crystal chemistry of some tungsten oxides containing crystallographic shear planes. *Chemica Scripta* **14**, 147–59.

VEBLEN, D.R. (1991). Polysomatism and polysomatic series: A review and applications. *Amer. Mineral.* **76**, 801–26.

VEBLEN, D.R. AND BUSECK, P.R. (1979). Chain width order and disorder in biopyriboles. *Amer. Mineral.* **64**, 687–700.

VEBLEN, D.R. AND BUSECK, P.R. (1981). Hydrous pyriboles and sheet silicates in pyroxenes and uralites: intergrowth microstructures and reaction mechanisms. *Amer. Mineral.* **66**, 1107–34

WEGNER, M.W. AND CHRISTIE, J.M. (1985). Chemical etching of amphiboles and pyroxenes. *Phys. Chem. Minerals* **12**, 86–89.

8 Energetics and mineral stability I – basic concepts

One of the central themes in mineralogy is the study of mineral behaviour, which describes the response of a mineral structure to a change in the physical or chemical environment. The structure responds so that its free energy always tends to a minimum value. In the process the structure may distort and change its symmetry; it may undergo a major reconstructive transformation to a new polymorph, or alternatively, occupancies of the cation sites may become more ordered or less ordered. Whatever the nature of the response, the direction of change is always in the direction of a free energy minimum, i.e. towards an equilibrium state, and in this chapter we will outline the thermodynamic concepts involved in defining mineral equilibrium in simple systems.

On a larger scale, when we consider an assemblage of minerals in a rock, it too responds to changes in the physical and chemical environment by reactions between the minerals. Cation interchange may alter their relative compositions, or alternatively completely new mineral assemblages may form. For any given bulk composition, changing the temperature and pressure conditions defines a sequence of different equilibrium states consisting of different assemblages of coexisting minerals. The study of the way mineral assemblages in rocks respond to changing conditions in the Earth is a central theme in the field of metamorphic petrology, and in having to deal with a large number of possible mineral reactions in various rock compositions, the thermodynamic procedures used there are somewhat different to those we will describe in this chapter. However, before we can say anything about the thermodynamics of reactions between minerals we need to know the free energies of the individual mineral phases, and how they change with temperature and pressure. In mineralogy therefore, we are mainly concerned with the processes which take place within the mineral and the energy changes involved. This thermodynamic data becomes the input to the large databases which can be called upon to define the equilibrium mineral assemblage at any pressure and temperature.

In this chapter we will be concerned with defining what is meant by the free energy of a mineral. We will then ask how this free energy changes if the mineral undergoes a transformation of some kind, and how we might measure these changes. Many minerals can have a wide range of possible compositions through cation substitutions, and in the following chapter we will look at the energetics of such solid solutions and the criteria which define the extent of possible solid solution. Cation substitutions also lead to the possibility of ordered or disordered occupancy of atomic sites in a structure, and cation ordering on cooling is another important process in minerals.

Although we will mainly be concerned with processes within minerals, i.e. *intramineral* processes, rather than the *intermineral* reactions between minerals in rocks, the separation of the two topics is somewhat arbitrary. The continuing refinement of the often approximate estimates of pressure and temperature obtained from the compositions of natural mineral assemblages (i.e. geobarometry and geothermometry) depends on improving our understanding of the energy of intramineral processes. The magnitudes of the energies involved in intramineral processes are on a similar scale to the energies involved in intermineral reactions. Therefore processes such as cation ordering in one mineral, can have a very marked effect on the stability of the whole mineral assemblage. Furthermore, the experimental determination of the free energy of an individual mineral is not usually made in isolation, but is derived from its equilibria with other minerals whose free energy is known.

First we will briefly review the basic thermodynamic functions we need to describe mineral energetics, particularly the criteria which define an equilibrium state. A full treatment of even basic thermodynamics is not possible here, and reference is given in the bibliography to more rigorous texts on the topic. Our aim is to outline the main concepts and procedures, sufficient to cover the examples of mineral behaviour described in later chapters.

8.1 Some basic thermodynamic concepts

The *internal energy*, U, of a mineral structure is the sum of the potential energy stored in the interatomic bonding and the kinetic energy of the atomic vibrations. Adding more heat increases the kinetic energy (and hence the temperature), and consequently increases the internal energy. However, if the crystal is allowed to expand it does some work on its surroundings, and so the total change in the internal energy is expressed as

$$dU = dQ - PdV \qquad (8.1)$$

where dQ is the change in heat content, and PdV is the amount of work done in the expansion, P being the pressure and dV the change in volume. This is effectively a statement of the First Law of Thermodynamics, i.e. that the total energy is conserved.

For reasons explained below it is convenient to define another energy function termed the *enthalpy* H as

$$H = U + PV \qquad (8.2)$$

where P is the pressure and V is the volume.

A fundamental property of any material is its *heat capacity* C, which describes the amount of heat dQ involved in changing the temperature of a mole of material by dT.
Thus
$$C = dQ/dT \qquad (8.3)$$

There is a difference between the definition of the heat capacity of a system at constant volume and the heat capacity at constant pressure and this is derived in the following way.

Dividing the terms in eqn. (8.1) by dT gives

$$dU/dT = dQ/dT - p\,dV/dT$$

If the volume is held constant then $dV/dT = 0$, and

$$\left(\frac{\partial U}{\partial T}\right)_V = \left(\frac{\partial Q}{\partial T}\right)_V \equiv C_v \qquad (8.4)$$

where C_v is the heat capacity at constant volume. Differentiating eqn. (8.2)

$$dH = dU + pdV + Vdp$$

Substituting eqn. (8.1) gives

$$dH = dQ + Vdp$$

Dividing the terms by dT gives

$$dH/dT = dQ/dT - Vdp/dT$$

If the pressure is held constant then $dp/dT = 0$. and

$$\left(\frac{\partial H}{\partial T}\right)_P = \left(\frac{\partial Q}{\partial T}\right)_P \equiv C_p \qquad (8.5)$$

where C_p is the heat capacity at constant pressure.

In experiments with solids it is easier to consider constant pressure rather than constant volume and hence enthalpy changes are easier to measure than energy changes. For a system at constant pressure the heat input is equal to the enthalpy change dH and the enthalpy is thus analogous to the internal energy.

In solids $(C_p - C_v)$ is very small and becomes significant only at high temperatures, $> \sim 1500\text{K}$.

It can be shown that

$$C_p - C_v = TV\alpha^2/\beta \qquad (8.6)$$

where α is the thermal expansion coefficient and β is the compressibility. α and β reflect the response of the bond angles and bond lengths to changes in T and P.

At low temperature, as T approaches 0 K, both C_p and C_v approach zero. The way in which the heat capacity varies with temperature will be discussed in a later section, but has the general form shown in Figure 8.1. The enthalpy change between temperature T_1 and T_2 is given by integrating eqn. (8.5) to give

$$\Delta H = \int_{T_1}^{T_2} C_p dT$$

and the overall enthalpy at any temperature T_1 is

$$H = H_o + \int_0^{T_1} C_p dT \qquad (8.7)$$

where H_o includes the enthalpy due to the potential

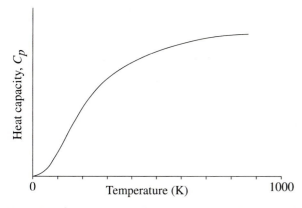

Figure 8.1. General form of the variation of specific heat with temperature.

energy of the crystal at 0 K as well as the zero point vibrational energy. The integral includes the kinetic energy contribution up to temperature T_1, plus changes in the potential energy due to expansion and tilting of bonds.

8.2 Enthalpy changes in mineral transformations and reactions

The absolute values of many thermodynamic quantities cannot be obtained and therefore we are generally concerned with differences between different states. It is convenient to define a *standard state*, at which the value of the particular quantity is specified at a reference temperature and pressure. For solids the standard state is usually 298.15 K and 1 atmosphere. The standard state enthalpies of the elements are arbitrarily assigned a value of zero, and enthalpies of binary compounds are specified as an enthalpy of formation ΔH_f^o, from the elements. For ternary oxide minerals it is often more convenient to define the standard state enthalpy relative to the component binary oxides, under the standard state conditions. (The superscript 'o' refers to the standard state).

For example, in tabulations of enthalpies of formation, ΔH_f^o for MgO is given as -601.5 kJ mol^{-1}. This is the enthalpy change for the reaction

$$Mg + 1/2O_2 \Rightarrow MgO$$

at 1 atmosphere and 298K.

For the reaction $Si + O_2 \Rightarrow SiO_2$ (low quartz) under the same conditions, the enthalpy change ΔH_f^o is -910.7 kJ mol^{-1}. The negative sign in each case

means that the enthalpy of the right hand side of the equation is smaller, and hence heat is given out in the reaction.

For the ternary compound Mg$_2$SiO$_4$ (olivine structure), ΔH_f^o from the oxides is given as -56.69 kJ mol^{-1}, which is the enthalpy change for the reaction

$$2MgO + SiO_2 \Rightarrow Mg_2SiO_4$$

at 1 atmosphere and 298K.

The enthalpy of formation of the compound Mg$_2$SiO$_4$ from the elements according to the reaction

$$2Mg + Si + 2O_2 \Rightarrow Mg_2SiO_4$$

can be found by combining the enthalpy changes for the above three reactions i.e. $- (2 \times 601.5) - 910.7 - 56.69 = - 2170.39$ kJ mol^{-1}. The principle of the addition of enthalpies (Hess' Law) makes it possible to calculate unknown enthalpy changes from combinations of reactions.

To determine the enthalpy at a different temperature T we need to find the change in enthalpy of the substance between 298K and T, using the expression

$$H_T = H_{298} + \int_{298}^{T} C_p \, dT$$

which assumes that we know the variation in the heat capacity as a function of temperature. We will return to this point in a later section.

When a mineral changes its structure in a polymorphic transformation, or one mineral assemblage changes to another there will be a change in the enthalpy ΔH. If the enthalpy is reduced in such a reaction, ΔH is negative and the process is termed *exothermic* i.e. heat is evolved; when ΔH is positive the process is *endothermic* i.e. heat is absorbed.

For example, the standard state enthalpy of tridymite (SiO$_2$) is ΔH_f^o for the reaction

$$Si + O_2 \Rightarrow SiO_2 \text{ (tridymite)}$$
$$\Delta H_f^o = -907.5 \text{ kJ mol}^{-1}.$$

The equivalent reaction for the formation of quartz is

$$Si + O_2 \Rightarrow SiO_2 \text{ (low quartz)}$$
$$\Delta H_f^o = -910.7 \text{ kJ mol}^{-1}.$$

To determine the enthalpy change for the transformation from tridymite to quartz these two reactions are combined, taking care with the signs of the reactions. Thus:

SiO_2 (tridymite) $\Rightarrow$ Si $+$ O_2 $\Rightarrow$ SiO_2 (low quartz)
$$\Delta H = +907.5 - 910.7 = -3.2 \text{kJ mol}^{-1}$$

The transformation is therefore exothermic.

The reverse reaction: SiO_2 (low quartz) $\Rightarrow$ SiO_2 (tridymite) at the standard state conditions has an equal but opposite enthalpy change to the forward reaction, i.e. $+3.2$ kJ mol^{-1}, and is endothermic.

In seeking to define a criterion for stability, and hence for the direction of change in a mineral transformation, one might think that the enthalpy must be minimised. This is *not* however the case, as is shown from the following example.

Calcium carbonate $CaCO_3$ exists in two polymorphic forms, calcite and aragonite. The standard state enthalpy of calcite is -1207.37 kJ mol^{-1}, while that of aragonite is -1207.74 kJ mol^{-1}. The transformation from aragonite to calcite at 25°C and 1 atm. pressure therefore involves an *increase* in enthalpy of 370 joules, i.e. the reaction is endothermic. However, at 25°C it is calcite which is known to be the stable polymorph of $CaCO_3$, illustrating the point that a reduction in enthalpy is not a sufficient criterion for stability.

An additional criterion must be introduced to be able to define the relative stability of minerals and predict the direction of a mineral reaction. This criterion involves the quantity known as *entropy*.

8.3 Entropy and disorder

When a mineral changes from one structure to another it exchanges heat with its surroundings. The *entropy* is defined as the quantity which measures the change in the state of order associated with this process. The overall entropy change is the sum of the entropy change in the mineral (i.e. the system under consideration) and the entropy change in the surroundings, i.e.

$$dS = dS_{\text{system}} + dS_{\text{surr.}}$$

For a reversible reaction, i.e. one which passes through a continuous sequence of equilibrium states, $dS = 0$, but for any natural reaction proceeding towards equilibrium, $dS > 0$, according to the Second Law of Thermodynamics. This means that if a mineral becomes more ordered in a transformation process, and hence reduces its entropy, the heat liberated must increase the entropy of the surroundings by a greater amount, to satisfy the second law.

The entropy change is defined by

$$dS > dQ/T \tag{8.8}$$

where dQ is the amount of heat exchanged by the system at temperature T. In a system free to exchange heat with its surroundings the change in entropy of the system is related to the enthalpy change by the relation

$$dS > dH/T$$

noting that at constant pressure $dQ = dH$. Thus the criterion for a mineral transformation or reaction to proceed is that

$$dH - TdS < 0$$

If $dH - TdS = 0$ no further change is possible, i.e. the system is at equilibrium. If $dH - TdS$ is > 0 the reaction will not proceed.

The quantity $(dH - TdS)$ therefore can be used to define a criterion for the direction of change in a mineral reaction and a definition of equilibrium. This quantity is known as the change in the *Gibbs free energy*, dG of the system. Thus

$$dG = dH - TdS$$
or $$G = H - TS \tag{8.9}$$

The equivalent criterion when a system is considered at constant volume rather than at constant pressure would be that $(dU - TdS)$. The equivalent free energy is known as the *Helmholtz free energy*, dF of the system. Thus at constant volume

$$F = U - TS \tag{8.10}$$

Since constant pressure experiments with solids are far easier to perform than constant volume experiments, the Gibbs free energy is much more useful than the Helmholtz free energy. Whenever the term free energy is used in this book, it refers to the Gibbs free energy, G.

Entropy is not an intuitively straightforward concept and its formal definition does not explain its physical basis. It was introduced in classical thermodynamics to explain the practical problem of the direction of heat flow in heat engines and the fact that a 100% efficiency could not be theoretically achieved, but its atomistic basis had to await the advent of statistical thermodynamics.

8.3.1 Configurational entropy

In the statistical definition, according to Boltzmann, the entropy of a system in a given state is related to the probability of the existence of that state. The "state" in this context refers to a particular distribution of atoms and their vibrational energy levels. Mathematically this is expressed as

$$S = k \ln \omega \qquad (8.11)$$

where ω is the probability that a given state will exist and k is Boltzmann's constant (1.38×10^{-23} JK^{-1}). The probability is related to the state of disorder or randomness in the structure which may be expressed statistically by the number of different ways in which atoms can arrange themselves in that state.

For example, consider a distribution of atoms of A and B on a simple cubic lattice (Figure 8.2) which contains a total of N atomic sites. To determine the entropy of a completely random distribution of A and B atoms we need to calculate the number of ways in which the atoms can be arranged. If the atomic fraction of A atoms is x_A and of B atoms is x_B, there are $x_A N$ atoms of A and $x_B N$ atoms of B to be distributed over N sites. The number of such arrangements, ω, is determined from elementary statistics as

$$\omega = \frac{N!}{(x_A N)! \, (x_B N)!}$$

The entropy associated with this disorder is therefore

$$S = k \ln \omega = k \ln \frac{N!}{(x_A N)! \, (x_B N)!}$$

When the number of sites N is very large, as is the case in a mole of a mineral, we can simplify this expression by using Stirling's approximation:

$$\ln N! = N \ln N - N$$

Thus
$$S = -Nk(x_A \ln x_A + x_B \ln x_B)$$

For a mole of sites N is Avogadro's number (6.02×10^{23} mol^{-1}) and $Nk = R$, the gas constant (8.31 J mol^{-1}K^{-1}).

Hence
$$S = -R(x_A \ln x_A + x_B \ln x_B)$$

By taking N as Avogadro's number we have limited the calculation to considering only one structural site over which substitution of A and B occurs. In a complex mineral structure there may be more than one site per formula unit over which disorder can occur and the more general form of the above expression is

$$S = -nR(x_A \ln x_A + x_B \ln x_B) \qquad (8.12)$$

where n is the number of sites on which mixing occurs. This expression is known as the *entropy of mixing*. Since x_A and x_B are both fractions, S is always positive as shown in Figure 8.3 which gives the general form of the curve for the entropy of mixing ΔS as a function of the atomic fraction x_B of B atoms.

In general, when the entropy is due to atomic disorder it is known as configurational entropy and is calculated using eqn.(8.12) above, as illustrated in the following example:

In cordierite $Mg_2Al_4Si_5O_{18}$ there are 4 Al atoms and 5 Si atoms distributed over 9 tetrahedral sites in the formula unit (Section 6.8.4). If the distribution

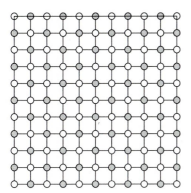

Ordered

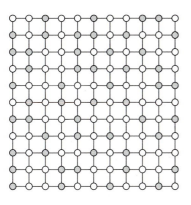

Disordered

Figure 8.2. Ordered and disordered distributions of equal numbers of A and B atoms (shaded and unshaded) on a simple cubic lattice.

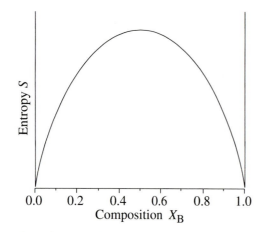

Figure 8.3. The general form of the configurational entropy S, associated with the disorder of A and B atoms, as a function of composition. The maximum entropy occurs when the atomic fraction of A and B atoms is equal, i.e. $X_A = X_B = 0.5$.

was totally random over all 9 sites, the configurational entropy would be

$$S = -nR \left(x_A \ln x_A + x_B \ln x_B \right)$$

$$= -9 \times 8.31 \left(\tfrac{4}{9} \ln \tfrac{4}{9} + \tfrac{5}{9} \ln \tfrac{5}{9} \right)$$

$$= 51.39 \text{ J mol}^{-1} \text{ K}^{-1}$$

Consider however the case where the 9 tetrahedral sites are subdivided into 3 T_1 sites and 6 T_2 sites and the disorder is described as 1 Si and 2 Al atoms disordered over the 3 T_1 sites, and 4 Si and 2 Al atoms disordered over the 6 T_2 sites. The entropy must now be calculated for each type of site and summed as follows:

For T_1: $S = -3 \times 8.31 \left(\tfrac{1}{3} \ln \tfrac{1}{3} + \tfrac{2}{3} \ln \tfrac{2}{3} \right)$

$$= 15.87 \text{ J mol}^{-1} \text{ K}^{-1}$$

For T_2: $S = -6 \times 8.31 \left(\tfrac{4}{6} \ln \tfrac{4}{6} + \tfrac{2}{6} \ln \tfrac{2}{6} \right)$

$$= 31.74 \text{ J mol}^{-1} \text{ K}^{-1}$$

giving a total configurational entropy of
47.61 J mol^{-1} K^{-1}

For a completely ordered distribution of Al and Si where the occupancy of each atomic site is specified, the configurational entropy is zero (because $x \ln x$ is 0 when $x = 0$ or 1).

8.3.2 Electronic entropy

Electronic configurational entropy arises when electrons in unfilled shells can be distributed over a number of degenerate orbitals. In the transition metals, with unfilled $3d$ orbitals, this form of entropy is an important contribution to the total entropy. For example, the Ti^{3+} ion has only one $3d$ electron and in regular octahedral coordination this occupies one of the three t_{2g} orbitals (see section 4.5). The number of ways of distributing this electron over 3 degenerate orbitals is $-3R\tfrac{1}{3}\ln\tfrac{1}{3} = 9.13 \text{ JK}^{-1}\text{mol}^{-1}$.

In a distorted octahedral site the degeneracy between the t_{2g} orbitals is removed (Figure 4.28) and the electron occupies the lowest energy orbital. There is therefore no electronic entropy associated with this configuration.

As another example we can compare the cases of Fe^{3+} and Fe^{2+} in an undistorted octahedral site. In Fe^{3+} the five electrons all have unpaired spins and occupy each of the five $3d$ orbitals, i.e.

$$\uparrow \ \uparrow \ \uparrow \qquad \uparrow \ \uparrow$$
$$t_{2g} \qquad\qquad e_g$$

The electronic entropy in this case is therefore zero. For Fe^{2+} in an undistorted octahedral site, the six $3d$ electrons are distributed so that the extra electron may be spin-paired with any of the three t_{2g} electrons, one of which could be

$$\uparrow \ \downarrow \ \uparrow \ \uparrow \qquad \uparrow \ \uparrow$$
$$t_{2g} \qquad\qquad e_g$$

The electronic entropy associated with this state is therefore $-3R\tfrac{1}{3}\ln\tfrac{1}{3}$.

When a change in the structure of a mineral results in a change in the coordination of a transition metal ion, the electronic configuration also changes. The thermodynamic consequences may include an electronic entropy change as well as an enthalpy change due to a change in the crystal field stabilization energy.

8.3.3 Vibrational entropy

Another important source of entropy in a crystal is due to the disorder associated with the lattice vibrations. As we have discussed in Chapter 4, the energy of lattice vibrations is quantised and each quantum of vibrational energy known as a phonon. When the amplitude of atomic vibrations increases, we speak of an increase in the number of phonons in the crystal. The phonon spectrum defines the number of phonons in each frequency range and the function describing this frequency distribution is termed the phonon density of states.

The vibrational entropy arises by considering the number of ways of distributing the atoms over the vibrational energy levels which exist in a crystal. This is very much more difficult to calculate than configurational entropies because the number of energy levels is not as simply defined as the number of crystallographic sites. However, the vibrational entropy may be calculated from the heat capacity which defines the way in which an increase in the heat content in the crystal is related to an increase in temperature. This depends on the way in which the extra phonons are distributed over the available vibrational energy levels.

The relation between the vibrational entropy and the heat capacity C_p can be seen by considering that for a reversible process $dS = dQ/T$. From the definition of the heat capacity, $C_p = dQ/dT$, it follows that $dS/dT = C_p/T$ and hence

$$S = \int \frac{C_p}{T} \, dT$$

Figure 8.4 shows the variation of the function (C_p/T) with temperature. The entropy at any temperature T_1 is given by the expression

$$S = S_o + \int_0^{T_1} \frac{C_p}{T} \, dT \qquad (8.13)$$

where S_o is the entropy at zero K.

A simple statement of the Third Law of Thermodynamics is that at 0 K the entropy of a perfect crystal is zero, all atoms occupying the ground state, hence $S_o = 0$. However if a material has a disordered atomic distribution at a higher temperature, rapid cooling may freeze in this state, even down to 0 K where the thermodynamically favoured state would be the one with the minimum internal energy, i.e. complete order. Such residual disorder, and the associated *zero-point entropy*, as it is called, is due to the fact that at low temperatures the rate of atomic diffusion becomes vanishingly small. Al,Si disorder in aluminosilicates is an example. A measurement of the heat capacity in such a case records only the vibrational contribution to the entropy and the configurational contribution must be separately assessed.

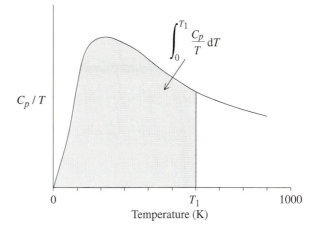

Figure 8.4. The vibrational entropy at a particular temperature T_1 is the shaded area under the curve of C_p/T against T between $T = 0$ and $T = T_1$.

8.4 The Gibbs free energy G, and equilibrium

In the previous section we concluded that the condition for stability is that the free energy G is a minimum, and hence the criterion for a mineral transformation to take place is that the change in the Gibbs free energy ΔG, defined as $\Delta G = \Delta H - T\Delta S$ must be negative. Referring back to the example of the transformation of aragonite to calcite (section 8.2):

$$\text{aragonite} \Rightarrow \text{calcite}$$
$$\Delta H^o = +370 \text{ joules at 25°C and 1 atm.}$$

The entropy of aragonite and calcite under the same conditions is 88 J mol^{-1} K^{-1} and 91.7 J mol^{-1} K^{-1} respectively. Thus for the transformation

$$\text{aragonite} \Rightarrow \text{calcite}$$
$$\Delta S^o = +3.7 \text{ J mol}^{-1} \text{ K}^{-1} \text{ at 25°C and 1 atm.}$$

Therefore ΔG for the aragonite $\Rightarrow$ calcite transformation equals $\Delta H - T\Delta S = +370 - (298 \times 3.7) = -732.6$ joules. Calcite thus has the lower free energy and is the stable polymorph of calcium carbonate at 25°C and 1 atm. pressure.

The way in which the free energy G changes as a function of temperature and pressure is shown in Figure 8.5(a,b). We can derive the relations between G, T and P from the basic definitions of the thermodynamic functions as follows:

$$G = H - TS$$

Substituting $H = U + PV$ into this equation gives

$$G = U + PV - TS$$

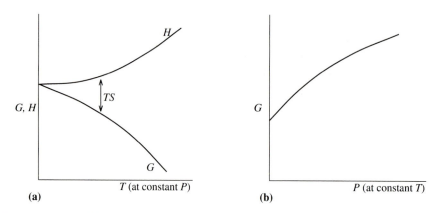

Figure 8.5. (a) The change in enthalpy H, and free energy G as a function of temperature, at constant pressure. The difference between the enthalpy and free energy is TS, according to the definition $G = H - TS$. (b) The change in free energy G as a function of pressure, at constant temperature.

Differentiating gives

$$dG = dU + PdV + VdP - TdS - SdT.$$

From the first law, $dU = dQ - PdV$, and for a reversible process (which is always in equilibrium) $dQ = TdS$. On substitution

$$dG = VdP - SdT \qquad \text{at equilibrium} \quad (8.14)$$

Therefore at constant pressure:
$$(\partial G/\partial T)_P = -S \qquad\qquad (8.15)$$

and at constant temperature:
$$(\partial G/\partial P)_T = V \qquad\qquad (8.16)$$

Thus Figure 8.5(a,b) shows the way in which the free energy of a mineral structure changes when it maintains equilibrium with a changing T and P.

In a simple polymorphic transformation occurring as a function of temperature, the G–T curves for the two structures intersect at some temperature T_c as shown in Figure 8.6. Above T_c the phase labelled β has a lower free energy and hence is more stable than phase α, while below T_c, α is the more stable phase. At the transformation temperature T_c the free energies of the two phases are equal, and hence $\Delta G = 0$. A transformation from one phase to the other is therefore accompanied by an abrupt change in the enthalpy given by $\Delta H = T\Delta S$, which is the *latent heat of transformation* (Figure 8.6).

If we consider both temperature and pressure stability fields for two polymorphs α and β, the free energy curves become surfaces in G–T–P space (Figure 8.7(a)) and their intersection defines the equilibrium between α and β. The slope of this

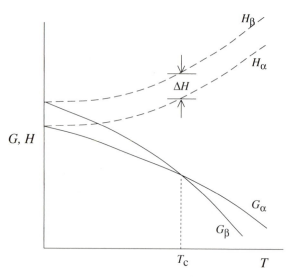

Figure 8.6. The variation of free energy G and enthalpy H of two phases α and β as a function of temperature. The enthalpy difference ΔH, between the phases shows little variation with temperature. If however phase β has a higher entropy than α, their free energy curves will cross at some temperature T_c. The stable phase is α below T_c, and β above T_c.

intersection line when projected on the P–T plane, i.e. dP/dT is given by the *Clapeyron relation* which arises from the equilibrium relation $\Delta G = 0$. At equilibrium, from eqn. 8.14,

$$\Delta VdP = \Delta SdT$$

and hence

$$dP/dT = \Delta S/\Delta V \qquad\qquad (8.17)$$

Figure 8.7(b) shows a typical Clapeyron curve for a polymorphic transformation. A linear slope implies

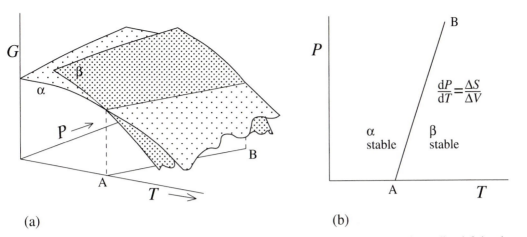

Figure 8.7. (a) Free energy surfaces for two phases α and β in pressure – temperature space intersect along a line defining the equilibrium between them. The projection of this line on the P–T plane is shown as AB. (b) The equilibrium between α and β defined by the line AB in the P–T plane.

that both ΔV and ΔS are independent of temperature and pressure, which in many reactions involving only solids of constant composition, is a reasonable assumption.

Note on units: When ΔS is measured in J mol^{-1} K^{-1}, and ΔV is in m^{3}mol^{-1} the pressure must be given in pascals (Pa). However in experimental petrology the pressure is still usually measured in bars and the P,T slope is given as bars deg^{-1}. In this case if ΔS is measured in J mol^{-1} K^{-1}, dP/dT comes out in J m^{-3} K^{-1} which is equivalent to Pa K^{-1}.

(1 pascal = 10^{-5} bars). Alternatively, the volume unit may be converted using the conversions 1m^3 = 1J.Pa^{-1} or 1 cm^3 = 0.1 J bar^{-1}.

8.4.1 Reversible and irreversible processes. Metastability

The two G–T curves in Figure 8.8(a) indicate that, in order to maintain equilibrium, phase β should transform at temperature T_c to phase α on cooling, as shown by the path marked with arrows. Along this path the system moves through a continuous sequence of equilibrium states and the entropy change $\Delta S = 0$. Along this path the transformation is reversible in that the same path will be followed on cooling and on heating.

However, in practice transformations do not occur at the thermodynamically defined equilibrium temperature, because at T_c the free energy of both phases is the same. Only when there is a reduction in free energy will there be a free energy drive to

transform β to α. Therefore some degree of undercooling, ΔT, will be required, as shown in Figure 8.8(b). The amount of undercooling depends on the structural changes which are required in the transformation. As mentioned already in Section 7.1.4 there is always an activation energy barrier associated with atomic diffusion and reorganisation. If the activation energy for the formation of the low temperature phase is small, the amount of undercooling required to nucleate the phase will also be small. A high activation energy implies a greater degree of undercooling in order to provide the necessary 'driving force' as it is often termed.

When a transformation occurs at a temperature below T_c on cooling, it is irreversible and will take place at a different temperature ($>T_c$) on heating. The amount of superheating ΔT, is smaller than the amount of supercooling, since increasing the temperature increases the atomic mobility. In irreversible transformations the actual temperature at which a transformation takes place depends on the cooling and heating rate i.e. on the kinetics of the process, and this will be discussed in Chapter 12. However, it is worth noting throughout any discussion of thermodynamics that while a reduction in free energy is a necessary condition for a transformation or reaction to occur, the actual behaviour will depend on the rate of the processes involved.

One effect of the decreasing mobility of atoms at lower temperatures is that structural changes become more difficult, despite the increasing free energy drive at greater undercooling. In an extreme

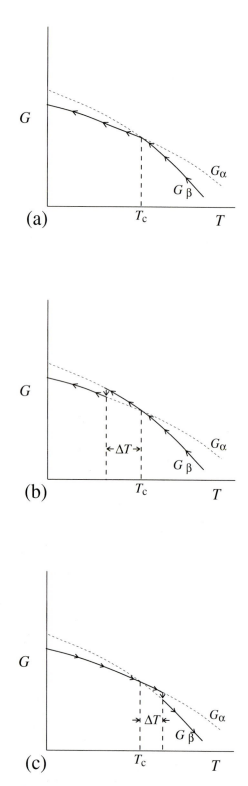

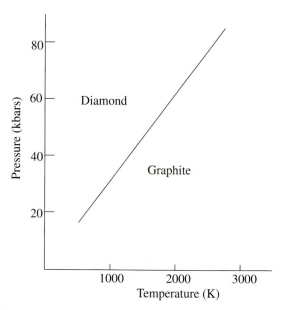

Figure 8.9. The stability fields of diamond and graphite as a function of temperature and pressure.

Figure 8.8. Free energy curves for two phases α and β intersecting at the equilibrium temperature T_c. (a) The full line shows the equilibrium cooling path with the transformation from β to α taking place at T_c. The same path would be followed in reverse on heating α. (b) In practice the transformation from β to α requires some degree of undercooling ΔT. The reverse transformation from α to β will not follow the same path but will transform at a temperature above T_c as shown in (c).

case nothing at all might happen. A well-known example is the case of diamond and graphite. Diamonds which have grown within their stability field deep in the Earth's crust (Figure 8.9) and been rapidly brought to the near surface in kimberlite rocks are preserved, out of equilibrium, indefinitely. Their rate of transformation to graphite at room temperature is insignificantly small, despite the fact that the reduction in free energy would be about 3 kJ mol^{-1}. Diamond is said to be in a metastable state in that although the structure occupies a local free energy minimum, there is a lower value of the free energy separated from it by a large activation energy (Figure 8.10).

An important outcome of the large undercooling required to enable some transformations to proceed at a significant rate even on a geological timescale, is that if a complex mineral structure fails to transform to the thermodynamically stable state, it may undergo a structurally easier transformation to some other phase. This is shown schematically in Figure 8.11, where the dashed G–T curve of the γ phase represents a structure which is not thermodynamically stable at any temperature, but may be kinetically more accessible than the α phase. If the β ⇒ γ transformation is easier (has a lower activation energy) than the β ⇒ α transformation, the β ⇒ γ transformation defines a *metastable equilibrium* at temperature T_2 in Figure 8.11.

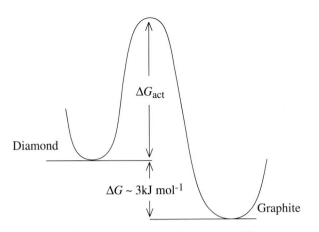

Figure 8.10. Diamonds do not transform to graphite at room temperature because although graphite is thermodynamically more stable there is a large activation energy barrier ΔG_{act}, for the transformation.

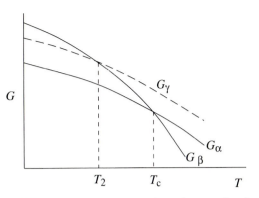

Figure 8.11. Free energy curves for three phases α, β and γ as a function of temperature. Only α and β have stability fields, below and above the equilibrium temperature T_c. Although γ does not have a stability field a transformation from β to γ at the metastable equilibrium temperature T_2 is possible on cooling if the transformation to the stable α phase is impeded.

An example of such behaviour is the case of the clinopyroxene to orthopyroxene transformation in low-Ca pyroxenes, such as pigeonite, discussed in Section 6.4.1. The high pigeonite structure is not thermodynamically stable at low temperatures and should transform to the orthopyroxene structure. This however involves a major structural reorganisation and can only take place in very slowly cooled rocks. If orthopyroxene fails to form, an alternative metastable transformation to low pigeonite takes place. Recall that the high–low transformation in pigeonite is a rapid displacive transformation which does not involve any bond breaking.

Another well known example of such metastability is in the phases of SiO_2. The high temperature phases cristobalite and tridymite both have distorted low temperature structures which are not thermodynamically stable relative to quartz. However, both low cristobalite and low tridymite can persist indefinitely in rocks crystallised at high temperatures and cooled at a rate too fast for the stable sequence of transformations high cristobalite ⇒ high tridymite ⇒ quartz to occur (see Section 6.8.1).

There is of course nothing unusual about the persistence of metastable states in natural rocks. In the literal sense all crystalline rocks are metastable and are slowly weathering to more stable clay minerals and muds. Their preservation over geological time is an indication of the sluggish kinetics of such processes at surface temperatures. However, in our discussions of mineral transformations we will be primarily concerned with higher temperature pro-

cesses where the atomic mobility is sufficient to allow structural responses to changing physical conditions.

Another concept which is useful to note is that of *partial equilibrium* when some of the variables in a system have reached equilibrium while others remain in a non-equilibrium state. For example in a mineral assemblage undergoing metamorphism some minerals will reach equilibrium before others due to the fact that reaction rates are not all the same. Furthermore in a reaction such as A + B ⇒ C + D chemical equilibrium between the phases may be established before the most stable structural state of the individual phases is achieved. The reaction and transformation sequence is controlled by the rates of the processes involved, and while every transformation reduces the free energy, the system as a whole may never reach the lowest possible free energy.

8.4.2 First- and second-order phase transitions

Structural phase transitions from one polymorph to another can be classified by the behaviour of thermodynamic quantities such as entropy, heat capacity, volume etc. through the transition. In Chapters 11 and 12 we will discuss a classification based on the structural mechanism involved, and see how the classifications relate to one another. Here we briefly outline the thermodynamic classification.

Phase transitions occur in response to a change in the external environment. At the equilibrium temperature (or pressure) at which the transition theoretically takes place, the free energies of the two polymorphs are equal, and hence there is no discontinuity in the free energy G on passing from one structure to another. However, in *first-order phase transitions* the first derivatives of the free energy $\partial G/\partial T$ and $\partial G/\partial P$ are discontinuous. Since $\partial G/\partial T = -S$ and $\partial G/\partial P = V$, first-order phase transitions are characterized by discontinuous changes in entropy and volume at the critical temperature. There is also a discontinuity in the enthalpy, as shown in Figure 8.6, corresponding to the latent heat of the transition. At the transformation temperature the specific heat becomes infinite, since the addition (or subtraction in an exothermic change) of heat serves to convert more of one phase into another, rather than to change the temperature.

In *second-order phase transitions* the first derivatives of the free energy are continuous, but the second derivatives $\partial^2 G/\partial T^2$ and $\partial^2 G/\partial P^2$ are discontinuous. The enthalpy change is continuous and so there is no latent heat associated with second order transitions. Since

$$\partial^2 G/\partial T^2 = -\partial S/\partial T = -C_p/T$$

and $$\partial^2 G/\partial P^2 = -V\beta; \quad \partial^2 G/\partial T\partial P = V\alpha$$

the discontinuities occur in the specific heat capacity C_p, the compressibility β and the thermal expansion α.

Figure 8.12 shows the changes in thermodynamic properties through first- and second-order phase transitions. The use of this classification obviously depends on being able to measure the appropriate properties in a mineral undergoing a phase transition. Usually first-order transitions are easy to detect if the discontinuities are large, by measuring lattice parameters as a function of temperature (or pressure), or by calorimetric methods which directly measure the enthalpy change. Second-order transitions are more difficult to detect, especially as the changes involved are generally small. Measuring properties such as C_p near the transition temperature is experimentally difficult since the variation with temperature can be large. There are also many transformations in minerals which cannot be studied directly in this way, since the changes involved are far too sluggish to be measurable on a laboratory time-scale.

Although this classification between first- and second-order transitions appears to be quite clear-

cut, in practice we should regard these as extreme cases. Even first-order transformations show continuous changes before T_c is reached, so-called 'premonitory' effects, and the discontinuity at T_c may be quite small. In second-order transitions most of the structural change takes place before T_c is reached, and T_c represents the temperature at which the changes are effectively complete. There are cases where transitions are referred to as 'nearly second order' and the changeover from first to second order is difficult to define.

Despite these problems we will refer to first- and second-order transitions frequently throughout the rest of this book. One important connection between the thermodynamics of a transition and the symmetry changes involved is made in Landau theory (Section 8.6) in which a formal procedure determines whether a second-order transition is possible by comparing the symmetry elements of the two structural states.

8.5 Determining thermodynamic quantities

Up to this point we have described some of the concepts leading to the definition of the free energy of a mineral under standard state conditions and how this free energy changes as a function of temperature and pressure. To be able to use thermodynamic data to solve geological problems we need to have a data base which tabulates values of enthalpy and entropy of end-member mineral phases in the standard state, and also defines the variation of their heat capacity with temperature. From this data enthalpies and entropies and hence free energies can be calculated at other temperatures. To determine the pressure dependence we also need to know molar volumes as well as values of the thermal expansion coefficient α and the compressibility β.

There are various ways of determining thermodynamic data. The volume properties are relatively easy to measure by measuring lattice parameters as a function of temperature and pressure using X-ray diffraction. The thermal properties are more difficult to measure and here we can only briefly refer to the main methods which we list under three headings: (i) direct experimental measurements, (ii) phase equilibrium studies and (iii) theoretical methods.

One of the perennial problems of determining the equilibrium state between minerals is that, since equilibrium is defined as $\Delta G = 0$, we are always

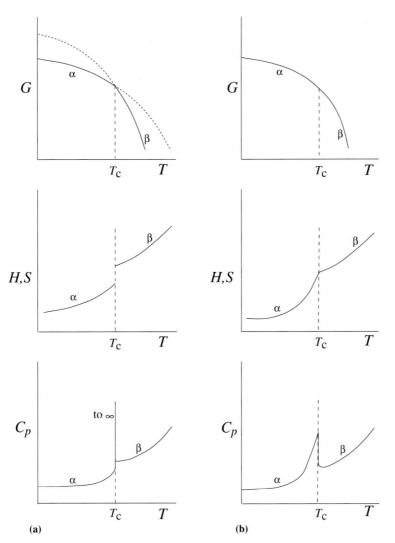

Figure 8.12. (a) The changes in the thermodynamic properties free energy (G), enthalpy (H), entropy (S) and specific heat (C_p) as a function of temperature for a first- order transformation between two phases α and β. (b) The same for a second-order transformation.

involved in small *differences* between thermodynamic quantities. These quantities are typically large numbers, and even when their relative errors are small, the result of subtraction is a small number with a relatively large error. Thermodynamic measurements therefore, must be made with very high precision if they are to be useful in accurately calculating equilibrium.

8.5.1 Direct experimental measurements

Direct measurements of heat capacity over a range of temperatures can be used to calculate both the enthalpy and entropy, assuming that the sample achieves equilibrium at the temperature of each

measurement. Thus the effect of any sluggish reequilibration process, such as Al,Si ordering and disordering may not be included in the heat capacity measurement. The entropy calculated from such measurements is the so-called *third-law entropy* which neglects any configurational effects due to frozen-in disorder.

Heat capacity measurements can be made using a number of different approaches:

1. At low temperature, heat capacities can be measured with very high precision (better than 0.2%) from liquid helium temperatures (4.2K) to around 400K using *adiabatic calorimetry*. The sample is thermally insulated from its surroundings and the rise in temperature recorded as a result of electrical heating for a fixed time at a measured current and voltage. Such measurements need relatively large amounts of sample (~10 gm) and this can be a serious limitation in cases of phases which are difficult to synthesise.

2. At higher temperatures (up to 1000K) heat capacities can be determined by *differential scanning calorimetry* (DSC) which measures the different amounts of power required to keep two sample holders at the same temperature. One sample is the unknown, the other a standard whose heat capacity is known very precisely, and the difference in the power is used to calculate the heat capacity of the unknown. The uncertainties in this method (±1%) are larger than with adiabatic calorimetry at low temperature, although less than 20mg of sample is required.

Another widely used but older method of heat capacity determination is *drop calorimetry*, where a measured mass of sample is heated to the required temperature and then dropped into a calorimeter operating at room temperature. The heat evolved as the calorimeter and its contents cool back to room temperature is measured.

The temperature dependence of heat capacity measurements above room temperature is often represented by an empirical relation of the type

$$C_p = a + bT + cT^{-2} + dT^{-1/2} \qquad (8.18)$$

and tabulations of thermodynamic data may quote values of the empirical constants a, b, c, d.

Heat capacities are approximately additive, allowing the estimation of an unknown value by a summation method. For example, grossular $Ca_3Al_2Si_3O_{12}$ may be written as 3 wollastonite + corundum i.e. $3 \times CaSiO_3 + Al_2O_3$, and hence the specific heat of grossular can be approximated as $3C_p(\text{woll.}) + C_p(\text{cor.})$.

3. An important experimental technique for measuring enthalpies is to use *heat of solution calorimetry* which measures the heat of solution in a suitable solvent. Most mineral data have come from various types of acid calorimeters in which the sample is dissolved in strong acid. However aluminosilicates dissolve only slowly, even in hydrofluoric acid, and high temperature molten salt calorimeters are now more commonly used. Molten lead borate which has an operating temperature of between 600°C and 1200°C is the preferred solvent and is able to dissolve most aluminosilicate minerals relatively

quickly. The heat of solution changes the temperature in the reaction vessel and this change in temperature is compared with that in a twin chamber which acts as a control. The heat of solution itself is not of interest, but differences between samples reflect differences in their enthalpy at the temperature at which the measurement is made (usually between 600°C and 1000°C). After calibration with known substances, this method is capable of determining enthalpies with a precision of around 1%.

8.5.2 Phase equilibrium studies

Determining the equilibrium temperatures and pressures for reactions between minerals has been a major source of thermodynamic data for the individual mineral phases. Here we will illustrate the principles involved and some issues which arise from such studies with a simple example which involves only solid phases. We will consider the reaction

$$\text{albite} \Rightarrow \text{jadeite} + \text{quartz}$$
$$NaAlSi_3O_8 \Rightarrow NaAlSi_2O_6 + SiO_2$$

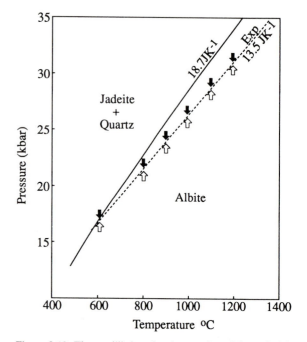

Figure 8.13. The equilibrium for the reaction albite ⇒ jadeite + quartz. The solid line represents the calculated equilibrium from measured thermodynamic properties, assuming that the albite has complete Al,Si disorder and hence a configurational entropy of $18.7 JK^{-1}$. The dashed line is the direct experimental determination of the equilibrium, the arrows bracketing the equilibrium from both directions. The slope of this line indicates a configurational entropy of $13.5 JK^{-1}$ in the albite. (Experimental data from Holland, 1980)

for which the equilibrium Clapeyron curve is shown in Figure 8.13.

The first requirement for phase equilibrium studies is to demonstrate equilibrium by reversing the reaction. The reaction is monitored in the forward direction (usually by X-ray diffraction of samples quenched from the high pressure, high temperature experiment), and then in the reverse direction to define as narrowly as possible the equilibrium pressure and temperature. Since reaction rates decrease dramatically as the equilibrium is approached, the equilibrium point can never be precisely located, and the best estimate is a bracketed line in P–T space.

Along the equilibrium line, the free energy of the albite is equal to the sum of the free energies of jadeite + quartz, i.e. $\Delta G = 0$. From eqn. (8.14) the equilibrium condition can be expressed as

$$\Delta G_{P,T} = \text{O} = \Delta G^{\circ}_{1,298} + \int_1^P \Delta V \mathrm{d}P - \int_{298}^T \Delta S \mathrm{d}T$$

Thus, at equilibrium

$$\Delta G^{\circ}_{1,298} = \int_{298}^T \Delta S \mathrm{d}T - \int_1^P \Delta V \mathrm{d}P$$

where $\Delta G^{\circ}_{1,298}$ is the difference in the standard state (1 atmosphere and 298K) free energies of the products and reactants, and the integrals define the way in which this free energy difference changes as a function of temperature and pressure respectively.

The pressure integral can be determined exactly if the compressibility β and the thermal expansion α are known. When this is not the case, estimates based on empirical relationships between volume, pressure and temperature for various structure types can be used. Alternatively, *in reactions involving only solids,* compressibility and expansion effects are sometimes ignored since the volume decrease when the whole assemblage is compressed from 1 bar to the pressure P at 298K, tends to be cancelled by the volume increase as the temperature is raised from 298K to the temperature T. The integral is then determined directly from the known volumes of the phases at 1 atmosphere and 298K and

$$\int_1^P \Delta V \mathrm{d}P = \Delta V(P - 1)$$

When P is large (generally kbars) this can be approximated by $P\Delta V$.

The temperature integral $\int_{298}^T \Delta S \mathrm{d}T$ depends on the change in the heat capacities of the phases with temperature, and this estimate depends on the power series used to describe this dependence. However, as pointed out in Section 8.3.2, heat capacity measurements can only be used to calculate third-law entropies and any configurational effects must be separately assessed.

Here we come to the first important issue raised by an experiment on the albite $\Rightarrow$ jadeite + quartz reaction, and that is the effect of Al,Si disorder. The Clapeyron relation indicates that the slope of the equilibrium reaction line depends on the total entropy change, which in turn depends on the degree of order in the albite. Since accurate heat capacity data are available for disordered albite as well as jadeite and quartz any reduction in the entropy of albite due to Al,Si order may in principle be detected from the experimental slope of the equilibrium line. Alternatively, by using albite which is known to be disordered, the standard free energy change for the reaction albite $\Rightarrow$ jadeite + quartz can be determined. These data can in turn be used to determine the free energy of an unknown phase from combinations of other equilibria involving minerals in common.

(A more detailed discussion and a worked example of the thermodynamic treatment of this reaction is contained in the Appendix to this chapter.)

Another issue which arises in such an experiment is the effect of any composition change. The composition of quartz is virtually pure SiO_2, but the compositions of both jadeite and albite can vary as part of a solid solution. The effect of solid solution on the free energy of one of the phases can be determined from experimental equilibria if all the other variables are fixed. Both solid solution effects and order–disorder have very important consequences on the thermodynamics of mineral reactions and each will be considered in the following chapter.

Extraction of thermodynamic data from phase equilibrium experiments always involves a large number of different experiments and one approach is to use data from one set of experiments to define properties of minerals used in other more complex reactions. The potential problems of error propagation have led to other statistically based methods where a large number of experimental equilibria (>120) are simultaneously considered and the best fitting set of thermodynamic parameters obtained by a least squares method. This provides an internally consistent data set which distributes the uncertainties in any individual experiment. It is also possible to recognise experiments which produce results which are inconsistent with the rest of the data set.

8.5.3 Determining the heat capacity from lattice vibrations

The direct measurement of the heat capacity C_p has several important disadvantages. First, the accuracy required is experimentally difficult to achieve, especially in calorimeters operating at higher temperatures. Second, the experimental measurement is dependent on the sample used, and includes all effects of impurities, defects as well as grain size. These extrinsic effects can make it difficult to obtain an intrinsic value of the heat capacity of the pure end-member phase. Another perennial difficulty is that the chosen mineral end-member cannot be always obtained in a pure enough form or in sufficient quantity.

Another approach to the problem is that the heat capacity C_v can, in theory, be determined if we know the frequencies of all the lattice vibrations, i.e. if we know the phonon spectrum. As outlined in Section 4.4 the complete description of the lattice vibrations of a crystal is given by the dispersion relations which relate the frequencies ω, of the normal vibrational modes, to the wavevectors $\mathbf{k}$ from $0 \leqslant \mathbf{k} \leqslant \pi/a$ where a is the lattice constant. This describes the frequencies of the phonon modes from the longest to the shortest wavelengths in various directions in the crystal. From the dispersion relations a function termed the phonon density of states $g(\omega)$, which describes the frequency distribution, i.e. counts how many different modes there are in each frequency range, can be numerically determined.

The internal energy U is directly related to the integral of the phonon density of states function over the whole frequency range i.e.

$$U \propto \int g(\omega)\mathrm{d}\omega$$

The way in which the lattice vibrations are excited with temperature is described by Bose–Einstein statistics which need not concern us in detail here. This describes the temperature dependence of the inter-

nal energy U, from which the heat capacity C_v can be determined by differentiation,

$$C_v = (\partial U/\partial T)_v$$

C_p can be calculated from C_v from the expression

$$C_p = C_v + TV\alpha^2/\beta$$

as described in Section 8.1.

Thus if the density of states function $g(\omega)$ is known, then all the thermodynamic parameters can be calculated. Inelastic neutron scattering experiments (Section 4.4) can be used to determine the phonon dispersion curves and hence $g(\omega)$ by a numerical summation procedure. However this is experimentally very demanding and requires large single crystals.

Another more approximate, but experimentally much easier method is to combine a more limited amount of experimental data on vibrational modes, with a theoretical model. First the vibrational spectrum of the mineral is determined by infra-red and Raman spectroscopy. This only gives information about the very long-wavelength optic phonons, i.e. at $\mathbf{k} \sim 0$ and the rest of the dispersion relation must be estimated from some theoretical model. However, the optic phonon frequencies are not very dependent on the wavevector $\mathbf{k}$ and reasonable approximations can be made.

The largest variation of the phonon frequencies with wavevector occurs for the three acoustic phonons. At $\mathbf{k} = 0$, these have frequency zero (see Figure 4.22) and a slope which can be determined from the elastic constants of the material. Separate experiments using Brillouin scattering or ultrasonic velocity measurements must be carried out to determine the slopes of the acoustic branches. However, even when the full elastic constants are not known various approximations can be made from other material constants.

Modelling the phonon density of states function can be done with various levels of sophistication which require different amounts of input experimental data. Although this approach has been applied to relatively few minerals it does have the potential to provide increasing amounts of thermodynamic data. Clearly, no single method is used in isolation and comparisons between the results of experimental and modelled heat capacities helps to refine the approximations which need to be made when dealing with complex structures.

Another point to remember is that the heat capacity describes the *change* in enthalpy (eqn. (8.7)) and entropy (eqn. (8.13)) and hence the change in free energy of a phase as a function of temperature, i.e.

$$G = G_o + G_{vibr.}$$

where G is the free energy at any temperature, G_o is the free energy at 0 K, and $G_{vibr.}$ is the change in free energy due to the vibrational energy. To determine the relative stability of two phases we still need to know the absolute difference in the free energy between them at 0K, in addition to the temperature dependent part defined by the heat capacity. To obtain this we need to determine experimentally one temperature and pressure at which the two phases are in equilibrium and hence $\Delta G = 0$. Then $\Delta G_o = -\Delta G_{vibr.}$ which can be determined if the heat capacities and their temperature and pressure variation are known.

Note again that heat capacity determinations allow the calculation of third-law entropies only, and any configurational entropy contribution must be separately assessed.

8.5.4 Determining thermodynamic parameters from computer simulations

Computer simulations of minerals and their properties represent the ultimate departure from the traditional experimental approach to one in which the 'experiment' is carried out in a computer supplied with information about the constituent atoms and the forces between them. The prospect of being able to compute the thermodynamic properties of materials under conditions of temperature and pressure where experiments are difficult or impossible, coupled with the advent of supercomputers has spurred the development of computer simulations of minerals. This development can be outlined by describing the various computational techniques which come under this general heading and their applications to the determination of the stability of minerals.

(i) The static lattice energy calculation

In a static lattice calculation the aim is to determine the minimum energy configuration of a given set of atoms by minimising the potential energy due to the interatomic bonding. We begin with a particular set of atomic coordinates and an *interatomic potential*

model which expresses the potential energy of this structure as a function of the atomic coordinates. By adjusting the atomic coordinates and the cell dimensions the computer program finds the minimum energy for a given topology and symmetry. There are various strategies which make this calculation more efficient, such as calculating the first derivative of the energy with respect to the parameters being varied and finding the steepest energy descent.

Defining a suitable interatomic potential is the main challenge in this type of calculation. Computer simulation studies were first developed with strongly ionic materials for which the starting point is a model which assumes that the forces between ions can be described by a Coulombic attraction. The energy of interaction between two ions i and j is then given by

$$U_{ij} = q_i q_j e^2 / r_{ij}$$

where q is the ionic valence, e is the electronic charge, and r_{ij} is the interionic distance. The total energy is then summed over all such pairs and becomes

$$U_{coul.} = 1/2 \sum q_i q_j e^2 / r_{ij}$$

the factor 1/2 appearing because in the summation all interactions are counted twice. In some models the valence need not be the formal charge on the ion but can take non-integer values.

When the interatomic distance r_{ij} gets sufficiently small, the electron clouds of neighbouring atoms begin to overlap and interact with one another. This causes the electrons to jump into higher energy levels raising the energy very rapidly with increasing overlap. The result is a very short-ranged repulsive force. This interaction is not easy to model, but an exponential energy term with the general form $A_{ij} \exp(-r_{ij}/B_{ij})$, where the A and B parameters are empirically determined (see below) is commonly used.

Another attractive interaction is the van der Waals force which exists in all solids including those where the atoms are electrically neutral. The most important term in this interaction is inversely proportional to the sixth power of the interatomic distance, i.e. has the form $C_{ij} r_{ij}^{-6}$, where C_{ij} is another variable parameter. The result of these three interactions is shown by the potential model illustrated schematically in Figure 8.14 .

The next level of sophistication in developing a realistic model is to include the effects of polarization of the charge clouds around each ion. The distortion of the charge cloud from a simple spherical distribution comes about whenever moving char-

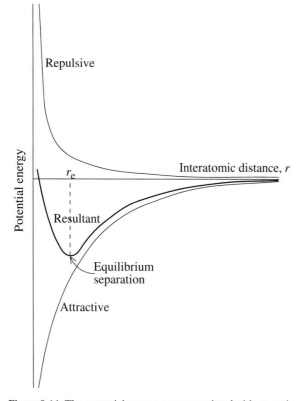

Figure 8.14. The potential energy terms associated with attractive and repulsive forces between atoms, as a function of interatomic distance. The resultant energy defines an equilibrium interatomic separation r_e.

ges interact with local electric fields due to neighbouring charges. This generates a temporary dipole moment and adds an extra contribution to the energy of the crystal. This polarisability is usually described in terms of the *shell model* in which the ions are assumed to be made up of a rigid core containing the nucleus and the inner electrons, surrounded by a shell of loosely bound outer electrons (Figure 8.15). The shell and the core are then held together by a simple harmonic force, i.e. the energy is proportional to the square of the distance between the centres of core and shell. The total interaction energy between two ions is then the sum of six separate interactions: core(1) – shell(1), core(1) – core(2), core(1) – shell(2), core(2) – shell(1), core(2) – shell(2), shell(1) – shell(2). This summation takes the form $1/2 \Sigma kd^2$ where k is the term corresponding to the spring constant, and d is the separation between core-shell centres.

The covalency of the interatomic bonding in silicate minerals requires that an additional term which describes the fact that electrons are not localised around the atomic core and that the bonding is directional. When the atoms move therefore, the surrounding electron distribution must also move. The energy associated with this is difficult to calculate, but can be modelled by bond-bending interactions which constrain the bond angle θ of an O–Si–O bond such that the interaction energy increases as $(\theta - \theta_o)^2$. The summation of this

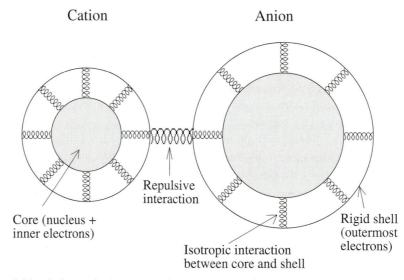

Figure 8.15. The shell model for the interaction between two ions. The nucleus and inner electrons form the rigid core of the ion, while the loosely bound outer electrons form the shell. The forces between the cores and shells are described by springs which behave in a simple harmonic way.

interaction between three atoms i,j,k introduces a term such as $1/2\Sigma\, k_{ijk}(\theta - \theta_o)^2$, where again k_{ijk} is a variable parameter.

The sum of all of these interaction terms describes the way in which the potential energy varies with interatomic distance, and hence constitutes the interatomic potential model. There is no overall consensus on the forms of the potential model to be used, or the best way of determining the unknown parameters in the interaction terms. The most common approach is to fit the terms empirically to reproduce the known atomic positions, lattice parameters, and static physical properties such as elastic constants, and dielectric constants. The best fit to experimentally determined values defines the potential model, which can then hopefully be applied to unknown structures. The aim is to be able to develop a set of interatomic potentials which can be applied to a wide range of different mineral structures i.e. a transferable potential, and not just one which works well in one particular structure.

This technique does not incorporate temperature and it is effectively a simulation of the internal energy at 0 K. However, much can be learned about the energies of real and hypothetical structures, as well as the energy associated with various types of defects using this relatively straightforward simulation method.

(ii) The lattice dynamics calculation

This technique is a natural development of the static calculation, and uses the same potential models to determine the frequencies of the atomic vibrations when the atoms are displaced from their equilibrium positions. In the *harmonic approximation* it is assumed that each atom undergoes a simple harmonic motion about its equilibrium position, which remains fixed. Since the vibrational frequencies, which are determined by the strength and nature of the bonding between atoms, are directly related to the heat capacity, a dynamical simulation makes the quantitative link between macroscopic thermodynamic properties and the interatomic forces.

The possible vibrational modes of the structure can be determined by a detailed consideration of the space group symmetry and the frequencies of these modes can be computed by the lattice dynamics program. The frequency of a particular vibrational mode, or phonon, depends on its wavelength in the crystal, and initially the calculation can be made for very long wavelengths, around $\mathbf{k} = 0$. The calculated frequencies can be then compared with infra-red and Raman spectra. Calculating the frequencies for a range of wavelengths leads to a description of the dispersion relations which can be compared with the results of inelastic neutron scattering experiments, where these are available. Figure 8.16 shows an example of a set of phonon dispersion curves for forsterite calculated from this type of simulation. The phonon density of states function from which the thermodynamic properties are derived is numerically determined from the dispersion curves.

The harmonic approximation, which fixes the equilibrium interatomic distance, does not directly include the effects of temperature, and therefore does not predict thermal expansion of the lattice. Since the effects of temperature on thermodynamic properties are of prime interest, some way of incorporating temperature must be included in a lattice dynamical calculation. This is achieved by doing the calculations for the different values of the lattice parameters which correspond to their values at different temperatures. This simulates the effect of temperature by assuming that changes in the phonon spectra are due to changes in the unit cell size. This is known as the 'quasi-harmonic' approximation.

The success of computer simulation of the lattice dynamics and hence physical properties of minerals is measured by the extent to which it can reproduce the experimental data on well studied mineral structures. This is a rapidly growing area at the forefront of mineralogical research with continuous improvements in the development of empirical potential models, as well as in theoretical methods of determining interatomic potentials. Undoubtedly, lattice dynamical simulations will play an increasing role in the determination of thermodynamic parameters, as well as furthering our understanding of their atomistic basis.

(iii) Molecular dynamics simulations

The molecular dynamics simulation is different in principle to the static and dynamic lattice simulations outlined above in that it directly includes the effects of temperature on atomic motions, and it can be applied equally as well to liquids and non-crystalline solids as it can to crystalline materials. The basic idea of the technique sounds very simple. A given number of particles (atoms, molecules) whose positions and initial velocities are specified,

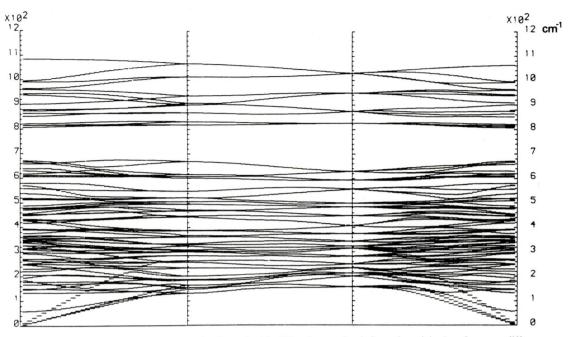

Figure 8.16. Calculated phonon dispersion curves for forsterite Mg_2SiO_4 along a circuit from the origin, $\mathbf{k} = 0$, to two different corners of the Brillouin zone, and back to the origin. (From Price *et al.*, 1987.)

are placed in a fictitious 'box', i.e. constrained by some boundary conditions. From a given model of the forces between these particles, the classical Newton's laws of motion are solved for the entire assembly at successive time-steps which are chosen to be smaller than the time of an atomic vibration. The positions and velocities of the particles are continuously updated and provide a description of the evolution of the system with time. Since the initial start up configuration of the particles is random, the early stages of the simulation determine the equilibrium positions. Further time-steps then generate the data on particle motions which are used to determine the dynamical properties of the system.

Molecular dynamics simulations are well developed in organic molecular physics where the forces between molecules can be reliably simulated using well-tested potential models. In complex mineral structures the technique is in its early stages and is by no means routine. The forces between atoms in minerals are described by the same potential models as discussed above, and as with all simulations, the success of the simulation depends on the availability of realistic models.

The fact that the temperature can be directly incorporated into the system via the kinetic energy of the particles is a major advantage of the tech-

nique, and can therefore provide direct information on atomic motions at high temperatures. Another very important advantage of molecular dynamics simulations is that *anharmonic interactions* between atoms can be incorporated in the calculation whereas this is not possible within the lattice dynamical simulation. Anharmonic interactions between atoms are responsible for many properties such as thermal expansion (i.e. the interatomic distances change with temperature), and the temperature dependence of elastic constants and phonon frequencies.

The change in phonon frequency with temperature may lead to a structural phase transition. In some crystalline materials the frequency of one of the vibrational modes present in the high temperature form may decrease as the temperature falls, eventually reaching zero. At this point the crystal is unable to sustain the corresponding atomic displacements and will yield to a distortion forming a lower symmetry structure. The mode whose frequency falls is termed a *soft mode* since the crystal becomes *soft* against the corresponding distortion. Structural changes due to such instabilities in the atomic vibrations are termed *soft-mode transitions*, and will be discussed again in Chapter 12.

The ability to model structural phase transitions is

an important advantage of molecular dynamics simulation, but against this is the fact that the method is computationally very expensive and requires many hours of supercomputer time. The rapid developments in the power of modern computers, especially with parallel array processors, coupled with improved potential models may make molecular dynamics simulations more routine in the future.

8.6 The Landau theory of phase transitions

One of the important aspects of determining thermodynamic properties of a mineral is to be able to evaluate the effect of a structural phase transition on the overall free energy. To determine the change in free energy associated with a structural distortion for example, both the enthalpy change and the entropy change would need to be measured by some calorimetric method. As mentioned above, heat capacity measurements to the precision required are experimentally very demanding. In a sluggish transformation such as Al,Si ordering the enthalpy change can be measured by heat of solution calorimetry, but then the configurational entropy change has to be separately assessed. Calculations of the thermodynamics of structural changes involve *microscopic* models which attempt to describe the variation in free energy from interaction energies between atoms.

In this section we outline a very different approach to determining free energy changes, based on the fact that the changes in *macroscopic* properties such as the strain, optical birefringence, average site occupancy etc. must also be related in some way to the changes in thermodynamic properties. Although the changes in macroscopic properties must clearly be related to microscopic interactions, this connection does not need to be made in Landau theory. If the mathematical form of the relationship between the change in free energy and the change in some macroscopic property is known, easily measurable properties can provide quantitative thermodynamic data. It is important to note that here we are only dealing with *changes* in free energy superimposed on the normal temperature dependent variation i.e. the excess free energy associated with the phase transition.

Central to Landau theory is the concept of an *order parameter* which describes the course of a

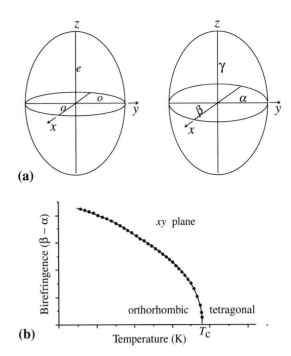

(a)

(b)

Figure 8.17. (a) In the transformation from a tetragonal to orthorhombic structure the optical indicatrix changes from uniaxial to biaxial. (b) In the *xy* plane the birefringence changes from zero to a finite value ($\beta - \alpha$). The change in birefringence is directly proportional to the order parameter Q.

phase transition. The Landau order parameter Q is not only confined to transitions which involve cation order–disorder, but is a generalized concept not restricted to any specific type of transformation. The order parameter Q is related to the change in some macroscopic property through the phase transition, and is scaled so that it is assigned a value of 0 in the high temperature form, and 1 in the low temperature form. Q therefore describes the deviation of the low temperature phase from that of the high temperature phase. The temperature evolution of the order parameter describes the thermodynamic character of the phase transition.

For example, in a transition from a high temperature tetragonal structure to a low temperature orthorhombic structure, the birefringence in the *xy* plane changes from zero to some finite value. The distortion of the structure can be represented by the distortion of the optical indicatrix, from a circular cross-section in the *xy* plane of the tetragonal structure, to an elliptical section in the orthorhombic structure (Figure 8.17(a)). The change in birefringence in this case gives a measure of the progress of the transformation, and is directly proportional to the order parameter Q. The order parameter may change smoothly or have a discontinuity at the critical temperature T_c of the phase transition (Figure 8.17(b)).

The measured physical property is not necessarily directly proportional to Q. In the example above, if the change in birefringence during the transition is measured in the xz plane it is proportional to Q^2, rather than Q. The reason for this is that strict symmetry rules define the form of the order parameter in relation to the symmetry change. The definition of the Landau order parameter has a rigorous theoretical basis in terms of the symmetry relations between the high and low temperature forms, and standard tables (see bibliography) list the correct form of the order parameter and its relationship to certain physical properties for a given change in symmetry. Most commonly measured properties scale as either Q or Q^2.

As another example we consider the case of the cubic ⇒ tetragonal transition which occurs in perovskite on cooling, as discussed in Section 5.2.5. In SrTiO$_3$ the transition is due to the rotation of TiO$_6$ octahedra in opposite directions about the [001] axis, as shown in Figures 5.23 and 8.18. The rotation angle φ increases smoothly as a function of temperature from 0 in the high symmetry form to 2° in the

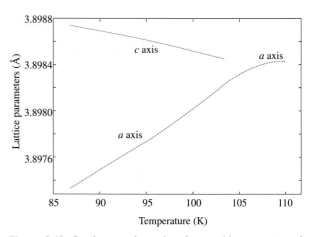

Figure 8.19. In the transformation from cubic to tetragonal SrTiO$_3$ there is a change in the shape of the unit cell, as indicated by the change in lattice parameters. This change in shape defines a spontaneous strain which can be used to define the extent of the transformation.

tetragonal structure, and can be used directly as a measure of the order parameter Q. The direct measurement of this angle as a function of temperature requires a full structure determination and it is easier to measure some macroscopic property which also changes through the transition. One such property is the change in the shape of the unit cell (Figure 8.19).

This change in shape due to the phase transition is expressed in terms of the *spontaneous strain* ε, which is defined in terms of the lattice parameters. It is an excess quantity which is always measured relative to an undistorted cell at the same temperature and is therefore additional to the normal effects of thermal expansion. In the case described here, if the cubic phase has lattice parameter a_o while the tetragonal phase has parameters a and c, the spontaneous strain can be expressed as

$$\varepsilon = \frac{a - a_o}{a_o}$$

and is proportional to Q^2. Both the form of the spontaneous strain, and its relationship with Q are dependent on the symmetry change involved in the transition.

One of the important criteria describing the symmetry change in a phase transition is whether the transition changes the translational symmetry of the lattice. If there is no change in the translational symmetry and the unit cell merely changes its shape, no extra reflections will appear in the diffraction

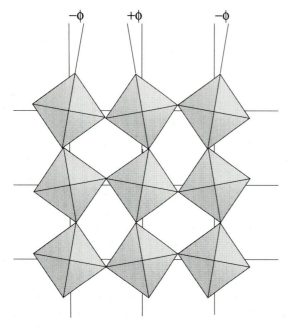

Figure 8.18. In the low temperature structure of SrTiO$_3$ perovskite, alternate TiO$_6$ octahedra rotate in opposite senses through an angle φ. In the high temperature cubic structure, φ = 0. The change in the value of φ can be used as a measure of the progress of the transformation i.e. the order parameter Q.

pattern. It is usual to describe this in reciprocal lattice terms, by describing any change in the Brillouin zone of the crystal (see Section 4.4) . For our purposes, all this means is that if no extra points appear in the reciprocal lattice, the transition does not involve a change in the boundaries of the Brillouin zone. Transitions of this type are termed *zone centre transitions*. On the other hand, if there is a change in the reciprocal lattice as would occur if the translational symmetry changed (i.e. change in the lattice type, or the formation of a superlattice), the boundaries of the Brillouin zone would be affected. This is termed a *zone boundary transition*.

In general, for a zone centre transition, $\varepsilon \propto Q$ while for a zone boundary transition, $\varepsilon \propto Q^2$. The cubic $\Rightarrow$ tetragonal transformation in $SrTiO_3$ involves a reduction in translational symmetry and the formation of a superlattice (see Figure 5.23(a)), and hence the spontaneous strain is proportional to Q^2.

Landau theory describes those phase transitions in which the low temperature form is derived from the high temperature form by a loss of some of the symmetry elements, which implies that the basic structures of the two forms are topologically similar. The low temperature form may be a distorted version of the high temperature form, or it may have an ordered cation distribution while the high temperature form is disordered. Formally, the symmetry of the low temperature form must be a subgroup of the high temperature symmetry. It is this underlying symmetry relationship that allows the order parameter Q to be mathematically defined and related to the thermodynamic quantities we are seeking to measure.

Once the relationship between the measured property and the order parameter Q is defined, the next step is to describe how the excess free energy G is related to the order parameter Q. Note again that the *excess* free energy is the excess over that which the high form would have if the phase transition did not occur. We will however use the notation G, S etc. for these excess quantities, rather than $\Delta G, \Delta S$ etc.

8.6.1 The Landau free energy

Landau proposed that the excess free energy due to a phase transition can be described as a polynomial expansion of the order parameter Q:

$$G = \alpha Q + \frac{1}{2}AQ^2 + \frac{1}{3}bQ^3 + \frac{1}{4}BQ^4 + \ldots$$

where α, A, b, B etc are coefficients which may or may not depend on material properties or on extensive variables such as temperature and pressure.

The equilibrium behaviour of Q through the phase transition is determined by minimising G with respect to Q, and therefore

$$\frac{dG}{dQ} = 0 \quad \text{and} \quad \frac{d^2G}{dQ^2} > 0$$

In the high symmetry form which is stable above some critical temperature T_c, the order parameter $Q = 0$. The equilibrium criteria above can only be satisfied if the linear term is absent and A is positive. If A is negative, Q must be bigger than 0 and the low temperature form becomes stable. As the temperature falls therefore, the sign of A crosses over from positive to negative at $T = T_c$. This temperature dependence of A is expressed as a linear function of T, such that

$$A = a(T - T_C)$$

where a is another constant. The other coefficients b, B, C etc are assumed to be independent of T.

The free energy expansion then becomes

$$G = \frac{1}{2}a(T - T_c)Q^2 + \frac{1}{3}bQ^3 + \frac{1}{4}BQ^4 + \ldots$$

$$(8.19)$$

The form of this free energy expansion is also determined by the nature of the symmetry change involved in the transition. Another aspect of Landau theory is that it predicts the symmetry conditions necessary for a phase transition to be thermodynamically second-order, i.e. one in which the order parameter Q can vary continuously between 0 and 1. In the above free energy expansion, the condition that $dG/dQ = 0$ at $T = T_c$ is satisfied by two values of Q, $Q = 0$ and $Q = -b/B$ (if terms above Q^4 are insignificant). This is not consistent with a continuous change in Q, and therefore the third-order coefficient b must be zero; similarly all other odd-order coefficients are zero. If odd-order coefficients are present in the expansion, the transformation must be discontinuous i.e. thermodynamically first-order. The symmetry changes which allow odd-order terms in the Landau free energy expansion are tabulated in standard texts (see bibliography).

When odd-order terms are absent the excess free energy reduces to

$$G = \frac{1}{2}a(T - T_c)Q^2 + \frac{1}{4}BQ^4 + \frac{1}{6}cQ^6 + \ldots$$
(8.20)

Normally two (or three) terms of this expansion are adequate to describe the free energy changes from experimentally measured values of Q. Although this expression allows Q to change continuously between 0 and 1 it does not preclude the possibility of a discontinuity in Q, depending on the values of the coefficient B.

Three cases are usually considered:

1. when B is positive and the sixth-order term is negligibly small, the expansion describes a thermo-dynamically second-order phase transition. The variation of Q with temperature at equilibrium can be found from

$$\frac{dG}{dQ} = a(T - T_c)Q + BQ^3 = 0$$

$$\therefore Q = \left[\frac{a}{B}(T_c - T)\right]^{1/2}$$

By definition $Q = 1$ at $T = 0$, therefore $\frac{a}{B} = \frac{1}{T_c}$ and

$$Q = \left[(T_c - T)/T_c\right]^{1/2}$$
(8.21)

The variation of Q with temperature is shown in Figure 8.20.

Substituting the value of Q into the free energy expansion in eqn.(8.20) (neglecting the sixth-order term) we obtain a temperature dependence for the excess free energy

$$G = -\frac{a^2}{4B}(T - T_c)^2$$
(8.22)

The temperature dependence of the excess free energy as a function of the order parameter Q is shown in Figure 8.21. At temperatures below T_c a continuous free energy pathway exists to the minimum value, which moves to larger values of Q as the temperature falls.

Other excess thermodynamic quantities can be derived directly from eqn. (8.22). Thus the excess entropy due to the phase transition is given by

$$S = -dG/dT$$

$$= \frac{a^2}{2B}(T - T_c) = -\frac{1}{2}a\,Q^2$$
(8.23)

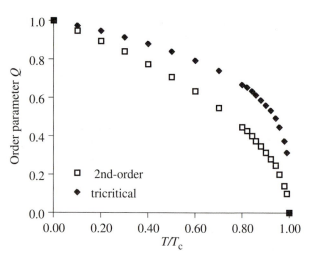

Figure 8.20. A comparison of the variation of the order parameter Q with increasing temperature for a second-order phase transition and a tricritical phase transition. Above $T = T_c$ the order parameter is zero in the high temperature form.

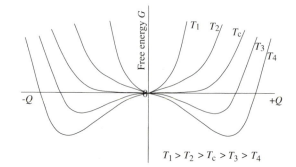

Figure 8.21. The excess free energy G as a function of the order parameter Q for a second-order transition. The positive and negative values of Q refer to two alternative geometric possibilities for a transformation, such as twin related distortions or antiphase related cation ordering. Above T_c the free energy increases for any non-zero value of Q, while below T_c a continuous free energy pathway leads to the free energy minimum.

The excess enthalpy is

$$H = G + TS$$

$$= -\frac{1}{2}aT_cQ^2 + \frac{1}{4}BQ^4$$
(8.24)

and the excess heat capacity is

$$C_p = T.\frac{dS}{dT}$$

$$= \frac{a^2}{2B}T$$
(8.25)

These simple expressions for the excess thermodynamic functions due to a second-order phase transition can be evaluated once the coefficients a and B have been determined. This can be done experimentally by measuring T_c and the value of a thermodynamic quantity at one other temperature. Once the coefficients are known a full description of the thermodynamics of the mineral can be determined over a wide temperature interval. We will discuss the extent to which experimental measurements fit this simple theory in more detail below, but first we return to two other possible solutions to the Landau expansion in eqn. (8.20).

2. when B is negative, both a and c must be positive, the sixth-order term being required to obtain a free energy minimum in the low temperature phase when $Q > 0$. The equilibrium condition yields

$$\frac{dG}{dQ} = 0 = a(T - T_c)Q + BQ^3 + cQ^5$$

from which

$$Q^2 = \frac{-B \pm [B^2 - 4ac(T - T_c)]^{1/2}}{2c} \text{ or } Q = 0 \tag{8.26}$$

This expression describes a discontinuity in Q as a function of temperature, as shown in Figure 8.22 and hence the transition is thermodynamically first order.

The temperature dependence of the free energy (Figure 8.23) also shows this jump in the order parameter from $Q = 0$ to $Q = Q_o$ at the transformation temperature, here labelled T_{tr}. The reason for relabelling the transformation temperature in this

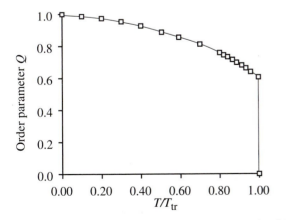

Figure 8.22. The variation of the order parameter Q with increasing temperature for a first order-transition. In this example there is a jump in the order parameter from $Q = 0.6$ to $Q = 0$ at the equilibrium temperature $T = T_{tr}$.

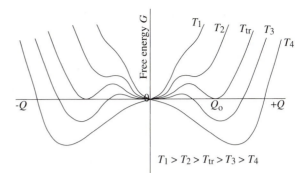

$$T_1 > T_2 > T_{tr} > T_3 > T_4$$

Figure 8.23. The excess free energy G as a function of the order parameter Q for a first-order transition. At the transition temperature $T = T_{tr}$ there is a jump in the order parameter from $Q = 0$ to $Q = Q_o$, accompanied by a free energy barrier. At sufficiently low temperatures continuous pathways towards equilibrium become available.

way is that the free energy curve which shows the equilibrium between the high temperature $Q = 0$ phase and the low temperature phase with $Q = Q_o$ occurs at a temperature higher than T_c. In Figure 8.23 this is the curve labelled $T = T_{tr}$.

From the above equations, the equilibrium transformation temperature T_{tr} is given by

$$T_{tr} = T_c + \frac{3}{16} \frac{B^2}{ac} \tag{8.27}$$

At T_{tr} the order parameter Q jumps from 0 to Q_o where Q_o is given by

$$Q_o = \pm \left[\frac{-4a(T_{tr} - T_c)}{B} \right]^{1/2} \tag{8.28}$$

The size of the discontinuity Q_o indicates how strongly first-order the transition is. The more negative the value of B, the larger the first-order step and the greater the difference between T_{tr} and T_c.

The other values of the excess thermodynamic quantities can be derived from these equations in the same way as in the case of second-order transitions. The model predicts discontinuities in entropy and enthalpy as expected for a first-order transition, and a latent heat L given by

$$L = \frac{1}{2} a Q_o^2 T_{tr} \tag{8.29}$$

3. The final case we will consider is the special case when $B = 0$. This intermediate case is called 'tricritical' and corresponds to the boundary between a first-order transition (B negative) and a

second-order transition (B positive). When $B = 0$, the Landau expansion must include the sixth-order term with c positive to produce a minimum in the low temperature form, and hence the excess free energy is given by

$$G = \frac{1}{2} a (T - T_c) Q^2 + \frac{1}{6} c Q^6 \qquad (8.30)$$

Repeating the minimisation procedure as in the cases above gives the equilibrium variation of Q with temperature as

$$Q = \frac{a}{c} (T_c - T)^{1/4}$$

and

$$T_c = \frac{c}{a} \qquad (8.31)$$

Therefore

$$Q = [(T_c - T)/T_c]^{1/4} \qquad (8.32)$$

The variation of Q with T is continuous between $Q = 0$ and $Q = 1$, as shown in Figure 8.20.

The form of the variation of Q with T should be compared with that for second-order transitions. In both cases the general relationship can be written

$$Q = [(T_c - T)/T_c]^{\beta}$$

where β is termed the "critical exponent" and describes the variation of the order parameter with temperature for continuous transitions. For the two ideal cases we have considered, $\beta = 1/2$ for a second order transition and $\beta = 1/4$ for a tricritical transition.

The other excess thermodynamic quantities can be readily calculated as for the second-order case giving the excess entropy as

$$S = -\frac{1}{2} a Q^2, \qquad (8.33)$$

the excess enthalpy as

$$H = -\frac{1}{2} a T_c Q^2 + \frac{1}{6} c Q^6 \qquad (8.34)$$

and an excess heat capacity of

$$C_p = \frac{aT}{4\sqrt{T_c}} (T_c - T)^{-1/2} \qquad (8.35)$$

Although tricritical transitions may seem to be a very special case they are by no means exceptional in

minerals. Al,Si ordering transitions in alkali and plagioclase feldspars, cation ordering in omphacite, as well as the displacive transition in pure, ordered anorthite at low pressure, all appear to behave in this way.

8.6.2 Determining the Landau coefficients

If the experimentally observed behaviour of the order parameter Q conforms to the predictions of Landau theory, all of the excess thermodynamic functions can be calculated once the coefficients a, B and c have been determined. The measured change in lattice parameters due to a phase transition can be used to calculate the spontaneous strain, from which the temperature dependence of the order parameter can be determined. Figure 8.24 shows an ideal

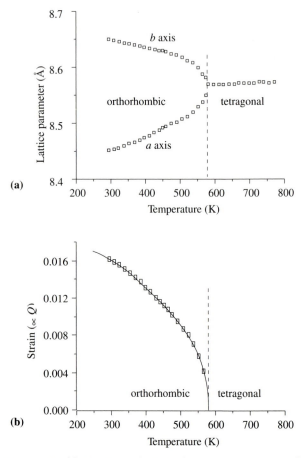

Figure 8.24. (a) The change in the lattice parameters associated with a typical tetragonal to orthorhombic transition on cooling. In this case the change in shape of the unit cell in the *ab* plane involves a contraction of one axis and an expansion of the other. The *c* axis remains constant, as does the overall volume. (b) The variation of the strain with temperature. The data points all lie along the line predicted for a second-order, zone centre transition.

example in a tetragonal to orthorhombic transition. In this case the spontaneous strain ε is directly proportional to Q, and the temperature dependence of Q conforms within experimental error to the predictions of Landau theory for a second order phase transition, i.e. $Q = [(T_c - T)/T_c]^{1/2}$.

Another example of the measurement of the temperature dependence of Q is provided by the phase transition in calcite $CaCO_3$, which occurs at 1260K and involves the orientational disordering of CO_3 groups. Below this temperature the planar CO_3 groups vibrate about a fixed orientation, alternate layers of CO_3 groups pointing in opposite directions (Figure 9.22), while in the high temperature form they are free to rotate and become equivalent. The symmetry reduction on ordering is from $R\bar{3}m$ to $R\bar{3}c$, with doubling of the c axis length, which introduces extra reflections into the diffraction pattern. As calcite is heated therefore, the intensity of these extra superlattice reflections decreases as the CO_3 groups disorder, and provides a measure of the order parameter Q. It can be shown from symmetry arguments that the intensity $\propto Q^2$. Since the experimental plot of (intensity)^2 against temperature is linear (Figure 8.25), it follows that $Q \propto (T_c - T)^{1/4}$ and the transition is tricritical. The value of T_c is determined from such an experiment.

From the equations in the previous section, it is clear that to determine the Landau coefficients, we need to know T_c as well as the excess value of a thermodynamic quantity such as the heat capacity or the enthalpy at at least one other temperature. For displacive transitions, the value of the heat capacity as a function of temperature can be measured by differential scanning calorimetry. In the case of Al,Si order-disorder transitions which are very slow to achieve equilibrium, the usual method is to prepare a series of equilibrated samples at various temperatures, quench the samples to room temperature thereby freezing in the Al,Si distribution, and then measure the differences in the heat of solution (Section 8.5.1).

For the example of the orientational disorder of CO_3 groups in calcite, neither of these methods is experimentally applicable since the transition is not quenchable, and scanning differential calorimeters cannot operate at temperatures above about 1000K. Drop calorimetry (Section 8.5.1) carried out at a series of increasing temperatures enables the excess enthalpy to be determined, and fitted to a tricritical Landau model, as found from the order parameter behaviour. The total excess enthalpy due to orientational disorder is found to be -10 kJ mol^{-1} (Figure 8.26). Substituting this value, and the value of T_c into eqn. (8.34):

$$H = -\frac{1}{2}aT_cQ^2 + \frac{1}{6}cQ^6$$

and noting that

$$T_c = \frac{c}{a} \quad \text{(eqn. (8.31))}$$

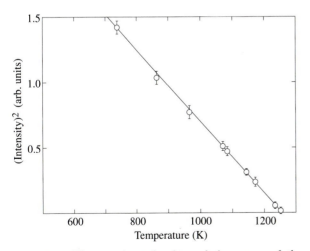

Figure 8.25. The experimental values of the square of the intensity of a superlattice reflection (associated with the $R\bar{3}m$ to $R\bar{3}c$ transition due to orientational ordering of CO_3 groups in calcite) with temperature. Complete disorder is achieved at ~1250K when the superlattice reflection disappears. The linear dependence indicates that the transition is tricritical. (Data from Dove and Powell, 1989.)

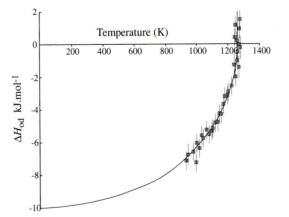

Figure 8.26. The excess enthalpy associated with CO_3 orientational disorder as a function of temperature, determined by drop calorimetry. (After Redfern *et al.*, 1989.)

gives

$$H = -\frac{a}{3} T_c = -\frac{c}{3} \text{ for the total enthalpy}$$

change from $Q = 0$ at 1260K to $Q = 1$. This gives values of the Landau coefficients as $a = 24$ J mol^{-1} K^{-1}, and $c = 30$ kJ mol^{-1}. Using these values the temperature dependence of the other excess thermodynamic quantities can be readily calculated.

8.6.3 What do the Landau coefficients mean?

During a phase transition there are several different contributions to the free energy change, all of which are incorporated in the Landau coefficients. For example, in the excess enthalpy part of the free energy expansion, which for the second-order case is given by eqn. (8.24), the contribution of the volume change V due to the transition, and the pressure P is included in the coefficients a and B. If we need to evaluate the effects of pressure on a transition, or if the volume change is significant, it is possible to separate these effects and include them explicitly in the Landau coefficients. Similarly, the effect of the strain on the free energy , as well as the effect of a change in composition can also be independently evaluated by including them in the coefficients.

(i) The effect of pressure

One of the postulates of Landau theory which we assumed in the free energy expansion above is that only the coefficient of the Q^2 term is temperature dependent. The same postulate applies to the pressure dependence, and so only the Q^2 term is involved. The free energy expansion for a second-order transition then becomes

$$G = \frac{1}{2} a(T - T_c)Q^2 + \frac{1}{4}BQ^4 + \ldots + \frac{1}{2} a_v P Q^2$$

where a_v is a coefficient describing the pressure dependence. Rearranging the terms gives

$$G = \frac{1}{2} a[T - (T_c - \frac{a_v}{a}P)] Q^2 + \frac{1}{4}BQ^4 + \ldots \quad (8.36)$$

This is in effect equivalent to substituting a new value for the transition temperature T_c^* to restore the temperature dependent part of the expansion to

$$G = \frac{1}{2} a(T - T_c^*)Q^2 + \frac{1}{4}BQ^4 + \ldots$$

with

$$T_c^* = T_c - \frac{a_v}{a}P \quad (8.37)$$

T_c is the transition temperature at zero pressure, and T_c^* is the transition temperature at pressure P. When experiments on phase transitions are carried out at a fixed temperature and the pressure is varied, the equilibrium pressure P_c for the transition is given by

$$T_c^* = T_c - \frac{a_v}{a} P_c$$

and thus

$$P_c = \frac{a}{a_v}(T_c - T_c^*) \quad (8.38)$$

Equation (8.37) predicts that the effect of pressure on a second-order phase transition is to induce a linear change in the equilibrium transition temperature. If the excess volume is positive, the sign of a_v will be positive and an increase in pressure will decrease the transition temperature.

(ii) The effect of the spontaneous strain

A change in shape of the unit cell through a phase transition can also be included. If a change in the order parameter Q induces a spontaneous strain e the two are coupled together. The excess free energy expansion due to this strain is given by

$$G_{str.} = d\varepsilon Q + e\varepsilon Q^2 + \ldots + f\varepsilon^2 \quad (8.39)$$

where d and e are two new Landau coefficients (only the first two terms in the series need to be considered), and $f\varepsilon^2$ is the elastic energy contribution (Hooke's Law: the elastic energy is proportional to the square of the strain). The terms εQ and εQ^2 are referred to as linear and quadratic coupling terms.

The total free energy change is then given by

$$G = \frac{1}{2} a(T - T_c)Q^2 + \frac{1}{4} BQ^4 + \frac{1}{6} cQ^6 + \ldots$$
$$+ d\varepsilon Q + e\varepsilon Q^2 + \ldots + f\varepsilon^2 \quad (8.40)$$

At equilibrium there must be no residual stresses in the crystal, and therefore the free energy must be at a minimum with respect to the strain, i.e. $dG/d\varepsilon = 0$.

We consider two cases:

(i) For a zone centre transition (i.e. no change in the translational symmetry of the lattice), it is generally assumed that the linear coupling term between the strain and the order parameter is sufficient, and hence $e\varepsilon Q^2$ is negligible. In this case, $dG/d\varepsilon = 0$ gives

$$\varepsilon = -\frac{d}{2f}Q \qquad (8.41)$$

When this is substituted back into the original Landau expansion it becomes

$$G = \frac{1}{2}a\left(T - T_c - \frac{d^2}{2af}\right)Q^2 + \frac{1}{4}BQ^4$$
$$+ \frac{1}{6}cQ^6 + \dots \qquad (8.42)$$

Just as in the case of pressure, strain has the effect of changing the transition temperature to T_c^* where

$$T_c^* = T_c + \frac{d^2}{2af} \qquad (8.43)$$

(ii) When a transition does involve a change in the translational symmetry (i.e. a zone boundary transition), only the quadratic coupling term is significant, i.e. $d\varepsilon Q$ is neglected. Repeating the above procedure for this case gives

$$\varepsilon = -\frac{e}{2f}Q^2$$

and

$$G = \frac{1}{2}a(T - T_c)Q^2 + \frac{1}{4}\left(B - \frac{e^2}{f}\right)Q^4$$
$$+ \frac{1}{6}cQ^6 + \dots$$

In this case it is the fourth-order term which is affected, and as we discussed in Section 8.6.1, the sign of the fourth-order coefficient defines whether the transition is first- or second-order. Thus a transition which may be second-order when there is negligible strain, could become first-order if the strain was significant.

(iii) The effect of composition

A common observation is that phase transitions in minerals which are end-members of a solid solution are increased or decreased in temperature by a change in the composition. The effect of substituting impurity atoms into the crystal is to generate a local strain field around the substituting atom. When sufficient atoms are substituted the strain fields coalesce to produce what is in effect an extra strain term in the free energy expansion. The magnitude of this effect will depend on the amount of impurity added, given by X, the mole fraction of the second component. The effect of this strain on the excess free energy of a phase transition will depend on Q, and so the two are coupled in much the same way as in the previous case.

As before, the effect of coupling is to add a second power series to the free energy expansion giving

$$G = \frac{1}{2}a(T - T_c)Q^2 + \frac{1}{4}BQ^4 + \frac{1}{6}cQ^6 + \dots$$
$$+ mXQ + nXQ^2 + \dots$$

In order to reproduce the observed behaviour the linear coupling term must be negligible, giving

$$G = \frac{1}{2}a\left(T - T_c + \frac{2n}{a}X\right)Q^2 + \frac{1}{4}BQ^4$$
$$+ \frac{1}{6}cQ^6 + \dots$$

As previously, we can write a new transition temperature T_c^* where

$$T_c^* = T_c - \frac{2n}{a}X \qquad (8.44)$$

8.6.4 Coupling between two order parameters

In complex mineral structures, the response of the structure on cooling may involve more than one phase transition. In particular we refer to the situation in framework aluminosilicates, where the open structure tends to distort on cooling, while the Al,Si atoms tend to order on the tetrahedral sites. We have discussed this situation in Section 6.8. The classic case is that of albite $NaAlSi_3O_8$, which undergoes a displacive transition from monoclinic to triclinic symmetry ($C2/m$ to $C\bar{1}$)) and also undergoes an Al,Si ordering process. If the displacive transition was suppressed (as is the case in $KAlSi_3O_8$) Al,Si ordering would produce the same symmetry change. It is well known that one transition is influenced by the other. For example, fully ordered albite when heated could not transform to monoclinic by a displacive transition until the Al,Si atoms became completely disordered.

The influence of one transition on the other is described by a coupling of the order parameters for the displacive transition (Q) and the ordering transition (Q_{od}) respectively. The physical origin of this coupling comes from the fact that both transitions involve a change in the shape of the unit cell, i.e. a strain, and this strain which has a long-range influence over the whole crystal communicates between the displacive and the ordering process. The coupling between the order parameters is therefore via the strain ε. Since both Q and Q_{od} can vary over quite a wide temperature range, this coupling can be significant even if the equilibrium transformation temperatures for the two transitions are several hundreds of degrees apart, as is the case in albite.

The free energy expansion for the overall structural change must incorporate both order parameters Q and Q_{od}. The way in which this coupling is described in the free energy expression depends on the symmetry change associated with each transition, but the simplest case is when neither phase transition involves any change in the translational symmetry (i.e. both are zone centre transitions). As in Section 8.6.3(ii) above, only linear coupling between the order parameters and the strain e needs to be considered in this case. When both order parameters are linearly coupled to the strain we refer to *bilinear coupling*. This coupling introduces terms εQ and εQ_{od} into the Landau expansion which becomes

$$G = \frac{1}{2} a_1(T - T_c)Q^2 + \frac{1}{4}B_1Q^4 + \frac{1}{6}c_1Q^6 + \ldots$$
$$+ \frac{1}{2} a_2(T - T_c)Q_{od}^2 + \frac{1}{4}B_2Q_{od}^4 + \frac{1}{6}c_2Q_{od}^6 + \ldots$$
$$+ d_1\varepsilon Q + d_2\varepsilon Q_{od} + \ldots + f\varepsilon^2$$

Although this expression looks cumbersome it is merely the addition of two separate power series plus two linear coupling terms between the strain and each order parameter. As in eqn.(8.40) where the coupling between order parameter and strain was first introduced the $f\varepsilon^2$ term describes the Hooke's Law contribution to the elastic energy.

Exactly the same procedures as described above must be followed to derive the equilibrium conditions, i.e. the crystal must be stress free, so that $dG/d\varepsilon = 0$. Differentation of the above expression gives

$$\frac{dG}{d\varepsilon} = d_1Q + d_2Q_{od} + 2f\varepsilon = 0$$
$$\therefore \quad \varepsilon = -\frac{d_1Q + d_2Q_{od}}{2f}$$

Substituting this value of ε back into the free energy expansion gives the equilibrium excess free energy as

$$G = \frac{1}{2} a_1 \left(T - T_c - \frac{d_1^2}{2a_1f} \right) Q^2 + \frac{1}{4} B_1Q^4$$
$$+ \frac{1}{6} c_1Q^6 + \ldots$$
$$+ \frac{1}{2} a_2 \left(T - T_{cod} - \frac{d_2^2}{2a_2f} \right) Q_{od}^2 + \frac{1}{4} B_2Q_{od}^4$$
$$+ \frac{1}{6} c_2Q_{od}^6 + \ldots - \frac{d_1d_2}{2f} QQ_{od} \qquad (8.45)$$

Notice the effect of this coupling on the second-order coefficients of Q. The temperatures T and $T_{c\,od}$ are the temperatures of the separate transitions if we ignored any coupling. Just as in the previous section we can see that the coupling changes the second-order coefficients, and hence the temperature of the transition. We can write that

$$T_c^* = T_c + \frac{d_1^2}{2a_1f}$$

and

$$T_{c\,od}^* = T_{c\,od} + \frac{d_2^2}{2a_2f}$$

where now we have taken the coupling into account. The temperature T_c^* is now the equilibrium transition temperature for the displacive transition if the crystal is completely disordered i.e. $Q_{od} = 0$, and the temperature $T_{c\,od}^*$ is the equilibrium ordering transition temperature in a crystal where the displacive transition is suppressed (i.e. $Q = 0$).

The coefficient of the term QQ_{od} describes the direct coupling between the two order parameters and may itself be temperature dependent if the elastic constants (present in the f term) vary with temperature. We can replace this coefficient by λ and write a simplified expression for the expansion

$$G = \frac{1}{2} a_1 \left(T - T_c^* \right) Q^2 + \frac{1}{4} B_1Q^4 + \frac{1}{6} c_1Q^6 + \ldots$$
$$+ \frac{1}{2} a_2 \left(T - T_{c\,od}^* \right) Q_{od}^2 + \frac{1}{4} B_2Q_{od}^4$$
$$+ \frac{1}{6} c_2Q_{od}^6 + \ldots \lambda QQ_{od} \qquad (8.46)$$

The further condition of equilibrium is that the free energy must be at a minimum with respect to both Q and Q_{od} and hence $dG/dQ = dG/dQ_{od} = 0$.

Applying this to the above expression gives

$$a_1(T - T_c^*) Q + B_1Q^3 + c_1Q^5 + \ldots \lambda Q_{od} = 0$$

and

$$a_2(T - T_{c\,od}^*) Q_{od} + B_2Q_{od}^3 + c_2Q_{od}^5 + \ldots \lambda Q = 0.$$

To determine the coefficients and equilibrium values of Q it is necessary, in a case such as albite, to follow Q as a function of temperature (by measuring the lattice parameters for example) under conditions where Q_{od} is fixed at known values. Since Q_{od} described Al,Si ordering which is a very sluggish process compared to the instantaneous displacive transition, measuring Q as a function of temperature can be done without Q_{od} varying within the time scale of the experiment.

The specific example of albite will be considered in more detail Chapter 12, but here we summarize briefly the general effects of bilinear order parameter coupling of this type:

1. The result of the coupling of two order parameters is that there is one phase transition, and not two separate transitions, and both order parameters are involved in the transition.

2. The variation of Q and Q_{od} with temperature is different than would be expected if only one order parameter was necessary.

3. The equilibrium transition temperature is different from either T_c^* or $T_{c\,od}^*$. In the case of albite the resultant transition temperature is higher than both, i.e. if a crystal with any Al,Si order is heated, the displacive transition to monoclinic symmetry will require a higher temperature than if the crystal was disordered. Similarly a crystal which has undergone a distortion due to the displacive transition will have a higher Al,Si ordering temperature relative to an undistorted crystal. This can be understood intuitively since both processes have the same effect on the crystal symmetry. The coupling therefore tends to stabilize the low temperature, ordered and distorted structure.

One of the consequences of order parameter coupling is that, once the coupling behaviour is understood, an experiment measuring one order parameter can be used to deduce the value of the other order parameter. This has important applications in a case such as albite where one of the order parameters (Q) is due to a rapid and easily measurable transition, while the other (Q_{od}) is due to a very sluggish transition which cannot be easily studied on a laboratory timescale. The nature of the displacive transition (i.e. the variation of Q with T) can be used to measure the degree of Al,Si order. Clearly, to be able to do this the form of the coupling has to be verified. As we see below, other forms of coupling are also possible.

Other forms of coupling

By analogy with the treatment in Section 8.6.3(ii), we can envisage other forms of coupling via the strain. When both transitions involve a change in the translational symmetry (i.e. both are zone boundary transitions), the coupling with the strain is quadratic, i.e. involves terms such as εQ^2, and hence an equivalent set of equations can be built up using εQ^2 and εQ_{od}^2 as the coupling terms. This is termed *biquadratic coupling*. An example of such a case is anorthite, in which the Al,Si ordering transition ($C\bar{1}$ to $I\bar{1}$) involves a doubling of the unit cell, and the displacive transition ($I\bar{1}$ to $P\bar{1}$) are both zone boundary transitions. Further reference to this will be made in Chapter 11.

Another possibility is the coupling between a zone centre transition and a zone boundary transition when the appropriate coupling terms would include εQ and εQ_{od}^2 (or vice versa), and the coupling would be described as *linear-quadratic*. The algebraic methods in these cases are the same as previously, but will not be described in detail here. Further reference can be found in the bibliography.

8.6.5 The applicability of Landau theory to silicate minerals

We conclude this chapter with some brief observations on the general applications of Landau theory, and some comments on its success in describing phase transitions in minerals. One of the assumptions of Landau theory in its simplest form, is that the order parameter is defined macroscopically, and therefore represents the average value for the whole crystal. For a transition which involves a distortion of the unit cell, it assumes that all unit cells in the crystal are correlated and distort in the same way. For a cation ordering process, Q in Landau theory means the same as S in the Bragg–Williams approximation (Section 8.4.2) and any short-range order cannot be simply dealt with in this way.

The assumption of long-range correlations breaks down near the critical temperature T_c when local fluctuations in Q become so large that they exceed the average value for Q in the crystal. This interval in which Landau theory breaks down is known as the Ginsburg interval, but in framework silicates it appears to be very small i.e. less than a few degrees and not a serious limitation to the application of the theory. The success of Landau theory in describing transitions in many framework aluminosilicates is partly due to the fact that the the flexibility of the framework allows large strains to develop as a result of atomic displacements or ordering processes. These strains, which extend over at least 10–100 unit cells, correlate the changes in the order parameter over relatively large distances.

Another case where the assumption of long-range correlations may break down is when the symmetry change involves a loss of translational or point group symmetry and hence allows the possibility of domain structures such as antiphase or twin domains (Sections 6.3.4 and 6.3.5). Under non-equilibrium conditions such inhomogeneities in the order parameter can become stabilized and grow in amplitude. These local gradients in Q will lead to a measured macroscopic order parameter which is not the same as the order parameter within the domains. Development of Landau theory along these lines is beyond the scope of this book, but further references can be found in the bibliography.

Appendix
A worked example : The thermodynamics of the reaction

Jadeite + Quartz $\Rightarrow$ Albite
$NaAlSi_2O_6 + SiO_2 \quad NaAlSi_3O_8$

from calorimetric measurements and experimental phase equilibria

In this example we describe the basic methods of manipulating thermodynamic data for a simple reaction involving only solid phases. There are a number of possible variations of these procedures, depending on the type of data available. Here we assume that we have calorimetric data on the variation of heat capacity C_p with temperature for the three pure phases low albite, jadeite and quartz. The low albite is assumed to have complete Al,Si order. These data yield the standard state entropies ($S°_{298}$) for the three phases, as well as the variation in entropy with temperature.

The entropy is calculated from the heat capacity data using the expression

$$S_T = S_o + \int_0^T \frac{C_p}{T} \, dT \quad \text{(see eqn. 8.13)}$$

or in this case

$$S_T = S°_{298} + \int_{298}^T \frac{C_p}{T} \, dT$$

where S_T is the entropy at temperature T.

The heat capacity data for each phase are usually expressed as a function of temperature using the relation $C_p = a + bT + cT^{-2} + dT^{-1/2}$, where a,b,c and d are constants. These values are given in the Table below. The entropy change from 298K to 1000K has been calculated from the above equation as follows:

$$S_{1000} - S°_{298} = \int_{298}^{1000} \frac{C_p}{T} \, dT$$

$$S_{1000} - S°_{298} = \int_{298}^{1000} \frac{a}{T} + b + cT^{-3} + dT^{-3/2} . \, dT$$

$$= \left[a \ln T + bT - \frac{cT^{-2}}{2} - 2dT^{-1/2} \right]_{800}^{1000}$$

$$= a \, (\ln 1000 - \ln 298) + b(1000 - 298)$$

$$- \frac{c(1000^{-2} - 298^{-2})}{2} - 2d(1000^{-1/2} - 298^{-1/2})$$

	S_{298}	a	$b \times 10^{-5}$	c	d	$\int_{298}^{1000} \frac{C_p}{T} \, dT$	S_{1000}
	$kJ.K^{-1}$						$kJ.K^{-1}$
Albite	0.2074	0.4520	−1.336	−1276	−3.954	0.3232	0.5306
Jadeite	0.1335	0.3011	1.014	−2239	−2.055	0.252	0.3855
Quartz	0.0415	0.1044	0.607	34	−1.070	0.0735	0.115

Substitution of the constants a,b,c,d for each phase gives the entropy changes tabulated below:

Data for low albite, jadeite and quartz

The molar volumes at 298K, converted to units of kJ $kbar^{-1}$ using the conversion 1 $cm^3 = 0.1$ kJ $kbar^{-1}$ are given below:

Albite: 10.01
Jadeite 6.040
Quartz 2.269

Data from experimental phase equilibria

Experiments have been carried out on the reaction

$$\text{Jadeite + Quartz} \Rightarrow \text{Albite}$$

by mixing together jadeite, quartz and high albite and reacting them at various temperatures and pressures. (The standard reversal technique is to put all phases together in reacting proportions, to avoid nucleation problems.) The reverse reaction was carried out by changing the pressure and temperature conditions. The results are shown in Figure 8.13. Over a considerable P,T interval the equilibrium boundary is a straight line, the slope of which has been measured from the graph as

$$\frac{dP}{dT} = 0.0263 \pm 0.002 \text{ kbar K}^{-1}.$$

From the Clapeyron relation

$$\frac{dP}{dT} = \frac{\Delta S_r}{\Delta V}$$

where ΔS_r is the entropy change for the reaction at a particular temperature, and ΔV is the difference in the volume of the product and the reactants, i.e. volume change in the reaction. As explained in Section 8.5.2, the thermal expansion and compressibility tend to cancel one another and ΔV is unlikely to

change very significantly over this temperature range from its value at 298K. The fact that the equilibrium boundary is a straight line therefore indicates that ΔS is also approximately constant over this interval.

The $P–T$ slope allows an estimate for ΔS_r from the Clapeyron relation:

$$\Delta S_r = \Delta V . \frac{dP}{dT}$$

$$= \{10.01 - (6.040 + 2.269)\} \times 0.0263$$
$$= 1.701 \times 0.0263$$
$$= 0.0447 \pm 0.0034 \text{ kJ K}^{-1}$$

Comparison with calorimetric data for the entropy change

The entropy change for the reaction, ΔS_r, at 1000K can be determined from the tabulated calorimetric data above:

$$\Delta S_r = 0.5306 - (0.3855 + 0.115)$$
$$= 0.0301 \text{ kJ K}^{-1}$$

The fact that this entropy change is considerably smaller than the value of 0.0447 kJ K^{-1} determined from the experimental reaction indicates that the albite formed in the reaction had a higher entropy than the calorimetrically determined value of 0.5306 kJ K^{-1}. Note that the calorimetric value refers to low albite, and heat capacity measurements as a function of temperature do not include the configurational entropy, as discussed in Section 8.3.3. Assuming that all of the extra entropy involved in the reaction is due to Al,Si disorder in the albite, the configurational entropy contribution in the albite is given by

$$\Delta S_r = 0.0447 = S_{Ab} - (0.3855 + 0.115)$$
$$S_{Ab} = 0.5452 = S_{vib.} + S_{config.} = 0.5306 + S_{config.}$$
$$S_{config} = 0.0146 \text{ kJ K}^{-1}$$

If the albite which formed in the reaction had been totally disordered, its configurational entropy would be given by

$$S_{\text{config}} = 4R \left[\frac{1}{4}\ln\frac{1}{4} + \frac{3}{4}\ln\frac{3}{4} \right] = 0.0187 \text{ kJ K}^{-1}$$

From this we can conclude that the albite in the experimental reaction is almost 80% disordered.

The enthalpy change ΔH_r for the reaction at temperature T, and at 298K, determined from the experimental equilibrium

From the experimental data, we may choose any convenient equilibrium reaction point, e.g. $T = 1000$K (727°C) and $P = 19.75 \pm 0.25$ kb. At equilibrium

$$\Delta G_{P,T} = \Delta H_{1,T} - T\Delta S_{1,T} + P \cdot \Delta V_{P,T} = 0$$

Therefore $\quad \Delta H_{1,T} = T\Delta S_{1,T} - P \cdot \Delta V_{P,T}$

Assuming $\Delta V_{P,T} = \Delta V_{1,298}$, and using the value of $\Delta S_{1,T} = 0.0447 \pm 0.0034$ kJ K^{-1} from the experimental data,

$$\Delta H_{P,T} = 1000 \times 0.0447 - 19.75 \times 1.701$$
$$= 11.105 \pm 3 \text{ kJ}$$

This is the value of the enthalpy change for the reaction at 1000K. To reduce this to the standard state value at 298K, we may use the relation

$$\Delta H^{\circ}_{r} = \Delta H_{1,T} - \int_{298}^{T} \Delta C_p dT \quad \text{(see eqn 8.7)}$$

To evaluate the integral we once again use the expression $C_p = a + bT + cT^{-2} + dT^{-1/2}$, where a, b, c and d are constants listed in the data Table.

$$\int_{298}^{1000} C_p dT = a(1000 - 298) + \frac{b}{2}(1000^2 - 298^2)$$
$$- c(1000^{-1} - 298^{-1}) - 2d(\sqrt{1000} - \sqrt{298})$$

Substituting values of a, b, c, d gives:

$$\int_{298}^{1000} C_p dT$$

Low Albite	194.77 kJ
Jadeite	151.70 kJ
Quartz	44.49 kJ

from which the change in enthalpy of the reaction between 298K and 1000K, i.e.

$$\int_{298}^{1000} \Delta C_p dT = 194.77 - (151.70 + 44.49)$$
$$= 1.42 \text{ kJ mol}^{-1}$$

$$\Delta H^{\circ}_{r} = \Delta H_{1,T} - \int_{298}^{T} \Delta C_p dT$$

$$= 11.105 + 1.420$$
$$\Delta H^{\circ}_{298} = 12.525 \pm 3.0 \text{ kJ}$$

The entropy change ΔS_r for the reaction at temperature T, and at 298K, determined from the experimental equilibrium

We have already determined the entropy change for the reaction at 1000K from the slope of the equilibrium reaction line, i.e. $\Delta S_r = 0.0447 \pm 0.0034$ kJ K^{-1}.

To reduce this to the standard state value at 298K we use the expression

$$\Delta S_{r,1000} = \Delta S^{\circ}_{298} + \int_{298}^{1000} \frac{\Delta C_p}{T} dT$$

The values of the integral for each of the three phases has already been calculated and is tabulated above. Therefore, for the reaction

$$\int_{298}^{1000} \frac{\Delta C_p}{T} dT = 0.3232 - (0.252 + 0.0735)$$

$$= -0.0023 \text{ kJ. K}^{-1}$$

Therefore the standard state entropy change for the reaction is given by

$$\Delta S^{\circ}_{298} = 0.0447 + 0.0023$$
$$= 0.0470 \pm 0.0034 \text{ kJ.K}^{-1}$$

The standard free energy change for the reaction, $\Delta G^{\circ}_{1,298}$, determined from the experimental equilibrium

From the determined values of the standard enthalpy and entropy we can now determine the free energy change for the reaction at 1 atmosphere and 298K using the expression

$$\Delta G_{1,298} = \Delta H_{1,298} - T\,\Delta S_{1,298}$$
$$= 12.525 - 298 \times 0.0470$$
$$= -1.481 \text{ kJ mol}^{-1}$$

	ΔH_{soln} (kJ mol^{-1})
Ordered albite	84.69 ± 1.34
Disordered albite	71.81 ± 1.34
Jadeite	88.02 ± 0.67
Quartz	−4.26 ± 0.21

A comparison with heat of solution data ΔH_{soln} for ordered and disordered albite, jadeite and quartz

Heats of solution have been determined at 700°C (973K) for both ordered and disordered albite and pure jadeite and quartz. These are tabulated below:

From this data the following enthalpies of reaction at 700°C can be calculated:

(i)   Jadeite + Quartz ⇒ Disordered albite
$$\Delta H = (88.02 - 4.26) - 71.81$$
$$= 11.95 \pm 1.51 \text{ kJ mol}^{-1}$$

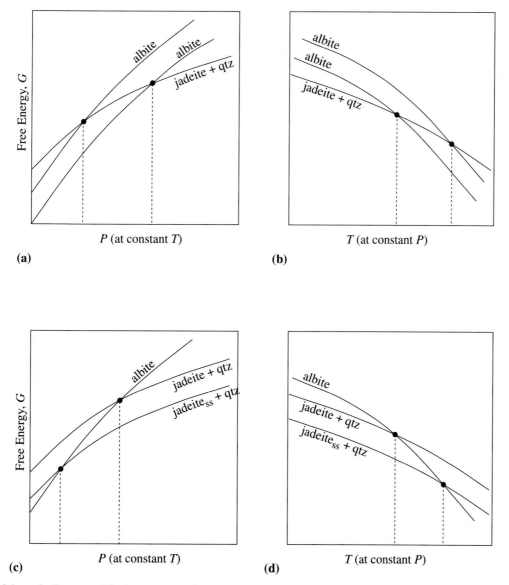

Figure 8.27. Schematic diagrams of the free energy for jadeite + quartz, and albite, illustrating how a change in the free energy of one of the reactants displaces the equilibrium boundaries. In (a) and (b) an increase in the free energy of the albite (by increasing the degree of Al,Si order relative to the equilibrium value) decreases the equilibrium pressure and increases the temperature of the reaction. In (c) and (d) a decrease in free energy of jadeite (by forming a stable solid solution with another pyroxene component) reduces the equilibrium pressure and increases the temperature of the reaction.

(ii)   Jadeite + Quartz ⇒ Ordered albite
$$\Delta H = (88.02 - 4.26) - 84.81$$
$$= 1.05 \pm 1.51 \text{ kJ mol}^{-1}$$

By comparison, the equivalent enthalpy change, determined from the phase diagram, for the reaction at 1000K (727°C) is 11.1 ± 3.0 kJ mol^{-1}.

The heat of solution measurements, combined with the heat capacity data, can be used to determine the thermodynamics of the reactions, but the errors involved are likely to be greater than those derived from experimental equilibria.

The effect of changing composition in albite and jadeite

Although quartz usually occurs as a pure phase, natural albite and jadeite generally contain some calcium. i.e. albite is part of the albite – anorthite solid solution ($NaAlSi_3O_8 - CaAl_2Si_2O_8$), and jadeite is part of a solid solution between jadeite and diopside ($NaAlSi_2O_6 - CaMgSi_2O_6$). The effect of mixing these other components will be to change their free energy. The free energy changes are due to both the enthalpy and entropy terms associated with mixing and these are considered in some detail in the following chapter. In the case of jadeite – diopside there is an additional complication due to the fact that intermediate compositions (omphacites, see Sections 6.4.1 and 9.6) undergo cation ordering.

In general terms, the effect of changing the free energy of one of the phases is to displace the equilibrium boundaries. This can be seen in a schematic way in Figure 8.27, which shows the intersection of free energy curves for the reactants (jadeite + quartz) and the product (albite), and how changes in the relative free energies affect these intersection points.

Bibliography

Thermodynamics and experimental methods:

NORDSTROM, D.K AND MUNOZ, J.L. (1985). *Geochemical thermodynamics.* The Benjamin/ Cummings Publishing Co., Inc.

HOLLOWAY, J.R. AND WOOD, B.J. (1988). *Simulating the Earth* Unwin Hyman.

Lattice Vibrations and Heat Capacity:

KIEFFER, S.W. (1985). Heat capacity and entropy: systematic relations to lattice vibrations. In:

Microscopic to macroscopic. Reviews in mineralogy **14** Mineralogical Society of America.

SALJE, E. AND WERNECKE, CHR. (1982). The phase equilibrium between sillimanite and andalusite as determined from lattice vibrations. *Contrib. Mineral. Petrol.* **79**, 56–67.

Computer simulations and thermodynamics:

CATLOW, C.R.A. Computer modelling of silicates.

PRICE, G.D. AND PARKER, S.C. The Computer simulation of the lattice dynamics of silicates.

DOVE, M. Molecular dynamics simulations in the solid state sciences

All of the three papers above appear in: *Physical properties and thermodynamic behaviour of minerals* Ed. E.K.H. Salje Nato ASI Series C225 D. Reidel Publishing Co. (1988)

Landau Theory:

SALJE, E.K.H. (1990). *Phase transitions in ferroelastic and co-elastic crystals* Cambridge University Press.

CARPENTER, M.A. (1992). Thermodynamics of phase transitions in minerals: a macroscopic approach. In: *Stability of minerals* G.D. Price and N.L. Ross (Eds) Chapman & Hall.

References

DOVE, M.T. AND POWELL, R.M. (1989). Neutron diffraction study of the tricritical orientational order/disorder phase transition in calcite at 1260K. *Phys. Chem. Minerals* **16**, 503–7.

HOLLAND, T.J.B. (1980). The reaction albite = jadeite + quartz determined experimentally in the range 600 – 1200°C. *Amer. Mineral.* **65**, 129–34.

PRICE, G.D., PARKER, S.C. AND LESLIE, M. (1987). The lattice dynamics and thermodynamics of the Mg_2SiO_4 polymorphs. *Phys. Chem. Minerals* **15**, 181–90.

REDFERN, S.A.T., SALJE, E. AND NAVROTSKY, A. (1989). High temperature enthalpy at the orientational order-disorder transition in calcite: implications for the calcite/aragonite phase equilibrium. *Contrib. Mineral. Petrol.* **101**, 479–84.

9   Energetics and mineral stability II – solid solutions, exsolution and ordering

The existence of solid solutions plays a major role in determining the composition of minerals. The range of composition over which a mineral structure is stable depends very much on the temperature, high temperatures allowing a greater flexibility in atomic substitution. In this chapter we will first describe how the formation of solid solutions affects the free energy of a mineral, and hence its thermodynamic relationship with other minerals in a rock. We then consider how this free energy changes as the temperature is decreased. In this way we can build up a thermodynamic description of the behaviour of solid solutions, leading to an understanding of how solid solutions behave on cooling. Again the emphasis is on the tendency for disorder at high temperatures, giving way to order at low temperatures. The concepts discussed in this chapter are not only very important in minerals which have formed at high temperatures and then cooled, but also in low temperature minerals where solid solutions and disordered states can exist metastably.

9.1 Solid solutions

Most natural mineral groups can exist over a range of chemical compositions which are generally specified in terms of fixed composition end-members. Thus olivine $(Mg,Fe)_2SiO_4$ can have any composition between the end-members forsterite (Mg_2SiO_4) and fayalite (Fe_2SiO_4) by substituting Mg and Fe on the M1 and M2 octahedral sites. Olivine is therefore a *substitutional solid solution*. In the description of mineral structures in Chapters 5 and 6 there were many examples of such solid solutions. Where one element substitutes for another in a structure the amount of substitution depends on the ionic radius. Elements of similar size such as Mg^{2+} and Fe^{2+} are very likely to substitute for one another: a general

rule of thumb is that if the size difference between the ions is less than around 15% a wide range of substitution is possible. In some mineral groups the amount of cation substitution is more restricted and defined by compositional limits which are temperature dependent – at higher temperatures the compositional range is greater. For example, when Na^+ substitutes for K^+ in alkali feldspars the considerable difference in size introduces strain and limits the degree of solid solution. At high temperatures however, the solid solution is virtually complete.

Although relative cation size is a useful indicator of the possibility of substitutional solid solution, the amount of mismatch which can be tolerated in a solid solution does depend on the ability of the rest of the structure to bend bonds and accommodate. For example the degree of solid solution due to Ca $\Leftrightarrow$ Mg substitution is very limited in CaO $-$MgO even at high temperature, while in $CaCO_3 - MgCO_3$ it is fairly extensive, and is complete in high temperature grossular–pyrope garnets $(Ca_3Al_2Si_3O_{12} - Mg_3Al_2Si_3O_{12})$. As the matrix structure in which the substituting cations are sited becomes more complex, the possibility of bond bending becomes greater, allowing a greater flexibility in the degree of solid solution.

A common phenomenon is that of coupled substitution when cations of different valence are interchanged. When Al^{3+} substitutes for Si^{4+} the need to maintain charge balance requires a substitution of some other pair of ions such as Ca^{2+} and Na^+. Thus in the plagioclase feldspars the coupled substitution is written:

$$Al^{3+} + Ca^{2+} \Leftrightarrow Si^{4+} + Na^+$$

which describes the solid solution between the end-members albite $(NaAlSi_3O_8)$ and anorthite $(CaAl_2Si_2O_8)$.

Another less common way in which minerals can continuously vary their composition is by *omission solid solution*. In the pyrrhotites, $Fe_{1-x}S$, cation vacancies result in a range of compositions from FeS to Fe_7S_8, the overall charge balance in the structure being maintained by converting some of the Fe^{2+} ions to Fe^{3+} (Section 5.2.4). Another example is that of mullite where a compositional range from at least $3Al_2O_3.2SiO_2$ to $2Al_2O_3.SiO_2$ is achieved by a substitution of Al^{3+} for Si^{4+} and the creation of oxygen vacancies to maintain charge balance (Section 6.3.3).

Finally, in *interstitial solid solutions*, structural sites which are not normally occupied may be used to broaden the compositional range. The composition of tridymite SiO_2 can be varied towards nepheline $NaAlSiO_4$ by using the large channel sites for Na^+, while substituting Al^{3+} for Si^{4+} in the tetrahedra (Section 6.8.2). Complete solid solution between these end-members is not possible due to geometric constraints.

The type of solid solution in a mineral cannot be simply inferred from the relative abundances provided by a chemical analysis. In a case such as pyrrhotite, the formulae FeS_{1+x} and $Fe_{1-x}S$ could both have the same Fe:S ratio, although the first implies an interstitial solid solution with excess S, while the second is an omission solid solution (Fe deficient). However, density measurements can distinguish between an interstitial solid solution (the density increases relative to the stoichiometric end-member) and an omission solid solution (the density decreases). In complex minerals containing a number of cations, a knowledge of the structure and the known likely substitution schemes can provide a good guess at the occupancy of the various sites, although structural or spectroscopic analysis may be necessary.

In this section we are concerned with the way in which an understanding of the thermodynamics of solid solutions can help to explain the ranges of chemical composition found in minerals. Substitutional solid solutions are by far the most common and we will be mainly concerned with these. Initially we will assume that the solid solution exists across the whole compositional range from one end-member to the other. This requires that the end-members have the same structure.

9.1.1 The entropy of a solid solution

The existence of a solid solution implies a degree of disorder and the random substitution of one element for another will increase the entropy relative to that in the pure end-member. Since we are only concerned with the contribution to the entropy due to the solid solution we will refer to this as the *excess entropy*, relative to the end-member.

We have already derived (Section 8.3.1 and Figure 8.3) the entropy of mixing ΔS for the simple substitution of two elements A and B on a single site in a structure as

$$\Delta S = -R \left(X_A \ln X_A + X_B \ln X_B \right)$$

where X_A and X_B are the atomic fractions of A and B respectively. Note that in the derivation we had to assume a completely random mixing of A and B in the solid solution. This assumption may not always be justified.

If we apply the equation to the case of olivine $(Mg,Fe)_2SiO_4$, which is the most quoted example of an ideal solid solution, and we assume that the M1 and M2 sites can be considered together, the expression becomes

$$\Delta S_{mix} = -2R \left(X_A \ln X_A + X_B \ln X_B \right)$$

to take into account the two M sites per formula unit. However there is a slight preference for Fe to occupy the M1 site (see Section 6.3.1) so that even here the assumption of total randomness may not be completely justified. As an illustration of the method, if for example in a composition $FeMgSiO_4$, 60% of the Fe was in the M1 site, the distribution of Fe,Mg could be written:

$$(Mg_{0.4}Fe_{0.6})_{M1} (Mg_{0.6}Fe_{0.4})_{M2} SiO_4$$

and the entropy calculation would be carried out separately for each site as follows:

$$M1: \Delta S_{mix} = -R \left(X_A \ln X_A + X_B \ln X_B \right)$$

where X_A and X_B now refer to the atomic fraction on the M1 sites.

$$\Delta S_{mix} = -R \left(0.4 \ln 0.4 + 0.6 \ln 0.6 \right)$$
$$= 0.673R$$

The value for M2 is the same, and therefore the total

$$\Delta S_{mix} = 2 \times 0.673R = 1.346R$$

By comparison, if we assume complete randomness over both sites the entropy is larger and is given by

$$\Delta S_{mix} = -2R(0.5 \ln 0.5 + 0.5 \ln 0.5)$$
$$= 1.386R$$

In the derivations which follow we will assume mixing on one site per formula unit.

9.1.2 The enthalpy of a solid solution

To determine the free energy of mixing in a solid solution, and hence to be able to explain the compositional range over which a solid solution will be stable, we next need to describe the change in enthalpy which results from atomic substitution. To do this we need a model which describes the energy of the interactions between the substituting atoms A and B, assuming that the extra energy terms arise due to the interactions between A and B.

In the simplest model we can assume that these extra energy terms, described by the enthalpy of mixing, ΔH_{mix}, arise only from the interactions between nearest neighbour A and B atoms. Suppose that each A atom has z nearest neighbour B atoms and vice versa i.e. both A and B have a coordination number z. If the total number of A and B atoms is N, then the total number of nearest neighbour pairs is $1/2Nz$ (the factor $1/2$ appears because the number of bonds is $1/2$ the number of atoms).

We can define the energy of interaction between A and B atoms by defining three interaction terms:

(i) ω_{AA} is the interaction energy between nearest neighbour A–A atoms.
(ii) ω_{BB} is the interaction energy between nearest neighbour B–B atoms.
(iii) ω_{AB} is the interaction energy between nearest neighbour A–B atoms.

The enthalpy of mixing is then calculated by determining the number of each of these three types of bond in the solid solution. In this model we assume that A and B are completely randomly mixed and the usual statistical methods apply. If the atomic fraction of A atoms is X_A and of B atoms is X_B:

the probability of A–A nearest neighbours = X_A^2.
the probability of B–B nearest neighbours = X_B^2
the probability of A–B nearest neighbours = $2X_AX_B$.

The total interaction energy is therefore:

$$H_{total} = \frac{1}{2} Nz (X_A^2 \omega_{AA} + X_B^2 \omega_{BB} + 2X_AX_B \omega_{AB})$$

$$= \frac{1}{2} Nz (X_A \omega_{AA} + X_B \omega_{BB} + X_AX_B [2\omega_{AB} - \omega_{AA} - \omega_{BB}])$$

The first two terms in this expression i.e. $\frac{1}{2} Nz X_A\omega_{AA}$ and $\frac{1}{2} Nz X_B\omega_{BB}$ are the enthalpies due to $A - A$ and $B - B$ interactions in the appropriate proportions of the pure phases A and B respectively. Adding these two terms would give the enthalpy of a mechanical mixture of A and B as shown in Figure 9.1(a). The *excess enthalpy* due to the formation of the solid solution is the third term i.e.

$$\Delta H_{mix} = \frac{1}{2} Nz X_AX_B[2\omega_{AB} - \omega_{AA} - \omega_{BB}] \quad (9.1)$$

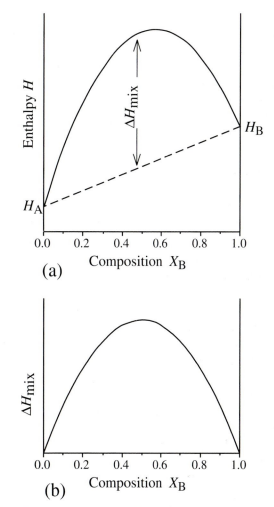

(a)

(b)

Figure 9.1. (a) The enthalpy of the two end-members A and B of the solid solution is given by H_A and H_B respectively. The enthalpy of a mixture of A and B lies along the dashed straight line between H_A and H_B. The enthalpy of a solid solution between A and B also includes an enthalpy of mixing term ΔH_{mix}, here shown to be positive. The curve gives the resultant enthalpy. **(b)** The enthalpy of mixing term ΔH_{mix} as a function of the composition across the solid solution.

The sign of the term $[2\omega_{AB} - \omega_{AA} - \omega_{BB}]$ determines whether the enthalpy of the solid solution is higher or lower than that of a mechanical mixture of the two components.

When $[2\omega_{AB} - \omega_{AA} - \omega_{BB}] = 0$, the energy of two A–B bonds is equal to that of an A–A bond plus a B–B bond. In other words $\Delta H_{mix} = 0$ and there is no difference between a solid solution and a mechanical mixture. Since there is no net gain or loss in energy in forming a solid solution the solid solution is said to be *ideal*, with the internal energy independent of the distribution of atoms.

When $[2\omega_{AB} - \omega_{AA} - \omega_{BB}] > 0$, $\Delta H_{mix} > 0$ and the energy of A–B bonds is greater than the sum of A–A and B–B bonds. Thus the formation of the solid solution raises the enthalpy relative to that of a mechanical mixture and forming the solid solution is endothermic. The maximum value of the ΔH_{mix} term occurs when $X_A = X_B = 0.5$ (Figure 9.1(b)) and its magnitude in different solid solutions depends in a general way on the relative sizes of the substituting cations e.g. in the grossular – pyrope $(Ca_3Al_2Si_3O_{12}$ –$Mg_3Al_2Si_3O_{12})$ solid solution the maximum $\Delta H_{mix} \sim 9.1$ kJ mol^{-1}; in high albite – orthoclase $(NaAlSi_3O_8 - KAlSi_3O_8)$ $\Delta H_{mix} \sim 6.1$ kJ mol^{-1} while in enstatite – ferrosilite $(MgSiO_3 - FeSiO_3)$ $\Delta H_{mix} \sim 1.0$ kJ mol^{-1}.

When $[2\omega_{AB} - \omega_{AA} - \omega_{BB}] < 0$, $\Delta H_{mix} < 0$ and the energy of A–B bonds is less than that of the sum of A–A and B–B bonds. The formation of the solid solution lowers the enthalpy relative to that of a mechanical mixture. This means that A–B bonds are energetically favoured and there will be a tendency to maximise their number by the formation of an ordered compound.

9.1.3 The free energy of a solid solution

Having derived expressions for the entropy and enthalpy of our simple model of a solid solution we are now in a position to determine the free energy of mixing and hence the stability of solid solutions. Since the free energy change associated with the formation of a solid solution is given by

$$\Delta G_{mix} = \Delta H_{mix} - T\Delta S_{mix} \qquad (9.2)$$

it is clear that the entropy of mixing, which is always positive, will tend to reduce the free energy and hence always favour the formation of a solid solu-

tion. The limitation on the degree of solid solution therefore comes from the enthalpy of mixing term.

When the structure of the two end-members of the solid solution is the same, the free energy can be represented by a single curve across the whole composition range. The derivations for the entropy and enthalpy of mixing have also been derived in terms of a complete solid solution, a situation which is only possible when the end-members are isostructural. The conclusions we draw from the discussion which follows refer to complete solid solutions, although the principles will also apply in the case of more restricted solid solutions between end-members with different structures.

We will consider two cases.

(i) The ideal solid solution, when $\Delta H_{mix} = 0$

The ideal solid solution is stabilized over the whole compositional range by the entropy term and

$$\Delta G_{mix} = -T\Delta S_{mix} = RT(X_A \ln X_A + X_B \ln X_B)$$

as shown in Figure 9.2(a). The effect of temperature is merely to change the depth of the free energy minimum. Virtually no mineral solid solutions are truly ideal although Fe,Mg substitution involves only a small positive enthalpy contribution.

The total free energy of an ideal solid solution will also depend on the free energies G_A and G_B of the end members A and B respectively. A mechanical mixture of A and B has a free energy

$$G_1 = X_A G_A + X_B G_B \qquad (9.3)$$

where X_A, X_B are the molar fractions of A and B and G_A and G_B are their molar free energies.

The total free energy of a solid solution is then given by:

$$G_T = G_1 + G_{mix}$$

For an ideal solid solution therefore:

$$G_T = X_A G_A + X_B G_B + RT(X_A \ln X_A + X_B \ln X_B) \qquad (9.4)$$

as shown in Figure 9.2(b).

(ii) The regular solid solution, when $\Delta H_{mix} > 0$

If the enthalpy of mixing, defined by the pairwise interactions between A and B nearest neighbours, is positive, and the increase in entropy arises totally from the configurational effects of random mixing (ΔS_{mix}), it is termed the *regular solution model*.

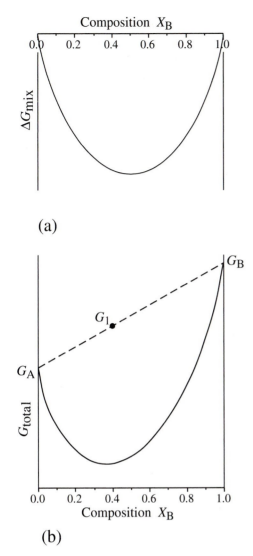

(a)

(b)

Figure 9.2. (a) The free energy of mixing ΔG_{mix} of an ideal solid solution between end-members A and B, as a function of composition. **(b)** The total free energy of the solid solution is given by $G_1 + G_{mix}$ where G_1 is the free energy of a mixture of A and B, and lies on the dashed straight line joining the free energies of the end-members.

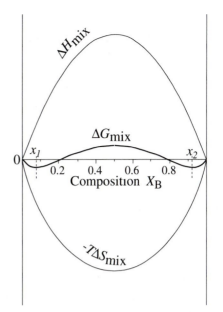

Figure 9.3. The free energy of mixing ΔG_{mix}, of a regular solid solution is the resultant of the enthalpy of mixing ΔH_{mix}, in this case shown to be positive, and the term $-T\Delta S_{mix}$ which is always negative. In the situation shown here, the resultant free energy of mixing is negative for dilute solutions but becomes positive in the intermediate compositions. The two minima in the free energy curve occur at compositions x_1 and x_2 respectively.

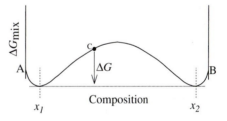

Figure 9.4. The ΔG_{mix} curve taken from Figure 9.3, showing the common tangent drawn to the two minima at compositions x_1 and x_2. For any composition of the solid solution between x_1 and x_2, the minimum free energy is given by the coexistence of two separate phases with compositions x_1 and x_2 respectively. The free energy reduction associated with this phase separation for a bulk composition C is shown here as ΔG.

The free energy of mixing, ΔG_{mix}, of a regular solid solution is shown in Figure 9.3 where schematic curves for ΔH_{mix} and $-T\Delta S_{mix}$ are drawn, together with the resultant free energy $\Delta G_{mix} = \Delta H_{mix} - T\Delta S_{mix}$. Depending on the relative values of ΔH_{mix} and $-T\Delta S_{mix}$ the free energy of mixing may still be negative throughout the whole composition range if the temperature/entropy contribution outweighs the enthalpy increase. This is more likely at high temperatures where the $T\Delta S_{mix}$ term will be increasingly dominant. At lower temperatures the $T\Delta S_{mix}$ term becomes smaller and the situation shown in Figure

9.3 will arise. This shows a ΔG_{mix} curve which is no longer always concave upward but reverses its slope resulting in two minima at the compositions labelled x_1 and x_2.

The significance of this upward inflexion in the shape of the ΔG_{mix} curve is that solid solution compositions between X_1 and X_2 have a higher free energy than a mixture of two phases, one with composition X_1 and the other with composition X_2. The free energy of a such a mixture would lie along the straight line between the minima, shown in Figure 9.4 by the common tangent. In a later section

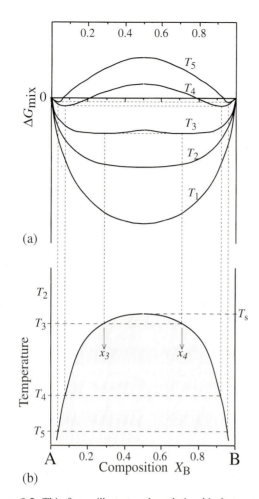

Figure 9.5. This figure illustrates the relationship between the temperature dependence of the free energy curves and the equilibrium phase diagram. **(a)** The variation in free energy of mixing ΔG_{mix} as a function of temperature in a regular solid solution in which the enthalpy of mixing is positive. As the temperature is reduced from T_1 to T_5 the decreasing influence of the entropy term results in free energy of mixing curves which develop two minima. **(b)** The locus of these minima defines the solvus curve which describes the extent of equilibrium solid solution as a function of temperature. Above T_s a complete solid solution is stable but within the shaded region, equilibrium involves the coexistence of two separate phases.

we will demonstrate in more detail why the common tangent construction defines two coexisting compositions at equilibrium, but here we merely note that any solid solution with composition between X_1 and X_2 could reduce its free energy by *unmixing* into two phases, one richer in A, defined by composition X_1, the other richer in B, at composition X_2. On an atomic scale this involves a clustering of like atoms with the tendency to increase the total number of A–A and B–B bonds at the expense of A–B bonds. With $\Delta H_{mix} > 0$, i.e. $2\omega_{AB} > \omega_{AA} - \omega_{BB}$, this is the

type of behaviour we would expect in a system trying to minimise its energy.

The temperature dependence of the ΔG_{mix} curves (Figure 9.5) shows the decreasing influence of the $-T\Delta S_{mix}$ term as the temperature is reduced. At high temperatures the entropy dominates and the free energy curve is everywhere concave upwards i.e. a homogeneous solid solution is stable over the whole composition range. As the temperature falls, the $T\Delta S_{mix}$ term is only larger than ΔH_{mix} in dilute solutions. Thus there is a region at intermediate compositions, between the two minima of the ΔG_{mix} curve, in which the solid solution could reduce its free energy by unmixing into two separate compositions. As the temperature falls further this region broadens and only very dilute solid solutions are stable.

The downward slope of the ΔG_{mix} curve at small concentrations X_A or X_B arises because the slope dS/dX of the ΔS_{mix} curve is always greater than the slope dH/dX of the ΔH_{mix} curve at small values of X, and so small deviations from end-member compositions are always more stable than the pure end-members, even at low temperatures.

Depending on the magnitude of the ΔH_{mix} term, the regular solution model predicts that above a certain temperature a complete solid solution will be stable, while at decreasing temperatures the composition range of a stable solid solution becomes more restricted. This is usually illustrated on an *equilibrium phase diagram* which shows the stability range as a function of temperature and composition.

9.2 The equilibrium phase diagram for a regular solid solution – the solvus

The equilibrium phase diagram for a regular solid solution is drawn by plotting the locus of the minima of the ΔG_{mix} curves as a function of temperature, as shown in Figure 9.5. The result is a solubility line called the *solvus* which defines the limits of solid solution of A in B and B in A, and the field, shown shaded, in which two coexisting phases are more stable than the solid solution.

Above the critical temperature T_s the solid solution is stable over the whole composition range. Below T_s there is a miscibility gap within which the stable coexisting compositions are defined by the ends of horizontal *tie-lines* such as the dashed line which indicates that at temperature T_3, any solid

Figure 9.6. An optical micrograph of a pyroxene which grew as a solid solution with composition between augite and pigeonite, and which subsequently has exsolved into coexisting augite and pigeonite lamellae. The original crystal was twinned giving the final microstructure a 'herring-bone' appearance.

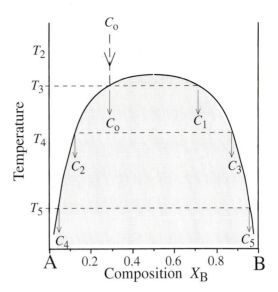

Figure 9.7. A solid solution with composition C_o meets the solvus at temperature T_3 as it cools. At T_3 it is in equilibrium with a composition C_1 determined from the dashed horizontal tie line. At successively lower temperatures the solid solution exsolves to two equilibrium coexisting compositions defined by the horizontal tie lines. At T_4 a bulk composition C_o consists of an intergrowth of $C_2 + C_3$; at T_5 the equilibrium coexistence is $C_4 + C_5$. The relative proportions of the two phases are such that the bulk composition remains constant.

solution with a composition between X_3 and X_4 will achieve equilibrium by unmixing into two separate phases with compositions X_3 and X_4 respectively. As the temperature falls, the miscibility gap widens until at low temperatures very little solid solution exists and the equilibrium consists of a phase mixture of almost pure A and pure B. The breakdown of a single phase solid solution into two phases is variously termed unmixing, phase separation, solid state precipitation, or exsolution (Figure 9.6). Exsolution is the usual term used in mineralogy.

The thermodynamically predicted behaviour of a high temperature regular solid solution as it cools is illustrated by considering the sequence of processes which should take place in a single phase of composition C_o as it cools from above the miscibility gap, shown in Figure 9.7. At temperature T_3 the solid solution with composition C_o is in equilibrium with an infinitesimal amount of solid solution C_1 as shown by the dashed tie-line. If the high temperature solid solution is cooled to T_4, the free energy drive to unmix increases and, given sufficient time, a small amount of material of composition C_3 will nucleate, changing the composition in the rest of the solid solution to C_2, to maintain the mass balance. The phase of composition C_3 forms precipitates within the solid composition C_2.

Further reductions in temperature require that the compositions of the two coexisting phases change continuously as indicated by the tie-lines, until at temperature T_5 the most stable state is a phase mixture of regions of composition C_4 and C_5. During the unmixing process the total free energy remains constant since the system is at equilibrium, although the compositions and relative proportions of the two coexisting phases change with temperature. The method of determining the relative proportions of the two phases is called the *lever rule*, and is described in the Appendix to this chapter.

Although the regular solution model is very simple, it does predict the overall features of the behaviour of many mineral solid solutions. The existence of a miscibility gap which is temperature dependent, with a high temperature closure near the centre of the compositional range is a common feature of the phase equilibria of many mineral systems, although most are complicated by the existence of other phase transformations superimposed on the solid solution.

Before outlining some of the possible refinements of the regular solution model which would give a closer correspondence to reality, it is worth re-emphasising a point made in connection with polymorphic phase transformations. That is, that thermodynamics predicts what *ought* to happen, but the kinetics of the processes involved will decide what *will* happen. The predicted cooling behaviour of the high temperature solid solution described above is rarely achieved in solids, even if the model is correct. There are two main reasons for departures from equilibrium.

1. The first step in the unmixing process involves the nucleation of a second phase, e.g. the nucleation of a region of composition C_3 at temperature T_4 in the example above. This region will have different lattice parameters from the matrix phase and hence will introduce strain energy. This in turn suppresses the nucleation to lower temperatures until a sufficient free energy drive is produced by the undercooling to counteract the strain energy term.

2. The second problem is that the formation of the second phase requires the diffusion of considerable amounts of the B component to the nucleation site. Since solid state diffusion is slow, especially at low temperatures, the attainment of equilibrium by continuously readjusting the coexisting compositions on cooling is not always achievable, even on a geological time-scale. The problem becomes particularly serious in any experimental determinations of unmixing where the available time-scale is short.

It is clear that a description of the real behaviour of a solid solution must include time as one of the important parameters. The temperature at which nucleation takes place, and the degree to which equilibrium is achieved, depends on the cooling rate. The temperature of the solvus curve is also important – a solid solution with a small positive ΔH_{mix} will have a relatively low T_s where diffusion may be too slow to achieve any unmixing and a solid solution may persist metastably. Often techniques such as electron microscopy will provide evidence for fine-scale unmixing in minerals which appear to be single-phase solid solutions. We will return to the problems of kinetics in greater detail in Chapters 10 and 11.

9.2.1 Spinodal decomposition

Another prediction of the regular solution model is that in the region of the $G_{\mathrm{mix}} - X$ curve between the two inflexion points, where $d^2G/dX^2 = 0$, the solid solution is unstable to any fluctuations in composition (Figure 9.8(a)). In this region, fluctuations in composition about any point on the free energy curve result in a reduction in the overall free energy. The continued growth in amplitude of these fluctuations leads to the possibility of a continuous process of unmixing into regions of different composition. Eventually the stable coexisting compositions, defined by the common tangent to the $G(X)$ minima, will be reached. This continuous process of unmixing is termed *spinodal decomposition*.

The locus of the inflexion points (where the second derivative of the $G_{\mathrm{mix}} - X$ curve for different temperatures is zero, i.e. $d^2G/dX^2 = 0$) defines a *spinodal curve*, (Figure 9.8(b)) in the same way that the locus of the first derivative $dG/dX = 0$ defined the solvus. According to this model, spinodal decomposition is possible in any solid solution with a composition within the spinodal curve.

The position of the spinodal curve relative to the solvus can be calculated from the $G-X$ curve for the solid solution. In the regular solution model

$$T_{\mathrm{spinodal}} = 4\,T_{\mathrm{solvus}}\,X(1\text{-}X) \qquad (9.5)$$

where X is the molar concentration of one of the components. Thus at $X = 0.5$, $T_{\mathrm{spinodal}} = T_{\mathrm{solvus}}$ and the two curves touch, as in Figure 9.8.

At compositions outside the spinodal but within the solvus, such as composition c_1 in Figure 9.8(a), small fluctuations in composition result in an increase in free energy. The only way such a composition can move towards equilibrium is to nucleate a composition X_2 on the opposite side of the solvus line. This is a discontinuous process involving a large local compositional change and an activation energy barrier. Kinetically, the discontinuous nucleation and growth process is likely to be much slower than the continuous spinodal process, and hence within the spinodals the continuous mechanism will be preferred.

Spinodal fluctuations in composition create coex-

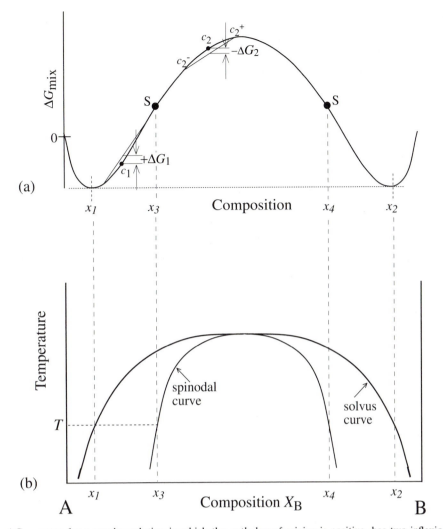

Figure 9.8. (a) The ΔG_{mix} curve for a regular solution in which the enthalpy of mixing is positive, has two inflexion points labelled S. Between these two inflexion points the free energy of the solid solution is reduced for any fluctuation in composition about the mean. For a solid solution with bulk composition c_2, compositional fluctuations between c_2^+ and c_2^- reduce the free energy by ΔG_2. On the other hand, for compositions outside the inflexion points, such as c_1, fluctuations increase the free energy. **(b)** The locus of the inflexion points S, as a function of composition, defines the spinodal curve, shown together with the solvus which is defined by the locus of the tangent points to the free energy curves.

isting regions with slightly different lattice parameters, creating local strains within the crystal. The tendency to minimise the strain energy results in a spatial organisation of the compositional fluctuations which is dictated by the symmetry of the elastic properties. The fluctuations are periodic and can be described as sinusoidal compositional modulations throughout the crystal (Figure 9.9(a)). The modulations are on a very fine scale, with a wavelength typically between 50Å and about 500Å, and so can only be observed by transmission electron microscopy (Figure 9.9(b)). Eventually, as the peaks and troughs in these waves become amplified they become the two discrete phases defined by the free energy minima.

The direction of the modulation corresponds to the elastically 'soft' direction(s) in the crystal. In a cubic crystal for example, this is usually <100> and the symmetry leads to three sets of mutually perpendicular modulations. In lower symmetry minerals only one or two compositional modulations are observed. When two sets of modulations coexist in a crystal, transmission electron microscopy shows a 'tweed' microstructure, very similar to that observed in the early stages of transformation twinning (see Section 7.3.6).

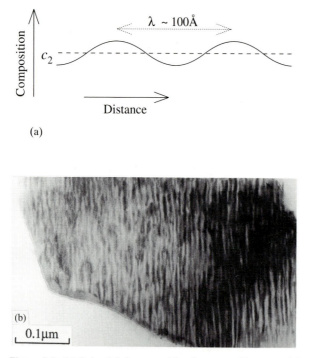

(a)

(b)

0.1μm

Figure 9.9. (a) Spinodal decomposition in a crystalline material involves the generation of sinusoidal fluctuations in composition about the mean composition c_1. The wavelength of these fluctuations is of the order of 100Å. **(b)** Transmission electron micrograph of spinodal decomposition during the early stages of exsolution in alkali feldspar. The compositional fluctuations produce fluctuations in local lattice spacings and result in periodic fluctuations in the diffraction contrast. The strong dark region is a bend contour.

(Note: The tweed microstructure in twinning involves sheared regions with rotated, but undistorted regions in between. In tweed microstructures formed by spinodal decomposition, the modulation is due to compositional differences which produce slight differences in lattice parameters. The modulations are therefore compressional rather than transverse, and the tweed microstructure consists of regions with composition oscillating about the average, with regions of average composition in between.)

Spinodal decomposition is a very common phenomenon in natural mineral solid solutions in which the process of equilibration has been arrested by a combination of the sluggish kinetics and a rapid cooling rate. We will return to a discussion of the mechanism and kinetics in the following chapter, but it is important to note that spinodal decomposition is not an equilibrium process, and therefore does not appear on an equilibrium phase diagram.

9.2.2 Improved models of mineral solid solutions

Although the regular solution model qualitatively predicts the general behaviour of many solid solutions, more accurate analysis of experimental data shows that there are many features which cannot be explained by such a simple model. In this section we look at some of the refinements which could be incorporated to improve it.

(i) The effect of local clustering

One of the rather artificial assumptions of the regular solution model is that even though the enthalpy of mixing is not zero, a random distribution of atoms is used to calculate the configurational entropy. In practice we would expect if ΔH_{mix} is positive then there would be a greater number of A–A, and B–B pairs than for a completely random distribution, as this would further reduce the enthalpy even at temperatures above T_c. On the other hand, such local *clustering* effects also decrease the entropy of mixing, and hence the free energy increases. An optimum situation will be attained, reflecting a balance between the enthalpy and entropy contributions to the free energy. In general, local clustering will exist even at temperatures above the solvus maximum T_s, and a statistically random situation would only be achieved at significantly higher temperatures. There are a number of ways, based on statistical thermodynamics, of estimating how the pair probabilities depend on the interaction energy, but this is beyond the scope of this book. We will discuss local short-range effects again in relation to ordering in Section 9.3.1.

(ii) Incorporating asymmetry

The regular solution model predicts a solvus which is symmetric about the mid-point composition, $X_A = 0.5$. Most solvi are asymmetric to a greater or lesser extent, reflecting the fact that the pair-binding interaction energies ω_{AA}, ω_{BB} and ω_{AB} are dependent on composition, i.e. on the local chemical and structural environment. In other words, it may be energetically less costly to substitute atoms of A into the B end-member than vice versa: substituting a smaller cation into a larger site may be easier than putting a larger cation into a site previously occupied by a smaller one. The system $MgO - ZnO$ provides a simple example. Both have the sodium chloride structure, but while MgO can tolerate a considerable solid solution towards ZnO (around 40% ZnO at 800°C), ZnO has virtually no solid solution towards MgO.

(iii) Data fitting methods of modelling solid solutions

When experimental data of the heat of mixing in complex mineral solid solutions has been measured (see Section 8.5), it can be fitted to an expression which includes a generalised interaction parameter Ω, without necessarily specifying the atomic origin of the interactions. In the regular solution model the the enthalpy of mixing is given by the expression:

$$\Delta H_{mix} = \Omega X_A X_B$$

where

$$\Omega = 1/2 \, Nz \, [2\omega_{AB} - \omega_{AA} - \omega_{BB}]$$

A sub-regular, or two-parameter fit can be made by applying the expression

$$\Delta H_{mix} = X_A X_B (\Omega_1 X_A + \Omega_2 X_B) \qquad (9.6)$$

where Ω_1 and Ω_2 are called Margules parameters. The choice of appropriate values of Ω_1 and Ω_2 can lead to an asymmetric solvus, the physical meaning of these Margules parameters is lost. More generalized expressions in which the data are fitted to a polynomial in X have also been used, but increasing the number of parameters does not necessarily provide physically realistic models.

9.3 The chemical potential and the activity of solid solutions

In describing regular solid solutions, we did not write out in full the expression for the ΔG_{mix} of a regular solution, preferring to use diagrams such as Figure 9.3 to illustrate its variation with composition and temperature. This is because the expression is rather too cumbersome. As in the case of an ideal solid solution the total free energy is the sum of the free energy of a mechanical mixture of A and B (G_1) and the ΔG_{mix} term. Therefore for a regular solution

$$
\begin{aligned}
G_T = G_1 + \Delta G_{mix} &= G_1 + \Delta H_{mix} - T\Delta S_{mix} \\
&= X_A G_A + X_B G_B + 1/2 \, Nz \, X_A X_B [2\omega_{AB} - \omega_{AA} \\
&\quad - \omega_{BB}] + RT(X_A \ln X_A + X_B \ln X_B)
\end{aligned}
$$

This is not a very easy expression to use and would become very much more unwieldy if we had more than two chemical components in the solid solution, as is often the case in mineral solid solutions. To simplify this we introduce a few new definitions:

(i) Chemical potentials in an ideal solid solution

First we return to the ideal solid solution in which the total free energy was given by the expression

$$G_T = X_A G_A + X_B G_B + RT(X_A \ln X_A + X_B \ln X_B)$$

This may be rewritten

$$G_T = X_A(G_A + RT \ln X_A) + X_B(G_B + RT \ln X_B)$$

The term $(G_A + RT \ln X_A)$ is called the *partial molar free energy* of A, i.e. describes the contribution to the free energy of the solid solution due to the X_A moles of A. Alternatively, it is termed the *chemical potential* of A in the solid solution, and denoted μ_A. Similarly for the second component B.

Thus we can write

$$\mu_A = G_A + RT \ln X_A \qquad (9.7)$$
$$\mu_B = G_B + RT \ln X_B \qquad (9.8)$$

and hence for an ideal solid solution

$$G_T = \mu_A X_A + \mu_B X_B \qquad (9.9)$$

The relationship between the free energy curve and the chemical potentials for an ideal solid solution is shown in Figure 9.10. A tangent to the free energy curve intercepts the free energy axes at μ_A and μ_B respectively, and hence the free energy at any composition can be found by the linear proportion in eqn.(9.9). (The form of this equation should be compared with eqn. (9.3) for the free energy of a mechanical mixture of A and B).

This description of the meaning of the intercept of the tangent to the free energy curve on the free energy axis, leads to a further point regarding the common tangent construction used in Section 9.1.3. There, the equilibrium compositions of two coexisting phases were defined by the common tangent. The common tangent construction means therefore, that the chemical potential of the component A is the same in both coexisting phases. Similarly the chemical potential of B is equal in both phases. This leads to another definition of equilibrium, i.e. that for a free energy minimum each component must have the same chemical potential in each coexisting phase.

(ii) Activities in a non-ideal solid solution

When the enthalpy of mixing is non-zero, the free energy change on mixing is greater or less than that in an ideal solid solution and therefore equations (9.7) and (9.8) defining the chemical potentials of components A and B in the solid solution must be modified to take this into account. However, it is mathematically convenient to retain the form of these equations and so we define a quantity known as the *activity*, *a* of a component in the solution such that

$$\mu_A = G_A + RT \ln a_A \quad (9.10)$$
$$\mu_B = G_B + RT \ln a_B \quad (9.11)$$

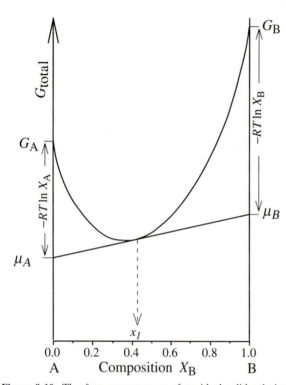

Figure 9.10. The free energy curve of an ideal solid solution, whose end-member components have free energies G_A and G_B respectively. The tangent drawn to this curve at any composition (in this case x_1) intercepts the free energy axes at μ_A and μ_B respectively, where μ is termed the chemical potential of the component in the solid solution.

where a_A and a_B are the activities of A and B in the solid solution. The activity therefore, behaves like an effective concentration of the component in the solution. If the enthalpy of mixing $\Delta H_{mix} > 0$, the stronger bonds between similar atoms increases their activity above their actual molar concentration. When $\Delta H_{mix} < 0$, the activity is lower than the molar concentration.

The relationship between the activity and the molar concentration is defined by the *activity coefficients* γ_A and γ_B as follows:

$$\gamma_A = a_A/X_A \qquad \gamma_B = a_B/X_B \quad (9.12)$$

This is shown graphically in Figure 9.11. For an ideal solid solution $a_A = X_A$ (Raoult's Law). A positive deviation from Raoult's Law with $a_A > X_A$ means that $\Delta H_{mix} > 0$ and indicates a tendency for unmixing; a negative deviation from Raoult's Law with $a_A < X_A$ means that $\Delta H_{mix} < 0$ and indicates a tendency towards ordering (i.e. increasing the number of A–B bonds). In nearly pure end-member compositions, Raoult's Law is obeyed even in non-ideal solid solutions due to the dominance of the entropy term.

9.4 Order–disorder in solid solutions

We return now to the case when the enthalpy of mixing in the solid solution is negative, i.e. $[2\omega_{AB} - \omega_{AA} - \omega_{BB}] < 0$, $\Delta H_{mix} < 0$ and the attraction between unlike atoms A and B in the solid solution is stronger than between like atoms. Both the enthalpy and the entropy terms favour the formation of a solid solution and the free energy curve is negative relative to the end-members throughout the whole compositional range (Figure 9.12).

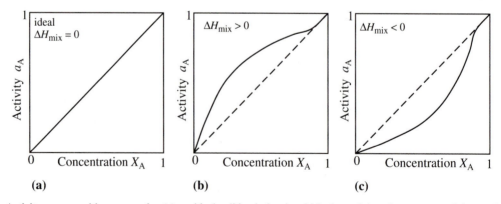

Figure 9.11. Activity – composition curves for **(a)** an ideal solid solution in which the activity of component A is equal to its molar concentration (Raoult's Law), **(b)** a solid solution in which the enthalpy of mixing ΔH_{mix} is positive, showing a positive deviation from Raoult's Law, except near pure A, and **(c)** a solid solution in which the enthalpy of mixing is negative, showing a negative deviation from Raoult's Law, except near pure A.

The tendency on cooling will be to increase the number of A–B bonds. At the 50:50 composition, where the numbers of A and B atoms are equal, this may lead to the formation of a fully ordered structure with alternating A and B atoms.

A number of concepts associated with the ordering process can be illustrated with reference to a simple solid solution in which mixing in the disordered state occurs on one set of equivalent sites, but ordering causes the sites to become non-equivalent, with A atoms on the sites labelled α, and B atoms on a crystallographically distinct set of sites labelled β (Figure 9.13).

This situation is termed *convergent ordering* (because the identity of the two sites α and β converges when they are randomly occupied), and can be distinguished from *non-convergent ordering* where the sites are crystallographically distinct even when randomly occupied. Si,Al ordering in K-feldspar is therefore an example of convergent ordering, whereas the distribution of Fe and Mg over the M1 and M2 sites in pyroxenes is non-convergent. Most of this section however, will be concerned with convergent ordering.

9.4.1 Short-range and long-range order

There are two different concepts used in describing the degree of order. First, *short-range order* is defined in terms of the proportion of atoms with the correct nearest neighbours i.e. the proportion of A atoms with B nearest neighbours.

The *short-range order parameter* $\sigma =$

$$\frac{q - q\,(\text{rand.})}{q\,(\text{max}) - q\,(\text{rand.})} \tag{9.13}$$

where q is the proportion of nearest neighbour A–B bonds, while $q\,(\text{max})$ is the maximum proportion of A–B bonds in the best ordered structure, and $q\,(\text{rand})$ is the proportion in a randomly disordered structure. At the composition AB, $q\,(\text{max}) = 1$, and $q\,(\text{rand}) = 0.5$, so that

$$\sigma = 2q - 1.$$

In the random structure $\sigma = 0$, while in the fully ordered structure $\sigma = 1$. Clearly, only the stoichiometric composition is able to achieve complete order.

Long-range order is defined in terms of the occupancy of the non-equivalent sites labelled α, and β (Figure 9.13). In the fully ordered structure at the composition AB, all α sites are occupied by A atoms and all β sites occupied by B atoms. The *long-range order parameter* **S** is defined so that the fraction of atoms on the correct sites is given by $1/2\,(1 + S)$ and hence the fraction on the incorrect sites is then

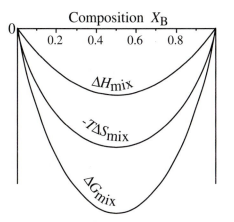

Figure 9.12. The free energy of mixing ΔG_{mix} of a solid solution in which the enthalpy of mixing ΔH_{mix} is negative. The free energy of mixing curve is negative throughout the whole compositional range.

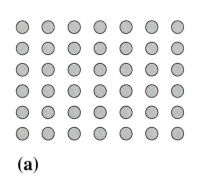

(a)

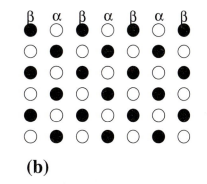

(b)

Figure 9.13. A schematic illustration of convergent ordering. Above a critical temperature, the sites over which ordering takes place are all equivalent, as shown in (a) where they are randomly occupied. Ordering distinguishes between the sites and the white atoms order into the sites labelled α, while the black atoms order into the sites labelled β.

$1/2 (1 − S)$. S is therefore the difference between the proportion of atoms correctly placed and incorrectly placed relative to the fully ordered structure. In the random structure $S = 0$, while in the fully ordered structure $S = 1$.

The onset of long-range order on cooling suggests a cooperative phenomenon in which site occupancies become correlated over the whole crystal. Before this happens however there will be considerable short-range order in which local ordered domains exist within the nominally disordered solid solution, although the domains are not correlated at high temperature. The argument is similar to that for clustering in solid solutions where $\Delta H_{mix} < 0$, and a balance between the reduced enthalpy and entropy must be reached to achieve a free energy minimum at every temperature. The evidence suggests that short-range effects persist in solid solutions at temperatures considerably above that where long-range order is lost. In solid solutions which have been cooled too quickly for long-range order to be achieved there may be significant short-range order. Any interpretation of thermodynamic measurements on such solid solutions must bear in mind the possibility of short-range order.

The detection of the degree of order in a solid solution can be made using X-ray diffraction methods. Long-range order of the type described above, where a single site becomes differentiated into non-equivalent sites due to ordering, reduces the symmetry of the structure. If there is a reduction in the translational symmetry, increasing the size of the unit cell or changing the lattice type, extra superlattice reflections will appear in the diffraction pattern. If the long-range order is well developed the superlattice reflections are sharp and well defined. Diffuse superlattice reflections suggest that the periodicity of the ordering is not perfect, increasing diffuseness indicating the loss of long-range order. By studying the diffuse X-ray scattering around the superlattice positions, the degree of short-range order can be measured.

Another result of the loss of translational symmetry is the presence of antiphase domains in the ordered structure (Section 7.3.4). At an antiphase boundary the meaning of 'correct site' and 'incorrect site' changes, since the displacement vector interchanges the α and β sites. This creates problems with the definition of long-range order; taken over the whole crystal even a single antiphase boundary can reduce the long-range order parameter S to zero, while not affecting the short-range order. Long-range order is defined in relation to the volume of the crystal which coherently scatters X-rays and produces the diffraction pattern. As the size of the antiphase domains becomes smaller than this volume, the superlattice reflections will become increasingly diffuse.

Ordering may also result in a loss of point group symmetry elements, resulting in the possibility of transformation twin domains (Section 7.3.5). The monoclinic – triclinic transformation in K-feldspar is a good example where the presence of submicroscopic twin domains produces a situation where the short-range order may be very high, but the long-range order is virtually zero.

9.4.2 A model for convergent ordering in a solid solution – the Bragg–Williams model

The aim in this section is to derive a model which can predict how the equilibrium degree of order S varies with temperature. To do this we need to evaluate the entropy and enthalpy of a solid solution in which $\Delta H_{mix} < 0$, in terms of the long-range order parameter S.

The model is very similar in principle to the derivation in Section 9.1, and makes the same assumptions regarding nearest neighbour interactions. We assume that the enthalpy of mixing ΔH_{mix} arises only from the interactions between nearest neighbour A and B atoms, and that each A atom has z nearest neighbour B atoms and vice versa i.e. both A and B have a coordination number z. The total number of A plus B atoms is $2N$, and ordering takes place over two sites α and β, which are equal in number i.e. N α sites and N β sites. There are thus Nz bonds from the α sites and the same number from the β sites. 'Correct' sites are defined as A atoms on α sites, and hence B atoms on β sites.

The model we describe is the well-known classical model for order–disorder in an AB alloy, and despite its simplicity does have many features common to more complex models. It is generally referred to in the literature as the Bragg – Williams model and has been successfully applied to predict the general form of phase diagrams in a number of mineral systems.

(i) The configurational entropy

From the definition of the order parameter S, the fraction of atoms on α sites which are A atoms is $1/2(1 + S)$, and the fraction which are B atoms is $1/2(1 - S)$. The entropy of mixing A and B atoms over the α sites is given by eqn. (8.12):

$$\Delta S_{mix} = -R\left\{ 1/2(1 + S) \ln 1/2(1 + S) + 1/2(1 - S) \ln 1/2(1 - S) \right\}.$$

The entropy of mixing over the β sites is identical, and therefore the total configurational entropy per mole is given by

$$\Delta S_{mix} = -2R\left\{ 1/2(1 + S) \ln 1/2(1 + S) + 1/2(1 - S) \ln 1/2(1 - S) \right\}.$$

$$\Delta S_{mix} = R\left\{ 2\ln 2 - (1 + S)\ln(1 + S) - (1 - S)\ln(1 - S) \right\} \qquad (9.14)$$

When $S = 0$, the solid solution has no long-range order and the entropy at composition $X_A = X_B = 0.5$, is therefore $\Delta S_{mix} = 2R \ln 2$. In the fully ordered state when $S = 1$, $\Delta S_{mix} = 0$, as we would expect.

(ii) The enthalpy of mixing

To evaluate the enthalpy we need to derive the number of each of the bond types A–A, B–B and A–B and multiply by the appropriate interaction energy. The total number of A atoms on α sites is $1/2N(1 + S)$. These A atoms on α sites give rise to $1/2N(1 + S)z$ bonds. Of these bonds some will be A–A bonds and some A–B bonds. The number of A–A bonds is given by multiplying this term by the fraction of A atoms on the neighbouring β sites.

$$n_{AA} \text{ bonds} = \frac{1}{2}N(1 + S) \cdot z \cdot \frac{1}{2}(1 - S)$$

$$= \frac{1}{4}Nz(1 - S^2).$$

The number of A – B bonds due to A atoms on α sites is therefore

$$n_{AB} \text{ bonds} = \frac{1}{2}Nz - \frac{1}{4}Nz(1 - S^2).$$

since $\frac{1}{2}Nz$ is the total number of bonds from α sites.

By the same reasoning the B atoms on α sites generate a number of B – B bonds given by

$$n_{BB} \text{ bonds} = \frac{1}{2}N(1 - S) \cdot z \cdot \frac{1}{2}(1 + S)$$

$$= \frac{1}{4}Nz(1 - S^2).$$

and the number of B – A bonds due to B atoms on α sites is then

$$n_{BA} \text{ bonds} = \frac{1}{2}Nz - \frac{1}{4}Nz(1 - S^2).$$

The total number of A–B, and B–A bonds (taken together, and now simply denoted n_{AB}) is then:

$$\text{Total } n_{AB} \text{ bonds} = 2\left\{ \frac{1}{2}Nz - \frac{1}{4}Nz(1 - S^2) \right\}.$$

$$= Nz - \frac{1}{2}Nz(1 - S^2).$$

$$= \frac{1}{2}Nz(1 + S^2)$$

We do not need to consider the bonds from the β sites as they have already been counted by the above procedure.

The total interaction enthalpy is thus given by the following:

$$\text{Enthalpy due to A–A bonds} = \frac{1}{4}Nz(1 - S^2) \cdot \omega_{AA}.$$

$$\text{Enthalpy due to B–B bonds} = \frac{1}{4}Nz(1 - S^2) \cdot \omega_{BB}.$$

$$\text{Enthalpy due to A–B bonds} = \frac{1}{2}Nz(1 + S^2) \cdot \omega_{AB}.$$

The total interaction enthalpy is therefore given by the sum:

$$H = \frac{1}{4}Nz(1 - S^2) \cdot \omega_{AA} + \frac{1}{4}Nz(1 - S^2) \cdot \omega_{BB} + \frac{1}{2}Nz(1 + S^2) \cdot \omega_{AB}.$$

$$= \frac{1}{2}Nz\left\{ \frac{1}{2}(\omega_{AA} + \omega_{BB})(1 - S^2) + \omega_{AB}(1 + S^2) \right\}$$

In the fully ordered form $S = 1$, and therefore the enthalpy is given by $Nz\,\omega_{AB}$. The enthalpy of mixing which is the difference between the partly disordered and the fully ordered states is therefore given by

$$\Delta H_{mix} = \frac{1}{2}Nz\left\{ \frac{1}{2}(\omega_{AA} + \omega_{BB})(1 - S^2) + \omega_{AB}(1 + S^2) \right\} - Nz\,\omega_{AB}.$$

which simplifies to

$$\Delta H_{mix} = \frac{1}{4}Nz\,[2\omega_{AB} - \omega_{AA} - \omega_{BB}](1 - S^2). \qquad (9.15)$$

(iii) The free energy of mixing

The free energy of mixing relative to the fully ordered state is given by the usual relation

$$\Delta G_{mix} = \Delta H_{mix} - T\Delta S_{mix}$$
$$= \frac{1}{4} Nz \left[2\omega_{AB} - \omega_{AA} - \omega_{BB}\right](1 - S^2)$$
$$- RT \{2 \ln 2 - (1 + S) \ln (1 + S) -$$
$$(1 - S) \ln (1 - S)\}$$
$$(9.16)$$

If $\Delta H_{mix} < 0$, i.e. $2\omega_{AB} < \omega_{AA} + \omega_{BB}$, the free energy will be reduced by ordering and at any temperature an optimum value of the degree of order will be attained, which represents a balance between the enthalpy reduction and the entropy reduction on ordering. This equilibrium will be a minimum in the curve of ΔG_{mix} versus the order parameter S, so that

$$\frac{\partial (G_{mix})}{\partial S} = 0$$

$$= RT\{\ln (1 + S) - \ln (1 - S)\}$$
$$- \frac{1}{2} z N[2\omega_{AB} - \omega_{AA} - \omega_{BB}] S = 0$$

i.e.

$$\ln (1 + S)/(1 - S) = \frac{1}{2} z [2\omega_{AB} - \omega_{AA}$$
$$- \omega_{BB}] S / k T$$
$$(9.17)$$

Since

$$\ln (1 + S)/(1 - S) = 2 \tanh^{-1} S$$

eqn (9.17) can also be written as

$$\tanh^{-1} S = \frac{1}{4} \cdot z [2\omega_{AB} - \omega_{AA} - \omega_{BB}] S / k T$$
$$(9.18)$$

This solution defines the equilibrium variation of S with temperature T and predicts that at a critical temperature T_c, S becomes zero. The value of T_c is given by

$$T_c = -2 [2\omega_{AB} - \omega_{AA} - \omega_{BB}]/k \qquad (9.19)$$

so that the lower the energy of an A–B bond relative to the average of A–A and B–B bonds the higher the disordering temperature T_c will be.

At any temperature T,

$$\ln (1 + S)/ (1 - S) = 2 S T_c / T$$

or

$$S = \tanh [S T_c / T] \qquad (9.20)$$

Figure 9.14 shows the variation of S with T which

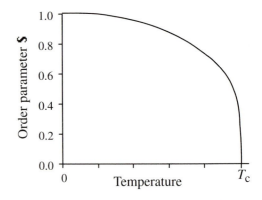

Figure 9.14. The variation of the long-range order parameter S with temperature, according to the Bragg–Williams model. Complete order ($S = 1$) is only achieved at 0K, and complete disorder ($S = 0$) exists above a critical temperature T_c.

illustrates the principal features of the Bragg–Williams model. As the temperature increases the degree of order remains quite large until the critical temperature is approached, after which disorder develops over a relatively small temperature range, becoming complete at the critical temperature T_c. This type of behaviour is typical of a cooperative phenomenon where the energy associated with disordering depends on the similarity or dissimilarity of the nearest neighbours and hence becomes progressively less as the amount of disorder increases.

Another feature of the Bragg–Williams model for ordering at the AB composition, is that the variation in the long-range order parameter is continuous and thermodynamically second order (as defined in section 8.4.2). In practice, this is not always the case, even in many simple alloy systems. The variation of S with T can be measured experimentally from the intensity of superlattice reflections, and the reality rarely conforms to the Bragg–Williams model. Taking into account interactions between second and even third nearset neighbours has often provided a better explanation for the observed variations of S with T, and can result in a model which predicts a first-order transformation.

Note that the Bragg–Williams model only describes the average or long-range order and cannot take into account short-range effects. The variation in the short-range order parameter with temperature (Figure 9.15) shows that although σ falls sharply at the critical temperature T_c, considerable short-range order persists even when the long-range order is zero.

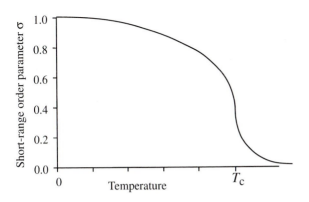

Figure 9.15. The variation of the short-range order parameter σ with temperature. Although there is a rapid fall in σ at T_c, significant short range order persists to higher temperatures.

9.4.3 Non-convergent ordering

In non-convergent ordering we are concerned with the distribution of cations over crystallographic sites which are non-equivalent. Often this type of ordering is termed intracrystalline partitioning. A simple example is the distribution of Fe and Mg ions over the M1 and M2 sites in pyroxenes. If there is an energetic preference for a cation to occupy one site rather than the other, it follows that the minimum enthalpy will be achieved by an ordered distribution e.g. the Mg^{2+} in M1 sites and the Fe^{2+} in M2 sites. However, at temperatures above 0K, the configurational entropy term will minimize the free energy of partially disordered distributions, the degree of order depending on the temperature and the enthalpy reduction associated with the ordering process.

This situation is very different from convergent ordering where the sites only become non-equivalent due to ordering. In the Bragg–Williams treatment of convergent ordering the enthalpy associated with the exchange of atoms between sites results from the interactions between pairs of atoms. The enthalpy change therefore depends on the degree of order and decreases as the disorder increases. In other words, the magnitude of the exchange energy depends on how similar the sites are, and goes to zero as the sites become statistically identical in the high temperature form. In non-convergent ordering there is no phase transition, and the exchange enthalpy ($\Delta H_{exch.}$) is fixed by the nature of the sites themselves and does not depend on the degree of order. Non-convergent ordering

produces a continuous change in the equilibrium cation distribution as a function of temperature. As the temperature increases, the entropy term becomes dominant, stabilizing an essentially random distribution, while at low temperatures the tendency to order is defined by the dissimilarity of the sites and hence the energy associated with the occupancy of each site by a specific atom.

In olivines the tendency for ordering Fe,Mg over the M1 and M2 sites is weak, although Ni,Mg and Co,Mg show a pronounced tendency to order (the transition ions preferring M1 in this case). Crystal field stabilization energies (Section 4.5) are often quoted as part of the stabilization of transition metal ions in certain geometric environments. In pyroxenes and amphiboles the tendency to order cations among the non-equivalent M sites is very much greater.

The thermodynamics of non-convergent ordering can be illustrated by a simplified account of the $MgSiO_3 - FeSiO_3$ orthopyroxene solid solution. The energetics of Mg,Fe ordering can be considered in two stages; first, the formation of a totally disordered solid solution of composition $(Mg_XFe_{1-X})_2Si_2O_6$ from the end members, followed by the exchange reaction of Mg and Fe between M1 and M2 sites.

The first stage can be described by the reaction:

$$XMg_2Si_2O_6 + (1-X)Fe_2Si_2O_6 \Rightarrow (Mg_XFe_{1-X})_{M1} (Mg_XFe_{1-X})_{M2} Si_2O_6$$

where the proportion of Mg and Fe on each site is the same as the overall Mg:Fe ratio. The configurational entropy associated with this totally disordered solid solution is given by

$$\Delta S_1 = -2R\{X \ln X + (1-X) \ln (1-X)\}$$

An assumption needs to be made at this point about the enthalpy of mixing for the formation of the disordered solid solution. For simplicity we will assume ideal mixing, i.e. a zero enthalpy.

The second stage in describing the distribution of Mg and Fe between M1 and M2 sites is the exchange reaction

$$(Mg^{2+})_{M2} + (Fe^{2+})_{M1} \Rightarrow (Mg^{2+})_{M1} + (Fe^{2+})_{M2}$$

This reaction is associated with an enthalpy ($\Delta H_{exch.}$ per mole) which we will assume to be constant, although later we will see that some temperature dependence of this enthalpy may have to be built into the model. The experimental observation that Fe^{2+} tends to prefer M2 sites suggests a negative $\Delta H_{exch.}$ for the forward reaction above. The formation of a more ordered solid solution can be described by the reaction

$$(Mg_xFe_{1-x})_{M1} (Mg_xFe_{1-x})_{M2} Si_2O_6 \Rightarrow$$
$$(Mg_{X+x}Fe_{1-X-x})_{M1} (Mg_{X-x}Fe_{1-X+x})_{M2} Si_2O_6$$

where x is the increase in the fractional occupancy of the M1 site by Mg (and M2 site by Fe), and hence measures the degree of order. The total enthalpy change on ordering, $\Delta H_{ord.}$, is then given by $x\Delta H_{exch.}$.

This ordering reaction decreases the configurational entropy to a new value given by

$$\Delta S_2 = -R\{(X+x)\ln(X+x)+(1-X-x)\ln(1-X-x)\}_{\text{for M1 sites}}$$
$$-R\{(X-x)\ln(X-x)+(1-X+x)\ln(1-X+x)\}_{\text{for M2 sites}}$$

This expression assumes that *within* M1 and M2 sites the Fe,Mg distribution is random.

The change in entropy on ordering is given by the difference between the configurational entropy for the partially ordered solid solution and the disordered solid solution, i.e.

$$\Delta S_{ord} = \Delta S_2 - \Delta S_1.$$

The change in free energy on ordering is then given by

$$\Delta G_{ord} = \Delta H_{ord} - T\Delta S_{ord}$$
$$= x\,\Delta H_{exch.} + RT\{(X+x)\ln(X+x) + (1-X-x)$$
$$\ln(1-X-x) + (X-x)\ln(X-x) + (1-X+x)\ln$$
$$(1-X+x) - 2X\ln X - 2(1-X)\ln(1-X)\}$$

At equilibrium,

$$\frac{dG_{ord}}{dX} = 0$$

$$\therefore \quad \Delta H_{exch.} + RT\ln\frac{(X+x)(1-X+x)}{(X-x)(1-X-x)} = 0$$

The term inside the logarithm is the equilibrium constant, K, for the reaction

$$(Mg^{2+})_{M2} + (Fe^{2+})_{M1} \Rightarrow (Mg^{2+})_{M1} + (Fe^{2+})_{M2}$$

where
$$K = \frac{X_{M1}^{Mg} X_{M2}^{Fe}}{X_{M2}^{Mg} X_{M1}^{Fe}} = \frac{(X+x)(1-X+x)}{(X-x)(1-X-x)}$$

and the X refers to the fractional occupancy of site M by the cation.

Thus, at equilibrium,

$$\Delta H_{exch.} = -RT\ln K \qquad (9.21)$$

Thus the exchange enthalpy uniquely defines the equilibrium cation distribution as a function of temperature. Conversely, if we know the equilibrium distribution at one temperature we can estimate the value of $\Delta H_{exch.}$.

The distribution of Fe and Mg between M1 and M2 sites has been determined in orthopyroxenes which have been isothermally heated for sufficiently long times to achieve equilibrium. The experimental methods used have generally been either complete X-ray structural refinements, or more popularly, Mossbauer spectroscopy which can quantitatively measure the ratio of Fe on M1 sites and on M2 sites. From such data the $\Delta H_{exch.}$ has been estimated at about 15 kJ mol^{-1} (for the $(Mg,Fe)_2Si_2O_6$ formula unit).

Using this value of the $\Delta H_{exch.}$ and the equations above, the equilibrium distribution of Fe between M1 and M2 sites can be calculated as a function of temperature and bulk composition (i.e. the Fe/(Fe+Mg) ratio). This is shown in Figure 9.16 as a set of equilibrium isotherms at 1000°C, 800°C, 600°C and 500°C. The dashed arrows show two examples of the expected ordering of Fe and Mg, for equilibrium cooling from 1000°C to 500°C, in orthopyroxenes of two different bulk compositions. Note that the model necessarily produces symmetric distribution curves around the composition Fe/(Fe+Mg) = 0.5.

Another way of expressing this relation is to plot $-\ln K$ against $1/T$ which gives a straight line with slope $\Delta H_{exch.}/R$ (Figure 9.17). Extrapolation of this plot to lower temperatures is only permissible if the ideal model described here remains valid.

One of the aims of this type of study is to determine the temperature of equilibration of natural orthopyroxenes by measuring the Fe,Mg distribution between M1 and M2 sites and referring it to a model. To obtain reliable results some slight adjustments need to be made to the model presented here. Although the real situation compares quite well with this simple model, there is some deviation from the symmetric distribution of Mg,Fe with bulk composition, especially at lower temperatures and at the Fe-rich end. To account for this asymmetry, the

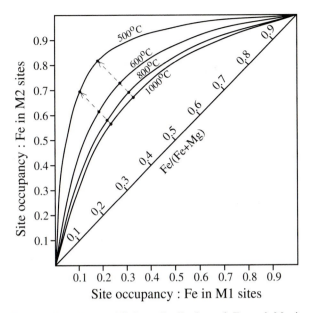

Figure 9.16. The equilibrium distribution of Fe and Mg in orthopyroxene $(Mg,Fe)_2Si_2O_6$, as a function of Fe content and temperature. The curves define the site occupancy of Fe in M2 and M1 at any composition at four different temperatures. For example, at 1000°C in an orthopyroxene with 50% Fe, 67% of the M2 sites and 33% of the M1 sites are occupied by Fe. The equilibrium occupancies as a function of temperature for compositions 50% and 40% Fe are shown by the dashed arrows.

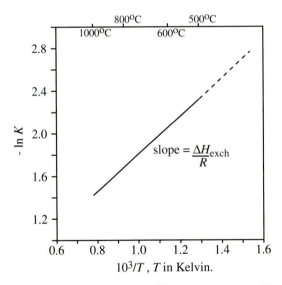

Figure 9.17. A plot of the equilibrium constant for Fe,Mg exchange between M1 and M2 sites in orthopyroxene as a function of inverse temperature, $1/T$. The slope of this line defines the enthalpy for the exchange reaction, ΔH_{exch}, equal to 15 kJ mole^{-1}.

assumptions which we made regarding ideal mixing of Fe,Mg in the disordered solid solution as well as *within* the M1,M2 sites has to be somewhat modified. This is usually done by introducing Margules

parameters (see Section 9.1.2) which modify the form of the enthalpy terms for both the mixing and the exchange reactions. Empirical equations which allow the enthalpy to depend on temperature or composition can produce a better fit between measured Fe,Mg distributions and those predicted from the thermodynamic model.

A similar approach to that described here can be applied in more complex cases when there are more than two cations mixing and ordering over more than two sites. In amphiboles for example, there are four M sites and so plenty of scope for complicating the equations, although Fe,Mg ordering appears to involve mainly M2 and M4 sites.

9.5 Phase diagrams for ordering – discontinuous and continuous

In this section we describe the general form of the equilibrium phase diagram when a solid solution between the two end members A and B tends to form an ordered AB compound. There are two possible phase diagrams depending on whether the transformation is thermodynamically discontinuous or continuous i.e. first-order or second-order. Although it is possible to derive free energy curves from the Bragg–Williams model, here we will use schematic diagrams which have the same general form as those predicted by such models. This avoids the need to be tied to any particular model, but does not affect the general conclusions.

1. In a first-order ordering transformation there is a discontinuity in the first derivative of free energy G, with respect to the long range order parameter $\boldsymbol{S}$ at equilibrium. The variation in the free energy as a function of $\boldsymbol{S}$ at a number of temperatures both above and below the ordering temperature T_{tr} is shown in Figure 9.18. At any temperature above T_{tr} an increase in the degree of order increases the free energy. At T_{tr} there is a non-zero value of $\boldsymbol{S}$ in equilibrium with $\boldsymbol{S} = 0$ indicating a discontinuous change in the degree of order. This suggests that regions with a certain degree of long-range order must grow within disordered material. As T decreases this discontinuity persists, but at a lower temperature T_4 it may disappear. At $T = T_4$ the disordered structure becomes unstable to any small fluctuations in the degree of order, and a continuous ordering pathway becomes available between $\boldsymbol{S} = 0$ and the equilibrium value. By analogy with spinodal

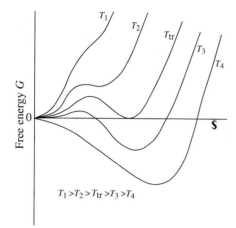

Figure 9.18. The variation in free energy G with the order parameter $\mathbf{S}$ for a first-order ordering transformation. Above the transformation temperature T_{tr} any degree of order increases the free energy. At T_{tr} disordered material ($\mathbf{S} = 0$) is in equilibrium with material with a degree of order $\mathbf{S}_1$. No intermediate values of $\mathbf{S}$ are stable, and hence there is a discontinuity at T_{tr}. At lower temperatures, this discontinuity is reduced, until at temperature T_4 the equilibrium order parameter can be reached by a continuous increase in $\mathbf{S}$.

decomposition (Section 9.2.1) the continuous ordering mechanism at temperatures such as T_4 is sometimes known as *spinodal ordering*.

For a first-order transformation, free energy curves for the disordered and ordered phases as a function of composition can be treated as distinct. As the temperature changes the two free energy curves change their relative positions as shown in Figure 9.19. Above T_{tr} the free energy curve for the disordered solid solution lies below the curve for the ordered phase over the whole compositional range. At T_{tr} the two free energy curves touch at the 50:50 composition (Figure 9.19(b)). Below T_{tr} the equilibrium is defined by common tangents between the curves. The two common tangents define four compositions c_1, c_2, c_3 and c_4 and five different compositional regions (Figure 9.19(c)):

(i) For compositions between pure A and c_1 the most stable phase is the disordered solid solution.

(ii) For compositions between c_1 and c_2 the most stable state is the coexistence of regions of disordered phase with composition c_1 and regions of ordered phase with composition c_2. This is referred to as a two-phase region.

(iii) For compositions between c_2 and c_3 the most stable state is the ordered phase.

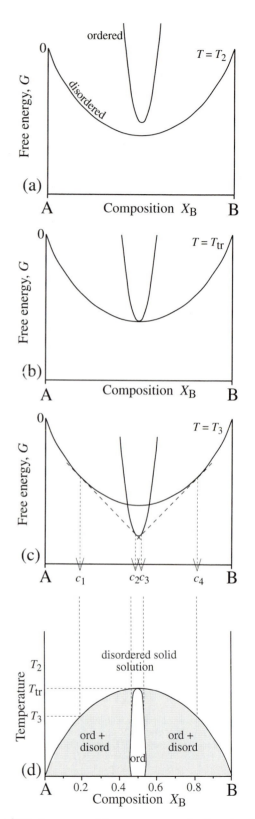

Figure 9.19. A sequence of free energy curves for the ordered and disordered phases in a first-order transformation, as a function of falling temperature (a) ⇒ (c). Below T_{tr} common tangents define two-phase regions (shaded) in the phase diagram (d). A narrow field of the ordered phase lies between these two phase regions.

(iv) For compositions between c_3 and c_4 the most stable state is the coexistence of regions of ordered phase with composition c_3 and regions of disordered phase with composition c_4.

(v) For compositions between c_4 and pure B, the most stable phase is the disordered solid solution.

These compositional fields as a function of temperature are displayed on the equilibrium phase diagram which plots the locus of the common tangent points (Figure 9.19(d)). The phase diagram shows a region of the ordered phase separated from the disordered phase on either side by two-phase fields where ordered and disordered phases coexist.

2. We now look at the same situation in the case of a second-order ordering transformation. The variation in the free energy as a function of the long-range order parameter S at a number of temperatures both above and below T_{tr} is shown in Figure 9.20. At all temperatures the equilibrium value of S occurs at the minimum in G, and increases continuously from zero as T decreases below T_{tr}. Furthermore there is always a continuous pathway between $S = 0$ and the equilibrium value, without the presence of any discontinuity in the degree of order.

Since there is no discontinuity between disordered and ordered regions, the variation of free energy with composition is represented by a single curve which changes its shape as the temperature falls, developing a deeper minimum at the centre (Figure 9.21(a),(b)). There are no two-phase fields in the phase diagram, and the boundary between disordered is represented by a single line (Figure 9.21(c)).

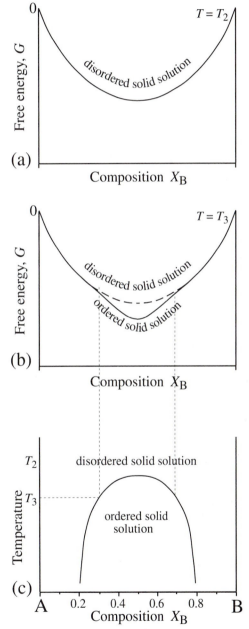

(a)

(b)

(c)

Figure 9.21. A sequence of free energy curves for the ordered and disordered phases in a second-order transformation, as a function of falling temperature (a) ⇒(b). The free energy curve for the ordered phase evolves continuously from that for the disordered phase, and the resultant phase diagram in (c) has no two-phase fields.

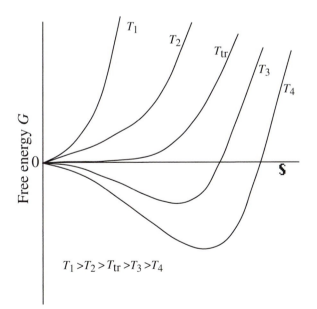

$T_1 > T_2 > T_{tr} > T_3 > T_4$

Figure 9.20. The variation in free energy G with the order parameter S for a second-order ordering transformation. Above the transformation temperature T_{tr} any degree of order increases the free energy. Below T_{tr} there is always a continuous pathway from $S = 0$ to the equilibrium value at the minimum value of G.

The lower value of T_{tr} at compositions away from the centre can be understood by considering that as the composition moves further from the ideal AB, the degree of order which can be achieved is reduced. Thus the maximum attainable value of **S** is reduced as is the value of T_{tr}.

9.6 Problems with simple nearest neighbour models – ordering and exsolution are not opposites

One of the major problems with the simple models of solid solutions we have been using throughout this chapter is that the ΔH_{mix} term is assumed to arise solely from the nearest neighbour interactions between the substituting atoms in the solid solution. This leads to the conclusion that a tendency to unmix on cooling, and a tendency to order are opposite extremes, depending on whether the interaction energy between unlike atoms, ω_{AB} is greater or less than the average of the interactions ω_{AA} and ω_{BB} between like atoms. Real mineral solid solutions however frequently show tendencies both to order and to unmix in the same system, and the two processes must not be regarded as being mutually exclusive.

For example, in the system calcite – magnesite ($CaCO_3 – MgCO_3$) an intermediate ordered compound dolomite, $(Ca,Mg)CO_3$, is superimposed on what would be a broad miscibility gap in the disordered solid solution (Figure 9.22(a)). The ordered dolomite structure is a superlattice of the calcite or magnesite structure, formed by ordering different cations into alternate cation layers between the layers of planar CO_3 groups (Figure 9.22(b)). In the disordered structure the cation sites are equivalent but become non-equivalent on ordering. On ordering, the symmetry is reduced from the rhombohedral space group $R\bar{3}c$ to $R\bar{3}$. A model of such a solid solution has to incorporate a tendency to separate Ca and Mg (the miscibility gap) with a tendency to order them into alternate layers.

This apparently contradictory behaviour is confirmed by calorimetric measurements of the enthalpy of formation of dolomite from calcite $CaCO_3$ and magnesite $MgCO_3$. For the reaction:

$$0.5CaCO_3 + 0.5MgCO_3 \Rightarrow Ca_{0.5}Mg_{0.5}CO_3$$

the enthalpy change to form ordered dolomite is $\Delta H = -5.74 \pm 0.25$ kJ mol^{-1}. For disordered dolomite

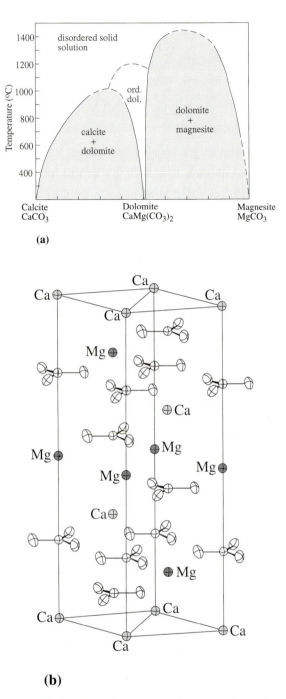

(a)

(b)

Figure 9.22. **(a)** The phase diagram for the system calcite–magnesite, with the ordered dolomite field at intermediate compositions. **(b)** The ordered structure of dolomite with alternating layers of Ca and Mg cations between the layers of planar CO_3 groups. In the hexagonal cell shown here $a = 4.8$Å, $c = 16.0$Å.

the enthalpy change for the same reaction is $\Delta H = +1.23 \pm 0.32$ kJ mol^{-1}. Thus the enthalpy of mixing is positive for the disordered phase, consistent with the existence of a solvus, but negative for ordered phase, as expected from the fact that an ordered

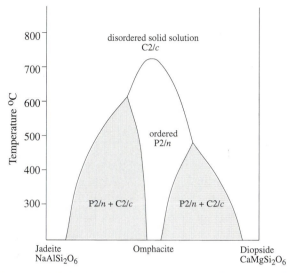

Figure 9.23. The phase diagram for the system jadeite – diopside, with the ordered omphacite field at intermediate compositions.

phase is observed. Such a situation cannot be treated using simple thermodynamic models (such as the regular solution and Bragg–Williams) which have only one energy of interaction parameter.

Not all carbonate systems with similar relative cation sizes have an intermediate ordered phase, e.g. no ordered phase exists in the systems $CoCO_3$ – $CaCO_3$ and $NiCO_3$–$CaCO_3$, although an ordered dolomite-type phase does exist in the system $CdCO_3$–$MgCO_3$.

Very similar behaviour occurs in the diopside – jadeite ($CaMgSi_2O_6$ – $NaAlSi_2O_6$) solid solution which is known to be non-ideal, and yet an intermediate ordered phase forms on cooling (Figure 9.23 and Section 6.4.1).

There are many other examples where unmixing and ordering are closely related and interdependent. It is clear therefore that physically realistic models cannot depend only on nearest neighbour effects. To understand the microscopic aspects of the real behaviour of mineral solid solutions we need more sophisticated models, although a detailed description is beyond the scope of this book. Most models are based either on pairwise interactions between second and third nearest neighbours, or interactions between local clusters of atoms, thus simulating the effects of short-range order which the simple models are unable to include.

While the macroscopic behaviour of any solid solution must be the result of the sum of both short-range and long-range interatomic forces, it seems that relatively small changes in composition, temperature or pressure may tip the balance between the various tendencies, favouring ordering under some conditions and unmixing in others. Where more realistic models have been applied (mainly in metal alloy systems) a complex sequence of processes is predicted, with a tendency to ordering at some temperatures, followed by unmixing at lower temperatures.

Another factor which has to be kept in mind when

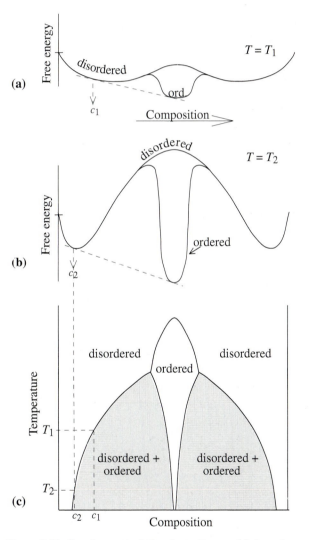

Figure 9.24. Development of the phase diagram (c) from free energy curves when a continuous ordering transformation is superimposed on a non-ideal solid solution during cooling, (a) ⇒(b). In this case the ordered phase has a stability field before the development of a miscibility gap in the solid solution. As the temperature falls the miscibility gap widens, defining two-phase regions on either side of the field for the ordered phase. The form of the phase diagram is very similar to that for calcite – magnesite, and jadeite – diopside.

attempting to unravel such behaviour is the role of kinetics in the observed processes. Ordering involves smaller diffusion distances than unmixing and therefore whenever both are possible, the faster kinetics will favour ordering followed by unmixing. In other words, the observed sequence of processes may be controlled by the kinetics rather than the thermodynamics of the system.

Given the shortcomings of free energy curves derived from microscopic models, we will proceed by using schematic free energy curves, not model-based, to explain the behaviour of systems such as those mentioned above.

9.7 Ordering superimposed on a non-ideal solid solution

In the system diopside–jadeite ($CaMgSi_2O_6$ – $NaAlSi_2O_6$) the solid solution is non-ideal, and therefore the free energy curve can be represented as shown in Figure 9.24(a, b). On cooling, cation ordering can be

represented by a minimum which develops in the free energy curve around the intermediate omphacite composition (see Section 6.4.1). As the temperature is reduced further, this minimum becomes deeper and narrower in composition. The continuous ordering process is consistent with experimental observations.

The phase diagram which can be drawn from these free energy curves is shown in Figure 9.24(c), and shares the general features of the real phase diagram of this system in Figure 9.23. At high temperature the ordered phase appears first, but the non-ideality of the solid solution results in two free energy minima on either side of the ordering minimum. The common tangents drawn in Figure 9.24(b) define the

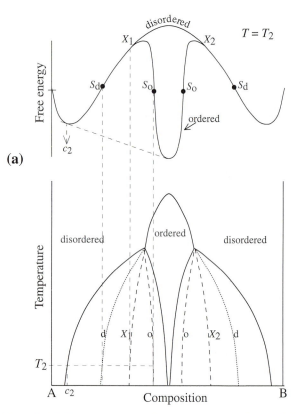

(a)

Figure 9.26. When continuous ordering is superimposed on a solid solution with a miscibility gap, metastable exsolution mechanisms may be explained by the positions of inflexion points S_o on the free energy curve for the ordered phase, and S_d on the free energy curve for the disordered solid solution. X_1 and X_2 are the points of contact between the ordered and disordered phases. The locus of inflexion points S_d define the position of the chemical spinodal within the disordered solid solution. This is shown as the dotted line labelled 'd' in the phase diagram. Additionally, between points S_o and X, compositional fluctuations due to the existence of the ordered phase, are stabilized. The locus of points S_o and X are shown dashed in the phase diagram, and define the two conditional spinodals lying within each miscibility gap. (After Carpenter, 1980.)

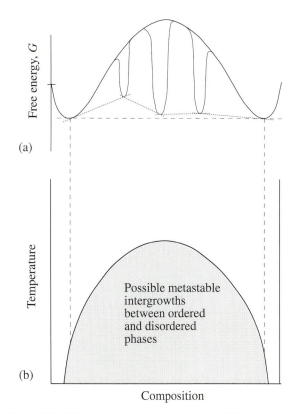

Figure 9.25. (a) Free energy curve for a miscibility gap in a disordered solid solution, superimposed on which are three metastable ordered phases. The equilibrium phase diagram is a simple solvus, although metastable intergrowths between ordered and disordered phases, shown by the common tangents in (a) are likely to form within the miscibility gap.

coexisting ordered + disordered two-phase regions of in the phase diagram. The narrowing of the compositional field of ordered omphacite at lower temperatures is a consequence of the sharpening of its free energy curve.

Another situation which can be envisaged is one in which ordered phases exist but are always metastable with respect to the unmixed phases (e.g. Figure 9.25(a)). The equilibrium phase diagram would be a simple solvus (Figure 9.25(b)) although the relatively favourable kinetics for ordering may result in ordered phases appearing in practice. Once the ordered phase has formed, the free energy drive for unmixing is reduced. Coupled with a large activation energy for diffusion this could result in the indefinite metastable persistence of the ordered phase.

9.7.1 The conditional spinodal

One of the consequences of superimposing ordering and unmixing in the same system, is the possibility of producing spinodal decomposition which is driven by ordering. Figure 9.26(a) shows how this can come about. In the continuous free energy curve between the non-ideal disordered solid solution and the ordered structure, there are inflexion points at S_o and S_d. Points 'S_o' lie on the free energy curve for the ordered phase, and points 'S_d' on the disordered solid solution. The free energy curve for the ordered phase meets the free energy curve for the disordered phase at points X_1 and X_2.

At compositions lying between the inflexion points S_d spinodal decomposition within the disordered solid solution is stabilized, as described in

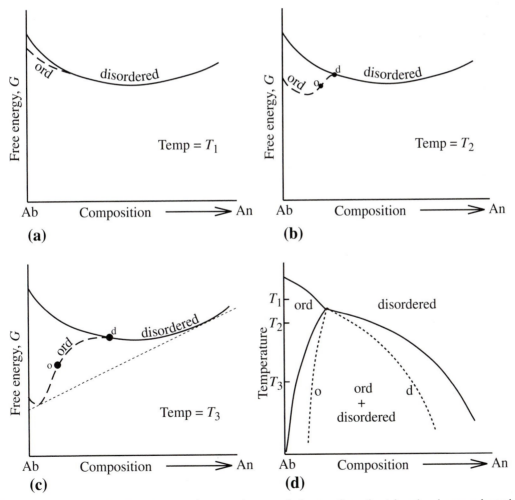

Figure 9.27. The development of the phase diagram when a continuous ordering transformation takes place in one end-member, as in the albite-rich plagioclases termed peristerites. The continuous ordering produces a minimum in the free energy curve (dashed line) and results in the formation of a miscibility gap between ordered and disordered phases. Within the two-phase region, the locus of the points o (the inflexion point on the ordered free energy curve) and the contact points d, as a function of temperature defines the position of the conditional spinodal (dotted line). (After Carpenter, 1981.)

Section 9.2.1. However, compositions lying between between points X and S_o can evolve via a different kind of spinodal decomposition in which the compositional fluctuations consist of an ordered and a disordered component respectively. Thus the fluctuations are both in composition and in the degree of order. The existence of the fluctuations depends on the existence of an ordered phase, and is in this sense 'conditional'.

Inflexion points lie on either side of the centre on the free energy curves, and their locus as a function of composition defines compositional fields on the phase diagram, in which different types of spinodal decomposition are possible (Figure 9.26). Normal compositional spinodal decomposition is possible between the loci of points S_d, and these are shown as dotted lines labelled 'd' on the phase diagram. Between the lines labelled X and 'o', conditional spinodal decomposition may also occur. For any given composition there are many different transformation sequences possible. For example, a disordered solid solution on cooling could first unmix into two components and then begin to order, or vice versa. Observations on natural omphacites confirm complex microstructures consistent with a finely balanced control over the actual behaviour under any particular set of circumstances.

A simpler example of a conditional spinodal is illustrated in Figure 9.27 which shows the free energy curve for a disordered solid solution at a sequence of decreasing temperatures T_1, T_2, T_3. One end-member continuously increases its degree of order with decreasing temperature, and this is represented by the deepening free energy curve for the ordered structure. A situation similar to this has been proposed for the albite-rich part of the plagioclase solid solution in which albite tends to order Al,Si on cooling.

Between the inflexion point labelled 'o' on the G curve for the ordered structure, and the point 'd' where the two curves meet, conditional spinodal fluctuations reduce the overall free energy. Again, their existence is due to ordering, and the fluctuations differ in both composition and degree of order. As the fluctuations grow in amplitude they approach the equilibrium situation shown by the dashed line in Figure 9.27(c). This intergrowth of ordered and disordered phases is shown in the phase diagram in Figure 9.27(d). In plagioclase feldspars this is known as the peristerite intergrowth (see Figure 6.57). If the peristerite intergrowth can be modelled in this way it would suggest that the phase separation is driven by ordering of the albite rich component.

A phase diagram with a very similar topology is that for hematite – ilmenite (Figure 12.67) in which ordering in the ilmenite end-member results in a miscibility gap.

Appendix
Phase diagrams in simple binary systems

The thermodynamic basis of phase equilibrium diagrams has already been discussed in this chapter in relation to the derivation of the solvus as well as for the case of an ordered intermediate phase. The purpose of this Appendix is to give some further examples of the consequences of the common tangent construction on the form of binary phase diagrams.

Here we will use a simple binary system with two components A and B. The 'components' in this context are the smallest number of independently variable constituents which may be used to express the composition of each phase. A phase is a chemically homogeneous and physically distinct part of the system, and each phase is separated from other phases by an interface. The definition of a component depends on the system we are describing. For example, we can describe the olivine solid solution in terms of the two components Mg_2SiO_4 and Fe_2SiO_4. If on the other hand we are describing the relationship between various magnesium silicates, such as forsterite Mg_2SiO_4 and enstatite $MgSiO_3$, we could do so within a binary system with two components MgO and SiO_2.

The starting point for understanding the form of phase diagrams is the free energy – composition curve (henceforth $G-X$ curve) of the phases involved. For an ideal solid solution (or an ideal melt) the $G-X$ curve is continuous with a minimum at the $X = 0.5$. This is a consequence of the entropy term on the free energy (Section 9.1). If a phase has a more restricted degree of solid solution, its $G-X$ curve is also narrower, the free energy rising rapidly as the extent of solid solution is exceeded. A phase which allows no solid solution is represented by a very sharp $G-X$ curve. The forms of these curves are shown in Figure 9.28.

When a number of phases coexist, the equilibrium between them is defined by the common tangent construction on their $G-X$ curves. This equilibrium condition was explained in Section 9.3 by the fact that the common tangent described the coexisting compositions for which the chemical potential of each component was the same in each phase. The way in which the free energy curves move relative to

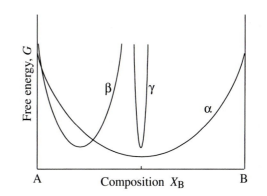

Figure 9.28. Schematic free energy – composition curves ($G-X$) for three phases α, β and γ within a binary system A – B. The α phase exists as a solution over the whole compositional range, while the composition over which the phases β and γ can exist is progressively more restricted, as shown by the narrower $G-X$ curves.

each other as a function of temperature defines the locus of the common tangent points, and hence the changes in the equilibrium between the coexisting phases. The different forms of phase diagrams are merely a consequence of the shapes of the $G-X$ curves and their relative positions with temperature.

Although in principle it is possible to calculate phase diagrams from thermodynamic data, in practice most are determined experimentally. The standard method is to make up charges of the appropriate composition and run them at a number of different temperatures (and/or pressures) for sufficient time to reach equilibrium. They are then quenched and the resulting phase mixture analysed. It is assumed that the quench retains the high temperatures phases, although any melt will now appear as a glass. Each phase is then compositionally analysed and plotted on the phase diagram.

Example 1. Crystallisation of an ideal solid solution

This can be illustrated with a very simple example in which a binary melt at high temperature cools to form a continuous solid solution at low temperature. In both phases (the melt, and the solid solution) the $G-X$ curves are broad and continuous (Figure

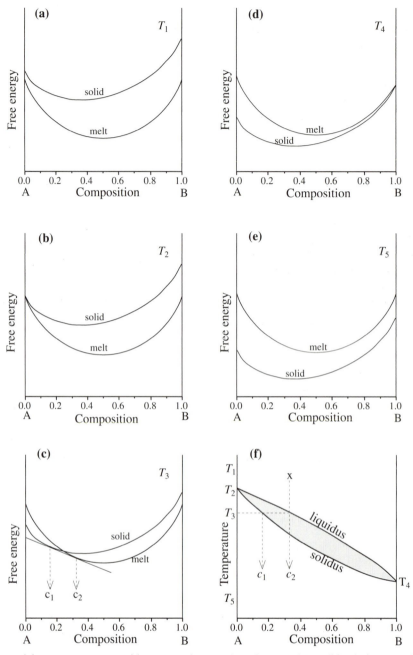

Figure 9.29. A sequence of free energy – composition curves for a melt and a complete solid solution as a function of decreasing temperature $T_1 \Rightarrow T_5$, and the resulting phase diagram (f).

9.29(a)). At high temperature, T_1, where the melt is stable over the whole composition range, its free energy curve is lower than that of the solid. On cooling however, the relative positions of the G–X curves gradually change, and at low temperatures (T_5) the curve for the solid solution will be the lower, when the solid solution is stable over the whole composition range (Figure 9.29(e)). At intermediate

temperatures the two curves must cross each other. Figure 9.29(b) shows the point where the two curves first touch $-T_2$, the melting point of pure A. In Figure 9.29(c), where the two curves cross each other the coexisting compositions of solid and liquid are defined by the common tangent. As the temperature decreases further, the tangency points move across the diagram towards the right until the last touching point which is the melting temperature T_4, of pure B (Figure 9.29(d)).

The resulting equilibrium phase diagram (Figure 9.29(f)) is a plot of the locus of the common tangency points and defines the coexisting compositions as a function of temperature. It can be used to define the equilibrium behaviour of a melt on cooling or of a solid solution on heating. As an example we can follow the equilibrium cooling of a melt of intermediate composition shown by the cross 'x' in Figure 9.29(f). On reaching the line termed the *liquidus*, the composition of the first solid which begins to crystallise is shown by the horizontal tie line, and is considerably richer in A than the bulk. Continued cooling produces more solid phase, but also keeps reequilibrating the composition of the solid and the melt so that at any point in the two-phase region the coexisting compositions are defined by the ends of the tie lines. The relative amounts of the two phases are given by a simple *lever rule* illustrated in Figure 9.30. At temperature T_4 all of the melt has solidified to form a homogeneous solid solution with a composition the same as that of the starting melt.

Just as in the discussion of the cooling of a solid solution through the solvus (Section 9.2) we must

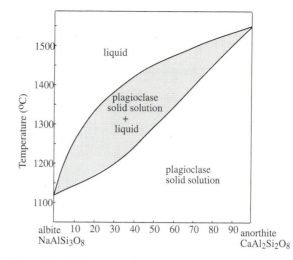

Figure 9.31. High temperature phase diagram describing equilibrium melting and crystallisation of the plagioclase solid solution.

note that the kinetics of the processes involved may not allow equilibrium to be maintained through a cooling process. The result of non-equilibrium cooling in the present case would be to produce zoned crystals with a core richer in A and a rim richer in B. This situation is quite common in the case of crystallisation of the high temperature albite–anorthite solid solution which is described by a phase diagram very similar to that in Figure 9.29. Anorthite has the higher melting point and therefore plagioclase feldspars with intermediate composition tend to have calcium-rich cores and sodium-rich rims (Figure 9.31).

If the $G–X$ curve for the melt has a deeper curvature than that of the solid, as shown in Figure 9.32(a), the resulting phase diagram will have a thermal minimum in the liquidus and solidus curves (Figure 9.32(b)). Similarly, if the solid has a deeper $G–X$ curve, the phase diagram will have a thermal maximum (Figure 9.32(c),(d)).

Example 2. Crystallisation where the two end-members have limited solid solution

If there are two separate free energy curves for the two limited solid solutions α and β, or a large maximum exists between them, the sequence of free energy curves with decreasing temperature is shown in Figure 9.33, together with the resulting phase diagram. Below the melting points of the two end-members the liquidus and solidus curves define the

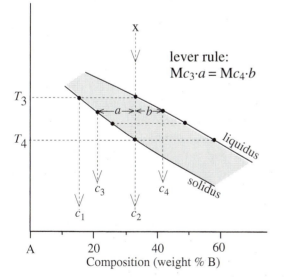

Figure 9.30. Illustration of the lever rule for determining the relative proportions of the coexisting phases at any temperature. The phase diagram is part of Figure 9.29(f). A melt composition 'x' meets the liquidus on cooling at temperature T_3, when it is in equilibrium with an infinitesmal amount of solid of composition c_1. At some lower temperature, the coexisting phases have compositions c_3 (solid) and c_4 (melt). To preserve the bulk composition 'x', the weights of the two phases present, denoted Mc_3 and Mc_4 respectively, are given by the lever rule.

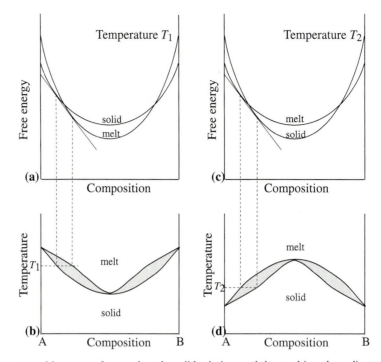

Figure 9.32. Free energy – composition curves for a melt and a solid solution, and the resulting phase diagrams when (a) the free energy curve for the melt has a deeper curvature than that for the solid and (b) the free energy curve for the solid has a deeper curvature than that for the melt.

two-phase regions (melt + α, melt + β) on either side of the phase diagram. A certain temperature must exist where the three G–X curves share a single common tangent, and hence are all in equilibrium. On the phase diagram this is temperature T_e and is known as the *eutectic temperature*. The point E in the phase diagram is the eutectic point at which the melt is in equilibrium with both α and β with compositions given by the ends of the horizontal tie-line. The eutectic reaction

$$\text{melt} \Rightarrow 2 \text{ solid phases}$$

describes the crystallisation of a melt at the eutectic temperature. Below the eutectic temperature the two-phase region of coexisting α + β occupies most of the phase diagram, with narrow single phase regions on either side.

Reducing the degree of solid solution allowed in the two end-members A and B narrows their free energy curves, and reduces the width of the single phase regions of the phase diagram. In the limit, when no solid solution exists between A and B, they contract to become 'line-phases' on the diagram, as shown in Figure 9.34.

The eutectic phase diagram can be thought of as a liquidus and solidus curve with a minimum such as that in Figure 9.32(a), intersected by a solvus (see Figure 9.5). This can be illustrated by an example which is directly relevant to the crystallisation of alkali feldspars from granitic melts at high and low water pressures (Figure 9.35). In a dry melt (low water vapour pressure) crystallisation occurs at a higher temperature, and the liquidus/solidus curves describing the crystallisation are separated from the miscibility gap at lower temperatures (Figure 9.35(a)). This crystallisation results in a single feldspar solid solution being formed. Subsequent cooling will result in phase separation within the solid so that the low temperature equilibrium state of the feldspars will be crystals which consist of a two-phase intergrowth i.e. perthites. At higher water pressures the melting temperatures are reduced by several hundreds of degrees, although the solvus is virtually unaffected. The solvus intersects the melting curves, and a eutectic is produced (Figure 9.35(b)). A melt of the appropriate composition could now solidify directly to two separate feldspar phases, Na-feldspar and K-feldspar which may

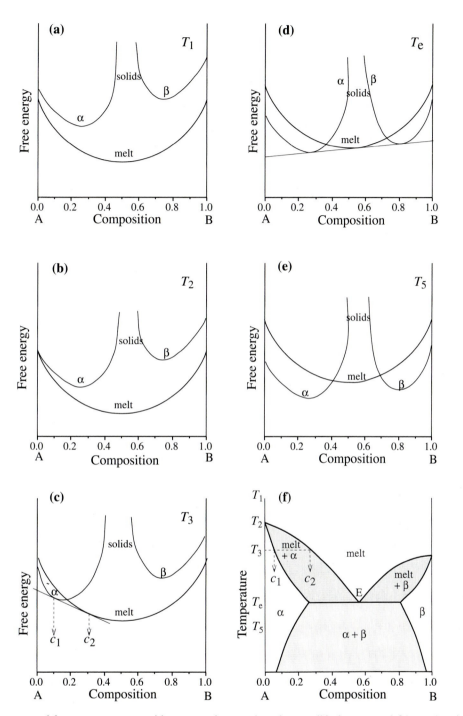

Figure 9.33. A sequence of free energy – composition curves for a melt and two solid phases α and β, as a function of decreasing temperature $T_1 \Rightarrow T_5$, and the resulting phase diagram (f). The free energy curves for the two solids could equally well be replaced by a solid solution with a miscibility gap.

themselves undergo further exsolution at lower temperatures. The presence of two alkali feldspars in a granite is therefore indicative of crystallisation at high water pressure.

Another type of binary phase diagram with only

limited solid solution of the two end members is shown in Figure 9.36, which can be derived by the intersection of a solvus with the solidus line of a phase diagram such as shown in Figure 9.29. Again a temperature T_p at which all three phases are in

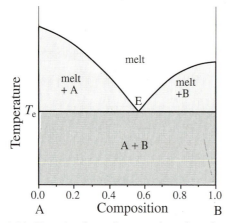

Figure 9.34. Eutectic phase diagram when there is no solid solution between the pure end members A and B. A melt with the eutectic composition E simultaneously crystallises A + B at the eutectic temperature T_e.

equilibrium occurs, and is known as the *peritectic temperature*. Similarly the peritectic point defines the composition and temperature where the equilibrium peritectic reaction

$$\text{melt} + \text{solid} \Rightarrow \text{solid}$$

takes place.

Example 3. Binary system with polymorphic phase transitions in the end members

We now consider a complete solid solution in which both end-members undergo first-order phase transitions on cooling. The change of structure will invariably change the extent of solid solution with the other end-member. The resulting phase diagram

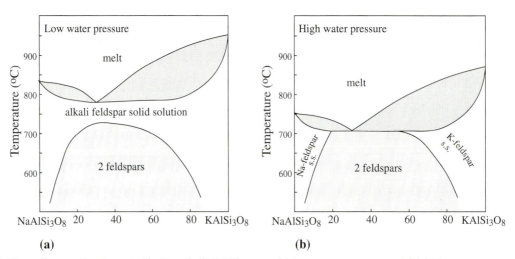

Figure 9.35. Phase diagrams for the crystallisation of alkali feldspars at (a) low water pressure, and (b) high water pressure. Structural phase transitions in the end-members have been neglected for simplicity.

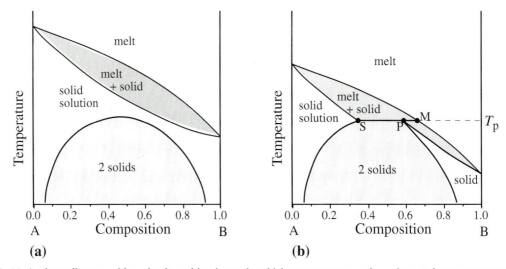

Figure 9.36. (a) A phase diagram with a simple melting interval at high temperatures and a solvus at lower temperatures. (b) The intersection of the solidus curve with the solvus generates a point P, termed the peritectic point, at which a solid of composition P (the peritectic composition) is in equilibrium with melt M + solid S. T_p is the peritectic temperature.

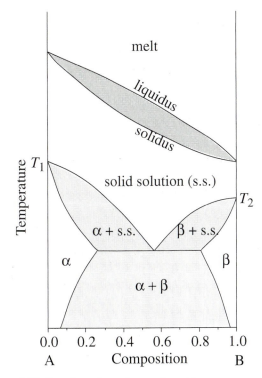

Figure 9.37. Complete phase diagram for a binary system with a high temperature solid solution in which the end-members A and B undergo polymorphic transformations to phases α and β at temperatures T_1 and T_2 respectively. The free energy curves for the solid solution, the α phase and the β phase are equivalent to those in Figure 9.33, except that here all phases are solid.

has the same topology as that in Example 2, since the complete solid solution can be represented by a continuous free energy curve similar to that of a melt, and the first-order structural transitions can be treated in the same way as melting and crystallisation.

The phase diagram (Figure 9.37) is therefore an all-solid version of Figure 9.33 with a temperature at which all three phases (the solid solution, α, and β) are in equilibrium. This temperature is termed the *eutectoid* temperature, and the eutectoid reaction

$$\text{solid solution} \Rightarrow 2 \text{ solid phases}$$

takes place at the eutectoid composition.

A solid-state phase diagram topologically similar to the peritectic diagram in Figure 9.36 is termed a *peritectoid*.

Example 4. More complex phase diagrams

The principal reaction points, namely, first-order transformations and melting, eutectics, eutectoids, peritectics and peritectoids have all been mentioned in the examples above. In addition, the concept of one-phase fields and two-phase fields in binary systems arises from the equilibrium criterion defined by the common tangent construction. More complex diagrams consist of combinations of these features which arise when one or more intermediate compounds exist in a binary system. The way in which the G–X curves for the solid and liquid phases move relative to one another as a function of temperature defines the form of the phase diagram.

For example, at high temperatures the system $MgO-SiO_2$ contains the following distinct solid phases : periclase, MgO; forsterite, Mg_2SiO_4; enstatite, $MgSiO_3$ and silica, SiO_2 as both cristobalite and tridymite (Figure 9.38). None of these phases has virtually any solid solution in this composition range and they appear as line phases on the diagram. In addition, the melt, which at high temperatures can exist as a single phase over the whole compositional range, becomes immiscible below about 2000°C at the silica-rich end. This miscibility gap is completely analogous to the solvus in a solid solution, and

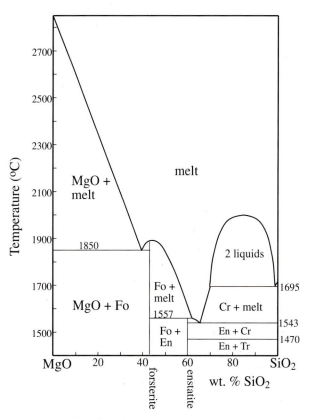

Figure 9.38. Phase diagram for the system $MgO-SiO_2$ containing the phases forsterite Mg_2SiO_4 and enstatite $MgSiO_3$.

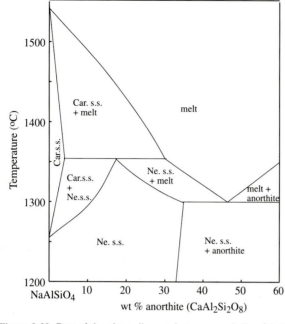

Figure 9.39. Part of the phase diagram between nepheline (NaAl SiO$_4$) and anorthite (CaAl$_2$Si$_2$O$_8$).

results in the separation of SiO$_2$-rich and more Mg-rich melts.

There are two eutectics in this system, one at around 1850°C where a melt composition ~40% SiO$_2$ crystallises to periclase + forsterite, and another at around 1540°C where a melt at ~65% SiO$_2$ crystallises to enstatite + cristobalite. Pure forsterite melts at ~1900°C to a melt of the same composition, i.e. it melts *congruently*, while enstatite melts *incongruently* at 1557°C to form forsterite plus a more silica-rich melt.

Between each single-phase region (in this case it is a line) on the phase diagram, there is a two-phase region, as dictated by the common tangent construction to the free energy curves. The transformation from cristobalite to tridymite is merely indicated by the horizontal lines between enstatite and SiO$_2$ marking the change in the coexisting phases.

When there is substantial solid solution, a polymorphic transformation has a pronounced effect on the form of the phase diagram. For example, in the system nepheline-anorthite (NaAlSiO$_4$– CaAl$_2$Si$_2$O$_8$), part of which is shown in Figure 9.39, the first-order transformation from carnegeite to nepheline as the temperature falls, increases the degree of solid solution, and leads to the peritectic (at ~1350°C and 20% anorthite). There is also a eutectic at ~1300°C and 45% anorthite.

Bibliography

COX, K.G., BELL, J.D. AND PANKHURST, R.J. (1981). *The interpretation of igneous rocks*. Allen and Unwin.

GORDON, P. (1968). *Principles of phase diagrams in materials systems*. McGraw-Hill.

HOLLAND, T.J.B. (1981). Thermodynamic analysis of simple mineral systems. In: *Thermodynamics of minerals and melts*. R.C. Newton, A. Navrotsky and B.J. Wood (Eds.) Advances in Physical Geochemistry Vol.1 Springer-Verlag.

LAUGHLIN, D.E. AND SOFFA, W.A. (1988). Exsolution, ordering and structural transformations: systematics and synergistics. In: *Physical properties and thermodynamic behaviour of minerals*. E. Salje (Ed.) NATO ASI Series C. D. Reidel Publishing Co.

NAVROTSKY, A. (1987). Models of crystalline solutions. In: *Thermodynamic modeling of geological materials*. H.P. Eugster and I.S.E. Carmichael (Eds) *Reviews in Mineralogy* Vol. 17. Mineralogical Society of America.

PORTER, D.A. AND EASTERLING, K.E. (1983). *Phase transformations in metals and alloys*. Van Nostrand Reinhold (UK) Co.Ltd.

POWELL, R. (1978) *Equilibrium thermodynamics in petrology. An introduction*. Harper and Row.

SWALIN, R.A. (1972) *Thermodynamics of solids*. 2nd Ed. Wiley.

WOOD, B.J. AND FRASER, D.G. (1976). *Elementary thermodynamics for petrologists*. Oxford University Press.

References

CARPENTER, M.A. (1980). Mechanisms of exsolution in sodic pyroxenes. *Contrib. Mineral. Petrol.* **71**, 289–300.

CARPENTER, M.A. (1981). A conditional spinodal within the peristerite miscibility gap in plagioclases. *Amer. Mineral.* **66**, 553–560.

10 Kinetics of mineral processes

In the two previous chapters we discussed how the thermodynamics of the various structural states of minerals governs the transformations from one state to another, as well as the reactions between minerals. However, although a negative free energy change is a necessary prerequisite for a mineral transformation to occur, it does not guarantee that it will do so at any measurable rate. The rate depends on the *mechanism* of the reactions involved and may bear no relation to the magnitude of the free energy reduction which drives the transformation. This is illustrated by the effect of catalysts on chemical reactions – they do not change the free energy of the phases, but do provide a faster mechanism. The commonplace observation that minerals stable at high temperatures and/or pressures are preserved outside their stability fields is due to the fact that the atomic mechanisms of bond-breaking and diffusion may be very slow.

If kinetics are determined by factors other than ΔG, we must consider the states through which a system passes during the transformation. Equilibrium thermodynamics does not describe these states and so we need some additional concepts to be able to deal with transformation rates. We have already touched on this topic in relation to point defects and diffusion in Section 7.1.4. In this chapter we will develop some basic ideas of kinetic theory in a slightly more formal way, and describe methods of extracting information from laboratory experiments on transformation rates. The ultimate aim of many experiments on mineral transformations is to be able to apply the data to geological time-scales.

At equilibrium, when $\Delta G = 0$, no transformation can take place. Some overstepping is always required to provide the free energy reduction, and the problem in geological processes is to determine how great this overstepping is. Given the long time-scales available for geological processes, a small free energy drive may be sufficient to achieve a transformation near equilibrium; on the other hand there is abundant evidence from both the textures of rocks and the microstructure of minerals that equilibrium has not been achieved. Laboratory experiments are carried out at higher temperatures, far from equilibrium, where reactions proceed at a measurable rate, and the data are extrapolated to lower temperatures, near equilibrium, where reaction rates are much slower. This is an exercise full of uncertainties, some of which will be outlined here.

10.1 The basis of kinetic theory

One way of dealing with the problem of the pathways through which a mineral system passes during a transformation from the initial to the final state, is to define an *activated state* which has some intermediate configuration. This is a 'quasi-equilibrium' approach in that we assign unique values of the thermodynamic functions to this activated state. The shape of the free energy curve (Figure 10.1) is a consequence

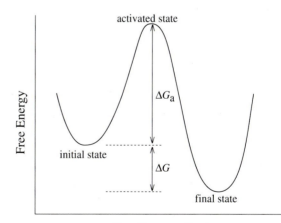

Figure 10.1. When a system passes from an initial state to a final state which has a lower free energy it passes through an intermediate activated state. The free energy reduction driving the transformation is ΔG, but to overcome the free energy barrier the system must first gain energy ΔG_a, the free energy of activation. The 'reaction co-ordinate' is any variable that defines the progress along the reaction path.

of the fact that if the initial and final states are in equilibrium (either stable or metastable), their free energies must be at minima, and any pathway from one to another must pass through a maximum.

The free energy of activation ΔG_a depends on the reaction pathway, and in heterogeneous reactions is independent of the thermodynamics of the initial and final states. The pathway with the lowest ΔG_a will be the one taken. A necessary condition for a transformation to take place at a measurable rate is that sufficient atoms have enough energy to achieve the transition state. This energy is supplied by thermal fluctuations. At all temperatures above 0 K atoms are in motion, collisions between them producing wide variations in the energy of individual atoms, with some having energies greatly in excess of the mean.

This concept of an activation energy barrier qualitatively explains a number of features of mineral reactions:

(i)   the persistence of metastable states, due to ΔG_a being very large compared to the mean free energy,
(ii) the fact that the reaction rate is independent of the driving force,
(iii) the effect of catalysts which change ΔG_a i.e. provide a reaction path with lower free energy of activation, and
(iv) the slow rate of many transformations, due to the fact that, at any one time, only a small number of the available atoms have sufficient free energy to overcome the activation energy barrier.

10.1.1 The activation energy (enthalpy) and entropy

The assignment of thermodynamic functions to the activated state includes values of internal energy U_a, volume V_a, enthalpy H_a and entropy S_a. As discussed in Section 8.4.1, we define a Gibbs free energy G_a for the activated state and write that

$$G_a = H_a - TS_a \qquad (10.1)$$

where the subscripts refer to the activated state.

For most solid state reactions at low pressures, the volume change is small and hence the change in enthalpy ΔH_a is approximated to the internal energy of the activated state. The term 'activation energy' is commonly used synonomously with activation

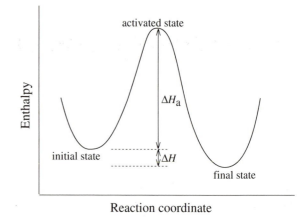

Figure 10.2. The enthalpy change for a transformation from an initial state to a final state. The overall enthalpy change is negative in an exothermic reaction (as shown here) or positive in an endothermic reaction. The thermal barrier to the process is defined by the activation energy ΔH_a.

enthalpy in such cases, although it is very important to note that this is not the same as the *free* energy of activation.

The activation energy is the difference between the internal energy (or enthalpy) of the transition state and the initial state (Figure 10.2). This curve is similar to the one for free energy (Figure 10.1) except that the overall enthalpy change ΔH for the transformation may be positive or negative. If ΔH is positive the reaction is endothermic (heat is taken in during the reaction), while if ΔH is negative the reaction is exothermic (heat is given out during the reaction).

Similarly, the entropy of activation ΔS_a is defined as the difference between the entropy of the activated state and the initial state. From the statistical definition of entropy (Section 8.3.1)

$$\Delta S_a = R \ln \frac{\omega_a}{\omega_I} \qquad (10.2)$$

where ω_a is the number of 'complexions' associated with the activated state, and ω_I is the number of 'complexions' associated with the initial state. By 'complexions' we mean all of the different ways in which the atoms can be spatially distributed, and how their thermal, electronic and other forms of energy are distributed over the permitted energy levels. The entropy change therefore includes changes in the configurational, electronic and vibrational entropy.

It is not easy to generalize about the sign or the magnitude of ΔS_a. However a large number of reaction paths results in a large entropy of activation and hence from eqn. (10.1), a lower free energy of activation ΔG_a. In general therefore, reactions between two states with a larger number of reaction pathways are faster. For example, even when the free energy changes are the same, evaporation is faster than condensation, melting is faster than crystallisation and disordering is faster than ordering.

In view of the requirement that an atom must acquire sufficient energy to overcome the activation energy barrier we need to determine how the thermal energy is distributed among the particles in the system. To do this we will merely quote a conclusion drawn from Boltzmann statistics:

In any system at equilibrium, the fraction f of the total particles having a thermal energy not less than ΔH_a is given by the expression

$$f = \exp\left(\frac{-\Delta H_a}{RT}\right) \qquad (10.3)$$

where ΔH_a refers to the activation energy per mole.

10.1.2 Rate of a single thermally activated process

For a process which involves only one basic atomic step, characterized by a unique activation energy ΔH_a, for example the diffusion of atoms from one site to another (see Figure 7.6) the rate equation can be formulated as follows:

The rate is proportional to:

(i) the frequency with which atoms attempt to jump from one site to the next, i.e. the vibration frequency v

(ii) the fraction of atoms with enough energy to surmount the activation energy barrier, i.e.

$$f = \exp\left(\frac{-\Delta H_a}{RT}\right)$$

(iii) the probability p that the atom which has the required energy satisfies some geometrical conditions i.e. it is jumping the right way. For a simple cubic distribution of atoms $p = 1/6$, since each atomic site is surrounded by 6 neighbouring sites. For more complex structures p is less, and for atomic diffusion involving the co-operative movement of

several atoms (Figure 7.5), p may be very small.

Therefore we may write that

$$\text{Rate} \propto p.v.\exp\left(\frac{-\Delta H_a}{RT}\right) = \frac{dy}{dt} \qquad (10.4)$$

where $y(t)$ is the ratio of the number of atoms/unit volume in the final state at time t, to the number of atoms per unit volume in the initial state at time $t = 0$, i.e. the fraction of atoms which have diffused or transformed from one state to the other.

To reconcile eqn. (10.4) with the dependence on the free energy of activation ΔG_a we note that the probability p is related to the entropy of activation ΔS_a in the following way:

$$p = \frac{\omega_a}{\omega_I}$$

i.e. the ratio of the number of complexions associated with the transition state to that associated with the initial state. The probability of a correct atomic jump is related to the number of possible pathways for the transformation. Since

$$\Delta S_a = R\ln\frac{\omega_a}{\omega_I}$$

$$p = \exp\left(\frac{\Delta S_a}{R}\right)$$

Therefore

$$\frac{dy}{dt} = v\exp\left(\frac{\Delta S_a}{R}\right)\exp\left(\frac{-\Delta H_a}{RT}\right)$$

Since

$$\Delta G_a = \Delta H_a - T\Delta S_a$$

$$\frac{dy}{dt} = v\exp\left(\frac{-\Delta G_a}{RT}\right)$$

More commonly we write

$$\frac{dy}{dt} = A\exp\left(\frac{-\Delta H_a}{RT}\right) \qquad (10.5)$$

where the pre-exponential factor A is known as the *frequency factor*. Equation (10.5) is known as the *Arrhenius equation*, first encountered in a slightly different form in Section 7.1.4.

A plot of $\ln\dfrac{dy}{dt}$ against $1/T$ is termed an *Arrhenius plot*:

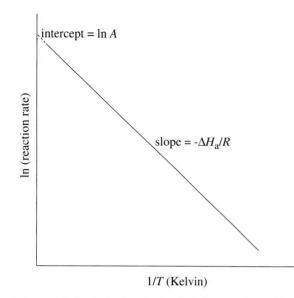

Figure 10.3. An Arrhenius plot in which the natural logarithm of the reaction rate is plotted against $1/T$. In a thermally activated transformation the plot is linear with a slope which is proportional to the activation energy.

$$\ln \frac{dy}{dt} = \ln A - \Delta H_a / RT \qquad (10.6)$$

and if both A and ΔH_a are independent of temperature, this is linear with a gradient $-\Delta H_a / R$ and an intercept on the rate axis of $\ln A$ (Figure 10.3).

The magnitude of ΔH_a describes the temperature dependence of the reaction rate while the pre-exponential A, which contains the entropy of activation term, determines the absolute reaction rate. When we consider that the rate of ordering or exsolution processes in natural minerals as they cool is controlled by an exponential function of this kind it is not surprising that metastable states are so commonly encountered. For example if the activation energy for a cation diffusion process is 240 kJ mol^{-1} at 1000K,

$$\exp\left(\frac{-\Delta H_a}{RT}\right) \sim \exp\left(\frac{-240000}{8.3 \times 1000}\right) = 2.8 \times 10^{-13}$$

while at 300 K

$$\exp\left(\frac{-\Delta H_a}{RT}\right) \sim \exp\left(\frac{-240000}{8.3 \times 300}\right) = 1.4 \times 10^{-42}$$

Therefore the rate of the process is about 10^{29} times faster at 1000 K than it is at room temperature. If the activation energy is halved to 120kJ mol^{-1}, the same calculation gives a factor of 10^{14} increase in the rate

at 1000 K compared to 300 K. We can see therefore since most solid state processes in minerals have activation energies between about 200 and 400 kJ mol^{-1}, a decrease in temperature of a few hundred degrees is enough to virtually stop a reaction which was taking place at moderate velocity at higher temperatures.

Figure 10.4 shows the effect of changing the activation energy ΔH_a and the pre-exponential factor A on the general form of the Arrhenius plots.

10.2 A more general theory of reaction rates

In most mineral transformations we do not have a reaction involving a single thermally activated process. In the most general case we might be measuring the rate of reaction of two minerals A and B to a new assemblage B + C. If we determined the temperature dependence of this rate, the activation energy we would derive would not necessarily refer to any specific atomic process, but would represent some overall activation energy which could be a combination of several processes.

In dealing with a reaction such as

$$A + B \Rightarrow C + D$$

we again assume the presence of an activated complex, written AB*, which is treated as another atomic species which is in equilibrium with the reactant even though its life-time is short (Figure 10.5). Therefore we can write the equilibrium reaction

$$A + B \Rightarrow AB^*$$

The equilibrium constant for this reaction can be written

$$K^* = \frac{C_{AB^*}}{C_A \cdot C_B}$$

where C_{AB^*}, C_A and C_B are the concentrations of the respective species (assuming unit activity coefficients).

The equilibrium constant is also defined thermodynamically by the free energy change per mole for the reaction $A + B \Rightarrow AB^*$:

$$K^* = \exp\left(\frac{-\Delta G_a}{RT}\right)$$

and therefore

$$\Delta G_a = -RT \ln K^*$$

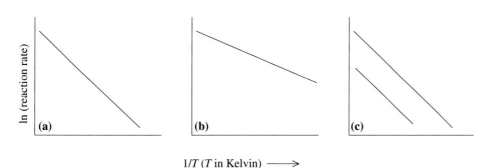

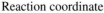

Figure 10.4. Arrhenius plots illustrating the effect of changing activation energy and pre-exponential factor A. (a) has a higher activation energy than (b). In (c) the activation energy is the same as in (a) but the different pre-exponential factors lead to different reaction rates.

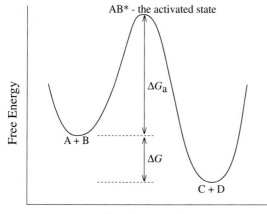

Figure 10.5. The change in free energy for the reaction A + B ⇒ C + D, where AB* is the activated state.

In this simplified account of the theory we assume that the rate of the transformation is entirely controlled by the number of activated complexes formed, and that the decomposition of the activated complexes to the products C + D is virtually instantaneous. Therefore the reaction rate can be equated with the number of activated complexes that decompose per unit time, which in turn depends on the concentration of complexes $C_{AB}{}^*$.

Thus

$$Rate \propto C_{AB}{}^*$$
$$\propto K^* C_A C_B$$

Since

$$K^* = \exp\left(\frac{-\Delta G_a}{RT}\right)$$

$$K^* = \exp\left(\frac{\Delta S_a}{R}\right) \exp\left(\frac{-E_a}{RT}\right)$$

We have now replaced ΔH_a by E_a which is an empirical activation energy and does not refer to any specific process. (Note that we have dropped the 'Δ' prefix from E_a merely for convenience.)

We can now write that the

$$Rate = c \exp\left(\frac{\Delta S_a}{R}\right) \exp\left(\frac{-E_a}{RT}\right). C_A.C_B \tag{10.7}$$

Thus the rate depends on the initial concentration of the reactants. As the concentration of the reactants changes during the course of the reaction, so will the rate of reaction. The usual way of expressing this is to write

$$Rate = k. f(C) \tag{10.8}$$

where k is termed the *rate constant* and describes the way the rate depends on the concentration of the reactants.

The rate constant $k = c \exp\left(\frac{\Delta S_a}{R}\right) \exp\left(\frac{-E_a}{RT}\right)$ (10.9)

or

$$k = A \exp\left(\frac{-E_a}{RT}\right) \tag{10.10}$$

10.2.1 The empirical activation energy E_a

The experimental determination of the rate constant k and hence the empirical activation energy E_a is discussed in the next section, but it is worth re-emphasising that most mineral reactions are heterogeneous, involving the nucleation and growth of new phases, separated from the parent phase by an interface. Such reactions involve at least two and often more consecutive steps, each with its own activation energy.

In sequential reactions the overall reaction rate is determined by the *slowest* step and hence an experimentally determined activation energy would most probably refer to this process. In this context we often refer to a 'rate-determining process' which controls the overall reaction rate. However, there may be the possibility that there are a number of *parallel processes* operating in a reaction, in which case the observed reaction rate would be determined by the *fastest* process.

For example, at low temperatures bulk diffusion in a polycrystalline material is dominated by surface diffusion along the grain boundaries because volume diffusion through the structure is too slow. As the temperature increases volume diffusion becomes faster, until above some temperature it becomes the dominant process. (Although surface diffusion also becomes faster, it is quantitatively less important at high temperatures, unless the grain size is extremely small). Figure 10.6 shows an Arrhenius plot in which the overall diffusion rate is dominated by two different processes at different temperatures.

An example of a sequential reaction is the growth of a new phase during an exsolution process. After the nucleation event, the subsequent growth depends on the rate of two processes: the diffusion of atoms to the growing nucleus, followed by its incorporation onto the surface of the new phase. Thus a diffusion process is followed by an interfacial process. If the activation energy for these two processes is different, but one is not overwhelmingly larger than the other, the overall empirical activation energy for growth will be some function of the two activation energies. We can illustrate such a sequential process with a mechanical analogy.

Figure 10.7 illustrates this mechanical analogy. Vessel A is maintained at a constant water level H_A

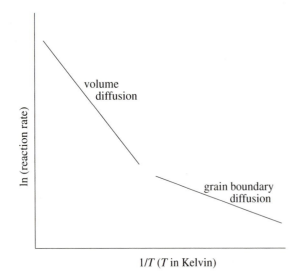

Figure 10.6. A schematic Arrhenius plot describing the temperature dependence of the overall diffusion rate. At low temperatures, transport is by grain boundary diffusion which has a low activation energy; at higher temperatures volume diffusion with a higher activation energy becomes the dominant diffusion mechanism.

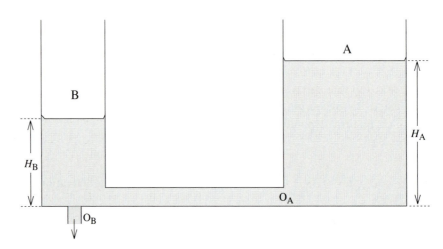

Figure 10.7. A simple analogy to illustrate the effect of two consecutive steps on the overall flow rate through the system. Water in tank A is maintained at a constant head H_A and flows through outlet O_A into vessel B, before flowing out through O_B. The height H_B depends on the relative sizes of O_A and O_B.
 (i) If $O_B \gg O_A$, $H_B = 0$ and the flow rate is controlled by the size of O_A.
 (ii) If $O_B \ll O_A$, $H_B = H_A$ and the flow rate is controlled by the size of O_B.
 (iii) For any intermediate case, such as shown here, the flow rate through O_B depends on both O_A and O_B. (After Burke, 1965.)

from a reservoir. It can discharge through an outlet O_A into vessel B, which has an outlet O_B. If the size of O_B is very large compared with O_A no water will accumulate in B and the flow of water through O_B is simply the rate through O_A under a head of water H_A. On the other hand, if O_A is very large compared with O_B, the water will fill up in B until it reaches the same level as A. The rate of flow of water through O_B is then independent of O_A and depends only on the size of O_B. These are the two limiting cases.

For some intermediate situation, shown in Figure 10.7, the level in B will reach a steady state value H_B so that the flow through O_A under a head ($H_A - H_B$) equals the flow through O_B under a head H_B. Thus flow from the system will involve both O_A and O_B.

In applying this analogy to growth of a precipitate, O_A can be equated to diffusion through the solid, and O_B to the incorporation of atoms into the growing surface. When O_B is large the process is controlled by O_A i.e. *diffusion controlled*. An experimental determination of the empirical activation energy E_a would give a value for the diffusion process. If O_B is small the reaction is *interface controlled*, and the measured activation energy would refer to the atomic process of crossing the interface from the matrix to the growing precipitate. In the intermediate case the rates of both processes are important and a measured activation energy for growth would have no unambiguous significance.

In general, caution must always be exercised in attaching fundamental significance to any measured value of the activation energy of a transformation. It is a measure of the temperature dependence of the rate of a transformation carried out under a certain set of experimental conditions. Before applying this value to a different set of conditions, or extrapolating it to different temperatures it is necessary to demonstrate that the mechanism is still the same as that operating during the experiment.

10.3 Rate equations and the rate constant for heterogeneous reactions

As shown in eqn. (10.8) the rate of a heterogeneous reaction at fixed temperature, pressure etc. is a function of time, due to the fact that reaction rates depend on the concentration of reactants and these change continuously as the reaction proceeds. This problem is overcome by expressing the experimental results in terms of a *rate equation* which expresses the rate as a function of the concentration or the fraction of the transformation completed. From eqn. (10.8)

$$\frac{dy}{dt} = k \cdot f(y) \qquad (10.11)$$

where k is the *rate constant* and y is the fraction transformed. $f(y)$ is some function of y.

Experimentally it is y which is measured, not dy/dt which is the rate of change of y, so we must put this equation in a different form. Separation of the variables and integration gives

$$g(y) = kt \qquad (10.12)$$

where $g(y)$ is a further function of y. Eqn. (10.12) is known as an *integrated rate equation*.

In practice the function $g(y)$ is determined empirically by finding functions which describe the variation of y with time. When an equation is found which fits the observed data, the rate constant k can be found.

Rate equations are not always defined strictly in accordance with eqn.(10.11) and in many cases the usage can be quite loose. As we shall see in the rest of this chapter, equations which describe the fraction transformed as a function of time are also referred to as rate equations.

10.3.1 Empirical rate equations for heterogeneous reactions

Reactions which involve nucleation and growth of the product phase(s) within the parent phase are termed heterogeneous. Specific cases will be referred to in the following chapter, but here we deal in a general way with reactions which include exsolution, reconstructive polymorphic transformations, reactions between minerals and the breakdown of minerals.

Empirically it is found that the isothermal kinetics of a wide range of mineral reactions can be described by an equation of the general form

$$\frac{dy}{dt} = k^n t^{n-1}(1-y) \qquad (10.13)$$

where k is a rate constant, t is time, y is the fraction transformed and n is a constant which depends on the mechanism. Note that since k is now defined in

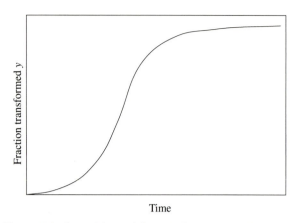

Figure 10.8. General form of the rate of a reaction with time for reactions which conform to the Avrami equation.

terms of both y and t, it is not a true rate constant in the same sense as in eqn. (10.11).

Separating the variables and integrating gives

$$\int \frac{dy}{1-y} = \int k^n t^{n-1} dt$$

$$\ln \frac{1}{1-y} = (kt)^n, \text{ where the term } 1/n$$

has been incorporated into the constant. This can also be written

$$y = 1 - \exp(-kt)^n \qquad (10.14)$$

The rate constant k has dimensions time^{-1}. In some kinetic studies rate equations do not have the dimensions time^{-1}. This can lead to confusion in comparing activation energies derived from different rate equations; an example is given in Section 10.4.1.

Equation (10.14) has a sigmoidal form, the reaction rate being small at the beginning, then increasing to a maximum, and finally decreasing to zero as the reaction goes to completion (Figure 10.8). Rate equations with the general form of eqn (10.14) are usually referred to as the *Avrami equations*.

To extract values of the rate constant k from eqn (10.14), the usual approach is to linearize the equation as follows:

$$y = 1 - \exp(-kt)^n$$
$$1 - y = \exp(-k^n t^n)$$
$$\ln(1-y) = -k^n t^n$$
$$-\ln\ln(1-y) = n\ln k + n\ln t \qquad (10.15)$$

Thus a reaction whose kinetics conform to this Avrami equation gives a straight line when $-\ln\ln(1-y)$ is plotted against $\ln t$. The value of n, which is

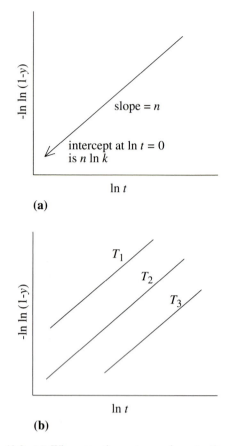

Figure 10.9. (a) When reaction rates conform to the Avrami equation a plot of $-\ln\ln(1-y)$ against $\ln$ (time) is linear. The slope is equal to the constant n, and the intercept on the y axis (at $\ln$ (time) = 0) provides a value of the rate constant k. **(b)** When the reaction is carried out at 3 different temperatures T_1, T_2 and T_3, a set of parallel lines indicates a constant value of n, suggesting that the reaction mechanism is the same. Each isothermal experiment provides a different value of the rate constant.

used as an empirical parameter used to compare reaction mechanisms, is derived from the slope of the line. The intercept on the y axis gives the value of $n \ln k$ from which the rate constant k is determined (Figure 10.9).

When the reaction is carried out at a series of temperatures T_1, T_2, T_3, the plot gives a set of parallel lines, if the mechanism does not change within this temperature range (Figure 10.9(b)). Such processes are termed *isokinetic*.

10.3.2 Examples of the application of the Avrami equation

(i) The decomposition of $BaCO_3$ to $BaO + CO_2$
Figure 10.10 shows the data for this decomposition reaction plotted on a $-\ln\ln(1-y)$ against $\ln t$ plot at

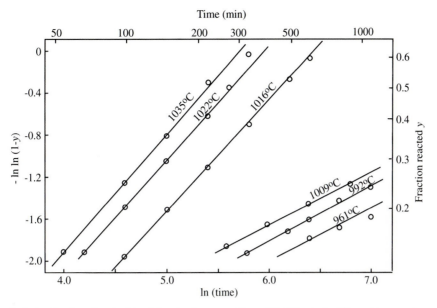

Figure 10.10. A plot of $-\ln\ln(1-y)$ against $\ln$ (time) for the decomposition of $BaCO_3$ at six different temperatures. The change in the slope of the lines between 1009°C ($n \sim 0.5$) and 1016°C ($n\sim1.0$) suggests a change in reaction mechanism from phase boundary control to diffusion control. (Data from Hancock and Sharp, 1972.)

six different temperatures. Such data can be obtained with a number of techniques e.g. by measuring the weight loss as a function of time at each temperature, or by monitoring the relative heights of a $BaCO_3$ and BaO X-ray diffraction peaks in the products.

The interpretation of this plot is that the reaction mechanism changes between 1009°C and 1016°C. At higher temperatures the value of n is very near to 1.0, while at the lower temperatures $n \sim 0.5$. Empirically it is found that for diffusion controlled reactions $n \sim 0.5$, whereas for phase boundary controlled reactions $n \sim 1.0$. This example illustrates this point well, although many isokinetic processes have been described with very different values of n.

The crystallisation of albite from analcite + quartz
When the reaction

$$NaAlSi_2O_6.H_2O + SiO_2 \Rightarrow NaAlSi_3O_8 + H_2O$$
$$\text{analcite} \qquad \text{quartz} \qquad \text{albite}$$

is carried out at 1 kbar pressure the data plotted in Figure 10.11 show the fraction of albite X_{Ab} in the run products as a function of time at four different temperatures. X_{Ab} may be equated with y, the fraction transformed. When the data at each temperature are fitted to the Avrami equation (eqn. (10.14)), the $-\ln\ln(1-y)$ against $\ln t$ diagram (Figure 10.12) gives a set

of four virtually parallel lines indicating that the reaction is isokinetic over this temperature range. Table 10.1 shows the values of n and k obtained from this plot at each temperature.

10.4 The experimental determination of the empirical activation energy E_a

(i) The rate constant method
This is the most direct method and follows from the method of fitting empirical rate equations. Isothermal experiments are carried out and the data of fraction transformed against time are treated as in the previous section. A value of the rate constant k is then determined at each temperature.

The value of E_a is then determined from an Arrhenius plot of $\ln k$ against $1/T$. Since

$$k = A \exp\left(\frac{-E_a}{RT}\right)$$

(eqn. (10.10)), this plot is linear with a slope of $-E_a/R$.

The disadvantage of this method is that the determined value of k depends on the empirical selection of the function $f(y)$ in the rate equation (eqn.(10.11)). The empirical activation energy E_a obtained this way may depend therefore on the

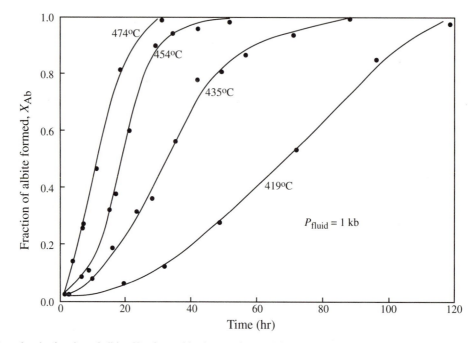

Figure 10.11. Data for the fraction of albite X_{Ab} formed in the reaction analcite + quartz $\Rightarrow$ albite + water as a function of time at four different temperatures. When $X_{Ab} = 1$ the reaction is complete. (from Matthews, 1980)

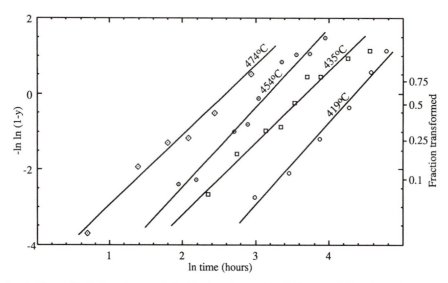

Figure 10.12. The data in Figure 10.11 plotted on a $-\ln \ln (1-y)$ against $\ln$ (time) diagram yields values of the rate constant k (from the intercept on the y axis) and the constant n (from the slope). These are listed in Table 10.1.

Table 10.1. *Values of n and the rate constant k, obtained from Figure 10.12*

Temperature	n	Rate constant k
419°C	2.15	1.25×10^{-2}
435°C	1.90	2.56×10^{-2}
454°C	2.07	4.08×10^{-2}
474°C	1.85	7.35×10^{-2}

choice of rate equation. A number of different forms of the rate equation may fit the data equally well, but result in different values of E_a. It is therefore essential to specify clearly the rate equation used in the determination.

A second disadvantage of this method is that it is not possible to ascertain whether E_a changes during the course of an isothermal transformation.

(ii) The 'time to a given fraction' method

This method makes the determination of the activation energy, E_a, independent of the function f(y) in the rate equation and also makes it possible to test whether E_a varies during the course of a transformation.

The method relies on the fact that the fraction transformed y and the time t are functionally related and so it is possible to make t the dependent variable. Beginning from the general rate equation

$$\frac{dy}{dt} = k.\,f(y) \quad (eqn.(10.11))$$

$$dt = k^{-1}\,f^{-1}(y)\,dy$$

The time t_Y to transform a given fraction $y = Y$ is then

$$t_Y = k^{-1}\int_{y=0}^{y=Y} f^{-1}(y)\,dy$$

If the function f(y) does not change over the temperature range studied, the integral above has a constant numerical value, and therefore

$$t_Y \propto k^{-1}$$

$$t_Y \propto A^{-1}\exp\left(\frac{E_a}{RT}\right)$$

Therefore

$$\ln t_Y = \text{const} - \ln A + E_a/R\left(\frac{1}{T}\right) \quad (10.16)$$

The time t_Y to the chosen value of the fraction transformed Y is determined from a series of isothermal experiments carried out at a number of temperatures. A plot of $\ln t_Y$ against $1/T$ is linear if E_a is independent of T. The slope of the graph is E_a/R, from which the value of E_a is obtained. To check for any variation of E_a during the course of the experiment the procedure is repeated for various values of the fraction transformed Y.

This method forms the basis for Time–Temperature–Transformation diagrams discussed in Section 11.1.7 where we return to the discussion of kinetics of specific processes.

10.4.1 Example 1: determining the activation energy for albite crystallisation from the reaction analcite + quartz ⇒ albite + water

In this example we will take the data for the crystallisation of albite according to the reaction

$$NaAlSi_2O_6.H_2O + SiO_2 \Rightarrow NaAlSi_3O_8 + H_2O$$

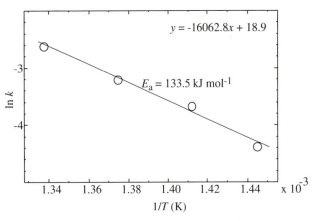

Figure 10.13. An Arrhenius plot using the four values of the rate constant k, obtained from Figure 10.11 (Table 10.1). The fitted straight line yields an empirical activation energy of 133.5 kJ mol^{-1}.

discussed in Section 10.3.2 and plotted in Figure 10.11 and examine various methods of determining values of the activation energy E_a.

In the rate constant method the data have to be fitted to a rate equation. Using the Avrami equation (eqn.(10.14)) the data can be plotted on a $-\ln\ln(1-y)$ against $\ln t$ diagram as in Figure 10.12. The values of the rate constant k at each temperature is tabulated in Table 10.1.

The empirical activation energy E_a is found from an Arrhenius plot of $\ln k$ against $1/T$, and is shown in Figure 10.13. The four data points are fitted by a straight line and from eqn.(10.10) the slope of this line is $-E_a/R$. The value of E_a is found to be 133.5 kJ mol^{-1}.

We next consider the situation if a different empirical rate equation had been chosen to fit the raw data of Figure 10.11. Occasionally, kinetic data can be fitted to much simpler rate equations with the general form

$$y = k'\,t^n \quad (10.17)$$

Note that this is not a rate equation in the same sense as eqn. (10.11) and moreover that in such a case the rate constant k' has dimensions (time)$^{-n}$. Values of E_a obtained from this equation cannot be compared with values derived from equations in which k has dimensions (time^{-1}.)

Figure 10.14 is a plot of the mole fraction of albite X_{Ab} against $t^{3/2}$. The data fall on straight lines through the origin and so the kinetics of albite

Table 10.2. *Values of the rate constant k, obtained from Figure 10.14 and used in the Arrhenius plot in Figure 10.15*

Temperature	$1/T$ (K)	k' (time)$^{-3/2}$	k (time)$^{-1}$	$\ln k$
474°C	1.338×10^{-3}	9.80×10^{-3}	4.58×10^{-2}	-3.08
454°C	1.375×10^{-3}	5.95×10^{-3}	3.28×10^{-2}	-3.41
435°C	1.412×10^{-3}	2.55×10^{-3}	1.86×10^{-2}	-3.98
419°C	1.445×10^{-3}	9.02×10^{-4}	9.33×10^{-3}	-4.67

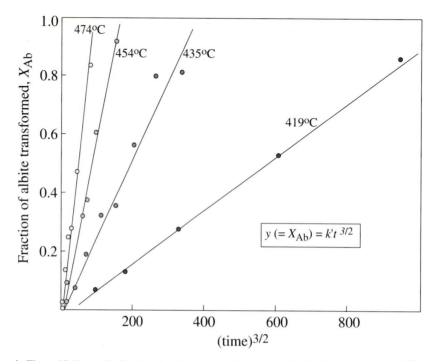

Figure 10.14. The data in Figure 10.11 can also be linearized by plotting the fraction of albite X_{Ab} against (time)$^{3/2}$. The slopes of these lines yield values of the rate constant k' which has dimensions (time)$^{-3/2}$ and are converted to (time)$^{-1}$ in Table 10.2.

crystallisation can be reasonably described by the equation

$$y \, (=X_{Ab}) = k \, t^{3/2} \qquad (10.18)$$

Values of k can be obtained directly from the slopes of these straight lines and converted to dimensions time^{-1} i.e. $k = (k')^{2/3}$ (Table 10.2). An Arrhenius plot of $\ln k$ against $1/T$ (Figure 10.15) gives a straight line from whose slope the value of E_a is calculated as 123.4 kJ mol^{-1}.

The time to a given fraction method can be illustrated by reading off the time, t_Y, to $X_{Ab} = 0.5$ from Figure 10.11 for each of the four temperatures and plotting $\ln t_Y$ against $1/T$ as shown in Figure 10.16 and Table 10.3. The slope of the straight line is $+E_a/R$ and gives a value of $E_a = 129.7$ kJ mol^{-1}.

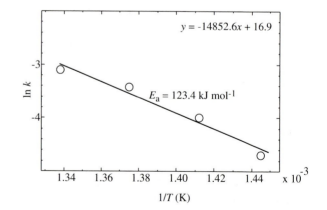

Figure 10.15. The slopes of the lines in Figure 10.14 yield values of the rate constants, which when converted to dimensions (time)$^{-1}$ (Table 10.2) are used in this Arrhenius plot. The fitted straight line yields an empirical activation energy of 123.4 kJ mol^{-1}.

Table 10.3. *Time taken to 50% completion, from the data in Figure 10.11*

Time to 50% Ab	ln (time)	Temperature	$1/T$
13.85 hrs	2.63	474°C	1.338×10^{-3}
18.1 hrs	2.90	454°C	1.375×10^{-3}
32.69 hrs	3.49	435°C	1.412×10^{-3}
68.1 hrs	4.22	419°C	1.445×10^{-3}

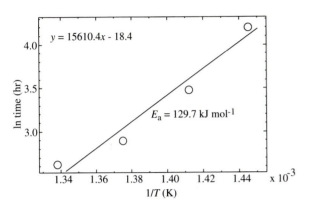

Figure 10.16. The data in Figure 10.11 plotted as ln (time) to 50% albite ($X_{Ab} = 0.5$) against $1/T$, according to the 'time to a given fraction method'. The fitted straight line (slope = E_a/R) yields an empirical activation energy of 129.7 kJ mol^{-1}.

These three values of the activation energy, the first two derived from different rate equations and the third derived independently of a rate equation, lie within the range expected from the experimental errors involved. One general observation which should be made regarding the use of rate equations is that to avoid the confusion which arises when comparing activation energies of different processes, rate equations should be formulated so that the rate constant k has dimensions (time^{-1}).

In nature, this albite forming reaction occurs during burial metamorphism of rocks at temperatures around 200°C. To apply the experimental rate data to natural rocks we need to extrapolate to these lower temperatures and determine the time taken for the reaction to take place. In the rate constant method, extrapolation of the Arrhenius plots leads to a value of k at 200°C which is then substituted into the rate equation. In the time to a given fraction method extrapolation can be made directly from the ln t_Y against $1/T$ plot.

However, extrapolation of experimental data is subject to a large number of uncertainties. First we assume that the mechanism of the reaction, and hence the rate equation and activation energy, remains unchanged over this temperature range. To some extent microscopic studies of the mechanism in experiments may be compared with the mineral textures to shed some light on this important problem. Secondly, the experiment rarely, if ever, reproduces the natural situation in terms of chemistry of the components and fluids, grain size of the particles and their state of compaction, effects of stress etc. All of these factors are known to affect the kinetics of heterogeneous reactions in experimental situations but their relative importance in controlling the kinetics of natural reactions is difficult to assess. Nevertheless, as more data become available, kinetic models of natural reactions will inevitably improve.

10.4.2 Example 2: The kinetics of coarsening of exsolution lamellae

Exsolution in minerals often takes the form of lamellar precipitates which coarsen with time. This process will be discussed in more detail in the following chapter, but here we are concerned with the rate laws which are used to describe the kinetics of coarsening and some of the potential problems when the dimensionality of the rate constant is not taken into account.

Figure 10.17 shows a series of TEM micrographs showing coarsening of a pigeonite – augite lamellar exsolution intergrowth. The thickness of the periodic lamellae can be described by a wavelength λ and experiments have been carried out to study the rate of coarsening as a function of time at different temperatures. Similar experiments have yielded data plotted in Figure 10.18 which shows a graph of the measured wavelength, λ, against (time)$^{1/3}$ in days, at three temperatures. The linear fit to the data suggests a rate law:

$$\lambda = \lambda_o + kt^{1/3}$$

Figure 10.17. A series of transmission electron micrographs from experiments to study the coarsening of a periodic pigeonite – augite intergrowth during exsolution. The micrographs illustrate three isothermal series of experiments at (a) 1000°C, (b) 900°C and (c) 800°C. The annealing times (increasing from left to right) in each isothermal sequence are 39hr, 359 hr, 1606 hr and 3886 hr. (Courtesy of G.L. Nord)

where λ_o is the initial wavelength and λ is the wavelength after time t. In this expression the rate constant k has dimensions length (time)$^{-1/3}$, i.e. $LT^{-1/3}$. The slopes of the straight lines in Figure 10.18 give values of k at three temperatures and a plot of ln k against $1/T$ is linear. In other words the Arrhenius law $k = A \exp\left(\dfrac{-E_a}{RT}\right)$ is obeyed and an empirical activation energy E_a can be obtained from the slope of the Arrhenius plot ($= -E_a/R$) in the usual way. The values of k derived from Figure 10.18 give $E_a = 413.8 \, \text{kJ mol}^{-1}$.

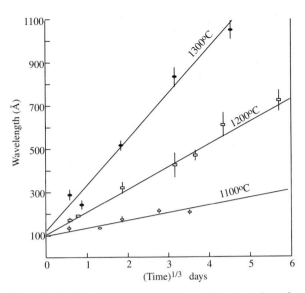

Figure 10.18. Experimental rate data on the coarsening of a periodic pigeonite – augite intergrowth as a function of (time)$^{1/3}$ at 1100°C, 1200°C and 1300°C. The linear fit suggests a rate law $\lambda = \lambda_o + kt^{1/3}$ where λ_o is the initial wavelength at $t = 0$. (Data from McCallister, 1978.)

We must next consider the meaning of this value of E_a. As we noted in the previous example, the dimensions of the rate constant k determine the value of E_a. If the rate equation for coarsening had been formulated in a different way, namely

$$\lambda = \lambda_o + (kt)^{1/3}$$

the rate constant would have dimensions (length^3) (time^{-1}) (L^{3}T^{-1}) and the values of k are equal to the (slope)^3 of the straight lines in Figure 10.18. The Arrhenius plot would then give an activation energy of $\frac{413.8}{3} = 138.9$ kJ mol^{-1}. With (time)$^{-1}$ in the dimension of k, this may be a more appropriate value for comparison with activation energies of other processes.

The physical interpretation of this empirical activation energy is not however clear. The coarsening of the augite–pigeonite intergrowth is dependent on the diffusion rate of Ca and Mg within the pyroxene structure, the Mg diffusing to the pigeonite lamellae and the Ca to the augite lamellae. It is tempting therefore to equate the measured activation energy for coarsening, to the activation energy for the diffusion process. However, as well as the uncertainties regarding the rate controlling atomic process which determines coarsening rate, there is also once again the problem of the definition of the rate constant and its dimensions. As we shall discuss

further in Section 11.2 the activation energy for diffusion describes the temperature dependence of the diffusion coefficient D, according to the Arrhenius law

$$D = D_o \exp(-E_a/RT) \quad \text{(see also Section 7.1.4)}$$

The definition of the diffusion coefficient (Section 11.2.1) gives it units m^{2}s^{-1} (i.e. dimensions L^{2}T^{-1}) which are different from the dimensions of the rate constant k used to determine either of the two values above for the activation energy for lamellar coarsening. It is therefore not correct to equate either of these measured values of activation energy to the activation energy for the diffusion process, even if the coarsening rate is controlled by diffusion.

Note also that the $t^{1/3}$ rate law is not always obeyed, especially at low temperatures, and values of the exponent n in the general rate equation $\lambda = \lambda_o + (kt)^{1/n}$ may vary between 1.5 and 5. If coarsening is impeded in any way (for example, by impurities on the interfaces), even higher values of n may be found.

10.4.3 Example 3: Coarsening of antiphase domains

Antiphase domains form during structural phase transitions and cation ordering transitions whenever the symmetry change associated with the transition involves a loss of some translational symmetry element (Section 7.3.4). In Section 7.3.4 we described the formation of antiphase domains during cation ordering in omphacite and during Al,Si ordering in anorthite. The antiphase domains coarsen with annealing time, (resulting in a sequence of microstructures similar to those in Figure 12.70) and the usual way of describing the rate law for the process is

$$\delta^n - \delta_o{}^n = k(t - t_o) \tag{10.19}$$

where δ_o is the initial domain diameter at time t_o and δ is the domain diameter after time t. The exponent n is a dimensionless constant and k is the rate constant $= A.\exp(-E_a/RT)$. When the rate equation is formulated in this way the rate constant k has dimensions L^{n}T^{-1}.

In the ideal case, such as antiphase domain coarsening in pure metals, the constant $n \sim 2$ and hence k has dimensions L^{2}T^{-1}, the same as that for the diffusion coefficient. Since antiphase domain coarsening is controlled by the diffusion rate, it is reasonable to equate the activation energy deter-

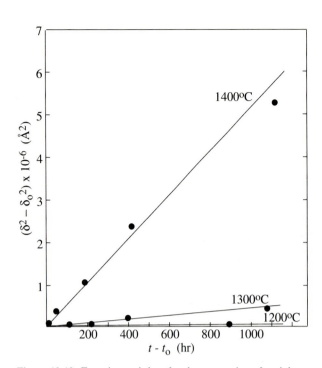

Figure 10.19. Experimental data for the coarsening of antiphase domains during Al,Si ordering in anorthite, $CaAl_2Si_2O_8$, where δ_o is the initial domain diameter, and δ is the diameter after time t. The slope of each line is equal to the rate constant k. (data from Carpenter, 1991.)

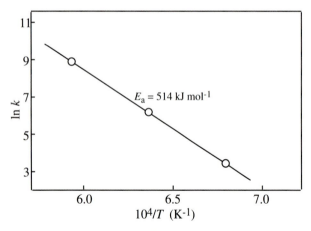

Figure 10.20. An Arrhenius plot using the values of the rate constant k obtained from Figure 10.19. The fitted straight line yields an empirical activation energy of 514 kJ mol^{-1}.

mined from the temperature dependence of k to the activation energy for diffusion.

In pure anorthite, antiphase domain coarsening is also well described by a rate law with $n = 2$, i.e. the data fall on a linear plot of $\delta^2 - \delta_o^2$ against $(t - t_o)$ (Figure 10.19). The slopes of these graphs at three different temperatures 1200°C, 1300°C and 1400°C

gives three values of k. An Arrhenius plot of $\ln k$ against $1/T$ (Figure 10.20) gives an activation energy E_a of 514 kJ mol^{-1}. This value is comparable to the activation energy for Al,Si diffusion in plagioclases.

In some experiments the value of n is found to be considerably greater than 2, and hence the coarsening rate is slower than that expected from diffusion. This was found to be the case for antiphase domain coarsening in omphacite for which $n \sim 8$. Such behaviour is usually explained by the dragging effects of impurity atoms and other defects which preferentially migrate to the antiphase domain boundaries.

10.5 The kinetics of continuous processes

So far in this chapter we have considered heterogeneous reactions where the products and reactants are sufficiently different in structure and/or chemistry such that the transformation process involves nucleation and growth. In the rest of this chapter we consider processes in which a continuous sequence of structural states exists between the initial state and the final state. For example, a change from a disordered cation distribution to a fully ordered state may take place by a gradual increase in the degree of order. In such a case we are not dealing with a situation where two separate free energy curves for the initial and final states are separated by some hypothetical activated state, and the formulation of the rate equations can be made differently.

Continuous transformations are described by a continuous free energy surface as a function of some external variable such as temperature, pressure etc. A sequence of such free energy curves as a function of temperature is shown in Figure 9.20 and defines the shape of a free energy surface in terms of the order parameter Q and the temperature. Such a surface is shown in Figure 10.21. At any temperature the equilibrium order parameter is that with the minimum free energy. Above T_c the equilibrium state is $Q = 0$, i.e. complete disorder. Below T_c the free energy surface forms two valleys, defining values of $+Q$ and $-Q$ which diverge with decreasing temperature. The positive and negative values of Q represent the ordered and anti-ordered states separated by antiphase domains, as discussed in Section 7.3.4.

Every point on this surface represents the free energy, due to ordering, of the mineral at a defined temperature and a particular state of order, and it is

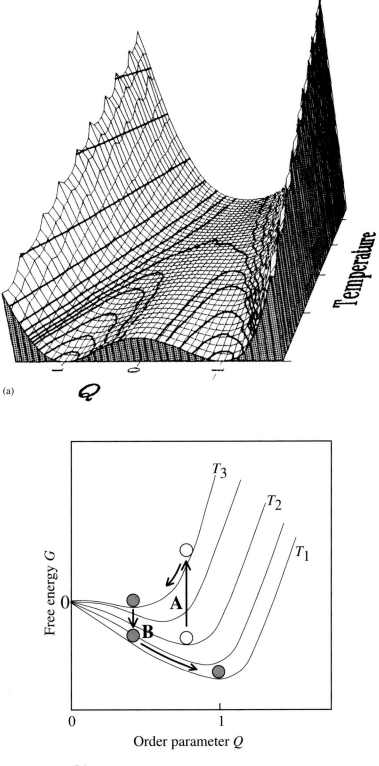

Figure 10.21. (a) The shape of the free energy surface as a function of order parameter Q and temperature T, for a continuous transformation process. The heavy lines are contours of free energy. (b) Free energy curves drawn from this surface at a number of different temperatures, decreasing from T_3 to T_1. Two different non-equilibrium paths are shown: (i) if a mineral with an equilibrium degree of order at T_2 is rapidly heated to T_3 it will begin to disorder down the maximum slope of the free energy curve (path A); (ii) a mineral in equilibrium at T_3 is rapidly cooled to T_1 it will tend to order, following path B.

assumed that during the ordering process the energy of the system stays on this surface.

Changes in the degree of order as a function of temperature are equivalent to moving from one point on the surface to another. For example, during equilibrium cooling the order parameter Q increases and the system moves along the free energy minima along the bottom of the energy valley. On the other hand, if a mineral with a certain degree of order is rapidly heated to a higher temperature, e.g. T_2 to T_3 in Figure 10.21(b), it will begin to disorder and the free energy will follow path A . If a relatively disordered mineral with $Q \ll 1$ is rapidly cooled e.g. T_3 to T_1, and held at this temperature, it will tend to order and the free energy will follow path B.

On the atomic scale, ordering and disordering involve local diffusion and the interchange of atoms. Although the number of possible pathways for these individual atomic interchanges may be almost infinite in number, statistically speaking the overall effect of the most probable of these steps determines the direction of change at the macroscopic level. Thus ordering along path B in Figure 10.21(b) involves atomic interchanges which on average reduce the overall free energy of the system. Each atomic interchange has a free energy of activation G_a and this can be thought of in terms of a frictional force on the free energy surface, restricting the rate at which the system can move towards a free energy minimum.

The task of identifying the most probable pathway along the free energy surface during ordering (or disordering) is a statistical one and first involves determining the probabilities of interchanges on a microscopic scale. These probabilities change with the local degree of order, so that when a group of atoms is ordered, further interchanges become less likely i.e. the probabilities change with time. These individual steps must then be summed, according to their assigned probabilities, over the whole crystal. The statistical methods used to do this need not concern us here and reference is given in the bibliography. The result of such a calculation is an expression for dQ/dt, the rate at which the macroscopic order parameter Q changes with time.

The equation which describes the passage of such a system across the free energy surface during an order–disorder process is termed the Ginzburg–Landau (G–L) rate law. In its simplest form

$$\frac{dQ}{dt} = \frac{-\gamma \lambda \exp(-\Delta G_a/RT)}{2RT} \cdot \frac{dG}{dQ} \quad (10.20)$$

where Q is the order parameter, γ and λ are constants depending on the mineral, ΔG_a is the free energy of activation for the process, and T is the temperature in kelvins. dG/dQ is the slope of the free energy surface. The rate at which the order parameter changes with time is therefore proportional to the slope of the free energy curve at that value of Q, modified by the usual exponential dependence on the activation energy.

To be able to use this equation to analyse experimental data, we first need to know the shape of the free energy surface. This is a problem of equilibrium thermodynamics and was discussed in the context of Landau theory in Section 8.6. The advantage of formulating the excess free energy of the transition in terms of Landau theory is that the order parameter Q can be determined experimentally by measuring the variation in a number of different properties which are related to the process operating. Thus cation ordering can be studied by measuring the variation in the spontaneous strain, or the birefringence, or excess enthalpy etc., by relating these properties to the order parameter Q. We do not therefore need to know the microscopic mechanism of the process to be able to formulate the rate equation.

To illustrate the use of the G–L equation (eqn.(10.20)) we will first use a simple example, and then point out some of the complications which can arise.

10.5.1 The rate of disordering in omphacite

Natural omphacite with ideal composition $(Ca_{0.5}, Na_{0.5})(Al_{0.5}, Mg_{0.5})Si_2O_6$ has an ordered cation distribution at low temperatures, with space group $P2/n$ (Section 6.4.1). Above about 860°C the cations are disordered within the M1 and M2 sites and the symmetry becomes $C2/c$. The diffraction pattern of the disordered structure has systematic absences of the type $h+k$ odd and therefore the process of disordering can be followed experimentally by measuring the decrease in intensity of these superlattice reflections.

The equilibrium behaviour of omphacite order–disorder appears to fit well with a tricritical model, i.e. intermediate between a first-order transition and a second-order transition (Section 8.6.1). The change in free energy due to ordering is therefore given by a Landau expansion of the type

$$G = \frac{1}{2} a (T - T_c)Q^2 + \frac{1}{6} c Q^6 \text{ (see eqn. (8.30))}$$

and therefore

$$\frac{dG}{dQ} = a (T - T_c)Q + cQ^5 \qquad (10.21)$$

The equilibrium values of Q over a range of temperatures have been determined experimentally, and the values of the Landau coefficients a and c obtained by measuring the excess enthalpy due to ordering, as explained in Section 8.6.2. The result of such experiments gives the following values: $a = 22.8$ J mol^{-1}K^{-1}, $c = 25\,900$ J mol^{-1} and $T_c = 1138$ K. We now have a fully defined expression for the variation of G with Q as a function of T, i.e. the free energy surface shown in Figure 10.21.

Having defined the equilibrium order–disorder thermodynamics in omphacite, we are now in a position to analyse the kinetics of the disordering process. As mentioned above, the intensity of the superlattice reflections can be used to monitor the disordering process. For example the intensity of the 050 reflection is zero in the disordered form, but present in the ordered form. The ratio of intensity of this reflection (I_{050}) relative to another reflection not affected by disordering (e.g. I_{060}) will therefore decrease during the disordering process. Thus if we start with an ordered crystal at time t_o with ratio $I_{050}/I_{060} = I_o$, and heat it above T_c for a time t the same ratio will fall to I_t. Thus I_t/I_o will fall from a value of 1 at time t_o to zero in the fully disordered form.

Scaled in this way the ratio I_t/I_o has the form of an order parameter. There is one further point to note which is that for a zone boundary transition, i.e. in which the translational symmetry changes, the intensity of the superlattice reflections vary as Q^2 and therefore the ratio I_t/I_o is proportional to Q^2/Q_o^2.

Figure 10.22 shows the experimental data for omphacite disordering, plotted as I_t/I_o ($= Q^2/Q_o^2$) against ln (time). To compare this data with the theoretical predictions we take the G–L equation (eqn.(10.20)) and substitute the form of the free energy surface given by eqn. (10.21) i.e.

$$\frac{dQ}{dt} = \frac{-\gamma \lambda \, exp(-\Delta G_a/RT)}{2RT} \cdot [a(T - T_c)Q + cQ^5] \qquad (10.22)$$

or

$$\int_{t_o}^{t} dt = \int_{Q_o}^{Q} \frac{-2RT}{\gamma \lambda \, \exp(-\Delta G_a/RT)} \cdot \frac{1}{[a(T - T_c)Q + cQ^5]} \, dQ \qquad (10.23)$$

which gives the time taken from t_o to t to change the order parameter from Q_o to Q.

Another factor that needs to be taken into account is that the free energy of activation ΔG_a may not be constant during the disordering process. Although the activation energy ΔH_a is generally constant, the entropy of activation ΔS_a which defines the number of microscopic reaction pathways, will change with the state of order. When a crystal is fully ordered, virtually any cation exchange decreases Q, whereas when the crystal is substantially disordered a smaller proportion of the possible interchanges will lead to a further decrease in Q. The coupling between ΔG_a and the order parameter Q is expressed as

$$\Delta G_a = \Delta G_o + \varepsilon Q^2$$

but if it is only the entropy of activation which is affected we may write

$$\Delta G_a = \Delta H_a - T(\Delta S_a + \varepsilon_s Q^2) \qquad (10.24)$$

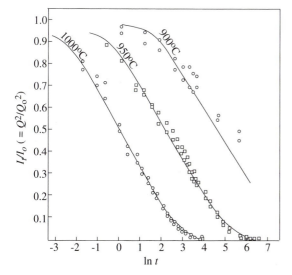

Figure 10.22. Experimental data for the rate of cation disordering in omphacite plotted as the change in the degree of order (measured as Q^2/Q_o^2) as a function of ln (time) at three temperatures. $Q^2/Q_o^2 = 1$ indicates complete order, while $Q^2/Q_o^2 = 0$ means complete disorder. (Data from Carpenter et al., 1990.)

where ε_s is the entropy part of the coupling constant ε.

Substituting eqn.(10.24) into eqn.(10.23) gives the final form of the G–L equation which must be solved to describe ordering and disordering kinetics:

$$\int_{t_o}^{t} dt = \int_{Q_0}^{Q} \frac{-2RT \exp(-\Delta S_a/R) \exp(-\varepsilon_s Q^2/R)}{\gamma \, \lambda \, \exp(-\Delta H_a/RT)[a(T - T_c)Q + cQ^5]} \, . dQ$$

(10.25)

Although this equation looks complicated, it can be solved numerically. The best fit to the experimental data is given by the solid lines in Figure 10.22. and gives an activation energy ΔH_a of 540 kJ mol^{-1} with $\varepsilon_s = 20$ J mol^{-1} K^{-1} and $\exp(-\Delta S_a/R)/\gamma\lambda = 9 \times 10^{-19}$ secs.

The significance of this fit between the experimental data and the calculated curve in Figure 10.22 is that it demonstates the validity of the G–L equation which is derived from a rigorous statistical analysis of the atomic interchanges. It also establishes the relationship between the equilibrium properties of the mineral i.e. the free energy surface, and the non-equilibrium behaviour. Furthermore, the equation can now be used in a predictive way to determine the kinetics of any change in the degree of order of omphacite.

Admittedly, eqn. (10.25) is not a simple equation to solve and it is worth pointing out that if we merely

wish to determine the activation energy of the disordering process from the experimental data, empirical methods such as described earlier in this chapter provide a quick and often satisfactory solution. The 'time to a given fraction' method discussed in Section 10.4 can be easily adapted to the case of omphacite ordering. Equal values of I_t/I_o in Figure 10.22 represent the same degree of order measured as a fraction ($I_t/I_o = 1$ represents complete order and $I_t/I_o = 0$ represents complete disorder). Therefore contours of I_t/I_o as a function of ln (time) and $1/T$ represent the time to taken to achieve a certain degree of disorder, as a function of temperature. These contours plot as straight lines (Figure 10.23) with slope equal to E_a/R. The value of E_a (which we may here equate to the term ΔH_a in eqn. (10.25)) obtained from the slope of these lines is ~560kJ mol^{-1}, in good agreement with the theoretical value obtained above.

Another method is to attempt to linearize the data in Figure 10.22 by finding a suitable empirical rate equation. Figure 10.24 shows that these data fall on straight lines with slope ~0.5 when plotted as ln ln (I_t/I_o) against ln (time). Expressing this as a rate law of the Avrami type

$$y = 1 - \exp(-kt)^n \text{(see eqn.(10.14))}$$

where y is the fraction of the process completed, the ratio (I_t/I_o) is equivalent to $1-y$, since when $(I_t/I_o) = 0$, $y = 1$ and when $(I_t/I_o) = 1$, $y = 0$. Thus the

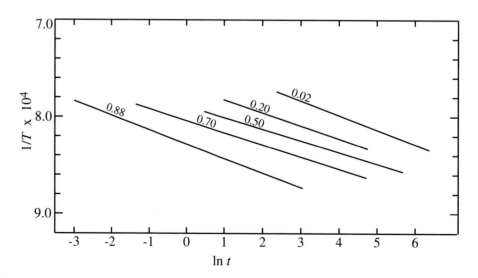

Figure 10.23. The data in Figure 10.22 plotted as a 'time to a given fraction' method. The time t to reach a particular value of the degree of order (given by Q^2/Q_o^2 in Figure 10.22) is measured at each temperature T. The lines drawn on this plot are contours of Q^2/Q_o^2 as a function of time and $1/T$. The approximately equal slope of each line indicates that the activation energy remains constant during the disordering process, and yields a value of the activation energy of ~560 kJ mol^{-1}.

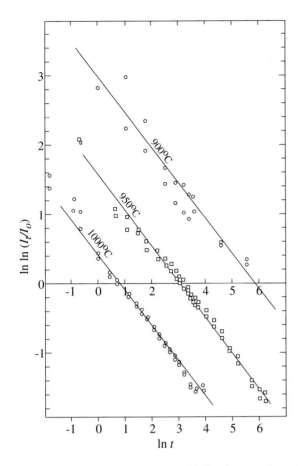

Figure 10.24. The data in Figure 10.22 fitted to an Avrami equation, $\ln (I_t/I_o) = -(kt^n)$. The slope and intercepts yield values of n and k at three temperatures. (After Carpenter *et al.*, 1990.)

Avrami equation becomes

$$(I_t/I_o) = \exp (-kt)^n$$
$$\ln (I_t/I_o) = -k^n t^n$$
$$-\ln \ln (I_t/I_o) = n \ln k + n \ln t$$

The straight lines in Figure 10.24 all have the same slope, indicating the same mechanism at each of the three temperatures. In this case $n \sim 0.5$. The values of the rate constant $k = A \exp (-E_a/RT)$, are obtained at each temperature from the intercepts, and an Arrhenius plot (Figure 10.25) is used to determine the activation energy E_a. The value of E_a found from the slope of this Arrhenius plot is $\sim$620 kJ mol^{-1}. This value is somewhat larger than the values obtained from the G–L equation, or the 'time to a given fraction' method which is independent of the rate equation chosen. This reflects the fact that different values for the activation energy can be

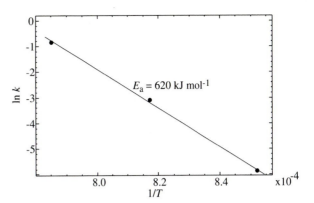

Figure 10.25. An Arrhenius plot using the values of the rate constant k determined from Figure 10.24. The fitted straight line yields an empirical activation energy of $\sim$620 kJ mol^{-1}.

obtained from the same data by using different methods.

10.5.2 The effect of spatial fluctuations in the order parameter: Al,Si ordering in anorthite

To be able to apply the G–L rate equation we need to know the form of the equilibrium free energy surface, i.e. the Landau free energy expansion as a function of Q and T. In the case of omphacite this is quite well known, and furthermore at equilibrium, the order parameter Q remains homogeneous throughout the crystal. As was mentioned in Section 8.6.5, this is not always the case and the assumption of long-range correlations in the order parameter Q may break down near T_c. When there are local fluctuations in Q, the form of the Landau free energy expansion has to contain terms which takes into account the energy due to gradients in Q.

This is the case for Al,Si ordering in anorthite. End-member anorthite, $CaAl_2Si_2O_8$ has the fully ordered $I\bar{1}$ structure at all temperatures up to the melting point (Figure 6.56) but when crystallised experimentally from a glass it grows with a substantially disordered Al,Si distribution and has space group $C\bar{1}$. During ordering, and hence transformation from $C\bar{1}$ to $I\bar{1}$, extra superlattice reflections appear in the diffraction pattern and the increase in their intensity with time could be used to monitor the increase in the order parameter Q. In omphacite, the superlattice reflections remain sharp and well-defined throughout the ordering and disordering process. In anorthite however, Al,Si ordering begins with the development of spatial fluctuations in the order parameter Q. These fluctuations cause the superlattice reflections to be diffuse, indicating a range of periodicities associated with the Al,Si distribution. Eventually, as the degree of Al,Si order increases the superlattice reflections become sharper, and the doubled c axis of ordered anorthite is established.

The outcome of this is that the Landau free energy expansion must contain extra terms due to these fluctuations. The general form of the Landau free energy becomes

$$G = \frac{1}{2} a (T - T_c)Q^2 + \frac{1}{4} bQ^4 + \frac{1}{6} cQ^6 + \\ \cdots \frac{1}{2} g (\nabla Q)^2$$

where $(\nabla Q)^2$ describes the variation of the order parameter with position in the crystal:

$$(\nabla Q)^2 = \left(\frac{\partial Q^2}{\partial x}\right) + \left(\frac{\partial Q^2}{\partial y}\right) + \left(\frac{\partial Q^2}{\partial z}\right)$$

where x,y,z are spatial co-ordinates. The final term in the energy expansion is the energy due to gradients in Q.

At present there is insufficient data to carry out the analysis of the kinetics of ordering in anorthite in the same way as for omphacite. However, kinetic data obtained from the variation in the enthalpy of solution of anorthite samples with increasing degrees of order, and from the variation in spontaneous strain during ordering, allow determination of the change of Q with time. Approximate solutions to the G–L rate equation can be found yielding an activation energy for Al,Si ordering of $\Delta H_a = 540$ kJ mol^{-1}. As more data become available and more complex ordering processes are understood more minerals will become amenable to the same level of analysis as omphacite.

10.5.3 Application of the G–L rate equation to spinodal decomposition

The Ginzburg–Landau equation quoted as eqn. (10.20) is expressed in its simplest form and makes a number of assumptions which do not hold in all continuous transformations. Although eqn.(10.20) can be successfully applied to omphacite disordering, it must be modified to apply it to other continuous processes, such as the onset of compositional modulations during spinodal composition. Spinodal decomposition is described by a free

energy surface rather similar to that in Figure 10.21, see for example the sequence of curves for the free energy of mixing (Figure 9.5), and should therefore be amenable to similar analysis.

Before this can be done we need to introduce two further concepts. One of the assumptions made in the derivation of eqn.(10.20) is that the *correlation length* ξ of Q is large compared with the *conservation length* ξ_c. These two concepts of length scales arise from the statistical analysis, but also have an important physical significance in transformation processes. The *correlation length* of Q is related to the distance over which a local change in Q will produce correlated changes in adjacent parts of the crystal. The value of the correlation length depends on the extent to which interatomic forces are propagated through the crystal. In framework silicate structures the elastic strains are a very efficient mechanism for communication between different parts of a crystal, ensuring a long correlation length. This point was also made in Section 8.6.5.

The *conservation length* is the distance over which an increase in one component in the crystal is compensated by a decrease in another. For example, in cation ordering an increase in the occupancy of a site by one cation is compensated by a decrease in the occupancy of this cation in an adjacent site. Thus the conservation length in this case is the distance between these neighbouring sites. In order/disorder transformations therefore, the correlation length is large compared with the conservation length and $\xi_c/\xi \sim 0$.

For the growth of spinodal modulations, the conservation length is the length scale of the compositional fluctuations, typically hundreds of ångströms, which is similar to the correlation length. Thus $\xi_c/\xi \sim 1$. The order parameter Q can be regarded as the amplitude of the compositional modulations, and their development with time as dQ/dt. The growth of this amplitude depends on the spatial gradient of the free energy surface and the equivalent of the G–L equation for spinodal decomposition becomes

$$\frac{dQ}{dt} = \frac{-\gamma \lambda \exp(-\Delta G_a/RT)}{4RT} \cdot \zeta_c^2 \nabla^2 \left(\frac{dG}{dQ}\right)$$

$$(10.26)$$

This equation describing the kinetics of spinodal decomposition is referred to as the Cahn equation. The factor $\nabla^2 (dG/dQ)$ describes the way in which the slope of free energy – order parameter surface, dG/dQ, varies with position in the crystal.

The application of this equation depends on knowing the form of the free energy curves with composition i.e. the free energy of mixing, and the conservation length ξ_c which is the wavelength of the compositional modulations.

Both equations (10.20) and (10.26) are special cases of a more general rate equation developed theoretically from the statistics of small changes in a crystal. Its description is beyond the scope of this book, but appropriate references for further reading are given in the bibliography.

Bibliography

BURKE, J. (1965). *The kinetics of phase transformations in metals*. Pergamon Press.

CARPENTER, M.A. AND SALJE, E. (1989). Time dependent Landau theory for order/disorder processes in minerals. *Mineral. Magazine* **53**, 483–504.

LASAGA, A.C. (1981). Rate laws of chemical reactions; Transition state theory; The atomistic basis of kinetics – defects in minerals. Chapters 1,4,7 in *Kinetics of Geochemical Processes*. Eds: A.C. Lasaga and R.J. Kirkpatrick Volume 8 Reviews in Mineralogy. Min. Soc. America.

SALJE, E. (1989). Towards a better understanding of time-dependent geological processes: kinetics of structural transformations in minerals. *Terra Nova* **1**, 35–44.

References

CARPENTER, M.A. (1991). Mechanisms and kinetics of Al,Si ordering in anorthite, I: incommensurate structure and domain coarsening. *Amer. Mineral.* **76**, 1110–19.

CARPENTER, M.A., DOMENEGHETTI, M.C. AND TAZZOLO, V. (1990). Application of Landau theory to cation ordering in omphacite II: kinetic behaviour. *Eur.J.Mineral.* **2**, 19–28.

HANCOCK, J.D. AND SHARP, J.H. (1972). Method of comparing solid state kinetic data and its application to the decomposition of kaolinite, brucite and $BaCO_3$. *Journ. Am. Ceram. Soc.* **55**, 74–77.

McCALLISTER, R.H. (1978). The coarsening kinetics associated with exsolution in an iron-free clinopyroxene. *Contrib. Mineral. Petrol.* **65**, 327–331.

MATTHEWS, A. (1980). Influences of kinetics and mechanism in metamorphism: a study of albite crystallisation. *Geochim. Cosmochim. Acta* **44**, 387–402.

11 Transformation processes in minerals I: exsolution

Minerals transform in many different ways as a result of changes in their physical and chemical environment. Attempts to understand transformation processes have led to classifications based on thermodynamic, structural or kinetic criteria, all of which provide a description of some aspect of the transformation. Two broad categories of transformations are those that involve changes in composition, such as unmixing of a solid solution, and polymorphic transformations in which there is a symmetry change but no change in composition. Often the latter are referred to as phase transitions, particularly when the various polymorphs are structurally related modifications.

Here we will not be overconcerned with the formalities of the various classifications, nor will we attempt to classify rigorously the mineral transformations we describe. Many aspects of transformation processes are common to different transformation types. For example in a thermodynamically first-order transformation (Section 8.4.2) two phases coexist at equilibrium, and are separated by an interface. The formation of the new phase involves a nucleation event, followed by subsequent growth. The structural relationship between the two phases controls the nature of the interface which in turn controls the way the new phase grows and its orientational and spatial relationship to the parent phase. This textural relationship between the two intergrown solid phases is referred to as the microstructure. This process of nucleation and growth can be described in general terms and is relevant to both exsolution processes as well as polymorphic transformations.

In this chapter we introduce general aspects of phase transformations and the factors which control the development of the microstructure. Although much of the discussion is relevant to both exsolution processes and polymorphic transformations, the emphasis here will be on examples of exsolution processes, with polymorphic transformations discussed in the next chapter. In some cases the examples will provide an opportunity for introducing some new concepts not yet mentioned in this book, while in other cases they will bring together ideas which have already been referred to in previous chapters.

The concept of the kinetic control of transformation processes underlies this chapter, and we will build on the more formal aspects of kinetic theory discussed in Chapter 10. It is vital to keep in mind that while the thermodynamics describes the relative stability of phases under any particular set of conditions, and so describes which phase *should* form, the actual behaviour also depends on the kinetics of the processes involved. When a transformation takes place under near-equilibrium conditions the rate is dominated by the thermodynamic driving force for the transformation, whereas far from equilibrium the role of kinetics of the atomic processes involved becomes increasingly important and may result in the formation of metastable phases not shown on an equilibrium phase diagram (Section 8.4.1). The product of a transformation and the resulting microstructure is a result of the compromise between the thermodynamics and the kinetics of transformation processes. This is illustrated in a very simple way in the description of the rate of nucleation of a new phase within a solid undergoing a first-order transformation, discussed in Section 11.1.

11.1 Homogeneous nucleation in solids

Most first-order and reconstructive transformations take place by a mechanism of *nucleation and growth*, where the new phase is initiated by a local fluctuation in composition and/or structure which develops and grows into the parent. The discussion which follows refers to the nucleation of a new phase in a solid solution in which the enthalpy of mixing,

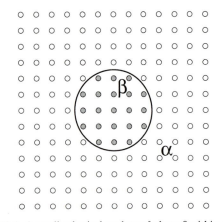

Figure 11.1. A small spherical nucleus of phase β within a matrix of the parent phase α. The nucleus may have a different composition and/or structure to the matrix; in this case the structure and lattice parameters of both phases are approximately the same, although their compositions are different.

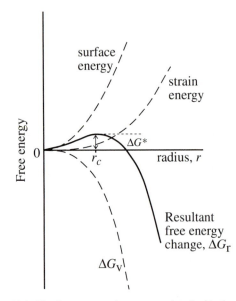

Figure 11.2. The free energy changes associated with the nucleation of a spherical nucleus, radius r. The free energy drive for nucleation ΔG_v reduces the free energy, while the surface and strain energy terms are positive. The resultant free energy change (heavy line) increases to a maximum at a critical value of the radius r_c, before decreasing with further growth. ΔG^* is the free energy of activation for the formation of a critical nucleus.

ΔH_{mix}, is positive and hence the tendency at low temperatures is to unmix into a two-phase inter-growth. The thermodynamics of this situation has been discussed in Chapter 9. Many of the concepts of nucleation are however directly applicable to other first-order transformations as well.

We begin with the simplest case by considering the nucleation of small spherical particles of phase β

within a matrix of the parent phase α (Figure 11.1). The driving force for this transformation arises from the fact that below the transformation temperature T_c, the formation of the new β phase lowers the total free energy by ΔG_v per unit volume of β phase. At the equilibrium temperature T_c, $\Delta G_v = 0$, while below T_c, ΔG_v is negative. The nucleation of the β phase however is opposed by two positive free energy terms which accompany the creation of an interface between the new phase and the matrix. The first is a surface energy term, σ per unit area of surface, which exists due to the surface tension between any two phases in contact. The second is a misfit strain energy term, ε per unit volume of the phase formed, which arises because the new phase will not in general have the same density and hence the nucleus will not occupy the same volume as did the parent.

The overall change in free energy, ΔG_r, on forming a spherical nucleus of radius r is therefore:

$$\Delta G_r = \Delta G_{volume} + \Delta G_{surface} + \Delta G_{strain}$$

$$= \tfrac{4}{3} \pi r^3 \Delta G_v + 4\pi r^2 \sigma + \tfrac{4}{3} \pi r^3 \varepsilon$$

$$= \tfrac{4}{3} \pi r^3 (\Delta G_v + \varepsilon) + 4\pi r^2 \sigma \qquad (11.1)$$

The positive and negative contributions to the free energy and the resultant free energy as a function of radius, is shown in Figure 11.2. This illustrates two important features of this classical nucleation theory. First, that the growth of a nucleus smaller than some critical radius r_c increases the overall free energy because the positive surface and strain energy terms are dominant, and second, that an activation energy barrier ΔG^* must be overcome before the nucleus can continue to grow with a reduction in free energy.

Subcritical particles are referred to as embryos, and correspond to the fluctuations in composition and/or structure which exist within the solid solution near and below T_c. Within the stability field of the high temperature phase such fluctuations can only have a transitory existence and rapidly disperse to be replaced by others elsewhere in the crystal. Within the stability field of the β phase these fluctuations or clusters become potential embyros of β. The formation of a viable critical nucleus depends on the chance event that an embryo acquires sufficient thermal energy to overcome the activation energy barrier and achieve the critical size.

The condition for the continued stable growth of embryos is that at the critical radius r_c

$$\frac{d[\Delta G_r]}{dr} = 0$$

$$\frac{d[\Delta G_r]}{dr} = 4\pi r_{c^2}(\Delta G_v + \varepsilon) + 8\pi r_c\sigma = 0$$

$$\therefore \quad r_c = \frac{-2\sigma}{(\Delta G_v + \varepsilon)} \quad (11.2)$$

Substituting this value of r_c back into eqn. (11.1) gives the activation energy for nucleation

$$\Delta G^* = \frac{16\pi\sigma^3}{3(\Delta G_v + \varepsilon)^2} \quad (11.3)$$

Our main interest is the temperature dependence of r_c and ΔG^*, and since both σ and ε are virtually independent of temperature, the controlling factor will be the temperature dependence of ΔG_v. As illustrated in Figure 8.6 the free energy varies approximately linearly with temperature (it would be exactly linear if both the enthalpy and entropy were temperature independent), and hence $\Delta G_v \propto (T_c - T)$. We can therefore write that

$$r_c \propto \frac{1}{T_c - T} \quad \text{and} \quad \Delta G^* \propto \frac{1}{(T_c - T)^2} \quad (11.4)$$

Figure 11.3 shows the variation in the overall free energy ΔG_r as a function of temperature, and also the way in which the activation energy for nucleation and the size of the critical radius decrease as the undercooling $T_c - T$ increases. At the transformation temperature T_c no nucleation is possible, as both ΔG^* and r_c tend to infinity as T approaches T_c; hence some degree of undercooling is always required. With increased undercooling, the activation energy and the size of the critical nucleus eventually become negligible.

The activation energy for nucleation ΔG^*, which controls the probability that a nucleus of the new phase will form, can also be reduced if the surface and/or strain energy terms are reduced. Any nucleation mechanism which decreases their value will be kinetically more favourable. Broadly speaking, two related factors are important: the structural relationship between the phases and the nature of the nucleation site. In the treatment above, no account is taken of the nucleation site – it is assumed that the nucleation of the β phase takes place at any defect-free point in the matrix, hence the term *homogeneous nucleation*.

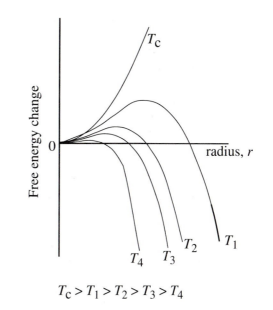

$$T_c > T_1 > T_2 > T_3 > T_4$$

Figure 11.3. The free energy change for nucleation as a function of temperature. At the equilibrium temperature for the transformation ΔG_v is zero and hence the only energy terms are positive. As the temperature falls below the transformation temperature the free energy drive becomes increasingly dominant relative to the surface and strain terms. Both the activation energy for the formation of a nucleus, and its critical radius fall.

11.1.1 Homogeneous nucleation – the coherency of the interface

One aspect of the relationship between the surface energy and the strain energy terms in the case of homogeneous nucleation is illustrated by considering the structural relationship between the precipitating phase and the matrix. *Heterogeneous nucleation*, where the reduction in the activation energy is achieved by nucleation on defects, will be considered in the next section.

(i) The nature of the interface

The structural relationship between the two phases determines the nature of the interface between them. If both phases have a similar structure the new phase may be oriented within the matrix so that there is good lattice matching across the interface. If the two structures are quite different it may be impossible to achieve any structural continuity across the inteface. The degree of lattice matching across an interface is described by the term *coherency*.

To illustrate some of the possibilities, Figure 11.4 shows four cases. In the first case (Figure 11.4(a)) both precipitate and matrix have identical struc-

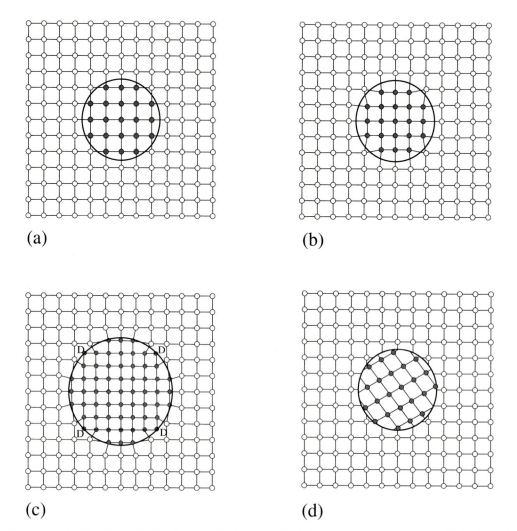

Figure 11.4. Four examples illustrating the structural continuity across the interface between two phases. **(a)** Perfect coherency with negligible lattice strain. **(b)** A coherent interface with strained bonds at the interface. **(c)** A semi-coherent interface with the strain relieved by dislocations at D. **(d)** An incoherent interface with no lattice matching.

tures, and the lattices match perfectly across the interface, so that the strain energy is zero. The surface energy term arises since the difference in chemical composition between the phases means that atoms at the interface of one phase will be partly bonded to different neighbours in the adjacent phase. The activation energy for nucleation however, is small. It is more likely however, that even when the structures are the same, the compositional difference between the precipitate and matrix will mean slightly different lattice parameters for stress-free crystals, and hence the precipitate will be strained in order to preserve the perfect coherency of the interface, as shown in Figure 11.4(b). The coherent interface maintains a low surface energy,

but the strain energy term raises the activation energy for nucleation.

When the structural difference between the phases is too large to be accommodated, it is energetically advantageous to relieve the coherency strains by generating dislocations at the interface. Although the strain energy is reduced, the surface energy is increased, and clearly a compromise needs to be achieved between the two, in order to minimise their sum. Such an interface is termed *semi-coherent* (Figure 11.4(c)). As the structural difference between the phases increases, the number of dislocations required at the interface also increases, until the strain fields around the dislocation cores overlap. At this point the interface is considered to be *incoherent* (Figure 11.4(d)). When there is no lattice matching, the interface has a low strain energy but a high surface energy due to the disordered atomic structure and dangling bonds, just as in a high-angle grain boundary.

(ii) The shape of the precipitate

Another factor which comes into play when considering coherent and semi-coherent interfaces is that the minimum free energy will be attained when the shape of the precipitate and its orientational relationship to the matrix are such as to reduce both surface and strain terms by achieving the best possible lattice matching. As most minerals are anisotropic some interfaces between the new phase and the matrix will have a lower surface energy than others. The new phase will tend to grow such that for any given volume, the total surface energy is minimised by preferential growth of those surfaces with minimum energy. This results in a definite morphology and orientational relationship between the phases. There are many examples of exsolution in minerals where the exsolved phase forms oriented plates or lamellae within the matrix phase. An example is shown in Figure 11.5 which is a high resolution electron micrograph of an exsolution lamella of pigeonite in augite. The lattice matching across the dominant interface is apparent.

In the early stages of nucleation and growth most exsolved phases probably have considerable coherence with the matrix. Continued growth tends to result in a progressive loss of coherence with an increased surface energy term and a less important strain energy term. The coarsening of a small plate-like lamella exsolving from a parent solid solution illustrates this point. If the shape of the growing plate is such that the ratio of its radius r, to its thickness t is constant, the surface energy term is proportional to t^2 and the strain energy term is proportional to t^3.

For a coherent precipitate the surface energy term is very small, and the energy term inhibiting nucleation will be proportional to t^3. For an incoherent precipitate the strain energy term is negligible, so that the energy inhibiting nucleation will be proportional to t^2. As shown in Figure 11.6, the t^3 function is smaller for small values of t, and hence we would

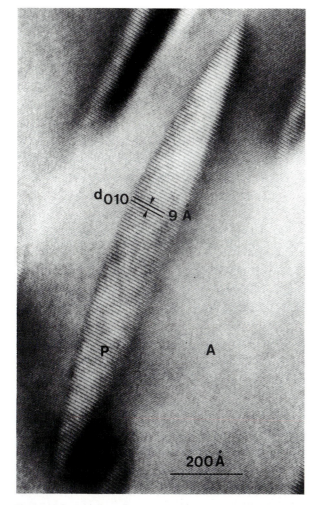

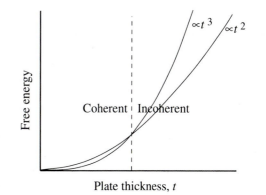

Figure 11.5. A high resolution transmission electron micrograph across the interface between a pigeonite lamella and the augite matrix. In the pigeonite, the 010 lattice fringes are imaged, but in the augite with a C-face-centred cell, 010 is absent and the lattice fringes are 020 (spacing 4.5Å). Note the perfect lattice matching across the interface. (Electron micrograph courtesy of W.F. Müller.)

Figure 11.6. The interfacial free energy terms for a growing lamella as a function of the thickness t. The strain energy is proportional to t^3 and the surface energy term is proportional to t^2. At small values of t, $t^3 < t^2$ and hence a strained coherent lamella is preferred to an incoherent lamella with a high surface energy. As the thickness increases, the situation is reversed and an incoherent lamella has a lower interfacial energy.

expect that in the early stages of nucleation a coherent interface would be most likely. The thickness at which the coherency is lost and an incoherent precipitate becomes more stable depends on the difference in structure between the precipitate and the matrix, as well as on the elastic constants of the two phases. A similar argument is also valid for precipitates with other morphologies, e.g. the loss of coherence of a spherical precipitate will occur as its radius increases.

A final point regarding the shape and orientation of the nucleating phase is the role of the *elastic anisotropy* of the parent phase. In the discussion above we emphasised the growth of minimum energy surfaces in controlling precipitate morphology. However, minimisation of the strain energy term must also be taken into account, and hence exsolution lamellae will tend to form on planes within the matrix that are normal to elastically 'soft' directions.

Examples of orientational relationships between exsolved phases and their host will be found throughout the rest of this chapter.

11.1.2 Heterogeneous nucleation

The positive energy terms associated with nucleation can be reduced if nucleation takes place on defects such as excess vacancies, dislocations, stacking faults, grain boundaries or free surfaces. In each case, if the creation of the nucleus destroys part of the defect, some of its energy will be released and so reduce the activation energy barrier. Defect energies increase in approximately the order given above, and so free surfaces and grain boundaries are the most likely sites for heterogeneous nucleation in a polycrystalline solid. Figure 11.7 shows transmission electron micrographs illustrating heterogeneous nucleation of pigeonite on dislocations in augite, and augite nucleation on antiphase boundaries in pigeonite.

Some of the features of heterogeneous nucleation can be illustrated by considering the nucleation of phase β at a planar grain boundary in the parent α phase (Figure 11.8). In this situation the surface energy is minimised if the nucleus has the shape of a spherical lens with a contact angle θ defined by the

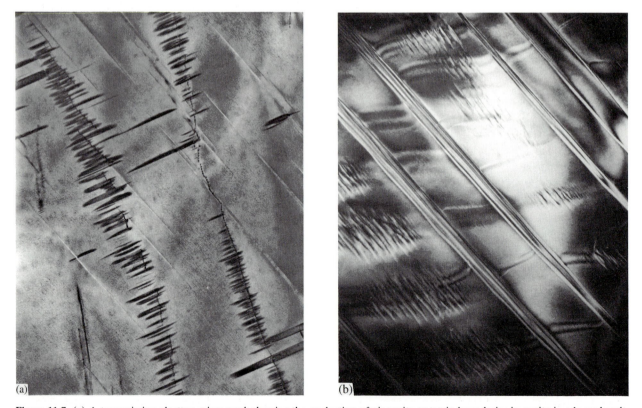

Figure 11.7. (a) A transmission electron micrograph showing the nucleation of pigeonite on grain boundaries in augite in a lunar basalt. (From Nord *et al.*, 1976). (b) Fine augite lamellae nucleated on antiphase domain boundaries in pigeonite. The long thin lamellae are a previous generation of exsolved augite and the antiphase boundaries in the pigeonite are imaged as S-shaped fringes. (From Carpenter, 1978.)

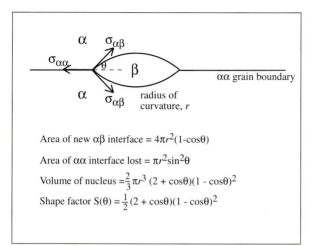

Area of new αβ interface = $4\pi r^2(1-\cos\theta)$

Area of αα interface lost = $\pi r^2\sin^2\theta$

Volume of nucleus = $\frac{2}{3}\pi r^3 (2 + \cos\theta)(1 - \cos\theta)^2$

Shape factor $S(\theta) = \frac{1}{2}(2 + \cos\theta)(1 - \cos\theta)^2$

Figure 11.8. Nucleation of a phase β on a grain boundary in the parent α phase. The shape with minimum surface area is a spherical lens with contact angle θ.

condition that the surface tensions balance in the surface plane. Thus

$$\sigma_{\alpha\alpha} = 2\sigma_{\alpha\beta}\cos\theta$$

or

$$\cos\theta = \frac{\sigma_{\alpha\alpha}}{2\sigma_{\alpha\beta}} \qquad (11.5)$$

The terms $\sigma_{\alpha\alpha}$ and $\sigma_{\alpha\beta}$ are the surface tensions (surface energies per unit area) between α–α and α–β surfaces respectively. Nucleation on the grain boundary creates new surfaces, whose energy per unit area depends on the $\sigma_{\alpha\beta}$ term. This is a positive energy contribution. However, part of the grain boundary is destroyed, and this results in a negative energy contribution dependent on the value of the $\sigma_{\alpha\alpha}$ term. Here we are assuming that both terms are isotropic and therefore not dependent on grain orientation.

If V is the the volume of the nucleus with radius of curvature r, ΔG_v is the free energy reduction per unit volume of β phase, $A_{\alpha\beta}$ is the created area of α–β interface, and $A_{\alpha\alpha}$ is the area of grain boundary destroyed, the overall change in the free energy on heterogeneous nucleation is

$$\Delta G_{het} = -V\Delta G_v - A_{\alpha\alpha}\sigma_{\alpha\alpha} + A_{\alpha\beta}\sigma_{\alpha\beta} \quad (11.6)$$

noting that the first two terms are both negative. As we are assuming an incoherent spherical nucleus we have ignored any strain energy terms. The equivalent expression for the homogeneous nucleation of a spherical particle is

$$\Delta G_{hom} = -V\Delta G_v - A_{\alpha\beta}\sigma_{\alpha\beta} \qquad (11.7)$$

To compare heterogeneous nucleation with homogeneous nucleation of a spherical nucleus with radius r we need to evaluate the relative areas and volumes. Geometrically this is somewhat awkward, but the values are shown in Figure 11.8. From these values it can be shown that

$$\Delta G_{het} = \Delta G_{hom}\, S(\theta) \qquad (11.8)$$

where $S(\theta)$ is a 'shape-factor' and equals $\frac{1}{2}(2 + \cos\theta)(1 - \cos\theta)^2$. For a given value of θ, $S(\theta)$ is a constant and so differentation of eqn.(11.8) with respect to r, in a manner analogous to that in the previous section, results in the same value of the critical radius r_c as for homogeneous nucleation. Thus the value of the critical radius is independent of the grain boundary and hence does not depend on the location of the nucleus. However, the activation energy barrier for heterogeneous nucleation ΔG^*_{het} is reduced by the factor $S(\theta)$ relative to that for homogeneous nucleation, i.e.

$$\Delta G^*_{het} = \Delta G^*_{hom}\, S(\theta) \qquad (11.9)$$

It can also be shown geometrically that $S(\theta)$ is equal to the ratio of the volumes of the lens shape and sphere in Figure 11.9, i.e. the relative volumes of the nucleus for heterogeneous and homogeneous nucleation and so obviously $S(\theta) < 1$. If the angle $\theta = 30°$ for example, $S(\theta) = 2.6 \times 10^{-2}$, and hence the reduction in the energy barrier for heterogeneous nucleation is significant.

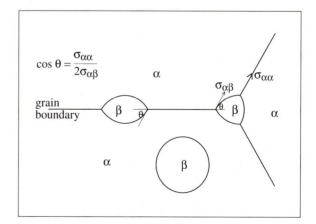

Figure 11.9. Nuclei of a phase β at a triple junction, a grain boundary and in the bulk of the crystal. In each case the radius of curvature for a critical nucleus is the same, but the relative volumes increase from grain junction ⇒ grain boundary ⇒ bulk.

Since the value of the critical radius for nucleation is independent of the location of the nucleus, the activation energy for nucleation on different types of grain boundaries depends only on $S(\theta)$ i.e. on the relative volumes of the critical nuclei. Figure 11.9 shows nuclei at a grain triple junction, a grain boundary and in the bulk of the crystal, all with the same radius of curvature r_c. As can be seen from this figure, the volume of the critical nucleus on a grain junction is smaller than for a nucleus on a grain boundary, which is in turn very much smaller than the volume of a critical nucleus in the bulk. The probability of nucleation will decrease in the sequence grain junction $\Rightarrow$ grain boundary $\Rightarrow$ bulk.

The effectiveness of a grain boundary as a nucleation site depends therefore on the relative surface tensions, and hence on $\cos\theta$ as given in eqn. (11.5). The activation energy barrier for nucleation on a grain boundary would only fall to zero if the ratio $\sigma_{\alpha\alpha}/\sigma_{\alpha\beta} \geq 2$, although it is always less than for homogeneous nucleation. The amount of undercooling required for heterogeneous nucleation is less than for nucleation in the bulk, and therefore nucleation is always expected to begin at grain junctions and boundaries. At large undercooling both heterogeneous and homogeneous nucleation can occur, although nucleation on grain boundaries would be earlier. Although stacking faults and antiphase boundaries can also act as sites for heterogeneous nucleation their effectiveness is considerably less than that of a high energy grain boundary.

11.1.3 The chemical solvus and the coherent solvus

In the thermodynamic discussion in Section 9.2, in which the solvus was derived from the temperature dependence of G–X curves, no mention was made of the mechanistic aspects of nucleation nor of the surface and strain energy terms which must be included. The solvus derived there is termed the *chemical solvus*, and refers to a situation when there is no activation energy barrier to nucleation. As we have seen, both homogeneous and heterogeneous nucleation necessarily involve some degree of undercooling and this depresses the solvus to lower temperatures. For homogeneous nucleation the undercooling will be greater, and the relevant solvus is termed the *coherent solvus*.

Figure 11.10 is a schematic phase diagram showing the relative positions of the chemical and coherent

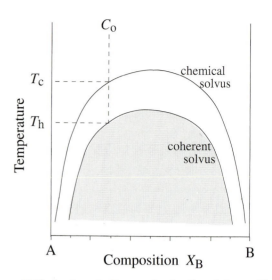

Figure 11.10. A schematic diagram showing the relative positions of the chemical solvus and the coherent solvus. Outside the coherent solvus but within the chemical solvus only heterogeneous nucleation can take place. Within the shaded region both heterogeneous and homogeneous nucleation are possible. A composition C_o reaches the chemical solvus at temperature T_c and the coherent solvus at T_h.

solvus. Exsolution by heterogeneous nucleation is possible at any composition and temperature within the chemical solvus and can take place very near to the equilibrium temperature given sufficient time, although some undercooling will always be required. Within the shaded region both heterogeneous and homogeneous nucleation are likely. During cooling, a solid solution may therefore undergo several episodes of exsolution; the first on grain boundaries early in the transformation, then on other defects such as dislocations and finally by homogeneous nucleation throughout the crystal.

11.1.4 The kinetics of nucleation

To appreciate the role of kinetics in determining transformation mechanisms in minerals, it is convenient to outline the general form of the temperature dependence of these processes. The classical theory of nucleation described here provides a simple example of the general principle.

The probability of forming critical-sized nuclei is exponentially dependent on the activation energy for nucleation and is given by the expression

$$\text{probability } p = \exp\left(-\frac{\Delta G^*}{RT}\right)$$

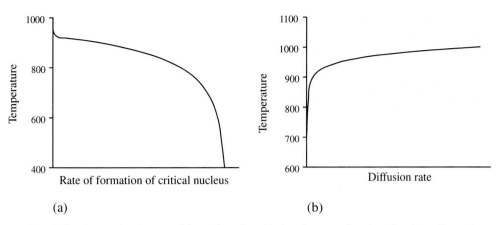

Figure 11.11. (a) The calculated curve for the rate of formation of a critical nucleus as a function of undercooling, using eqn.(11.10) and based on an equilibrium transformation temperature T_c, of 1000°C. **(b)** The calculated form of the curve for the diffusion rate D, using eqn.(11.11).

This standard expression for the Boltzmann distribution was first encountered in Section 7.1.4. The rate of formation of critical nuclei will be simply related to this probability function, so that

$$\text{Rate } c. = \exp\left(-\frac{\Delta G^*}{RT}\right) \qquad (11.10)$$

where c, the pre-exponential constant depends on the number of sites per mole at which embryos can form. This expression takes only the thermodynamic factors into account i.e. the driving force for the transformation ΔG_v and the strain and surface terms as given in eqn. (11.3). At the transformation temperature T_c, ΔG^* approaches infinity and the nucleation rate is zero. As shown in Figure 11.3 and eqn.(11.4), the activation energy G^* decreases with the amount of undercooling $T_c - T$, and the result is a rapid increase in the rate of formation of critical nuclei (Figure 11.11(a)).

No mention has yet been made of the atomic mechanism involved in the process, i.e. the rate at which atoms can diffuse to the nucleation site (in the case of exsolution), or the rate of atomic reorganisation (in the case of a polymorphic transformation). The atomic mechanism for these processes is also associated with an activation energy, E_A as discussed in Section 7.1.4. The temperature dependence of the diffusion rate is given by

$$D = D_o \exp\left(-\frac{\Delta E_A}{RT}\right) \qquad (11.11)$$

where D_o is a constant termed the frequency factor and depends on the vibrational frequency of the atoms, the distance of each atomic jump from one site to another and the diffusion mechanism. Since E_A is a constant, the term $\exp(-E_A/RT)$ decreases rapidly with temperature, becoming zero at 0°K (Figure 11.11(b)). In other words, the mobility of atoms is very strongly temperature dependent, and at low temperatures will inhibit any transformation which involves atomic diffusion.

The overall rate of nucleation, I, is the product of the thermodynamic term controlled by the driving force for nucleation, and the kinetic term dependent on atomic mobility, and is given by

$$\text{Rate of nucleation } I = A. \exp\left(-\frac{\Delta G^*}{RT}\right) \exp\left(-\frac{E_A}{RT}\right) \qquad (11.12)$$

where A is a mechanism-dependent constant. At temperatures near the transformation temperature T_c, the diffusion rate may be sufficiently high but the rate of nucleation is low because of the high value of ΔG^*. Thus at near equilibrium temperatures the rate of nucleation is controlled by the thermodynamic factors and less by diffusion kinetics. At low temperatures, the activation free energy barrier ΔG^* is negligible and a large driving force for nucleation exists, but the low diffusion rates again result in a low nucleation rate. The rate is now kinetically controlled. At some intermediate temperature the nucleation rate increases to a maximum, essentially a compromise between the thermodynamics and the kinetics.

The general form of the nucleation rate as a function of temperature is shown in Figure 11.12(a).

In practice it is more convenient to plot the time required for nucleation rather than the nucleation rate along the horizontal axis of such a plot. The resulting C-shaped curve (Figure 11.12(b),(c)) is characteristic of all diffusion controlled processes in solids.

11.1.5 Growth of precipitates – diffusion control and interface control

After the nucleation stage of a transformation has been accomplished, the growth of the new phase proceeds by the migration of this new interface between product and parent phases. Here a distinction has to be made between two different types of interface. We have already described interfaces as coherent or incoherent depending on the structural relationship between the phases; here we classify interfaces by the way in which they move. *Glissile interfaces* migrate by shearing the parent structure to form the product, as in many displacive transitions, and also in transitions controlled by the migration of shear-induced partial dislocations which generate stacking faults and mechanical twins (see Sections 7.2.8 and 7.3.5). No diffusion is involved in these transitions and their migration is relatively insensitive to temperature and termed *athermal*.

Non-glissile interfaces, on the other hand, grow by transfer of atoms across the parent–product interface and a nucleation barrier exists for their motion i.e. they are *thermally activated*. It follows that such interfaces may be quite mobile at high temperatures, but must become immobile at low temperatures. Most interfaces in mineral transformations are of this type, including those formed in all first-order exsolution and polymorphic transformations.

In a polymorphic transformation, where there is no change in composition, the rate of migration of the interface between the two polymorphs is controlled by the rate at which atoms of the parent phase can cross the interface and become part of the new phase. Such *interface-controlled growth* is characteristic of both displacive and reconstructive polymorphic transformations.

In an exsolution transformation the precipitated phase has a different chemical composition from the parent, and may also have a different structure. Growth of the precipitate requires long-range atomic diffusion through the crystal and this may control the rate at which the interface can migrate. Thus

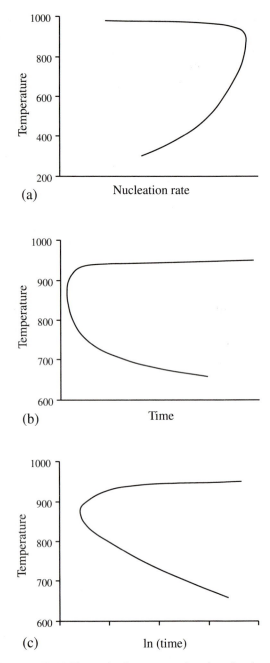

Figure 11.12. (a) The nucleation rate as a function of undercooling, as given by eqn.(11.12), which combines the effect of the thermodynamics and the kinetics, i.e. the product of the two functions in Figure 11.11(a) and (b). The calculations are based on an equilibrium temperature of 1000°C. **(b)** The time to form a nucleus as a function of undercooling. This is the same function as in (a) plotting time to nucleate = (nucleation rate)$^{-1}$. **(c)** Due to the exponential relationship between time and temperature, it is more convenient to plot the diagram using ln (time).

we have *diffusion-controlled growth*. There are circumstances however, when interface-controlled growth can dominate, even when the two phases have different compositions, if their structures are also different. When the interface is incoherent it does not in itself introduce any structural obstacle to growth, and the diffusing atoms can be incorporated in virtually any position. The migration of an incoherent interface is controlled by the rate at which the required atomic ingredients are supplied to it by diffusion. However, the incorporation of a new atom on a coherent or semicoherent planar interface results in an unstable, high energy situation and it is difficult for such interfaces to migrate (Figure (11.13)), even when diffusion is fast enough to supply the necessary atomic ingredients for the new phase. The growth of such precipitates may therefore be interface-controlled.

One result of the lower mobility of coherent and semicoherent interfaces relative to incoherent interfaces is the common observation of the lamellar morphology of exsolved phases in many mineral systems. In the early stages of nucleation and

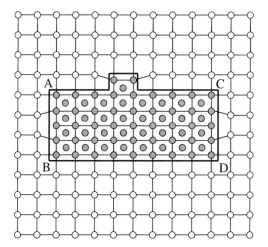

Figure 11.13. A section through a disc-shaped precipitate with coherent interfaces at the top and bottom, but semicoherent interfaces at the edges. The addition of an extra growth element on a coherent interface introduces relatively large surface and strain energy terms and is unfavourable compared to growth at the edges.

growth, small, coherent, disc-shaped precipitates often represent the minimum activation energy. These are bounded by planar coherent faces and curved incoherent edges. The migration rate of these edges is faster and so the precipitate grows along its length forming thin lamellar plates (Figure 11.14).

The thickening of these lamellae is usually achieved by the formation of a stepped interface as illustrated schematically in Figure 11.15(a), and shown in the high resolution electron micrograph in Figure 11.15(b). At each step in this semicoherent interface there is a dislocation. The interface can then grow perpendicular to itself by the lateral migration of these steps in the plane of the interface, although the problem of nucleating new steps restricts this mode of growth relative to that of the lamellar edges. Such stepped interfaces have been observed by electron microscopy in pyroxenes, amphiboles and feldspars.

Another function of these stepped interfaces is to control the degree of coherency between two phases. When the structural misfit becomes significant, large areas of coherent interface become impossible. The degree of coherency is increased by introducing these steps or ledges periodically along the interface. The presence of the steps changes the orientation of the macroscopic interface plane from some rational indices to some irrational plane misoriented by a few degrees from the coherent interface orientation as shown in Figure 11.15(a). The degree of misorientation can be controlled by changing the periodicity of the steps.

When exsolution lamellae grow during continuous cooling, the coherency strains between the lamella and matrix may change with temperature due to relative differences in the expansion coefficients. The ways in which such strain can be accommodated within exsolution lamellae in pyroxenes and alkali feldspars are discussed in the following section.

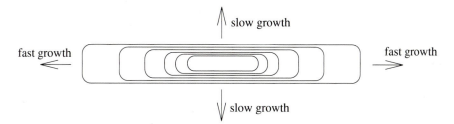

Figure 11.14. The development of a lamellar morphology for a disc-shaped precipitate as a result of slow growth normal to the coherent faces and fast growth normal to the incoherent edges.

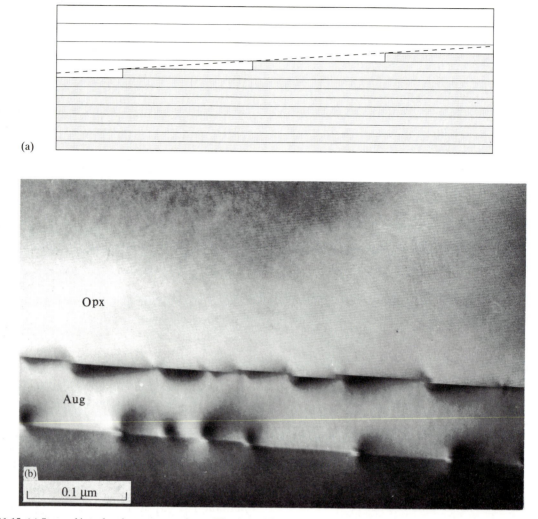

Figure 11.15. (a) Stepped interface between two phases. The dashed line is the average orientation of the interface, which is misoriented by several degrees from the local orientation of the interface. **(b)** Transmission electron micrograph of a stepped interface between an augite lamella (Aug) and the orthopyroxene (Opx) matrix. (From Champness and Lorimer, 1974.) See also Figure 7.42(d).

11.1.6 The accommodation of strain within exsolution lamellae

Example 1. Exsolution in the augite – pigeonite solid solution

In this section we are concerned with the way in which augite–pigeonite lamellae accommodate the strains associated with their coexistence as they exsolve from the parent clinopyroxene solid solution. First we need to bring together some of the ideas mentioned in previous chapters to describe the phase relations in the Ca,Fe,Mg pyroxenes (see Figure 6.12).

The general form of phase diagram for Ca,Fe,Mg pyroxenes is shown as a function of Ca content for a section across the pyroxene quadrilateral at an Fe content of about 30% ferrosilite (Figure 11.16). A broad solvus separates Ca-rich augite and Ca-poor pigeonite due to the non-ideality of the C2/c solid solution. At the Ca-poor end the reconstructive transformation from pigeonite to orthopyroxene results in a eutectoid below which orthopyroxene + augite are the stable phases. However, in all but the most slowly cooled rocks the pigeonite fails to transform to orthopyroxene and is metastably preserved to lower temperatures where it undergoes the displacive high–low pigeonite transformation (Section 7.3.4). Depending on the cooling rate therefore, the high temperature pyroxene solid solution may break down to an intergrowth of orthopyroxene +

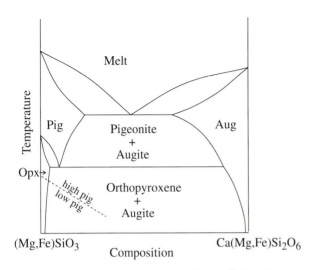

Figure 11.16. General form of the phase relations between Ca-rich and Ca-poor pyroxenes. The high ⇒ low pigeonite displacive transition is a metastable and shown as a dashed line.

augite (the stable state), or low-pigeonite + augite (a metastable state).

Here we consider the latter case when pigeonite exsolves from an augite-rich solid solution by a nucleation and growth process and continues to grow during cooling. During this process Ca,Mg and Fe diffuse through the silicate-chain substructure which can flex to accommodate the various cations, but is essentially preserved intact throughout the exsolution. Hence the pigeonite and augite crystallographic axes remain approximately parallel. The pigeonite grows as lamellae whose orientation within the augite host is determined by the minimum interfacial strain. This depends on the degree of lattice matching across the interface and the anisotropy of the elastic constants for the two phases. In the early stages the lamellae are coherent (as predicted from Figure 11.6) and both phases are elastically strained. The minimum interfacial energy occurs when pigeonite lamellae lie on (001) planes of the augite host. If the supersaturation is sufficiently great, i.e. the driving force for nucleation is high, other less favourable orientations may also be formed. Thus although the interfacial energy for lamellae on (100) planes is about twice as high as that for (001), this difference may be outweighed at high supersaturations.

The coherency strains associated with these early fine-scaled lamellae can be seen in electron diffraction patterns taken over an intergrowth (Figure 11.17(a),(b)). The diffraction pattern is oriented

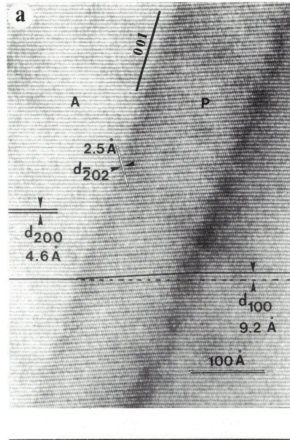

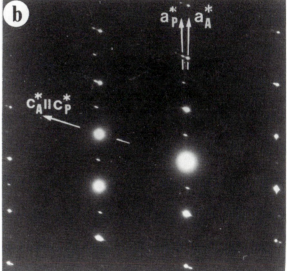

Figure 11.17. (a) A high resolution electron micrograph of a pigeonite lamella (extending from top right to bottom left of the picture) within augite. The interface is approximately parallel to (001). (200) lattice planes are imaged in the augite; in the pigeonite there is additional contrast for (100) fringes. Note the perfect coherency and the orientational change of the ($h00$) fringes across the interface. (b) The corresponding electron diffraction pattern across the intergrowth showing the difference in the β angle for augite and pigeonite. (Electron micrograph courtesy of W.F. Müller.)

correctly relative to the high resolution image which shows a single pigeonite lamella (P) within the augite matrix (A). This figure illustrates a number of important features of such intergrowths and the way they are studied. The topotaxy of the intergrowth (i.e. the sharing of common crystallographic orientations) results in the electron beam being parallel to the b axis of both phases. Hence the diffraction pattern is a superposition of the a^*c^* diffraction patterns of augite and pigeonite. The $h+k$ odd reflections are absent in the augite diffraction pattern due to the systematic absences of the C-centred lattice. Both patterns share a common c^* axis but the angle between the two a^* directions is due to the larger β angle in pigeonite. The $h00$ reflections of pigeonite and augite are joined by a streak of intensity which is due to the gradual bending of the (100) planes across the coherent (001) interface between them. This streaking indicates the degree of coherency.

In the image, the interface between the pigeonite and augite is approximately parallel to (001), since the interface is approximately normal to the c^* axes in the diffraction pattern. The ($h00$) lattice planes are imaged in both the pigeonite lamella and the augite matrix; in augite only the finer (200) fringes are imaged because 100 reflections are systematically absent, while in pigeonite, the coarser (100) fringes are also imaged. The lattice fringes are normal to the a^* directions of each phase. Note that the orientation of the lattice planes changes as they cross the augite–pigeonite interface, resulting in the observed angle between the a^* directions in the diffraction pattern.

As the lamellae grow they lose some coherency with the host and release some of the strain energy while increasing the surface energy term. This relaxation in the two structures allows their lattice parameters to move towards the 'strain-free' value they would have if unconstrained by the intergrowth, and hence introduces some lattice misfit across the interface. As the temperature falls the compositions and the lattice parameters of both phases change – the b lattice parameter varies only slightly across the solid solution while the a and c parameters, and the monoclinic angle β vary with both composition and temperature. To keep the lattice misfit at a minimum requires a slight adjustment to the orientation of the interface.

As shown in Figure 11.18(b), lamellae of

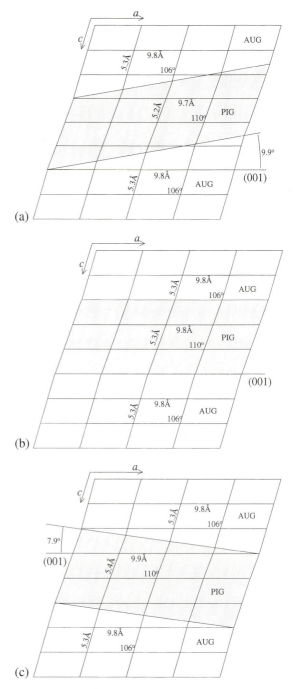

Figure 11.18. Scale drawings to illustrate the exact '001' phase boundary between a pigeonite lamella (shaded) and the augite matrix, depending on the relative a parameters of the unit cells. A variation in ± 0.1Å in the a parameter of the pigeonite changes the interface from $+9.9°$ to $-7.9°$ relative to the exact (001) boundary which is only strain-free when $a_{\mathrm{pig}} = a_{\mathrm{aug}}$. **(a)** $a_{\mathrm{pig}} < a_{\mathrm{aug}}$. **(b)** $a_{\mathrm{pig}} = a_{\mathrm{aug}}$ **(c)** $a_{\mathrm{pig}} > a_{\mathrm{aug}}$. (After Robinson *et al.*, 1977.)

pigeonite can only achieve an exact fit with the (001) planes of augite if the a lattice parameters are equal. If $a_{\mathrm{aug}} > a_{\mathrm{pig}}$ a better fit is obtained if the lamellae are inclined at a small angle (up to several degrees) to (001) away from the c axis (Figure 11.18(a)). Alternatively if $a_{aug} < a_{pig}$ the best fit is obtained when the lamellae are slightly inclined from (001) towards the c axis (Figure 11.18(c)). In these figures the lattices of the two phases are drawn to scale to show that an exact lattice match across the interface can be achieved by an interface orientation which is irrational but near to (001). The interface orientation is then referred to as '001'. To demonstrate the magnitude of the effect the figure shows how a variation of ± 0.1Å in the a lattice parameter of the pigeonite, while keeping the β angle the same, changes the interface from $+9.9°$ to $-7.9°$ relative to (001).

The same approach can be used to determine the exact orientation of pigeonite lamellae on (100) planes of the augite host. In this case the relative values of the c lattice parameters are the principal factor determining the deviation of the interface from (100). Again, the interface will only be exactly (100) if $c_{aug} = c_{pig}$. Using the geometry of coincident lattices as in Figure 11.18, it can be shown that a variation from $+0.1$Å to -0.1Å in the c parameter results in a variation from $+17.2°$ to $-15.0°$ from (100) planes, where the '+' and '−' angles refer to deviations towards and away from the positive a axis

respectively. These results are summarised in Figure 11.19.

Although the general principle of lattice matching is straightforward enough, the lattice parameters of natural clinopyroxenes are complex functions of composition, temperature, pressure and structural state, all of which can be involved in determining the orientation of the best-fitting augite–pigeonite phase boundary. It follows that the phase boundary angle in a natural intergrowth could be used as an indicator of the geological conditions under which it grew, as long as the variation of the lattice parameters with composition, temperature, pressure etc. are known. In general terms augite shows a gradual decrease in both a and c with decreasing temperature, but little change in β, while pigeonite shows more rapid decreases in a and c on cooling, especially towards the C2/c $\Rightarrow$ P2$_1$/c phase transition. At this transition the β angle in pigeonite abruptly decreases by about 1.5°. The net effect of these changes is that above the high-low pigeonite transition $a_{aug} < a_{pig}$ and $c_{aug} < c_{pig}$, while below the transition $a_{aug} > a_{pig}$ and $c_{aug} > c_{pig}$. This makes it possible to estimate the temperature at which exsolution took place.

Figure 11.20 is a high resolution electron micrograph of a '100' lamella of pigeonite in augite. The

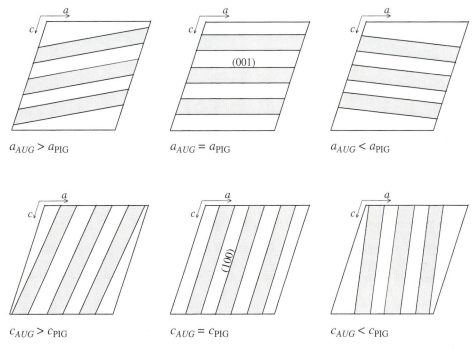

Figure 11.19. Relationship between the relative a and c lattice parameters and the orientation of exsolution lamellae (shaded) in clinopyroxenes. The relative a parameters affect the exact orientation of '001' lamellae, and the relative c parameters affect the exact orientation of '100' lamellae. (After Robinson *et al.*, 1977.)

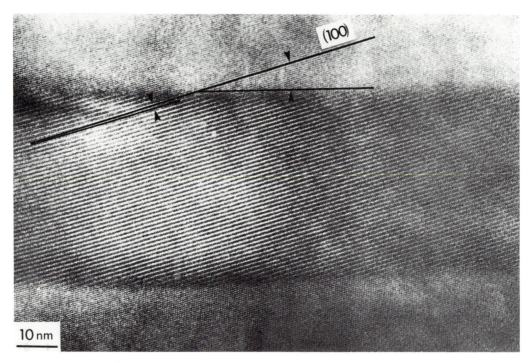

Figure 11.20. A high resolution transmission electron micrograph of a '100' pigeonite lamella in augite. The lattice fringes are (100) in the pigeonite, and (200) in the augite. The interface between the phases deviates by about 15° from the exact (100) orientation. (From Skrotzki *et al.*, 1991.)

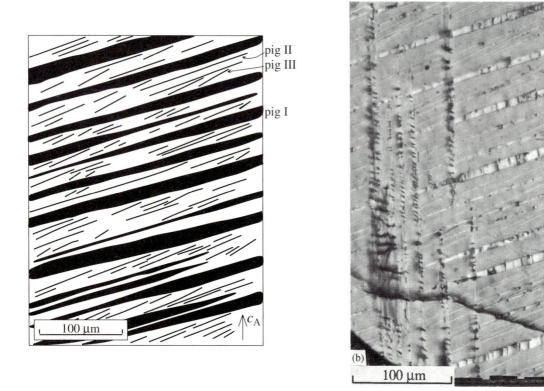

Figure 11.21. (a) Drawing of an optical micrograph of an augite grain containing three generations of '001' pigeonite lamellae, each at a slightly different angle to the exact (001) orientation, indicating three stages of exsolution at progressively lower temperatures. (b) Optical micrograph of a similar microstructure to that drawn in (a) The vertical features are orthopyroxene lamellae. (From Robinson *et al.*, 1977.)

pigeonite–augite interface is inclined at about 15° from the exact (100) lattice planes which are imaged in the pigeonite. This suggests that when the pigeonite lamella exsolved, $c_{aug} > c_{pig}$, a situation which indicates a relatively low exsolution temperature (± 500°C).

In an augite cooled from high temperatures pigeonite exsolution may nucleate on a number of different occasions over a large temperature interval resulting in several sets of '001' lamellae, each at a slightly different angle to the (001) planes of the host augite. This is shown in Figure 11.21.

We have not as yet addressed the question of how a pigeonite lamella exsolved above the high–low pigeonite transition and constrained within an augite host can manage to decrease its β angle by 1.5° as it cools through the transition. Electron microscopy

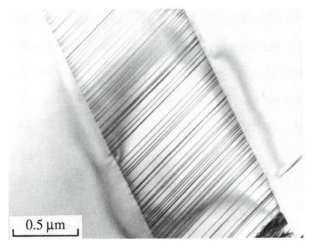

Figure 11.22. A transmission electron micrograph showing (100) stacking faults in a pigeonite lamella within an augite host. (From Robinson *et al.*, 1977.)

has shown that such pigeonite lamellae contain stacking faults on (100) planes (Figure 11.22). These faults relieve the strain by breaking the lamella up into segments which are small enough for the β angle to be able to change without changing the average angle of the interface. This is shown in exaggerated form in Figure 11.23 where each of the segments bounded by the stacking faults has the correct β angle while the stepped interface retains the original orientation. At the termination of the faults there must be a partial dislocation where the strain is locally concentrated. Thicker lamellae need more frequent stacking faults, whose incidence also increases as the β angle increases.

The detailed structure of such a stacking fault and the associated partial dislocations have been studied by high resolution transmission electron microscopy. Figure 11.24 shows two stacking faults in pigeonite terminating at the interface with augite. By carrying out Burgers circuits around the end of the stacking fault it is possible to establish that the partial dislocations are themselves further split into two dislocations with even smaller Burgers vectors, suggesting that the stacking fault is generated in a number of steps. Such observations make it possible to determine the sequence of atomic displacements which take place as the stacking fault propagates through the structure.

Another feature of (100) stacking faults in clinopyroxenes is that they locally change the arrangement of the silica chains from that in clinopyroxene to that in orthopyroxene (Sections 6.4.1 and 7.2.8). Note that pigeonite is metastable relative to orthopyroxene at the temperature of the high–low transition and therefore the energy associated with the

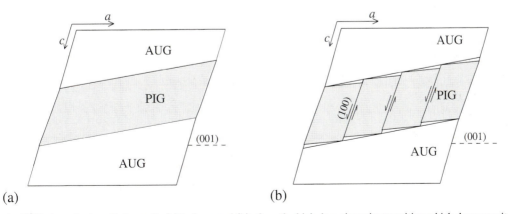

Figure 11.23. An '001' pigeonite lamella in augite (a) before, and (b) after, the high–low pigeonite transition which decreases its β angle by 1.5°. To accommodate this change, the pigeonite lamella undergoes shear displacements on (100) stacking faults. This results in a stepped interface while retaining its average orientation within the augite.

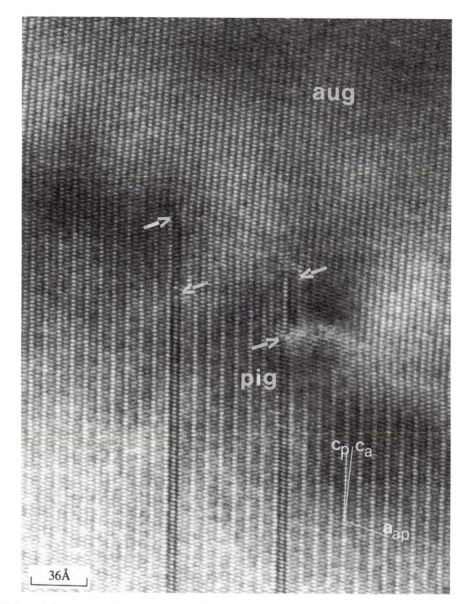

Figure 11.24. A high resolution electron micrograph showing the detailed structure of two stacking faults within pigeonite, terminating at a '001' interface with augite. The arrows point to dissociated partial dislocations at the end of the stacking faults. Note the step in the interface at each stacking fault, and the change in orientation of the *c* axes across the interface. The orientation of the micrograph is the same as in Figure 11.17. (From Livi and Veblen, 1989.)

stacking fault itself will be negative. The transformation from pigeonite to orthopyroxene may be promoted by the presence of stacking faults which are essentially unit-cell thin slabs of the orthopyroxene structure.

Example 2. Albite exsolution lamellae in alkali feldspars

The miscibility gap between Na-rich and K-rich alkali feldspars is shown in the phase diagram in

Figure 11.25. As discussed in Section 6.8.3, pure albite above 980°C is monoclinic C2/*m* but collapses to the triclinic C$\bar{1}$ structure on cooling, the temperature of the transition decreasing with K content. In the early stages of nucleation and growth exsolution of Na-rich feldspar from a K-rich host can take place entirely within the field of monoclinic feldspars. As the temperature falls and the composition of the Na-rich lamella moves towards albite it enters the stability field of the triclinic high-albite solid solu-

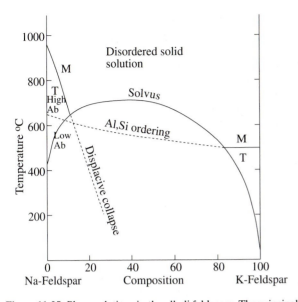

Figure 11.25. Phase relations in the alkali feldspars. The principal features are (i) the monoclinic ⇒ triclinic (M ⇒ T) displacive transition in Na-rich compositions; Al,Si ordering in the high albite leads to ordered low albite, (ii) Al,Si ordering resulting in a monoclinic ⇒ triclinic transition in K-rich feldspars, (iii) a solvus producing a miscibility gap between Na-rich and K-rich compositions. Slowly cooled intermediate compositions exsolve to two phase intergrowths.

tion. Thus we have a situation where a lamella of monoclinic Na feldspar constrained within a monoclinic K-rich host, attempts to collapse to a distorted triclinic structure. The interaxial angle α between the b and c axes changes from 90° in the monoclinic structure to about 93° in triclinic high albite.

The problem of how to accommodate the strain that would be generated at the lamellar interface is again solved by segmenting the lamella, as in the previous example, not by stacking faults however, but by albite twinning. The staggered interface produced by such twinning (Figure 11.26(a)) allows the average interfacial orientation to remain unchanged while the individual segments are able to transform to the distorted triclinic structure. The twinning produces domains distorted in opposite senses, so that the finer the twinning the more similar the 'average' structure is to the monoclinic state. This reduces the strain energy across the interface between the Na-rich and K-rich regions. A further reduction in strain energy is achieved by rounding the corners where the twin planes intersect the interface between the albite lamella and the matrix (Figure 11.26(b)).

The periodicity of the twinning in the albite-rich regions depends on their thickness. At some compositions the Na-rich and K-rich regions coarsen before the displacive transition. If the width of the Na-rich lamellae is variable, this is reflected in the twin periodicity (Figure 11.27), since thinner Na-rich lamellae require a higher twin frequency to accommodate the strain. Since the twin frequency is related to the lamellar width at the time of the displacive transition, the microstructure indicates whether the twinning developed before or after coarsening.

11.1.7 Time – Temperature – Transformation (TTT) diagrams

In isothermal experiments designed to measure the rate of a thermally activated transformation as a function of undercooling, the high temperature phase or solid solution is held at a constant temperature below the transformation temperature T_c. The progress of the transformation is monitored by quenching the sample to low temperature and determining the percentage transformation of the new phase. Repeating this procedure for various annealing times at different temperatures gives an overall Time–Temperature–Transformation (TTT) diagram which describes the progress of a transformation on a graph of temperature T against the logarithm of the time t. In Figure 11.28, the first curve represents the time and temperature at which the new phase is first detectable (at say 5% transformed) and the second curve the time at which the transformation is virtually complete (~95% transformed).

The measured time includes the effects of both nucleation rate and growth rate and since both are thermally activated processes the shape of the TTT curves is essentially the same as the C-curve in Figure 11.12(b). The exponential dependence on temperature means that at low temperatures transformations may never go to completion, even in geological times. This is borne out by the observation of the metastable persistence of high temperature phases at room temperatures.

An experimentally determined TTT curve can be used to obtain estimates of the two activation energy terms which control the transformation rate, namely an activation energy related to the thermodynamic driving force and an activation energy for the atomic

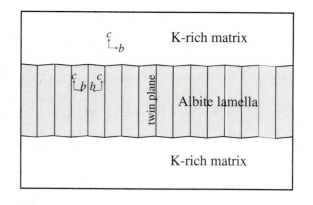

(a)

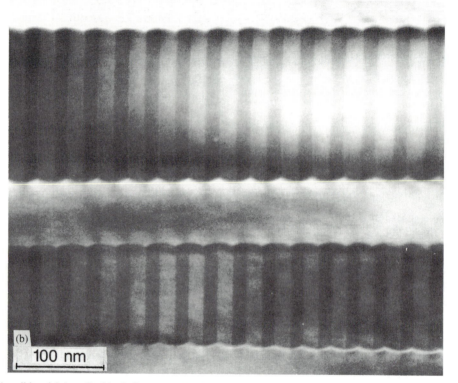

Figure 11.26. (a) An albite-rich lamella (shaded) may exsolve from a K-rich host, while both phases retain their monoclinic symmetry. When the albite undergoes the structural collapse to triclinic symmetry, the strain is accommodated by albite twinning, resulting in a zig-zag interface. (b) To reduce the strain at points of high curvature the interface becomes corrugated, with rounded corners as shown in the electron micrograph. (From Brown and Parsons, 1988.)

reorganisation. We can rearrange the basic rate equation (eqn.(11.12)) as follows:

Rate of nucleation

$$I = A \cdot \exp\left(-\frac{\Delta G^*}{RT}\right) \exp\left(-\frac{E_A}{RT}\right)$$

$$\ln I = \ln A - \left(\frac{\Delta G^*}{RT}\right) - \left(\frac{E_A}{RT}\right)$$

Since the time t for a given fraction to transform is

inversely proportional to the nucleation rate, we may write

$$\ln t = \frac{\Delta G^*}{RT} + \frac{E_A}{RT} - \ln A' \qquad (11.13)$$

Differentiating

$$\frac{d(\ln t)}{d(1/T)} = \frac{\Delta G^*}{R} + \frac{E_A}{R} + \frac{1}{RT}\left[\frac{d(\Delta G^*)}{d(1/T)}\right]$$

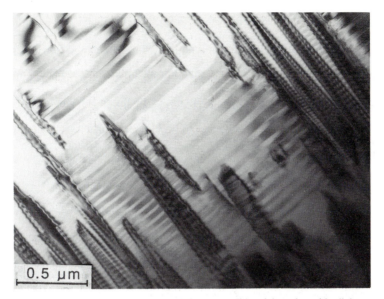

Figure 11.27. Transmission electron micrograph of an intergrowth between albite-rich regions (the lighter contrast twinned phase) and K-rich lenses. The albite twin periodicity varies with the local width of the Na-rich regions, showing that twinning developed after coarsening. (From Brown and Parsons, 1988.)

For large undercooling $\Delta G^* \sim 0$, and therefore

$$R\left[\frac{d(\ln t)}{d(1/T)}\right] = E_A \qquad (11.14)$$

Thus if we plot $\ln t$ against $(1/T)$ as shown in Figure 11.28(b), the activation energy term E_A may be found from the slope of the linear part of the curve at large undercooling. Note that the low temperature linear part of the TTT curve corresponds to the 'time to a given fraction' method of analysing the kinetics of phase transformations, as discussed in Section 10.4.

The linear part of the curve is described by the equation

$$\ln t = \frac{E_A}{RT} - \ln A' \qquad (11.15)$$

From equations (11.13) and (11.15)

$$\Delta G^* = RT(\Delta \ln t) \qquad (11.16)$$

This enables the free energy barrier ΔG^* to be determined at any temperature T by measuring the distance ($\Delta \ln t$) between the extrapolated straight line and the curve, as shown in Figure 11.28(b). When the curve becomes linear at increased undercooling, this energy barrier falls to zero.

The values of E_A and ΔG^* determined in this way are empirical activation energies for some combination of processes which control the overall transfor-

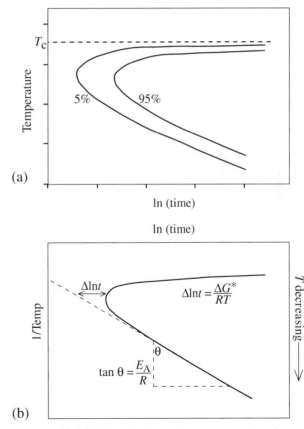

(a)

(b)

Figure 11.28. (a) Schematic Time – Temperature – Transformation (TTT) diagram describing the relative kinetics of a thermally activated transformation as a function of undercooling below the equilibrium temperature T_c. The two curves indicate the time taken achieve 5% and 95% completion of the transformation during isothermal heating. (b) The TTT diagram plotted as ln (time) against $1/T$ (kelvins).

mation rate. If the low temperature rate is diffusion controlled, then E_A may be the activation energy for diffusion. However, as we have mentioned above, the growth rate of a new phase may be interface-controlled in which case the measured E_A value would relate to the process of transfer of atoms across the interface between the two phases. Many mineral transformations are quite complex and may involve a number of different atomic processes, each with a different activation energy. The measured activation energy may be that for the slowest and therefore rate-controlling process, or it may be a function of the activation energies of a number of processes. This point has been discussed in more detail in Section 10.2.1.

11.1.8 TTT curves for heterogeneous and homogeneous nucleation

The depression of the coherent solvus for homogeneous nucleation relative to the chemical solvus for heterogeneous nucleation (Section 11.1.3) results in a TTT diagram as shown in Figure 11.29. This diagram refers to a composition C_o in Figure 11.10 for which the exsolution temperatures are T_c and T_h for heterogeneous and homogeneous nucleation respectively. The TTT curves for the two nucleation mechanisms will therefore be asymptotic to different maximum temperatures. This temperature difference depends on the relative magnitudes of the surface and strain energy terms involved in nucleation. Clearly, separate curves could be drawn for heterogeneous nucleation on grain boundaries, dislocations, and point defects, each with a slightly lower transformation temperature, but our present understanding of such details does not warrant this.

The slopes of the lower parts of the two curves are likely to be similar if atomic diffusion in both cases takes place through the bulk of the crystal. For heterogenous nucleation and growth, diffusion may also take place along grain boundaries and this would reduce the activation energy term and hence the slope of the low temperature part of the TTT curve. The relative positions of the two TTT curves along the time axis depend on the availability of suitable sites for nucleation. In Figure 11.29 homogeneous nucleation is shown to be a slower process than heterogeneous nucleation, although there have been experimental studies suggesting that at low temperatures homogeneous nucleation becomes

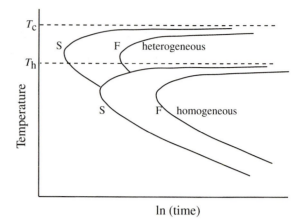

Figure 11.29. Schematic TTT diagram of the relative kinetics of heterogeneous and homogeneous nucleation, showing curves for the start (S) and finish (F) of the transformation. T_c and T_h are the temperatures of the chemical and coherent solvus respectively, for a given composition (cf. Figure 11.10).

faster than heterogeneous nucleation. Another factor which influences this is the concentration gradient of the atoms required to produce the exsolving phase. If heterogeneous nucleation on grain boundaries depletes the immediate surroundings of the necessary constituents, homogeneous nucleation in the bulk may be faster than the diffusion times required to supply grain boundary nuclei from the bulk of the crystal. We will return to the problems of concentration gradients and diffusion in Section 11.4 below.

Many of the factors which control the magnitude of the activation free energy for heterogeneous nucleation are not always predictable and depend a great deal on the precise nature of the mineral specimen (i.e. the distribution of grain boundaries, dislocations, impurities etc.). Experiments involving the onset of heterogeneous nucleation are therefore not always reproducible. Homogeneous nucleation, while less dependent on extrinsic properties will also depend on the atomic defects which control bulk diffusion and may vary with the thermal history of the specimen.

TTT diagrams are experimentally determined from isothermal experiments and hence indicate the relative times at which different processes will begin when the sample is annealed isothermally. In nature, we are concerned with minerals undergoing exsolution while cooling from high temperatures and it is usual to superimpose different cooling rates on TTT diagrams to show, in a schematic way, the temper-

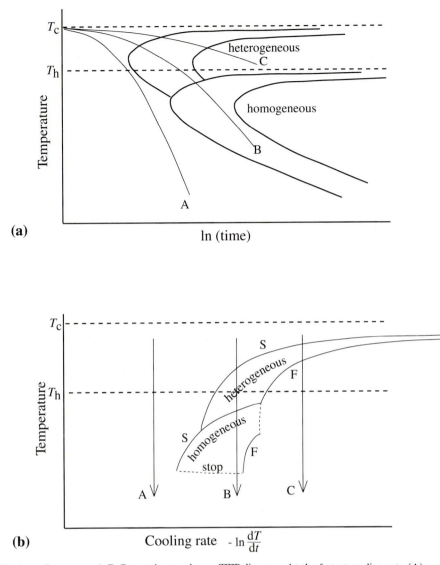

Figure 11.30. **(a)** Three cooling curves A,B,C superimposed on a TTT diagram. At the fastest cooling rate (A) no nucleation occurs; at cooling rate B the transformation begins by heterogeneous nucleation, followed by homogeneous nucleation before the transformation is complete; at cooling rate C, the transformation is achieved entirely through heterogeneous nucleation. **(b)** Cooling curves A and B superimposed on a Cooling rate – temperature – transformation (CTT) diagram (After Price, 1980.)

atures and times at which different processes will begin. An example is shown in Figure 11.30(a) which shows that for the slowest cooling rate (C) exsolution has gone to completion by heterogeneous nucleation and growth, while for the more rapid cooling rate (B) exsolution begins heterogeneously and is completed by homogeneous processes. At the fastest cooling rate (A) nuclei do not have time to form.

Strictly speaking it is not appropriate to use iso-thermally determined TTT curves for continuous cooling. An alternative would be to determine TTT curves for different cooling rates, as is done for metal alloy systems. A schematic diagram showing a cooling rate – temperature – transformation diagram for the onset of nucleation for the cooling rates A, B, and C is shown in Figure 11.30(b).

11.1.9 TTT curves with rising temperature

TTT curves are not restricted to situations involving the formation of new phases on cooling. The transformation from the low-temperature to the high-temperature state can also be shown. This is particu-

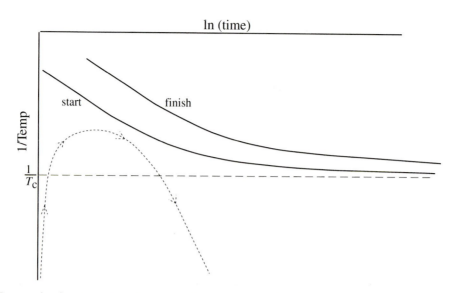

Figure 11.31. TTT curves for the start and finish of a transformation which takes place with rising temperature. T_c is the equilibrium transformation temperature. The dotted curve is a heating and cooling cycle which reaches a temperature above T_c but for which no transformation would be observed.

larly relevant to polymorphic transformations where we may be interested in the time taken to convert the low temperature phase to the high-temperature phase at various amounts of superheating ΔT above the equilibrium temperature T_c.

For small values of ΔT the rates of transformation will be dominated by the overall change in the free energy for the transformation. Near T_c the driving force will be low and the activation energy for nucleation will be very high. Consequently, transformation rates will be very sluggish, with infinite transformation times at T_c. With increasing temperature however, transformation rates will increase and will continue to increase as the diffusion rate also increases with temperature. The general characteristics of the TTT diagram associated with rising temperature are shown in Figure 11.31. Thus in a heating and cooling cycle over a relatively short period (as shown by the dashed line), no evidence of transformation will be observed at temperatures above T_c. Only in a longer-term heating and cooling cycle is it likely that the transformation will take place under near-equilibrium conditions. This type of diagram is relevant to polymorphic transformations and the behaviour of minerals during thermal and regional metamorphism.

The slope of the curve at temperatures far from equilibrium reflects the activation energy for diffusion. In Figure 11.32, a complete TTT diagram for a polymorphic reaction is shown, both for the low $\Rightarrow$

high transformation above T_c, and the high $\Rightarrow$ low transformation below T_c. The slopes of the two parts of the curve are the same at temperatures further from T_c, assuming that the diffusion mechanism is the same. Notice that the increased diffusion rate at high temperatures means that the amount of superheating required for a reaction to proceed will always be less than the amount of undercooling for the reverse reaction.

An example of an experimental metamorphic reaction which can be illustrated in this way is the breakdown of the layer silicate talc, $Mg_3Si_4O_{10}(OH)_2$ on heating. At high temperatures talc is not stable relative to the assemblage enstatite + quartz + vapour, but this decomposition reaction usually takes place via the intermediate formation of the amphibole, anthophyllite $Mg_7Si_8O_{22}(OH)_2$. With longer periods of heating, the anthophyllite ultimately breaks down to enstatite + quartz + vapour. Above about 750°C at 1kbar pressure the stable reaction is

$$\text{talc} \Rightarrow \text{enstatite} + \text{quartz} + \text{vapour}$$
$$Mg_3Si_4O_{10}(OH)_2 \Rightarrow 3MgSiO_3 + SiO_2 + H_2O \quad (1)$$

The intermediate reactions are:

$$\text{talc} \Rightarrow \text{anthophyllite} + \text{quartz} + \text{vapour} \quad (2)$$

$$7Mg_3Si_4O_{10}(OH)_2 \Rightarrow 3Mg_7Si_8O_{22}(OH)_2 + 4SiO_2 + 4H_2O$$

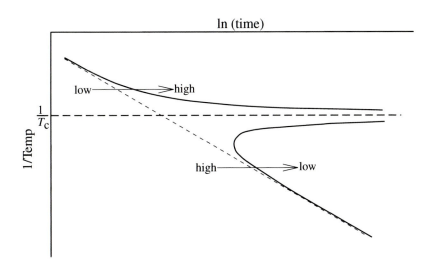

Figure 11.32. TTT curves for a transformation from the low ⇒ high temperature form above T_c, as well as the high ⇒ low transformation below T_c. The slopes of the linear parts of the curves at temperatures far from equilibrium are equal, indicating the activation energy for the transformation process. Near T_c both curves tend to infinity.

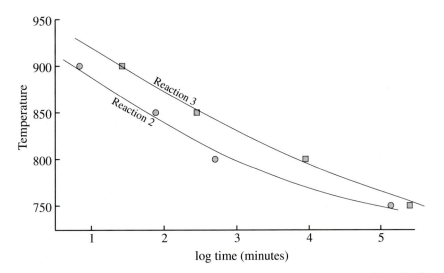

Figure 11.33. Experimental TTT curves for the reactions: talc ⇒ anthophyllite + quartz + vapour (Reaction 2), and anthophyllite ⇒ enstatite + quartz + vapour (Reaction 3). (Data from Greenwood, 1963.)

and

$$\text{anthophyllite} \Rightarrow \text{enstatite} + \text{quartz} + \text{vapour} \quad (3)$$
$$\text{Mg}_7\text{Si}_8\text{O}_{22}(\text{OH})_2 \Rightarrow 7\text{MgSiO}_3 + \text{SiO}_2 + \text{H}_2\text{O}$$

Data for the time taken for reactions (2) and (3) to be 50% completed is shown plotted on the TTT diagram in Figure 11.33. Several important points are illustrated here. First, for the reactions to occur at equilibrium temperatures (to which the curves tend at the lowest temperatures) requires very long experimental reaction times. Nearly six months are required before reaction (3) is 50% completed at 750°C. Second, at temperatures above 750°C where

enstatite + quartz + vapour is clearly the more stable assemblage, anthophyllite is the first phase to form in the breakdown of talc. Shorter term experiments could lead to the wrong conclusion that anthophyllite is the stable phase under these conditions.

Finally, the stability field of anthophyllite at 1kbar should be defined by the temperature interval between the two curves at the equilibrium temperatures for the reactions. The curves in Figure 11.33 are not conclusive on this point, but indicate that the stability of anthophyllite under these conditions can only be very limited, and the kinetics of forming anthophyllite within its stability field would be very sluggish. The most recent thermodynamic data on these reactions suggest that at 1kbar, anthophyllite is stable between about 685°C and 704°C.

11.2 Diffusion

Thermally activated transformations require atomic diffusion from one site in a mineral structure to another. We have already discussed several aspects of diffusion and its importance, both in terms of atomic mechanisms (Section 7.1.4) and the activation energy barrier and its effect on the transformation kinetics. The bulk diffusion of atoms results from thermal oscillations which may have sufficient amplitude to cause an atom to jump randomly from one site to the next. At high temperatures random fluctuations or clusters of atoms may exist throughout the structure, but if there is no free energy reduction associated with such clusters, they will not persist. As soon as some more stable state exists, certain atomic configurations or clusters will become more stable than others, potential gradients will be set up and hence there will be a greater tendency for atoms to jump in certain directions than in others. The net result is a macroscopic flow of atoms down a potential gradient. Such a statistical drift is ultimately responsible for the processes described above.

In this section we describe diffusion in terms of the mass transport, rather than the atomic mechanisms, and so define in a more formal way concepts such as the diffusion coefficient. The temperature dependence of the diffusion coefficient controls the rate of many mineral transformations as well as the microstructures which result from chemical gradients.

11.2.1 Atomic flux, Fick's laws and the diffusion coefficient D

Consider in Figure 11.34 a sequence of atomic planes in a structure, and assume that a given atom may jump with a frequency Γ from one plane to another, with jumps in either direction equally probable. For simplicity we will consider only a two-dimensional situation. If a concentration gradient exists normal to the planes we can write that the concentration of atoms in plane 1 is C_1 and that in plane 2 is C_2 with $C_1 > C_2$. C is measured in terms of

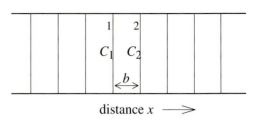

Figure 11.34. Atomic planes 1 and 2 with atomic concentrations C_1 and C_2 respectively. b is the interplanar spacing.

the number of atoms per unit volume, and so the number of atoms in plane 1 is therefore $n_1 = C_1 b$ and in plane 2, $n_2 = C_2 b$. In a small increment of time δt, the number of atoms leaving plane 1 is $n_1 \Gamma \delta t$. On average one half of these jumps is in each direction so that the number of atoms jumping from plane 1 to plane 2 is $\frac{1}{2} n_1 \Gamma \delta t$. Similarly the number of atoms jumping from plane 2 to plane 1 is $\frac{1}{2} n_2 \Gamma \delta t$. The flux J is the net flow of atoms which pass through the unit area in unit time, and is therefore given by

$$J = \frac{1}{2}(n_1 - n_2)\,\Gamma$$

Thus if $n_1 = n_2$ there is no net flux and the system is homogeneous. It is only when some inhomogeneity such as a nucleus of a new phase is stabilized that a net flux is established.

In terms of concentration

$$J = \frac{1}{2}b(C_1 - C_2)\Gamma$$

In practice the value of C for a given plane is not measurable, but if the concentration is measured at different points through the sample and is assumed to vary continuously with the distance x, we can write that the concentration gradient

$$\frac{\partial C}{\partial x} = -\frac{C_1 - C_2}{b}$$

and therefore

$$J = -\frac{1}{2}b^2\Gamma\,\frac{\partial C}{\partial x}$$

The flux J is now related to the experimentally measurable concentration gradient. The ratio of $-J/\partial C/\partial x$ is defined as the *diffusion coefficient D* (with units $m^2 s^{-1}$).

$$D = -\frac{J}{\partial C/\partial x} \quad \left(= \frac{1}{2}b^2\Gamma\right).$$

or

$$J = -D\frac{\partial C}{\partial x} \qquad (11.17)$$

Equation 11.17 is known as *Fick's first law* and simply states that the rate of flow is proportional to the concentration gradient. The minus sign indicates that the flow is from high to low concentrations. In an isotropic mineral the diffusion coefficient is independent of direction and there is only one value of D. For an anisotropic mineral the diffusion coefficient may vary by many orders of magnitude in different directions. The diffusion coefficient is a second rank tensor and must be described by the methods discussed in Chapter 2 (see Section 2.4).

Fick's first law describes a steady state situation in which the concentration at each point does not change as a function of time. More commonly we are concerned with cases where concentration does change with time. As a simple illustration we can consider the growth of an exsolution lamella rich in B atoms during cooling of an A–B solid solution. As the precipitate is richer in B than the solid solution, its formation will result in a depletion of B atoms within the matrix at the interface with the lamella. Atoms of B will diffuse from further away to make up this depletion only to be absorbed by the growing lamella. Thus the concentration of B around the lamella decreases with time until at equilibrium the gradient is zero throughout the matrix. The general form of the diffusion profiles around a lamella at times t_1, t_2 and t_3 during isothermal growth is shown in Figure 11.35.

To treat this more general situation we need a new equation which describes how C varies with both position and time. If we take a small part of one of the diffusion profiles of Figure 11.35 and consider points x_1 and x_2 a distance Δx apart (Figure 11.36(a)), we can determine the flux due to the changing concentration gradient along the profile. The flux J_1 at x_1 coming out of the region is higher than the flux J_2 going into the region. This is illustrated in Figure 11.36(b). If there are more atoms of B going out of the region than there are coming in, the concentration of B between x_1 and x_2 must decrease. The decrease in the number of atoms of B in this volume in a time increment δt is

$$(J_1 - J_2)\, A\, \delta t = A\, \Delta x\, \delta C \qquad (11.18)$$

where A is the cross sectional area through which the flux is flowing and hence $A\,\Delta x$ is the volume of the

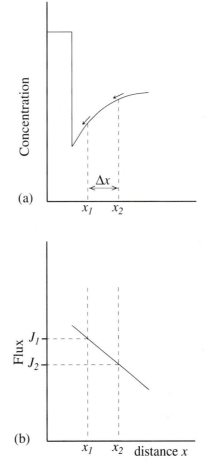

(a)

(b)

Figure 11.36. (a) Part of the diffusion profile in Figure 11.35, the arrows indicating the diffusion direction of B atoms. (b) The resulting atomic flux J as a function of distance x from the lamella.

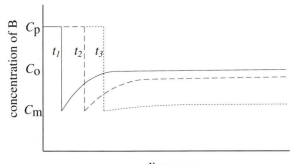

distance x

Figure 11.35. A sequence of composition profiles around a lamella at times $t_1 < t_2 < t_3$. C_p and C_m are the equilibrium concentrations of element B in the lamella and matrix respectively; C_o is the bulk composition of the original homogeneous solid solution.

region considered. Since J varies continuously with x, J_1 and J_2 are related by

$$J_2 = J_1 + \frac{\partial J}{\partial x} \Delta x \qquad (11.19)$$

for small values of Δx.

In the limit as $\delta t \Rightarrow 0$, we can combine eqns. (11.18) and (11.19) to give

$$\frac{\partial C}{\partial t} = -\frac{J}{\delta x} \qquad (11.20)$$

Substituting the equation for Fick's law (eqn. (11.17)) into eqn. (11.20) gives the relationship for the change in concentration with time as a function of the concentration gradient:

$$\frac{\partial C}{\partial t} = \frac{\partial}{\partial x}\left(D \frac{\partial C}{\partial x}\right) \qquad (11.21)$$

If D is assumed to be independent of the concentration and hence x we can write

$$\frac{\partial C}{\partial t} = D \frac{\partial^2 C}{\partial x^2} \qquad (11.22)$$

This equation is known as *Fick's second law*. The physical interpretation of this equation is quite straightforward. The term $\partial^2 C/\partial x^2$ is the curvature of the diffusion profile. Therefore whenever the concentration profile is concave downwards in a given region, the concentration decreases with time. This is the situation described by the set of diffusion profiles in Figure 11.35.

11.2.2 Solutions to Fick's laws and the determination of D

In order to apply Fick's Laws the equations must be solved for the appropriate experimental conditions. The theory of diffusion controlled processes such as nucleation and growth and spinodal decomposition consists of deriving and solving a diffusion equation modified by the thermodynamic requirements of the process (i.e. the free energy changes involved). The equations can also be modified to take into account elastic and interface parameters. Once the appropriate diffusion and thermodynamic parameters are known it is theoretically possible to predict the scale of the microstructures developed by certain thermal treatments and conversely to define the thermal history from a study of the microstructures.

In this section we will outline a form of the solution to Fick's equations for one important case which forms the basis for most experimental determinations of the diffusion coefficient D. This consists of setting up a diffusion couple in which a specimen with a certain elemental concentration C_1 is physically joined to another in which the concentration of this element is C_2. The details of this diffusion couple are important in distinguishing between several different diffusion coefficients. For example, if we make a couple by joining together olivine of composition Mg_2SiO_4 with olivine of composition Ni_2SiO_4, Mg will diffuse in one direction and Ni in the other until equilibrium is reached. One type of atom cannot move independently of the other and the movement of both species contributes to the rate of homogenization and hence to the value of D obtained from Fick's second law. A value of D obtained in this way is called the *interdiffusion coefficient*.

To determine the diffusion coefficient of Mg in pure Mg_2SiO_4 (the *self-diffusion coefficient*) it is necessary to set up a diffusion couple between two samples of Mg_2SiO_4 in which one of the samples has been prepared with an excess of the stable isotope Mg-25. Thus both samples are identical apart from this isotope difference, and the diffusion of the isotope into the specimen of lower concentration is then used to measure the diffusion coefficient. Strictly speaking this is the *tracer diffusion coefficient* (i.e. the diffusion of the isotope Mg-25) but in practice it is virtually identical to the self-diffusion coefficient.

The general form of the diffusion profiles for such a case as a function of time are shown in Figure 11.37. As expected from Fick's second law, the concentration rises steadily in those parts where the curvature of the profile is concave upwards ($\partial^2 C/\partial x^2$ is positive), falls in those parts where the curvature is concave downwards ($\partial^2 C/\partial x^2$ is negative), and remains constant where the curvature is zero. Experimentally, we measure the composition profile at various times and so obtain the concentration $C(x,t)$ as a function of both the distance from the boundary x and the time t. The solution to the diffusion equation which fits the experimental conditions and describes these profiles is

$$C(x,t) = \frac{C_1 + C_2}{2} + \frac{C_1 - C_2}{2} \operatorname{erf}\left(\frac{x}{2\sqrt{Dt}}\right)$$

$$(11.23)$$

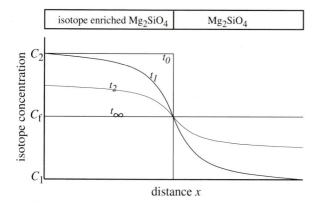

Figure 11.37. A diffusion couple between isotopically enriched Mg_2SiO_4 (isotopic concentration C_2) and standard Mg_2SiO_4 (isotopic concentration C_1). The final equilibrium distribution of the isotope at $t = \infty$ is C_f. Profiles of the isotope concentration as a function of time are shown for $t_0 \Rightarrow t_1 \Rightarrow t_2 \Rightarrow t_\infty$.

where 'erf' is the abbreviation for 'error function' which is a mathematical function values of which can be looked up in mathematical tables.

We are not concerned here with the details of this equation, except to point out one very important result. Our main interest is in determining the distance over which diffusion can operate in a given time, termed the *diffusion distance*. If we choose any composition between the start and finish concentrations (i.e. between C_1 and C_f or C_2 and C_f in Figure 11.37) and calculate the distance x from the boundary at which it will be found, we obtain the result that

$$x \approx \sqrt{D.t} \qquad (11.24)$$

where t is the time during which diffusion has been taking place. In the absence of a rigorous solution to Fick's laws, this result allows us to estimate the distance over which diffusion will affect the composition around a growing exsolution lamella, for example, in a given time. Since the scale of the microstructures associated with exsolution processes is controlled by diffusion and the cooling rate, this relationship allows simple quantitative estimates of D to be made in experimental systems where times are known and the separation of lamellae can be measured. Conversely, if the diffusion coefficient is known, the scale of the exsolution microstructure may be used to estimate the thermal history. Some aspects of the scale of exsolution microstructures are discussed in Section 11.5.

11.3 Spinodal decomposition

Spinodal decomposition was introduced in Section 9.2.1 in the context of the regular solution model which predicts that in a compositional region, defined by the inflexion points in the free energy curves, a solid solution can unmix by a continuous mechanism which does not require a discrete nucleation event. While a nucleation event can be considered as a fluctuation in composition which is very localised in extent but large in amplitude, spinodal fluctuations are very large in extent but small in amplitude.

A comparison of the compositional growth of spinodal fluctuations with nucleation and growth is shown schematically in Figure 11.38. As an example, we could consider that the diagram illustrates the development of an albite lamella due to exsolution from an alkali feldspar solid solution. Two features are illustrated in this figure. First, in the nucleation process local equilibrium is achieved in a single step, with the formation of a sharp interface separating

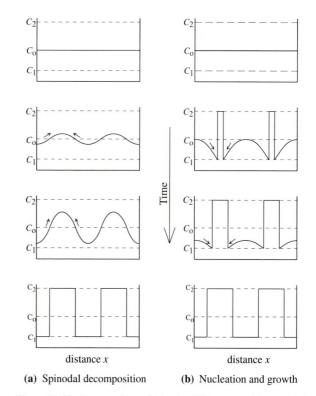

(a) Spinodal decomposition **(b)** Nucleation and growth

Figure 11.38. A comparison of **(a)** spinodal decomposition and **(b)** nucleation and growth, in terms of the development of compositional profiles with time. C_o is the initial composition of the solid solution and C_1 and C_2 are the equilibrium compositions of the exsolved two-phase intergrowth. The final lamellar intergrowth is the same in both cases.

two phases whose compositions near the interface closely approach the equilibrium values. Around the nucleus a concentration gradient develops due to local depletion of Na (in this case), and subsequent growth of the nucleus requires a continuous supply of Na from the matrix. The diffusion of Na is driven by the concentration gradient around the lamella.

In spinodal decomposition there is no distinct interface between the crests and troughs of the compositional modulations and in the early stages the solid solution remains a single phase. In order to build up the amplitude of the compositional fluctuations Na must diffuse up the concentration gradient. This 'uphill' diffusion is contrary to our general expectations but is driven by the reduction in free energy associated with forming compositional fluctuations within the spinodal (Section 9.2.1). As the fluctuations grow in amplitude and the compositions of the crests and troughs approach near-equilibrium values (within ~80% of the values defined by the coherent solvus), the compositional profiles square up and the resulting lamellae may be indistinguishable from those formed by nucleation.

A theoretical treatment of spinodal decomposition is beyond the scope of this book, but a key feature is that the diffusion coefficient is proportional to d^2G/dx^2 which is negative between the two spinodal points of the free energy curve. Thus the diffusion coefficient is negative and there is no barrier to the development of compositional fluctuations. Taking the diffusion term alone, the fluctuations would grow exponentially with time and the shorter the wavelength of the fluctuation, the faster it would grow. However the unrestrained growth of short wavelength fluctuations is controlled by two positive energy terms which would result. A short wavelength fluctuation with a large amplitude (i.e. a large compositional deviation from the mean) has very steep local concentration gradients which have two effects. First the concentration gradient itself is associated with a *gradient energy* term. This comes about because the close proximity of Na-rich and K-rich regions increases the number of Na–K interactions which raise the energy in a system where like neighbours are energetically preferred. The second term is a *strain energy* term which is due to the close coexistence of regions of different composition and hence unit cell size.

The outcome of these competing factors is a compromise which favours a particular value of the

wavelength for compositional fluctuations and allows this to grow in amplitude. The value of this wavelength depends on the relative values of the above terms and on the degree of undercooling below the spinodal curve.

Spinodal decomposition is an important process in the early stages of unmixing in mineral solid solutions between two end members which have the same structure, and hence with a continuous free energy curve between them. This includes solid solutions in virtually all of the important rock-forming mineral groups. In many cases spinodal modulations are observed in mineral solid solutions rapidly cooled from high temperatures, as in volcanic rocks. In other cases a combination of low temperatures for exsolution and sluggish diffusion rates means that unmixing has not been able to proceed beyond the spinodal stage even in minerals cooled slowly over geological time scales.

The scale of spinodal modulations (of the order of hundreds of ångströms) and their small compositional difference requires transmission electron microscopy for their observation. The modulations appear as periodic variations in diffraction contrast with diffuse boundaries (Figure 11.39(a),(b)). The diffraction contrast is due to the slight variations in lattice spacings associated with the compositional fluctuations. The electron diffraction pattern shows a single phase, i.e. one reciprocal lattice, but modified by the addition of *satellite reflections* around the principal diffraction spots. Since, as we discussed in Chapter 3, the reciprocal lattice must contain all the information about the structure, it must also contain the periodicity of the modulations. As expected in a reciprocal lattice, the satellite reflections are oriented at right angles to the modulations, and their distance from the main reflections is the reciprocal of the modulation wavelength. This is illustrated in Figure 11.39.

The orientation of the spinodal modulations is defined by the need to minimise the elastic strain energy and hence corresponds to elastically 'soft' orientations in the crystal. Where symmetry demands, compositional modulations form in two or three directions giving rise to 'tweed' microstructures in electron microscope images (Figure 11.39(b)). Even when not required by symmetry, more than one modulation may occur. For example in the earliest stages of spinodal decomposition of clinopyroxenes modulations form along both (001)

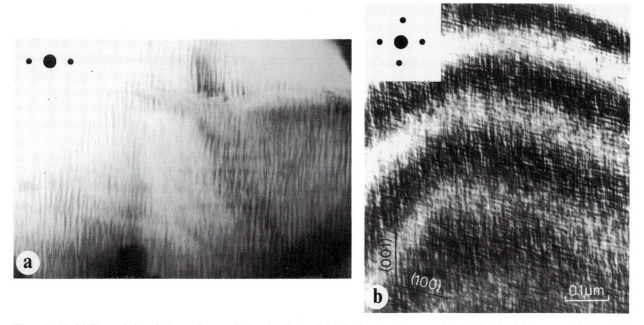

Figure 11.39. (a) Transmission electron micrograph showing the modulated microstructure associated with spinodal decomposition parallel to ($\bar{6}01$) planes in alkali feldspar. The inset shows the diffraction effects associated with this modulation. Each main reflection has a pair of satellites in a direction normal to the modulation. (b) The early stages of spinodal decomposition on both (001) and (100) planes in the augite-pigeonite solid solution. In the diffraction pattern, each main reflection has two pairs of satellite reflections (inset). (From Nord *et al.*, 1976.)

and (100) planes. As their growth develops the modulations in the (100) orientation eventually decay leaving only the (001) modulations which continue to coarsen. As noted in Section 11.1.6 the interfacial energy due to compositional variations on (100) planes is about twice as high as that for (001).

11.3.1 The chemical spinodal and the coherent spinodal

The inclusion of the strain and gradient energy terms into the discussion of the thermodynamics of spinodal decomposition requires some modification of the position of the spinodal curve on the composition – temperature phase diagram. In Figure 9.8 the spinodal curve was defined only on the basis of the locus of the inflexion points in the free energy curve. We will call this the *chemical spinodal* by analogy with the chemical solvus (Section 11.1.3). The position of the spinodal curve when the strain energy is taken into account will be depressed in temperature relative to the chemical spinodal, and is termed the *coherent spinodal*. The relative positions of the chemical and coherent spinodals, as well as the chemical and coherent solvus curves is shown in

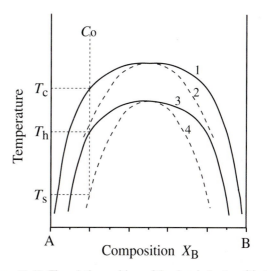

Figure 11.40. The relative positions of the chemical solvus (1), the chemical spinodal (2), the coherent solvus (3) and the coherent spinodal (4). A composition C_o, cooling from above the solvus temperature encounters the chemical solvus at T_c, the coherent solvus at T_h and the coherent spinodal at T_s.

Figure 11.40. The chemical spinodal has no practical application since there is no strain-free or heterogeneous spinodal process, whereas the chemical solvus corresponds to a strain-free heterogeneous nucleation process.

The depression of the coherent spinodal temperature relative to the chemical spinodal arises because of the increase in the free energy associated with the strain of coexisting compositional modulations. To attempt to evaluate this requires a modification to the model of the free energy of the solid solution e.g. the regular solution model, by including strain terms. For the case of an isotropic solid the elastic energy is given by an equation of the form

$$G_e = \eta^2 Y(X - X_o)^2 V$$

where $\eta = 1/a \, (da/dx)$ and a is the cell parameter (assuming a cubic unit cell). Hence η measures the way in which the lattice parameter varies with composition. Y is Young's modulus along the elastically 'soft' direction along which the spinodal modulations form, X_o is the average composition of the solid solution and X is the composition of the inhomogeneities. V is the molar volume. When this elastic term is added to the free energy of the solid solution $G(X)$ the total free energy of the solid solution becomes

$$G(X) + \eta^2 Y(X - X_o)^2 V$$

The addition of a positive energy term which varies as the square of the concentration moves the common tangent points inward and is responsible for a coherent solvus which lies totally within the chemical solvus. Similarly, the inflexion points move inwards and define the coherent spinodal within the chemical spinodal curve.

The difference in temperature between the chemical spinodal (T_{chem}) and the coherent spinodal (T_s) can be estimated by taking the second derivative of the total free energy with respect to composition hence defining the coherent spinodal for which

$$\frac{d^2G}{dX^2} + 2\eta^2 YV \leq 0 \qquad (11.25)$$

In general Young's modulus Y depends on direction in a crystalline material, and hence the temperature of the coherent spinodal will vary with the wavevector of the modulations. The maximum temperature will occur for the minimum value of $2\eta^2 YV$ (the 'soft' direction). To evaluate the difference between the temperature T_{chem} of the chemical spinodal and that T_s of the coherent spinodal we need to expand d^2G/dX^2 in terms of T (from the regular solution model of the solid solution, Section 9.1.3) and

substitute into eqn. (11.25), noting that at T_{chem}, $d^2G/dX^2 = 0$. The result gives

$$T_s - T_{chem} \sim -2\eta^2 \, YV/ \, (\partial G''/\partial T), \text{ where } G'' \text{ is } \frac{d^2G}{dX^2}.$$

Since $\partial G/\partial T = -S$, we can write

$$T_s - T_{chem} \sim 2\eta^2 \, YV/S'' \qquad (11.26)$$

where S'' is the second derivative of the molar entropy with respect to composition, i.e. d^2S/dX^2. Since $S'' < 0$, $T_s < T_{chem}$, as shown in Figure 11.40.

An example of how this result can be used in describing the spinodal decomposition of the Fe-Ti oxide spinel solid solution is discussed in Section 11.3.4.

11.3.2 TTT curve for spinodal decomposition

Although there is no activation energy barrier to the initiation of spinodal modulations, their development is controlled by atomic diffusion and by the fact that the value of the wavelength which is allowed to grow depends on the degree of undercooling below the coherent spinodal curve. At the coherent spinodal temperature T_s where the energy terms are all balanced the spinodal wavelength is infinite, and the rate of decomposition is zero. Just below the spinodal curve only very long wavelength and small amplitude modulations are possible. These can form only very slowly due to the long diffusion distances involved. As the temperature decreases, the increasing driving force provides the extra energy to allow increased gradient and strain energy terms and shorter wavelength fluctuations become possible. The decreased diffusion distance increases the rate at which these fluctuations can form. At much lower temperatures the growth of even short wavelength fluctuations becomes slow due to the influence of falling temperature on the diffusion rate.

The general characteristics of the TTT curve for spinodal decomposition (Figure 11.41) is therefore similar to that for nucleation – a C-shaped curve whose low temperature limb has a slope defined by the activation energy for diffusion. At each temperature below T_s the compositional modulations have a characteristic wavelength which remains constant with time as their amplitude increases. When a solid solution is cooled continuously there will be a range of wavelengths which can develop and the modulations are therefore less clearly defined. The satellite reflections in diffraction patterns will also not be as sharp, reflecting the range of periodicities present in the sample.

The relative kinetics of spinodal decomposition, homogeneous and heterogeneous nucleation can be shown on a schematic TTT diagram (Figure 11.42) in which the temperatures T_c, T_h and T_s refer to the temperatures at which the composition C_o in Figure 11.40 intersects the chemical solvus, coherent solvus and coherent spinodal respectively. Although spino-

dal decomposition is a faster process than nucleation and growth, it can only take place under non-equilibrium conditions when a solid solution is cooled too quickly for nucleation to be able to take place.

Although noting the reservations already made regarding drawing cooling curves on isothermal TTT diagrams (Section 11.1.8) it is nevertheless instructive to compare decomposition of a solid solution for the four cooling curves A,B,C and D in Figure 11.42. At the fastest cooling rate (A), the solid solution is quenched without any decomposition. Although such a specimen may have random compositional fluctuations sufficient time has not been allowed for periodic spinodal fluctuations to develop. Below a certain cut-off temperature T_f the solid solution will be preserved in this state indefinitely. At a slightly slower cooling rate, spinodal decomposition will be the only process able to operate. Spinodal modulations of this kind are often preserved in minerals from many terrestrial and lunar rocks. At cooling rate C, heterogeneous nucleation of the equilibrium phase will begin at relatively high temperatures, but the process will not have been completed before the curve for homogeneous nucleation is intersected. The rest of the exsolution process will then take place by this mechanism. At the slowest cooling rate D, exsolution will

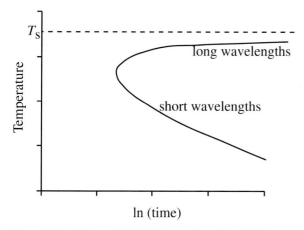

Figure 11.41. Schematic TTT diagram for spinodal decomposition.

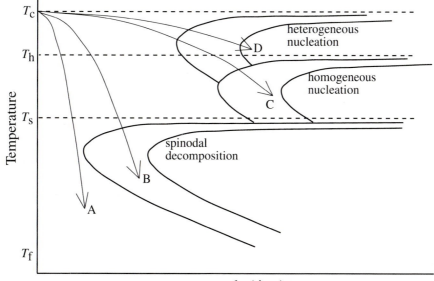

Figure 11.42. Schematic TTT diagram illustrating the relative positions of the curves for heterogeneous nucleation, homogeneous nucleation and spinodal decomposition. The temperatures T_c, T_h and T_s refer to the equilibrium transformation temperatures for a composition C_o in Figure 11.40. A, B, C, D are curves with decreasing cooling rates.

be entirely by the heterogeneous nucleation of the equilibrium phase.

11.3.3 Some cautionary thoughts on spinodal decomposition and nucleation models

Up to this point we have regarded nucleation and growth, and spinodal decomposition as very different and separate processes whose compositional and temperature regions of existence are defined by sharp lines in a phase diagram. These are simplistic models, useful for describing the principal features of these processes, but unrealistic in detail. For example, the classical nucleation theory introduced in Section 11.1 makes the assumption of a sharply defined interface for the nucleus, even when the undercooling is large enough for the critical radius r_c to be so small that the notion of a distinct interface must be quite artificial. The classical nucleation theory cannot deal with long-wavelength fluctuations in composition.

Our model of spinodal decomposition is based on the regular solution model of a solid solution. We have already pointed out some of the limitations of this model in Sections 9.2 and 9.6. As more rigorous models are used, especially when the atomic interactions driving the exsolution process are extended beyond nearest-neighbour pairs, the shape of the free energy curves becomes modified such that the inflexion points defining the spinodals disappear. In other words, the presence of the spinodal depends on the free energy model used. While the equilibrium phases are clearly defined thermodynamically, the spinodal itself is a non-equilibrium concept and depends on the kinetics of the decomposition process.

An important general principle is that under non-equilibrium conditions the path taken by a solid solution as it attempts to achieve equilibrium is irreversible and depends on the past thermal history of the sample. These paths cannot be uniquely defined in the same way that equilibrium states are. Nevertheless, we can recognise 'typical' decomposition mechanisms for which our simple models may serve as useful first approximations. Nucleation and growth is typical of solid solutions evolving just below the equilibrium temperature, while spinodal decomposition is characteristic of solid solutions unmixing at large supersaturations, i.e. at temperatures and compositions well inside the miscibility

gap. These are two extremes in a spectrum of exsolution behaviour which involves the development of compositional fluctuations which can vary in both amplitude and extent.

This behaviour is consistent with a wider general principle governing the equilibration of large numbers of interacting particles, namely that at small departures from equilibrium a system reduces its free energy through random large amplitude fluctuations (as in nucleation), while at large departures from equilibrium a system evolves by forming metastable periodic structures (as in spinodal decomposition). Further discussion along these lines may be found in articles in the bibliography.

11.3.4 Examples of spinodal decomposition in minerals

As mentioned above, many mineral solid solutions which have been cooled relatively quickly in nature, such as in volcanic eruptions, as well as experimentally studied solid solutions, show modulated structures when examined by transmission electron microscopy. These are consistent with spinodal decomposition, although there are very few studies in which experimental verification of the essential features of spinodal decomposition have been demonstrated. The three examples below illustrate some of these aspects of spinodal decomposition.

Example 1. Spinodal decomposition in the augite – pigeonite solid solution

Exsolution in the augite – pigeonite solid solution has already been discussed (Section 11.1.6) and the early stages of spinodal decomposition has been illustrated in the electron micrograph in Figure 11.39(b). We return to clinopyroxenes to describe an experimental verification of spinodal decomposition in isothermally annealed synthetic clinopyroxene in which compositional modulations of fixed wavelength increased in amplitude as a function of time. Furthermore the wavelength decreased with decreasing annealing temperature, in accordance with the theory.

The synthetic augite was prepared at 1200°C and quenched before annealing at a number of temperatures for periods up to 5 months to determine the TTT diagram for the decomposition process. All samples were studied by transmisssion electron microscopy. The starting material already contained

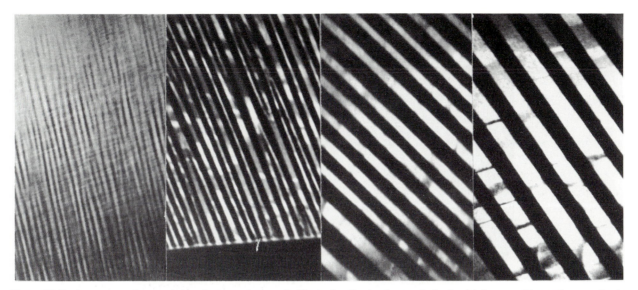

Figure 11.43. A sequence of transmission electron micrographs illustrating the development of spinodal decomposition and coarsening in experimentally annealed clinopyroxene. The annealing temperature was 1000°C for increasing times up to 5.5 months. (From Nord and McCallister, 1979; see also Buseck *et al.*, 1980.)

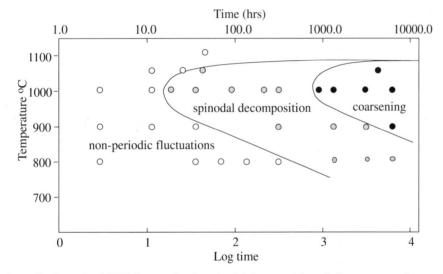

Figure 11.44. Experimentally determined TTT diagram for the spinodal decomposition of clinopyroxene of composition $Wo_{25}En_{31}Fs_{44}$ (from Nord and McCallister, 1979, see also Buseck *et al.*, 1980).

some weak non-periodic fluctuations and the onset of spinodal decomposition on annealing was defined as the development of periodic fluctuations with a clearly defined wavelength and satellite reflections in the diffraction pattern. As heating continues the diffraction contrast of the modulations increases, due to the increased strain as the amplitude increases. The final stage of the process at this scale is the beginning of coarsening of the lamellae when the periodicity of the lamellar microstructure

increases. Throughout this whole sequence the augite – pigeonite lamellae remain completely coherent with the host.

Figure 11.43 is a sequence of electron micrographs showing the sequence of microstructures, and Figure 11.44 is the experimentally determined isothermal TTT diagram. The form of the kinetics of decomposition is consistent with expectations from the theory.

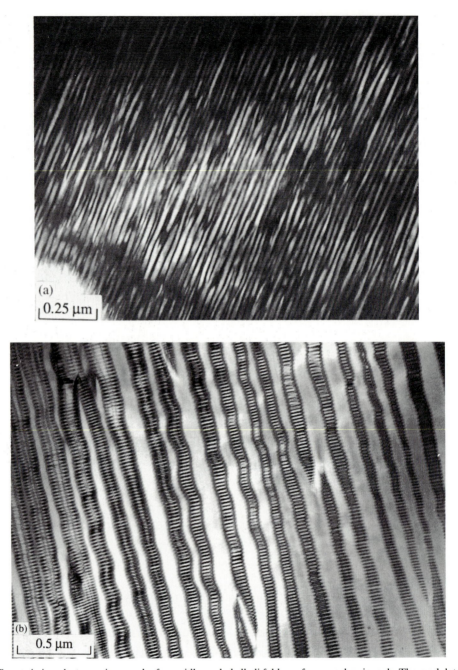

Figure 11.45. (a) Transmission electron micrograph of a rapidly cooled alkali feldspar from a volcanic rock. The modulated microstructure is consistent with spinodal decomposition (from Parsons and Brown, 1984). (b) With coarsening of the microstructure the Na-rich component develops periodic albite twinning, and the lamellae have a wavy interface (from Lorimer and Champness, 1973.)

Example 2. Spinodal decomposition and microstructural development in alkali feldspar solid solution

The alkali feldspar solid solution fulfills the conditions for spinodal decomposition by having a miscibility gap at intermediate compositions, separating two isostructural phases. The phase diagram is shown in Figure 11.25. Alkali feldspars from volcanic rocks appear homogeneous under an optical microscope, but transmission electron microscopy shows a typical spinodal texture (Figure 11.45(a)) which has also been reproduced by experiment. The modulations lie on $(\bar{6}01)$ which can be shown, using methods similar to those in Section 11.1.6 and

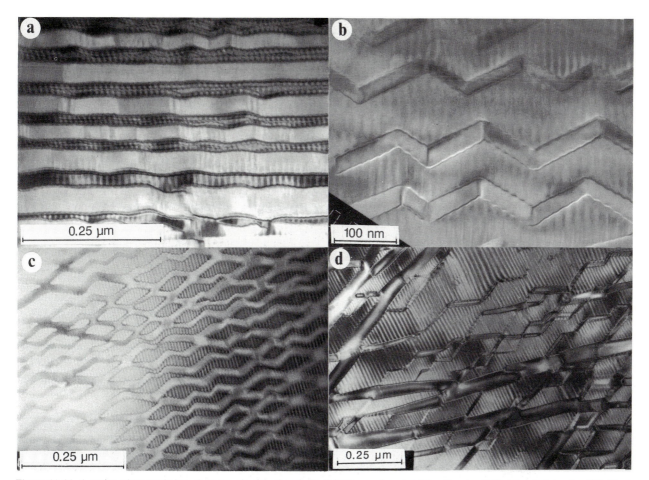

Figure 11.46. A series of transmission electron micrographs of alkali feldspar microstructures from a single igneous intrusion (Klokken syenite, Greenland) as a function of the distance from the margin. (a) Parallel and slightly wavy Na-rich and K-rich lamellae at a stage when both phases are locally triclinic. The twinning in the paler K-rich phase tends to correlate with the waves in the interface. (b) Zig-zag lamellae of low albite and microcline. (c), (d) Further coarsening breaks up the zig-zag albite lamellae into isolated albite lozenges. Note that the frequency of the twinning in the albite phase is independent of the width of the albite lamella, confirming that the twinning developed before coarsening. See also Figure 3.43. (From Smith and Brown, 1988.)

Figure 11.18, to be the plane of minimum strain between two monoclinic feldspars with slightly different cell parameters.

The alkali feldspar solid solution is not a simple solvus however, and the exsolution textures which develop by coarsening of the spinodal microstructure are complicated by the structural phase transformations which take place in the Na-rich and K-rich phases in the intergrowth. Aspects of the structural changes in the two end-members have been discussed in Sections 6.8.3 and 7.3.5, and here we look at these transformations again in a different context, namely the way in which the strain determines the microstructure of an intimate intergrowth. The microstructural development described in this section is rather generalised but consistent with a set

of samples taken from rocks with progressively slower cooling rates, such as would be found by sampling an igneous intrusion from the margin to the interior. It could also describe a sequence of microstructures developing from a homogeneous solid solution which had been rapidly cooled into the spinodal region to avoid nucleation but then allowed to coarsen under a slow cooling rate.

After development of the spinodal modulations the growth in amplitude and wavelength leads the Na-rich phase, which in the early stages has the undistorted monoclinic structure, into the stability field of the collapsed triclinic structure. Meanwhile, the K-rich component of the modulation remains monoclinic, since the larger K ion prevents the structural collapse. The Na-rich phase may to some

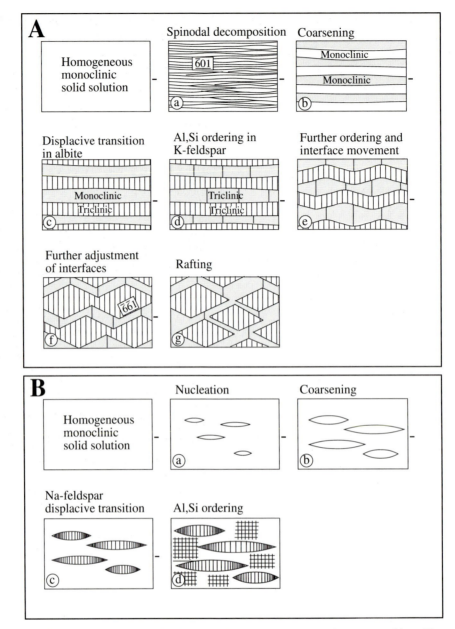

Figure 11.47. Schematic diagram illustrating the development of the microstructures in cryptoperthites. The upper sequence (A) summarizes the sequence under non-equilibrium conditions when spinodal decomposition occurs, the lower sequence (B) at near-equilibrium conditions with nucleation as the exsolution mechanism. (Based on Willaime *et al.*, 1976 and Brown and Parsons, 1984.)

extent be constrained from collapsing by the surrounding monoclinic intergrowth, although it eventually does transform, producing fine-scaled albite twins to minimise the strain on the boundary (Figure 11.45(b)). The situation is similar to that in Figure 11.26. Since the Na-rich phase has uniform thickness, the periodicity of the albite twins is also uniform and depends only on the degree of distortion from the monoclinic cell (i.e. the obliquity). With the fine scaled twinning the average strain

distribution throughout the lamella remains that of a monoclinic phase.

The next stage in the process involves Al,Si ordering in the triclinic Na-rich lamellae and in the monoclinic K-rich component. This involves a further increase in the interaxial angle α in the Na-rich component, and also the progressive distortion of the K-rich component to triclinic symmetry. When K-felspar is not constrained by its surroundings, the monoclinic to triclinic transformation results in a fine-scaled cross-hatched twinning (Section 7.3.5), but here, under the influence of the Na component, rather coarse albite twinning is the usual response to the interfacial strain. This produces triclinic regions in two twin-related orientations. The minimum strain interface can no longer remain the same as when both phases were monoclinic, and must also satisfy these two different twin orientations. The interface responds by changing its orientation to fit alternate K-feldspar domains, giving a wavy or zig-zag appearance to the interface (Figure 11.46).

Samples with a longer thermal history have more pronounced development of this zig-zag interface which is formed by the interface alternating on $(\overline{6}61)$ and its twin-related plane (Figure 11.46(b)). This is also in good agreement with the predicted interface of best fit for the two triclinic lattices. The zig-zags finally break up into lozenge-shaped islands of the albite-rich phase within a matrix of ordered microcline (Figure 11.46(c),(d)). Note that the albite twinning in the Na-rich phase is constant within these lozenge shapes, indicating that they formed within uniformly thick lamellae prior to the triclinic distortion in the surrounding matrix.

Throughout this sequence all intergrowths remain fully coherent and their morphology is totally controlled by the need to reduce the interfacial strain associated with the structural changes taking place, noting however that the time available dictates the extent of microstructural development. The sequence is summarised in schematic form in Figure 11.47. Clearly, the initial composition of the solid solution will determine the detailed nature of the intergrowths, although the same principles apply throughout.

These two-phase alkali feldspar intergrowths are known as *perthites*, and in these early stages of development are submicroscopic (hence *cryptoperthites*). The lozenge-shaped intergrowths are visible by optical microscope (*microperthites*), and further coarsening leads to intergrowths which can be easily seen by eye. When the coarsening has proceeded to that stage it is no longer possible to say whether the initial stages were by spinodal decomposition or by nucleation and growth.

Example 3. The decomposition of titanomagnetite solid solutions on cooling

Minerals with composition in the solid solution between magnetite (Fe_3O_4) and ulvöspinel (Fe_2-

TiO_4) are often termed titanomagnetites, and have the spinel structure (Section 3.2.5). At high temperatures the solid solution is complete, but below around 500°C unmixing begins, and slowly cooled natural titanomagnetites show a fine-scale two-phase intergrowth (cloth texture), just visible by high magnification optical microscopy (Figure 11.48). The phase diagram (Figure 11.49) is a simple solvus,

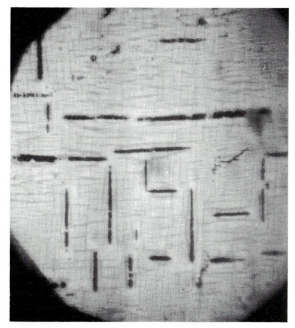

Figure 11.48. Reflected light micrograph showing exsolution microstructure in a natural spinel with composition in the field $MgAl_2O_4 - Fe_3O_4 - Fe_2TiO_4$. The coarser, dark lamellae are due to exsolution of $(Fe,Mg)Al_2O_4$ at high temperatures. The remaining magnetite – ulvöspinel solid solution exsolves at a lower temperature to form the fine 'cloth texture' in the background. The length of the scale bar is 15µm. (From Price, 1981.)

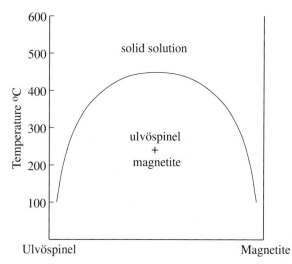

Figure 11.49. The approximate position of the solvus in the magnetite – ulvöspinel solid solution, determined from homogenisation experiments. (From Price, 1981.)

but its position is not known very precisely due to the difficulty of experimentally reproducing exsolution at such low temperatures on a laboratory time scale.

Natural titanomagnetites from more rapidly cooled dolerite rocks appear to be optically homogeneous, but electron microscopy shows a microstructure consistent with the later stages of spinodal decomposition (Figure 11.50). Spinodal modulations along the elastically soft <100> directions result in a three-dimensional tweed structure which forms small cubes of magnetite-rich phase surrounded by ulvöspinel-rich matrix. This fine distribution of isolated magnetite cubes has important effects on the magnetic properties of the host rock, a topic which is discussed in Section 12.6.8.

One aspect of the exsolution process in titanomagnetite is the extent to which the spinodal is depressed

Figure 11.50. Transmission electron micrograph of the microstructure in titanomagnetite from a dolerite rock. The texture is consistent with spinodal decomposition on the three {100} plane orientations, and results in small magnetite-rich cubes intergrown with ulvöspinel. (From Price, 1980.)

in temperature relative to the solvus. If the solid solution approximates to a regular solution model the temperature of the chemical spinodal relative to the solvus is given by

$$T_{chem} = 4\,T_{solvus}\,X\,(1-X)$$

where X is the mole fraction (Section 9.2.1). The suppression of the coherent spinodal relative to the chemical spinodal is given by eqn. (11.26):

$$T_s = T_{chem} + 2\eta^2\,YV/S''$$

The value of η can be determined from a linear interpolation between the lattice parameters of magnetite ($a = 8.40\text{Å}$) and ulvöspinel ($a = 8.54\text{Å}$), as can the molar volume V. The Young's modulus $Y \sim 2.1 \times 10^{-11}\text{Nm}^{-2}$. This leaves the determination of $S'' = d^2S/dX^2$ which is derived from a model of the cation distribution in the titanomagnetite solid solution.

The structure of the solid solution is however not very well understood. The two end members are both inverse spinels and the cation distribution between octahedral and tetrahedral sites may be written:

Magnetite $\quad[Fe^{2+}]^{IV}\,[Fe^{2+}Fe^{3+}]^{VI}\,O_4$
Ulvöspinel $\quad[Fe^{2+}]^{IV}\,[Fe^{2+}Ti^{4+}]^{VI}\,O_4$

where the superscripts IV and VI refer to tetrahedral and octahedral sites respectively. The cation distributions in intermediate compositions depend on the temperature but are difficult to determine by standard X-ray structure determination since Fe and Ti have very similar scattering factors. High temperature neutron diffraction is a possible technique in such cases, but must be carried out in an atmosphere with controlled oxygen partial pressure to maintain the correct Fe^{2+}/Fe^{3+} ratio. Three proposed cation distribution schemes are illustrated in Figure 11.51.

The cation distribution for the three models may be written as:

Model 1: $[Fe_{1-X}{}^{3+}\,Fe_X{}^{2+}]^{IV}\,[Fe^{2+}\,Fe_{1-X}{}^{3+}$
$\qquad\qquad\qquad Ti_X{}^{4+}]^{VI}\,O_4$
Model 2: $[Fe^{3+}]^{IV}\,[Fe_{1+X}{}^{2+}\,Fe_{1-2X}{}^{3+}\,Ti_X{}^{4+}]^{VI}\,O_4$
$\qquad\qquad\qquad\text{for } 0 \leq X \leq 0.5$
$\quad[Fe_{2-2X}{}^{3+}\,Fe_{2X-1}{}^{2+}]^{IV}\,[Fe_{2-X}{}^{2+}$
$\qquad\qquad Ti_X{}^{4+}]^{VI}\,O_4 \quad \text{for } 0.5 \leq X \leq 1.0$
Model 3: $[Fe_{1.2-X}{}^{3+}\,Fe_{X-0.2}{}^{2+}]^{IV}\,[Fe_{1.2}{}^{2+}$
$\qquad Fe_{0.8-X}{}^{3+}Ti_X{}^{4+}]^{VI}\,O_4 \quad \text{for } 0.2 \leq X \leq 0.8$

We can illustrate the consequences of different

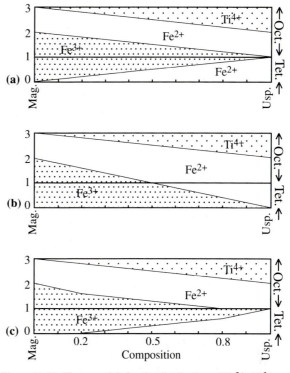

Figure 11.51. Three models for the distribution of Fe^{2+}, Fe^{3+} and Ti^{4+} among octahedral and tetrahedral sites in the magnetite – ulvöspinel solid solution. The models agree on the cation distribution in the end members, but differ in the relative proportions of Fe^{2+} and Fe^{3+} in the octahedral and tetrahedral sites. **(a)** the Akimoto model **(b)** the Néel model and **(c)** the O'Reilly–Banerjee model. (After Lindsley, 1976.)

entropy models on the suppression of the coherent spinodal by calculating the value of S'' for Model 1. Using the methods described in Sections 8.3.1 and 9.4.2, the configurational entropy for disorder over the one tetrahedral and two octahedral sites per formula unit is given by

$$S = -R\{(1-X)\ln(1-X) + X\ln X\}$$
for the tetrahedral sites, and
$$-R\{-2\ln 2 + (1-X)\ln(1-X) + X\ln X\}$$
for the octahedral sites

Thus

$$S = -2R\{-\ln 2 + (1-X)\ln(1-X) + X\ln X\}$$

from which

$$S'' = \frac{d^2S}{dX^2} = 2R\left[\frac{-1}{X(1-X)}\right]$$

For a composition $X = 0.5$, $S'' = -8R$. Substituting into eqn.(11.26) using the values for the other constants above gives a temperature suppression of 73.9°C for the coherent spinodal.

By comparison, the same calculation for model 3 gives

$$S = -R \{(1.2-X) \ln (1.2-X) + (X-0.2)\ln(X-0.2)$$
for the tetrahedral sites, and
$$-R \{\{- 2 \ln2 + 1.2 \ln1.2 + (0.8-X)\ln(0.8-X) - X \ln X \}$$ for the octahedral sites

Thus $S = R\{2 \ln2 - 1.2 \ln1.2 - (0.8-X)\ln(0.8-X) - X\ln X - (1.2-X) \ln (1.2-X) + (X-0.2)\ln(X-0.2)\}$

from which

$$S'' = \frac{d^2S}{dX^2} = R \left[\frac{2(X-1)}{(1.2-X)(0.8-X)} - \frac{2X-0.2}{X(X-0.2)} \right]$$

For a composition $X = 0.5$, $S'' = -10.1R$ giving a temperature suppression of 58.5°C for the coherent spinodal. Similar calculations for model 2 are more complex because of the discontinuity at $X = 0.5$, but near $X = 0.5$, $S'' \sim -21R$ giving a temperature suppression of only 28°C.

This example illustrates the importance of the free energy model of the solid solution in determining the temperature at which spinodal decomposition may occur. Had we assumed the most simplistic model of a binary regular solid solution with a random distribution of cations the temperature suppression would have been 147.5°C. Such a large temperature suppression is unlikely, since the low solvus temperature would mean that the coherent spinodal would occur below 350°C, a temperature too low for any significant solid state diffusion to take place. The fact that observable spinodal textures occur in natural minerals suggests a higher value for the spinodal. However, no experimental data against which these predicted temperature suppressions can be compared are available.

11.4 The nucleation of transition phases

In the previous section spinodal decomposition was presented as a process of non-equilibrium unmixing of a solid solution for which the two end members have the same structure. In this section we consider the likely metastable behaviour of a solid solution for which the low temperature stable phases have different structures, thus precluding the possibility

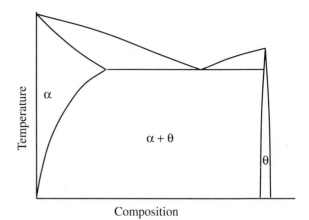

Figure 11.52. The general form of phase diagram in which a solid solution α coexists with a phase θ which has a limited solid solution and a different structure from the solid solution. Under equilibrium conditions the θ phase would exsolve from the α solid solution on cooling.

of spinodal decomposition. The schematic phase diagram in Figure 11.52 illustrates such a case, the stable θ phase precipitating from the α solid solution on cooling.

If the θ phase has a different structure to the solid solution the interface between the precipitate and the matrix will be incoherent with a high interfacial energy. The activation energy for nucleation will be high and homogeneous nucleation of the stable θ phase highly improbable. Under such circumstances a sufficiently slow cooling rate would allow heterogeneous nucleation on grain boundaries. If however the solid solution was cooled too quickly for heterogeneous nucleation to occur, we have the situation that either the solid solution will persist indefinitely in a metastable state, or some other non-equilibrium route may allow a reduction in free energy.

Observations of the behaviour of such systems show that a metastable route does often exist and involves the nucleation of a sequence of *transition phases* which have a structure (and in some cases composition) intermediate between the matrix and the stable phase. The structural similarity with the matrix allows the transition phases to have a coherent or semicoherent interface with the matrix, thereby reducing the activation energy for nucleation. Transition phases, being by definition less stable than the equilibrium θ phase, can only form when the supersaturation is high enough that their nucleation results in a free energy reduction.

The overall sequence of phases which nucleates

under non-equilibrium conditions, is one of a step-wise approach towards the equilibrium structure and composition, rather than a single-step nucleation of the stable phase. In this metastable route each step has a relatively small activation energy, but only reduces the overall free energy by a small amount. The single-step equilibrium behaviour involves a large activation energy barrier but also a large reduction in the overall free energy. This general principle has been known in physical chemistry for many years as the Ostwald step rule, and is illustrated in Figure 11.53.

Figure 11.54 shows a schematic free energy – composition diagram for this situation. The stable θ phase has a narrow free energy curve, consistent

with its narrow compositional range in the phase diagram (Figure 11.52). The two transition phases θ' and θ'' have higher free energies, again consistent with their intermediate structure and composition. The θ'' phase is shown as part of the same free energy curve as the a solid solution, indicating that they are isostructural. [This is not necessarily the case, nor are the compositions necessarily different as illustrated. All that is required is that the free energies progressively decrease from θ'' to θ' to θ, which could also be the case if the compositions of the three phases were all the same and only their structures were different.] The dashed lines show the common tangents between the precipitate phases and the solid solution, and the step-wise reduction in free energy of a composition C_0.

The relative free energies of the three phases determine the temperature and composition range in which each can form. Figure 11.54 is drawn to represent the free energy curves at some temperature T where C_1, C_2 and C_3 are the limiting compositions of the a solid solution which can coexist with the θ, θ' and θ'' phases respectively. When these compositions are transposed onto the phase diagram (Figure 11.55), they lie on the metastable solvi defining these coexisting compositions. A sequence of such free energy diagrams at different temperatures will also define the successively greater degrees of undercooling i.e. supersaturation, required to nucleate successively less stable phases.

The higher degrees of supersaturation needed to nucleate the θ, θ' and θ'' phases respectively, from a composition C_0, are shown on a TTT diagram in

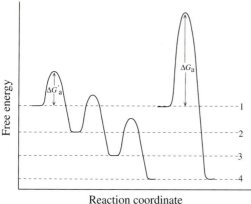

Figure 11.53. Illustration of the Ostwald step rule. The direct transformation from state 1 to state 4 involves a large activation energy ΔG_a and may be very sluggish. Transformation via a sequence of steps $1 \Rightarrow 2 \Rightarrow 3 \Rightarrow 4$ involves smaller activation energies and may be kinetically more favourable.

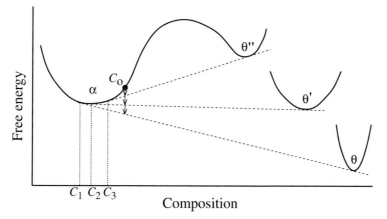

Figure 11.54. Free energy and composition changes associated with the precipitation of a sequence of phases θ'' and θ' before the nucleation of the stable θ phase. A composition C_0 of the a solid solution reduces its energy with each successive transformation, as shown by the arrows. The dashed common tangents indicate the compositions C_3, C_2 and C_1 in equilibrium with θ'', θ' and θ respectively.

Figure 11.56. At temperatures between T_1 and T_2 only the stable θ phase can form. At higher supersaturations, between T_2 and T_3, the nucleation sequence is θ' followed by θ. Below T_3 the supersaturation is great enough for the θ'' phase to nucleate first. The sequence $\theta'' \Rightarrow \theta' \Rightarrow \theta$ reflects the stepwise increase in the activation energy for nucleation and determines the relative positions of the noses of the TTT curves. The slopes of the low temperature part of the curves depend on the activation energy for diffusion through the matrix, and hence are all equal.

As each new phase in the transformation sequence nucleates, the preceeding phase begins to dissolve.

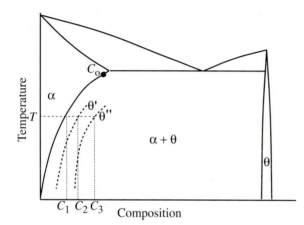

Figure 11.55. The phase diagram in Figure 11.49, on which the positions of the metastable solvi for the precipitation of the θ'' and θ' phases have been added. The composition C_o of the initial α solid solution, and the compositions C_3, C_2 and C_1 of the solid solution in equilibrium with θ'', θ' and θ respectively may be compared with those in the free energy – composition diagram in Figure 11.51.

The reason can be seen from the free energy curves which show for example that when the θ'' phase precipitates from the C_o starting composition, the matrix has composition C_1. When the next phase, θ', precipitates, the matrix composition moves to C_2 which is beyond the limiting composition with which θ'' can coexist. θ'' therefore redissolves in the solid solution. Similarly, when the equilibrium θ phase forms, the matrix composition must move to C_1 and the θ' phase dissolves.

The earliest precipitates in such a sequence are usually coherent platelets or discs which nucleate homogeneously throughout the solid solution. They may have the same structure as the matrix and merely represent segregations of the solute atoms within the solid solution. This type of coherent precipitate was first observed as segregations of copper atoms in aluminium in the system Al–Cu, studied by Guinier and Preston in the 1930s. They have subsequently been termed GP zones, and the term has been extended to analogous phases in other metallic systems and also in minerals.

11.4.1 Examples of transition phases in mineral exsolution

Example 1. The precipitation of augite from orthopyroxene

Earlier in this chapter (Sections 11.1.6 and 11.3.4) we discussed exsolution in the augite – pigeonite solid solution, and pointed out that we were dealing with a situation where the pigeonite had failed to transform to the more stable orthopyroxene struc-

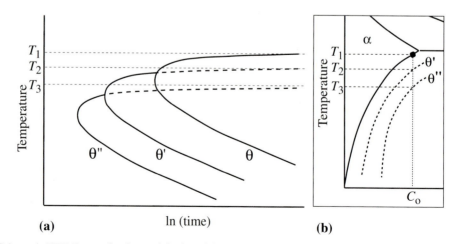

Figure 11.56. (a) Schematic TTT diagram for the precipitation of the metastable transition phases θ'', θ' relative to that of the equilibrium θ phase. The temperatures T_1, T_2 and T_3 may be related to the stable and metastable equilibrium temperatures in the phase diagram, part of which is inset in **(b)**.

ture. In very slowly cooled rocks we find that the pigeonite does transform to orthopyroxene on cooling. Since the orthopyroxene structure can hold even less Ca in solid solution than pigeonite, a Ca-rich phase will tend to precipitate from it. Such transformed pigeonites (often termed 'inverted' pigeonites) have a characteristic microstructure on an optical scale and illustrate a number of episodes of exsolution as shown in Figure 11.57. In this illustration the original pigeonite crystal was twinned, and on cooling, the exsolution of augite lamellae on (001) planes of the pigeonite formed the 'herring-bone' pattern. With further cooling, the whole pigeonite twin transformed to a single crystal of orthopyroxene. Further exsolution of augite from the orthopyroxene forms the lamellae on (100) planes of the host.

Orthopyroxene may also form as a primary phase in the early crystallization of basaltic magma, as well as in high grade metamorphic rocks, in which case augite may exsolve during cooling.

When dealing with the exsolution of augite from orthopyroxene, we must note that the difference in their structures does not allow the possibility of extensive solid solution between them, nor of spinodal decomposition as a mode of non-equilibrium exsolution. Transmission electron microscopy of orthopyroxenes from both lunar and terrestrial rocks show thin, coherent, disc-shaped precipitates around 1000 Å in diameter and <100Å thick, in addition to the equilibrium augite precipitates. These discs, which lie on (100) planes, have the characteristics of GP zones in that they are completely coherent with the orthopyroxene host, and show strain diffraction contrast indicating a slightly different lattice spacing (Figure 11.58).

When the precipitates are large enough to analyse by their X-ray emission spectra, they show Ca enrichment relative to the matrix, as well as an electron diffraction pattern which has main reflections the same as that of the orthopyroxene host (space group $Pbca$) but with some weak extra reflections. This indicates that the basic structure of the precipitates is the same as orthopyroxene, but that there is some reduction in symmetry. The extra reflections are of the type $h = 2n + 1$ which is consistent with a loss of the a glide plane of the orthopyroxene structure. This in turn indicates that instead of having two non-equivalent silica chains and two types of cation sites M1 and M2, there must

Figure 11.57. Optical micrograph of an inverted pigeonite from the Bushveld Complex. The sequence of processes leading to this observed intergrowth is : (i) growth of a twinned pigeonite crystal from the melt (ii) exsolution of augite on (001) giving the herringbone pattern (iii) transformation of pigeonite to orthopyroxene and (iv) further exsolution of augite from the orthopyroxene (small vertical lamellae).

be four distinct silica chains and four types of cation sites, as shown in Figure 11.59. From estimates of the chemical composition one in four of these cation sites would be Ca, and to be consistent with the symmetry they would occupy only the M2A or M2B sites.

In terms of the 18 Å (100) lattice planes of the orthopyroxene structure, this distribution of Ca can be described as 9Å half-cells alternately with and without Ca in the M2 sites (Figure 11.60). Such a model of the precipitates is consistent with a transition phase, having a composition between that of orthopyroxene and augite and a structure which is a variant of orthopyroxene.

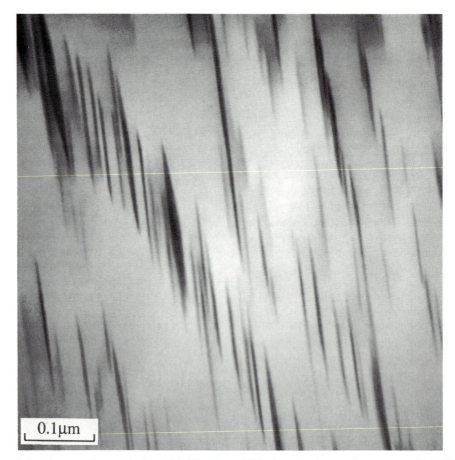

Figure 11.58. Transmission electron micrograph of parallel GP zones on (100) planes in a lunar orthopyroxene. (From Nord, 1980.)

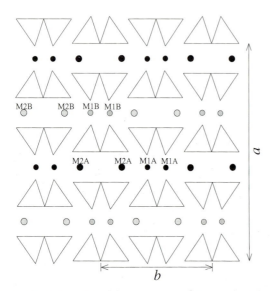

Figure 11.59. A model of the structure of GP zones in orthopyroxene, based on the electron diffraction and compositional data. The four non-equivalent silicate chains are labelled A1, A2, B1 and B2, and the M1 and M2 sites are also subdivided into M1A and M1B, M2A and M2B. The Ca atoms occupy either the M2A sites (filled circles) or the M2B sites (open circles). (After Nord, 1980; see also Smyth and Swope, 1990.)

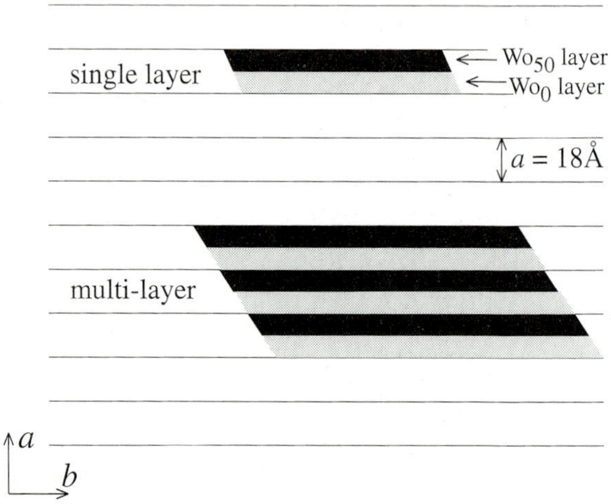

Figure 11.60. A model of the GP zones, based on Figure 11.59, in terms of the composition of alternate 9Å layers parallel to (100) planes of the orthopyroxene. If the Ca atoms are all in M2A or M2B sites, alternate 9Å layers have composition Wo_{50} and Wo_0, i.e. $Ca(Mg,Fe)Si_2O_6$ and $(Mg,Fe)SiO_3$ respectively. (After Nord, 1980.)

Example 2. Exsolution of Fe-rich precipitates from rutile

Rutile, TiO_2 has virtually no stable solid solution with the iron oxide phase hematite Fe_2O_3 at low temperatures, and their structures are quite different (Section 5.2). When natural rutile crystallises at low temperature it can frequently metastably accommodate up to several mole % Fe_2O_3 and transmission electron microscopy indicates an apparently homogeneous solid solution. When such rutile crystals are heated to provide the thermal activation required to enable them to achieve equilibrium, hematite is eventually exsolved. Metamorphosed iron-bearing rutiles which have undergone a period of post-crystallisation annealing, show similar fine-scale exsolution of an iron-rich phase.

This is another example of the exsolution of a structurally different phase in which transition phases could occur under non-equilibrium conditions. Experiments on natural iron-bearing rutiles have shown the existence of two transition phases. The first phase resembles GP zones, forming as coherent discs on the symmetrically equivalent (100) and (010) planes of the rutile structure (Figure 11.61). Structurally, they appear to be Fe-rich segregations within the rutile, straining the lattice but retaining the rutile structure. The second transitional phase, which forms after longer annealing times, is a modification of the rutile structure, with extra superlattice reflections in the diffraction pattern. The final stage is the growth of the equilibrium hematite phase.

11.5 The effect of cooling rate on the scale of exsolution textures

One of the common characteristics of exsolution textures in minerals is the uniformity of thickness and spacing of exsolution lamellae (Figure 11.62), and the observation that in slowly cooled rocks the scale of this exsolution is coarser than in more rapidly cooled rocks. When exsolution occurs by nucleation and growth, which does not in itself imply a periodicity, the spacing of nuclei is nevertheless

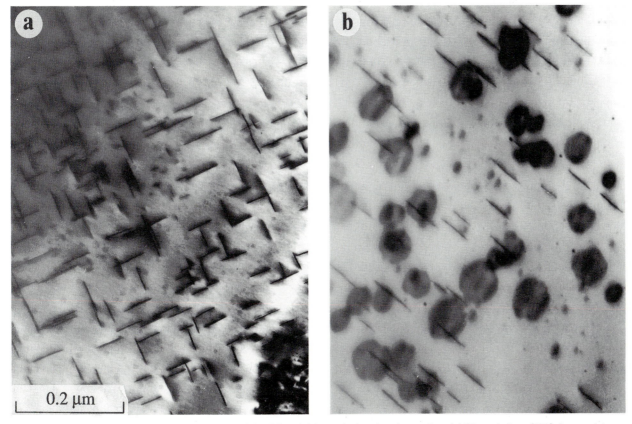

Figure 11.61. Transmission electron micrographs of Fe-rich precipitates in iron-bearing rutiles. (a) Viewed along [001] the precipitates appear as two sets of platelets, on (100) and (010) planes. (b) Viewed along [100] one set of platelets is face-on while the other set is edge-on. (From Putnis, 1978; see also Banfield and Veblen, 1991.)

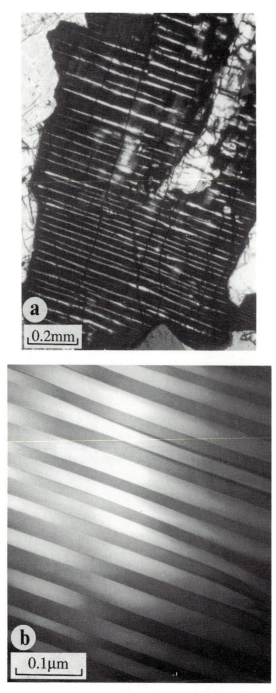

Figure 11.62. Exsolution lamellae are generally uniformly spaced, even when their spacing is many orders of magnitude different. (a) An optical micrograph of augite exsolution lamellae in pigeonite (b). A transmission electron micrograph of exsolution in orthoamphiboles with lamellae of gedrite in anthophyllite.

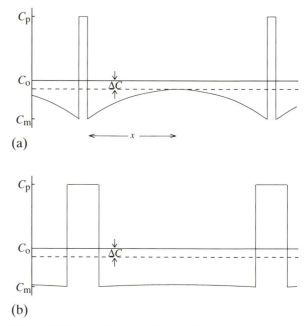

Figure 11.63. (a) Concentration profiles around exsolution lamellae which have formed by nucleation during cooling. C_p and C_m are the equilibrium concentrations of the lamella and matrix respectively, and C_o is the bulk composition of the initial solid solution. The diffusion distance x depends on the cooling rate, and defines the concentration profile around the lamella. If ΔC is the critical amount of supersaturation required for nucleation, the distance to the next lamella will be $2x$. **(b)** Continued growth of the lamellae reduces the concentration gradient between them, ultimately producing uniform equilibrium compositions C_p and C_m.

the TTT curves in Section 11.1.8). Nucleation will not begin until a critical degree of supersaturation is reached. At this stage the nucleation rate will be low and a number of isolated nuclei will form. Around each nucleus we can define a diffusion distance over which there is a depletion of solute in the solid solution. With very slow cooling and with a relatively high solvus temperature the diffusion distance ($\approx \sqrt{Dt}$) will be large, and within that distance the concentration of the solute will be reduced to a value less than the critical supersaturation (Figure 11.63(a)). Therefore within this distance no other nucleation event can take place unless the supersaturation increases. Beyond this diffusion distance other nuclei may form and establish their own depleted region around them. The establishment of a precipitate free zone around a lamella due to the sensitivity of nucleation rate to supersaturation at small values of the undercooling ΔT effectively determines the equilibrium spacing of nuclei in the very earliest stages of the exsolution process.

At slow cooling rates the diffusion rate down the

controlled by the effective diffusion distance (Section 11.2.2) for the specific cooling rate.

When exsolution occurs under near-equilibrium conditions, the initial nucleation event is determined very sensitively by the degree of undercooling (see

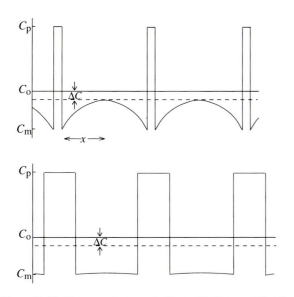

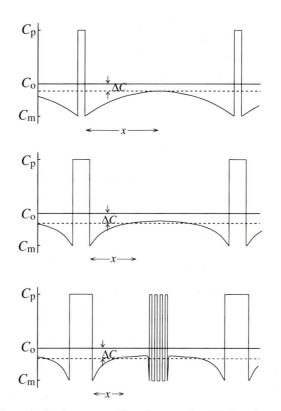

Figure 11.64. The same figure as in Figure 11.63 except that the cooling rate is faster and hence the diffusion distance x is smaller, resulting in closer spaced lamellae.

Figure 11.65. At more rapid cooling rates the diffusion distance decreases during growth, resulting in a stranded diffusion profile. At lower temperatures these stranded regions may become supersaturated with respect to the homogeneous nucleation of a second generation of exsolution lamellae, which form between the primary lamellae.

concentration gradients will be sufficiently rapid to maintain a depleted zone around each lamella so that the supersaturation between them will not increase above the critical value, and no further nucleation will occur. Steady growth of the early-formed nuclei will continue until the equilibrium concentrations of precipitate and matrix are reached (Figure 11.63(b)). If the initial cooling rate was faster, the time scale for nucleation would be less and the diffusion distance smaller. Under near-equilibrium conditions the same argument would apply as in the former case and no further nucleation would occur between the early formed nuclei. In this case however, the separation of the nuclei would now be smaller, as shown in Figure 11.64. The simple relationship $x \approx \sqrt{Dt}$ indicates that a change in one order of magnitude in the scale of the microstructure is equivalent to a change in the cooling rate of two orders of magnitude.

11.5.1 Stranded diffusion profiles

In the discussion above we assumed that the cooling rate was sufficiently slow that the diffusion distance remained constant and hence the whole region between lamellae was depleted in solute as the lamellae coarsened. In practice such a situation is only possible while temperatures are maintained very near to equilibrium. Even with slow cooling

rates the sensitivity of the diffusion coefficient to temperature, especially when the activation energy is large, results in a decreasing diffusion distance as the temperature falls. As the diffusion distance becomes smaller than the initial separation of the nuclei, the solute concentration in the regions between the lamellae becomes fixed and a stranded diffusion profile results.

These stranded regions may subsequently become supersaturated with respect to exsolution, and eventually a second generation of lamellae may form by homogeneous nucleation as the temperature continues to fall. If nucleation of the stable phase is kinetically impeded, supersaturations may become sufficiently high for metastable phases such as GP zones to form within the stranded profile. Typically, the second generation lamellae are separated from the initial lamellae by a precipitate-free zone in which the supersaturation was insufficient for nucleation. Figure 11.65 shows this situation sche-

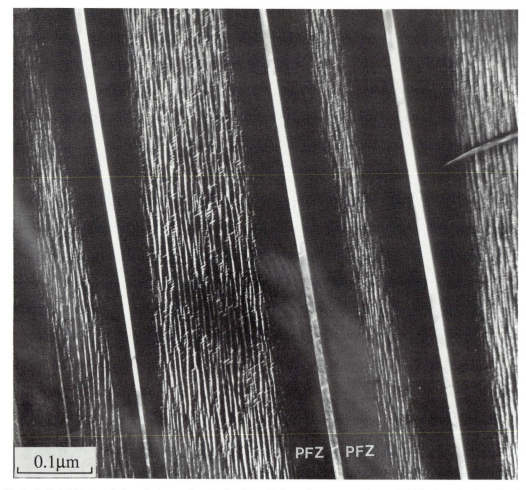

0.1μm

PFZ PFZ

Figure 11.66. A dark-field transmission electron micrograph of exsolution of two generations of '001' pigeonite lamellae in an augite matrix. The micrograph is taken using a diffraction spot of the type $h+k$ odd, i.e. from pigeonite only, and so the light lamellae are all pigeonite; the black matrix is augite. The three long lamellae are the first generation pigeonite. The second generation finer lamellae are separated from the first by a precipitate free zone (PFZ). (Micrograph courtesy of W.F. Müller.)

matically, and Figure 11.66 is a transmission electron micrograph showing second generation lamellae and a precipitate free zone around the primary exsolution lamellae of pigeonite in augite.

11.5.2 Exsolution textures and cooling rates in Fe–Ni meteorites

Iron–nickel meteorites show very characteristic and spectacular exsolution textures on a coarse scale (Figure 11.67) due to their very slow cooling in the cores of small planetary bodies. The reheating due to entry through the earth's atmosphere is very fast and has no significant effect on this microstructure. Since the diffusion rate of Fe and Ni as a function of temperature is well known, it is theoretically possible to use the scale of this exsolution texture and

the diffusion rate equations to make some estimates of the cooling rate of the core of the planetary body from which the meteorite originated.

The cause of the exsolution in Fe–Ni meteorites is shown in the phase diagram in Figure 11.68. Above 910°C an Fe–Ni solid solution with a cubic close-packed structure is stable. In meteorites this phase is termed taenite (γ-iron or austenite by metallurgists). In pure iron below this temperature there is a phase transformation to a body-centred cubic structure termed kamacite (α-iron, or ferrite), in which the solubility of Ni is considerably reduced. This results in a wide two-phase region in the phase diagram.

An iron meteorite with 15% Ni would enter the two-phase region at around 650°C which is a low temperature when considering diffusion rates in this system. As a general guide, diffusion rates are only

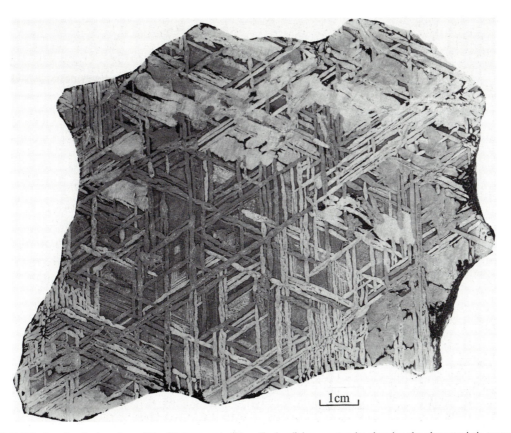

Figure 11.67. Polished and etched section of the Waingaromia (New Zealand) iron meteorite showing the characteristic texture produced by the exsolution of kamacite lamellae on {111} planes of the host taenite phase. (Courtesy of the Smithsonian Institution, Washington.)

significant on a laboratory time-scale at temperatures greater than two-thirds of the melting temperature, and so at these temperatures no significant exsolution could be induced experimentally. At the extremely slow cooling rates in meteorites however, kamacite begins to exsolve from the parent taenite. The lowest interfacial energy is when the kamacite occurs as lamellae on the close-packed {111} planes of the taenite. This results in four sets of symmetrically equivalent orientations and an octahedral pattern of lamellae which gives such meteorites the name *octahedrites*. The scale of the texture depends on the Ni content, which dictates the temperature of exsolution, and the cooling rate.

Even under the extremely slow cooling conditions of the parent bodies of meteorites stranded Ni-diffusion profiles show that equilibrium was not maintained throughout cooling. The sequence of diagrams illustrating the development of these profiles (Figure 11.69) should be interpreted together with the phase diagram which shows the coexisting kamacite and taenite compositions as a function of

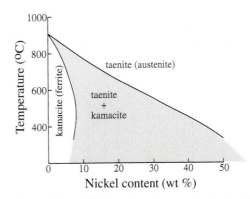

Figure 11.68. The equilibrium phase diagram for the Fe-rich part of the Fe–Ni system below 1000°C.

temperature at equilibrium. The diffusion profiles indicate a progressively decreasing diffusion distance so that at the cut-off temperature only the immediately adjacent part of the taenite matrix is in equilibrium with the kamacite lamellae.

The determination of cooling rates from the microstructure of iron meteorites is made either by

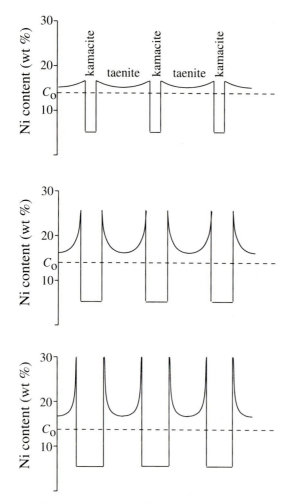

Figure 11.69. A sequence of diagrams illustrating the development of stranded M-shaped diffusion profiles in taenite during the growth of kamacite lamellae with falling temperature.

an analysis of the stranded diffusion profiles or from the final kamacite plate thickness, as both are ultimately related to the same factors. When the diffusion rates and their composition and temperature dependence are known it is possible to calculate the shape of stranded diffusion profiles for any given cooling rate and nucleation temperature. These calculated profiles may then be compared with the measured profiles from meteorites. Such analyses suggest that octahedrites must have cooled extremely slowly, losing only about 1 to 10°C every million years. This indicates cooling within a shield of poorly conducting material presumed to be the silicate crust of larger parent bodies. Such cooling rates are consistent with parent bodies 70 to 200 km in diameter, such as small planets in the asteroid belt.

Bibliography

BUSECK, P.R., NORD, G.L. AND VEBLEN, D.R. (1980). Subsolidus phenomena in pyroxenes. In: *Pyroxenes* C.T. Prewitt (Ed) Reviews in Mineralogy Vol 7. Mineralogical Society of America.

CAHN, J.W. Spinodal decomposition. *Trans.Met. Soc.AIME* **242**, 166–180.

CHADWICK, G.A. (1972). *Metallography of phase transformations* Butterworths.

CHAMPNESS, P.E. AND LORIMER, G.W. (1976). Exsolution in silicates. In: H-R. Wenk (Ed) *Electron microscopy in mineralogy*. Springer-Verlag.

CHRISTIAN, J.W. (1975). *The theory of transformations in metals and alloys*. Pergamon Press.

de FONTAINE, D. (1975). Clustering effects in solid solutions. In: *Treatise on solid state chemistry. Vol.5. Changes of State*. N.B. Hannay (Ed) Plenum Press.

GLANSDORFF, P. AND PRIGOGINE, I. (1971). *Thermodynamic theory of structure, stability and fluctuations*. Wiley.

PARSONS, I. AND BROWN, W.L. (1991). Mechanisms and kinetics of exsolution – structural control of diffusion and phase behaviour in alkali feldspars. In: *Diffusion, atomic ordering and mass transport*. J. Ganguly (Ed) Advances in Physical Geochemistry Vol.8. Springer-Verlag.

SHEWMON, P.G. (1963). *Diffusion in solids*. McGraw-Hill.

References

BANFIELD, J.F AND VEBLEN, D.R. (1991). The structure and origin of Fe-bearing platelets in metamorphic rutile. *Amer. Mineral.* **76**, 113–27.

BROWN, W.L. AND PARSONS, I. (1984). Exsolution and coarsening mechanisms and kinetics in an ordered cryptoperthite series. *Contrib. Mineral. Petrol.* **86**, 3–18.

BROWN, W.L. AND PARSONS, I. (1988). Zoned ternary feldspars in the Klokken intrusion: Exsolution microstructures and mechanisms. *Contrib. Mineral. Petrol.* **98**, 444–54.

BUSECK, P.R., NORD, G.L. AND VEBLEN, D.R. (1980). Subsolidus phenomena in pyroxenes. In: *Pyroxenes* C.T. Prewitt (Ed) Reviews in Mineralogy Vol 7. Mineralogical Society of America.

CARPENTER, M. A. (1978). Nucleation of augite at antiphase boundaries in pigeonite. *Phys. Chem. Minerals* **2**, 237–51.

CHAMPNESS, P.E. AND LORIMER, G.W. (1974). A direct lattice resolution study of precipitation (exsolution) in orthopyroxene. *Phil. Mag.* **30**, 357–62

GREENWOOD, H.J. (1963). The synthesis and stability of anthophyllite. *Journ. Petrology*, **4**, 317–51.

LINDSLEY, D.H. (1976). The crystal chemistry and structure of oxide minerals. In: *Oxide Minerals* D. Rumble III (Ed) Reviews in Mineralogy Vol. 3. Mineralogical Society of America.

LIVI, K.J.T. AND VEBLEN, D.R. (1989). Transmission electron microscopy of interfaces and defects in intergrown pyroxenes. *Amer. Mineral.* **74**, 1070–83.

LORIMER, G.W. AND CHAMPNESS, P.E. (1973). The origin and phase distribution in two perthitic alkali feldspars. *Phil. Mag.* **28**, 1391–99.

NORD, G.L. (1980). The composition, structure and stability of Guinier–Preston zones in lunar and terrestrial orthopyroxene. *Phys. Chem. Minerals* **6**, 109–28.

NORD, G.L., HEUER, A.H. AND LALLY, J.S. (1976). Pigeonite exsolution from augite. In: H-R. Wenk (Ed) *Electron microscopy in Mineralogy*. Springer-Verlag.

NORD, G.L. AND McCALLISTER, R.H. (1979). Kinetics and mechanism of decomposition in $Wo_{25}En_{31}Fs_{44}$ clinopyroxene. *Geol. Soc. Am. Abstracts* **11**, 488

PARSONS, I. AND BROWN, W.L. (1984). Feldspars and the thermal history of igneous rocks. In: *Feldspars and feldspathoids* W.L. Brown (ed) NATO ASI Series. Reidel.

PRICE, G.D. (1980). Exsolution microstructures in titanomagnetites and their magnetic significance. *Phys. Earth and Planet. Interiors* **23**, 2–12.

PRICE, G.D. (1981). Subsolidus phase relations in the titanomagnetite solid solution series. *Amer. Mineral.* **66**, 751–8.

PUTNIS, A. (1978). The mechanism of exsolution of hematite from natural iron-bearing rutiles. *Phys. Chem. Minerals* **3**, 183–97.

ROBINSON, P. AND JAFFE, H.W. (1971). Orientation of exsolution lamellae in clinopyroxenes and clinoamphiboles: consideration of optimal phase boundaries. *Amer. Mineral.* **56**, 909–38.

ROBINSON, P., ROSS, M., NORD, G.L., SMYTH, J.R. AND JAFFE, H.W. (1977). Exsolution lamellae in augite and pigeonite: fossil indicators of lattice parameters at high temperature and pressure. *Amer. Mineral.* **62**, 857–73.

SKROTZKI, W., MÜLLER, W.F. AND WEBER, K. (1991). Exsolution phenomena in pyroxenes from the Balmuccia Massif, NW-Italy. *Eur. J. Mineral.* **3**, 39–61.

SMITH, J.V. AND BROWN, W.L. (1988). *Feldspar Minerals 1*. 2nd Ed. Springer Verlag.

SMYTH, J.R. AND SWOPE, R.J. (1990). The origin of space group violations in lunar orthopyroxene. *Phys. Chem Minerals* **17**, 438–43.

WILLAIME, C., BROWN, W.L. AND GANDAIS, M. (1976). Physical aspects of exsolution in natural alkali feldspars. In: H-R. Wenk (Ed) *Electron microscopy in mineralogy*. Springer-Verlag.

12 Transformation processes in minerals II: structural phase transitions

In this chapter we discuss various aspects of the mechanism of transformations in minerals in which there is no change in chemical composition. Such *polymorphic transformations* are often described in terms of the degree of similarity between the structures, which in turn can be used to define the structural changes required to transform one to the other. Often a kinetic classification is also implied in this approach. For example a transformation between two very similar structures may only involve a distortion of the bonds such as that between the high and low silica polymorphs (Section 6.8.1). Such *displacive* transitions are generally fast and cannot be prevented from occurring even with very rapid cooling rates (i.e. they are unquenchable). Displacive transitions may be thermodynamically first- or second-order (Section 8.4.2).

On the other hand, a *reconstructive* transition involves a major reorganisation of the structure, with bonds being broken and new bonds formed. Transformations between the silica polymorphs (quartz ⇔ tridymite ⇔ cristobalite) are typical reconstructive transitions which have high activation energies and are kinetically very sluggish. Rapid cooling can easily quench the high temperature form which may persist indefinitely at low temperatures. The other classic example is the persistence of the diamond structure at atmospheric temperatures and pressures where graphite is the stable form of carbon. Reconstructive transformations are always thermodynamically first-order.

Order–disorder transitions may be slow, as in the case of substitutional disorder (e.g. Si,Al disorder in aluminosilicates) or fast, as in orientational disorder (e.g. the orientation of the CO_3 group in calcite – see Section 8.13.2 and Figure 8.38). Thermodynamically they may be first- or second-order.

12.1 Reconstructive polymorphic transitions in minerals

Reconstructive transitions involve a major reorganization of the crystal structure, achieved by breaking and reforming nearest-neighbour coordination bonds.

There need be no structural or symmetry relation between the two polymorphs, although the fact that the chemical composition remains the same will inevitably lead to some structural similarities. Typical of these transitions is the transformation between the two polymorphs of $CaCO_3$, aragonite ⇒ calcite, in which the coordination of Ca changes from 9 to 6. The transition from diamond to graphite involves a change in the bond type as well as carbon coordination.

In other cases, the primary coordination may remain the same and only the second-nearest neighbour coordination changes. If such transitions involve breaking nearest neighbour bonds they are still classified as, and have most of the features of, reconstructive transitions. For example, the silica polymorphs cristobalite, tridymite and quartz differ only in the arrangement of SiO_4 tetrahedra in the framework and the nearest neighbour coordination remains the same. However, the transformation from one to another requires breaking and reforming primary Si–O bonds.

Reconstructive transitions are always thermodynamically first-order (Section 8.4.2) and give rise to large discontinuities in cell volume, enthalpy and entropy. They typically have large activation energies and are kinetically very sluggish. In laboratory experiments, and even in nature, such transitions are associated with metastability and large degrees of superheating and undercooling. Even when the thermodynamic properties of the two polymorphs are known, thus enabling an equilibrium curve between them to be constructed, its relevance to real behaviour is always questionable.

Since the transformation mechanism involves nucleation and growth, many of the concepts regarding interfaces discussed in the previous chapter are directly applicable. The structural difference between the polymorphs makes heterogeneous nucleation the only likely transition mechanism and one of the consequences of this is that the measured activation energy depends very much on the defect density in the sample. Factors such as the grain size and shape, dislocation density, presence of trace amounts of impurities etc. can have a very significant effect on the transition, making any theoretical treatment virtually impossible. The mechanism may also depend on the experimental conditions, and factors such as the presence of non-hydrostatic stresses in the sample. These are not always easy to define in practice. Any catalytic effects which can change the mechanism of the transition, such as the presence of water for example, may also have a major influence on the progress of reconstructive transitions.

Thus although reconstructive transitions involve such a major change in structure they are much more difficult to define than the more subtle effects of displacive and order–disorder transitions. In this section we will describe reconstructive transitions in a number of important mineral groups, emphasising those aspects which illustrate the general features mentioned above.

12.1.1 Examples of reconstructive transitions in minerals

(i) The aragonite ⇔ calcite transition

The aragonite – calcite equilibrium phase diagram (Figure 12.1) shows that aragonite is the stable phase at high pressure and relatively low temperature in the earth's crust, such as is found in the blueschist metamorphic rocks in subduction zones. This stability field is consistent with the higher Ca coordination in aragonite and its greater density. The curvature of the Clapeyron slope is due to the progressive onset of CO_3 disorder in calcite as the temperature increases (Section 8.6.2). This increases both the entropy and the molar volume in a manner consistent with the observed curvature.

The metastability of aragonite in nature is demon-

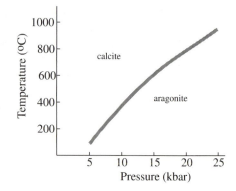

Figure 12.1. The equilibrium relations between calcite and aragonite, two polymorphs of $CaCO_3$. Aragonite is stable at high pressures, while the stable state at atmospheric conditions is calcite.

strated by its fairly common preservation in high pressure rocks which have been uplifted to the earth's surface, as well as the fact that aragonite also forms in low temperature sedimentary environments, i.e. within the stability field of calcite. These two manifestations of metastability are not unusual in polymorphism and, although related, may have different explanations. In the former case the sluggish kinetics of the aragonite to calcite transition prevents the transformation taking place within the time-scale of uplift, and in the latter case some kinetic factor enhances the low temperature nucleation and growth of aragonite relative to calcite in supersaturated solutions.

The nucleation and growth of sedimentary aragonites is frequently attributed to the role of Mg^{2+} in solution which may inhibit calcite nucleation or favour the aragonite. However, aragonite growth at 25°C and 1 atmosphere has also been demonstrated in Mg-free solutions. A possible explanation, which has general relevance, is that at high supersaturations the critical radius for nucleation of both calcite and aragonite is so small that the concept of surface and volume free energy used in classical nucleation theory is no longer valid. Assemblies of relatively few atoms need not have the same thermodynamic properties as the bulk, and the definition of the interface between 'crystal' and supersaturated solution requires an atomistic interpretation rather than a simple surface area dependence. When such factors are taken into account, a small nucleus of aragonite in a supersaturated solution may have a lower free energy than calcite, thus promoting its nucleation. Subsequent growth of aragonite may be kinetically easier than new nucleation of calcite even when the crystallite size makes calcite the more stable phase. The solid state transformation of the aragonite to calcite is completely inhibited at such low temperatures.

In aragonite which has grown within its stability field at high pressures and partially transformed to calcite during uplift, there is often a specific crystallographic relationship between the growing calcite nuclei and the aragonite host, with low-index planes in the calcite parallel to low-index planes in the aragonite. Despite the 7% increase in volume accompanying the transition there are some lattice spacings in calcite which are very similar to those in aragonite, allowing a degree of matching across the calcite – aragonite interface. A common feature of these relationships is the continuity of the Ca-atom positions across the interface, possible because in calcite the Ca^{2+} distribution approximates to cubic close-packing, while in aragonite it is approximately hexagonally close-packed. A coherent or semicoherent interface reduces the interfacial energy terms, and although the strain is increased, the balance favours the oriented growth. A number of different orientational relationships have been described, suggesting that the solid state transition mechanism may vary according to the conditions. The existence of such orientational relationships, termed *topotaxy*, illustrates that even in reconstructive transformations the interface exploits any structural relationships to reduce the total free energy.

Such observations have prompted many experimental studies of the kinetics of the aragonite to calcite transition as well as the reverse transition from calcite to aragonite. The apparent simplicity of a transition involving no compositional changes belies the complications caused by the unpredictability of nucleation processes in solids, as shown by the experimental results. Each published study produced different values of the overall rate of the transition as well as its temperature dependence, as shown by the range of activation energies from 150kJ/mole to 450kJ/mole! This is an enormous variation considering that the absolute rates are exponentially dependent on activation energy. It is generally agreed that the problem is due to the differences in factors such as grain size and shape, degree of crystallinity, and previous thermal and deformational history. It is difficult to quantify these

effects in either experiments or nature and yet nucleation kinetics is critically dependent on them.

The observation that it is possible to transform calcite to aragonite at room temperature by mechanical grinding is a further complication. A consequence of the relatively small difference in free energy between the phases is that the accumulated strain energy may raise the free energy of the calcite relative to both unstrained aragonite and unstrained calcite and the large shear stresses may promote some mechanism for the metastable transition to aragonite. The possible role of shear stress is mentioned again in Section 12.1.2 below.

In the presence of an aqueous solution the transformation from aragonite to calcite proceeds by a completely different mechanism of dissolution and reprecipitation via the fluid phase. The activation energy is low and the rate of transformation is far too rapid to be consistent with the preservation of aragonite on the time scale of metamorphic processes. This implies that preserved high pressure aragonites have been isolated from fluids during uplift.

It may seem from the above discussion that there is little likelihood of being able to apply any experimental data on the aragonite – calcite transition to understanding natural observations. However, if the unpredictable nucleation part of the process is isolated from the measurements, the subsequent growth rate behaves in a reproducible way. Using the assumption that growth rates measured under laboratory conditions (between 350°C and 450°C and 1 atmosphere pressure) can be extrapolated to the geological conditions during uplift ($\approx$150°C and 5kbars), which implies that the activation energy and hence transition mechanism remain the same, limits can be placed on the geological history of the sample.

The textures of natural partially transformed aragonites show many small calcite crystals throughout the aragonite, suggesting that the slow growth rate of calcite rather than its nucleation rate was responsible for the preservation of aragonite. Based on the size of these calcite crystals ($\approx 100\mu m$), the extrapolated experimental results suggest that the temperature during uplift could not have been much greater than 150°C as the rock passed into the calcite stability field. A higher temperature would have resulted in total transformation; a lower temperature would have produced no calcite growth at all.

Further general details of the relationship between experimental kinetic studies and their application to natural samples are discussed in Chapter 10.

(ii) Kyanite – andalusite – sillimanite

The polymorphic transitions in the aluminium silicate (Al_2SiO_5) minerals are probably the most notorious example of the difficulties encountered in determining an equilibrium phase diagram and then attempting to relate it to natural occurrences of these minerals. The general form of the phase diagram is shown in Figure 12.2(a) but there have been very significant deviations in the details, particularly the position of the invariant 'triple point'. The determination of the equilibrium phase relations has been particularly important in metamorphic petrology where the presence of a particular polymorph has long been used as an indicator of the metamorphic conditions during crystallisation of the rock.

The problems encountered in its determination can be attributed to two related factors. First, the uncertainties in the thermodynamics of the individual phases and second, to metastability both in the growth and the persistence of phases outside their stability field.

The structures of the phases, described in Section 6.3.3, have a number of features in common, although a transformation from one to another involves breaking Si–O and Al–O bonds, i.e. a reconstructive transition. The similarity of the structures is reflected in very small differences in the thermodynamic properties of the polymorphs. At equilibrium the free energy difference is zero, and

increases as the equilibrium boundary is overstepped. The energy difference depends on the relative slopes of the free energy curves as a function of temperature and pressure, i.e. on $\partial G/\partial T = -S$, and $\partial G/\partial P = V$ for each phase. The differences in the entropy and volume of each phase as a function of temperature and pressure therefore defines the free energy difference between the phases.

Entropy values have been estimated from heat capacity measurements and indicate that over a range of temperature, ΔS for andalusite – sillimanite is only about 4.4 J mol^{-1}K^{-1}, for sillimanite – kyanite it is 13.5 J mol^{-1}K^{-1} and for andalusite – kyanite ΔS is around 9.1 J mol^{-1}K^{-1}. When combined with the small temperature and pressure dependence of the molar volume differences, the result is that there are only very small free energy differences between the polymorphs even at temperatures and pressures considerably above or below the equilibrium values. The smallest free energy difference is between andalusite and sillimanite, and the uncertainty in the position of the equilibrium boundary between this pair contributes most to the uncertainty of the position of the triple point.

The small free energy difference between the polymorphs, combined with the necessity of a reconstructive transition to convert one to the other in the solid state, has a number of consequences which explain the problems in the determination of the equilibrium phase boundaries as well as the metastability in this system.

1. Phase equilibrium determinations where starting reactants are raised to various P, T conditions and the nature of the products determined will be

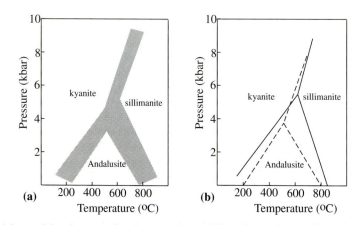

Figure 12.2. (a) The general form of the phase relations between the Al_2SiO_5 polymorphs. **(b)** Two experimental determinations of the phase diagram by Richardson *et al.* (1969) shown in solid lines, and by Holdaway (1971) shown in dashed lines.

extremely prone to any factors which affect the free energy. Strain energy in deformed materials, surface energies due to grain size, the presence of impurities and defects etc. all have magnitudes comparable to the free energy difference between the phases. For example, if the starting materials are finely ground, the surface energy terms may contribute around 400 J mole^{-1} and the defect energies introduced may be comparable. A change in ΔG of 400 J mole^{-1} would result in a shift of around 50°C in the position of the apparent equilibrium curve for kyanite – andalusite, and 300°C in the andalusite – sillimanite curve, making any petrological application meaningless.

Sillimanite often occurs as a fine grained acicular variety, termed fibrolite, and measurements of its heat capacity indicate a 4% increase in the entropy. This translates to an increase in the pressure and temperature of the triple point by around 2kbars and 200°C. The possible thermodynamic effect of Fe^{3+} substitution must also be taken into account in this context, although the effects of Al,Si disorder in sillimanite, once considered to be a likely source of excess entropy and enthalpy have now been discounted.

It is possible therefore that the various experimental equilibria are only relevant for the particular materials used, and their applicability to real conditions in nature is uncertain.

2. A second problem common in polymorphs with small free energy differences is the metastable crystallisation from reactants with high free energies, i.e. under conditions of high supersaturation. If we start with a glass or an oxide mix of Al$_2$SiO$_5$ composition and attempt to synthesise any of the polymorphs around their equilibrium P,T conditions, the ΔG for the synthesis reaction will be very large compared to the ΔG between the polymorphs themselves. This is illustrated schematically in Figure 12.3. The large free energy drive to nucleate any of the polymorphs means that which of them actually forms depends less on the thermodynamics than on kinetic factors such as preferential nucleation etc. Such synthesis reactions are always suspect as indicators of stability.

In metamorphic rocks, the aluminium silicates are usually formed by dehydration reactions such as:

muscovite + quartz ⇒ Al$_2$SiO$_5$ + K-feldspar + H$_2$O

which have a large ΔG compared to the ΔG between the polymorphs. Whether a stable or metastable

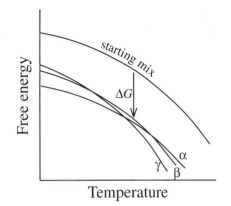

Figure 12.3. Schematic free energy – temperature curves for three polymorphs α, β and γ which have only small free energy differences, and the high energy starting synthesis mix. The free energy difference between the mix and any of the polymorphs is very much greater than that between the polymorphs, and the phase which is synthesised under any particular conditions depends more on the kinetics of the process than thermodynamic stability.

polymorph forms depends on their relative rates of nucleation and growth. In a heterogeneous system such as a rock, the role of preferential nucleation is difficult to evaluate, although it is unlikely that there would be a marked preference for the formation of the metastable phase and simultaneous suppression of the stable phase on a consistent basis. Over a fairly wide range of conditions both phases might be expected to occur, a situation borne out by natural assemblages. The overall correlation between the polymorph present in a metamorphic rock and the P,T conditions inferred from the rest of the mineral assemblage does support the qualitative use of the phase diagram.

Another factor which may be important in determining the phase which nucleates under non-equilibrium conditions has already been mentioned in the case of aragonite, i.e. the problem of defining the thermodynamics of nuclei when the critical radius consists of a relatively small assembly of atoms.

3. A third consequence of the small free energy difference between the polymorphs is the effect on the kinetics of transformation from one to the other. In a reconstructive transition small values of ΔG mean that very large departures from equilibrium will be required to reduce the activation energy for nucleation sufficiently to drive the reaction forward (eqn.(11.3)). The metastable persistence of phases outside their stability field is therefore to be expected. Furthermore, any evidence of a transformation from one phase to another in the solid state does not imply that the P, T conditions were anywhere near the equilibrium phase boundaries, rather the opposite. A rock containing all three polymorphs need not have ever been near the triple point, wherever it may be.

In nature, the disappearance of one polymorph and the growth of another as the metamorphic grade increases usually takes place by a series of reactions involving other minerals, rather than by the direct transformation of one polymorph to another. The extent of metastability at each stage is not known, but the role of fluids in catalysing metamorphic reactions is of primary importance.

Most of the methods discussed in Section 8.5 have been used in an attempt to obtain the 'correct' equilibrium phase diagram. Two of the most widely quoted experimental determinations using phase equilibrium studies are shown in Figure 12.2(b), despite the fact that the triple points differ by 1.75 kbars and 110°C. Another approach is to use well-characterized samples to measure the specific heat and volume as a function of temperature and so calculate the thermodynamic parameters needed to determine the phase diagram. This approach requires relatively large amounts of material to be able to obtain calorimetric data of sufficient accuracy, and also includes the effects of any impurities and defects. To avoid the experimental problems of surfaces, defects and impurities, specific heats can be calculated directly from the lattice vibrations (Section 8.5.3) to obtain a phase diagram for the ideal crystal structure. Finally the thermodynamic parameters can be calculated directly using a computer simulation of the structures (Section 8.5.4). While there is a general agreement in all of the methods, the relative merits of these various approaches is a matter of current debate.

(iii) The transformation of pigeonite to orthopyroxene

The microstructure of orthopyroxene which has transformed from pigeonite has been referred to in Section 11.4.1, without any discussion of the mechanism of the transformation except the indication that a twinned crystal of pigeonite transformed to a single crystal of orthopyroxene. There appear to be a number of different possible mechanisms for the pigeonite – orthopyroxene transformation, a feature which appears to be typical of many reconstructive transitions.

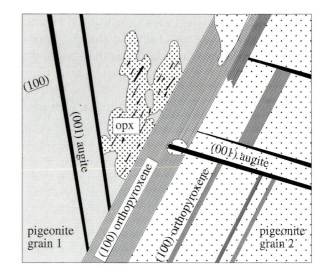

Figure 12.4. Drawing from an optical thin section from the Moore County meteorite showing two pigeonite grains partially transformed to orthopyroxene. In both grains the pigeonite has exsolved augite on (001) planes. In grain 2, orthopyroxene has grown topotactically on (100) planes in the pigeonite, possibly by an isothermal martensitic mechanism. This orthopyroxene provided a nucleation site for the massive transformation of parts of grain 1 into orthopyroxene. The orthopyroxene in grain 1 has no specific orientation relative to the host pigeonite, but has the same orientation as the orthopyroxene in grain 2. (Courtesy of G.L. Nord.)

The transformation from pigeonite to orthopyroxene may be isochemical, with augite subsequently exsolving from the orthopyroxene. In such cases the orthopyroxene may preserve some of the crystallographic features of the pigeonite, such as the silicate chain orientation, so that their c axes are parallel. Alternatively, there are many examples where there is apparently no crystallographic continuity between the orthopyroxene and the pigeonite. These two different mechanisms may operate at different cooling rates, with the probability that the latter mechanism is more likely at higher temperatures.

The case where the orthopyroxene has no special crystallographic orientation with respect to the parent involves a random nucleation event followed by growth of the incoherent interface which sweeps through the pigeonite (Figure 12.4). Such transformations in which there is no change in composition, and atomic diffusion is restricted to atomic jumps across an incoherent interface are often termed *massive*. At this high-energy interface the pigeonite structure is broken down and reassembled as orthopyroxene, the interface moving sufficiently fast to completely transform the pigeonite before any further nucleation has taken place. There are also cases where a cluster of pigeonite crystals has been transformed to a single crystal of orthopyroxene – evidently the grain boundaries need not represent a significant barrier to orthopyroxene growth. A very slow nucleation rate, yet a relatively fast growth rate suggests near equilibrium conditions for long periods.

Crystallographic continuity across the orthopyroxene – pigeonite interface on the other hand, may indicate that the initial nucleation was topotactic, the preservation of certain orientations and common lattice spacings reducing the overall interfacial energy. Another possibility is that the mechanism may involve some cooperative mechanism in which atoms move across the growing interface in an orchestrated way, by fixed displacements such as would be produced by the passage of partial dislocations. This mechanism is discussed in more detail in the next section (12.1.2).

The third mechanism of pigeonite transformation is a chemical decomposition reaction which may occur at high temperatures when the pigeonite composition lies near the eutectoid (see the phase diagram Figure 11.16). At the eutectoid (see Chapter 8 Appendix) the pigeonite breaks down to form the lower temperature equilibrium phases simultaneously by the reaction

pigeonite ⇒ orthopyroxene + augite

Such eutectoidal decompositions reactions usually begin by the heterogeneous nucleation of one of the phases followed by adjacent nucleation of the other. For example, nucleation of augite will deplete the immediate surroundings in Ca, thus promoting the nucleation of orthopyroxene. Repetition of this process produces a 'cell' which resembles a colony of alternating lamellae (Figure 12.5(a),(b)), and the whole colony advances into the parent phase (Figure 12.5(c)). Subsequent coarsening of this type of microstructure is thought to be responsible for inverted pigeonites in which irregular blebs of augite are intergrown with orthopyroxene. The augite and orthopyroxene are generally crystallographically related, sharing common c axes and with parallel (100) planes.

(iv) Quartz – tridymite – cristobalite

In our final example in this section we will add a few remarks regarding transformations between the silica polymorphs. As pointed out in Section 6.8.1, all three are framework silicates, with tridymite and cristobalite having similar structures based on different stacking arrangements of hexagonal rings of SiO_4 tetrahedra. Reconstructive transformations are required to transform one to the other. The difference in structure between quartz and the other polymorphs does not allow any coherency across an interface between them and a partly transformed crystal is likely to be an intergrowth of irregular grains within the host phase.

The tridymite–cristobalite structural relationship

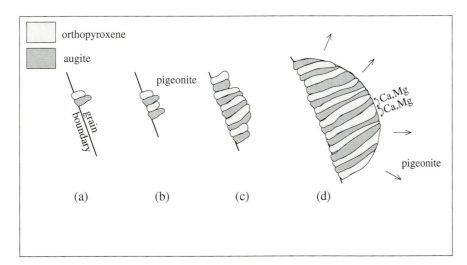

Figure 12.5. Mechanism of the eutectoidal breakdown of pigeonite to orthopyroxene + augite. **(a)** Adjacent nucleation of opx and augite at a grain boundary. **(b) to (d)** Further nucleation and growth of a 'cell' of alternating lamellae of opx and augite. The cell grows out into the pigeonite host and Ca,Mg interchange takes place between the lamellae at the growing interface.

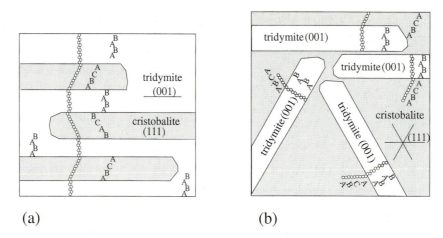

Figure 12.6. (a) In the topotactic transformation of tridymite ⇒ cristobalite the SiO$_4$ layers (denoted A, B etc) are parallel. The cristobalite lamellae may form in two twin-related orientations, with (111) planes of cristobalite parallel to (001) planes in the tridymite host. **(b)** In the cristobalite ⇒ tridymite transformation the tridymite (001) planes have four different possible orientations which preserve the topotactic relationship between the layers, three of which are shown here. The fourth orientation is parallel to the plane of the page in this section.

allows the possibility of a lamellar intergrowth in which the layers of SiO$_4$ rings are parallel. Observations of intimately interlayered slabs of the two polymorphs suggests that a transformation involving a cooperative mechanism at a coherent interface is possible. In practice both the massive and the topotactic transformations occur. The massive transformation produces randomly oriented nuclei of the new phase which grow to a granular microstructure within the parent phase. In the topotactic transformation the interface between tridymite and cristobalite is planar, with oriented lamellae such that {111} planes of cristobalite are parallel to the (001) plane in tridymite. The fact that there are four equivalent orientations of the {111} planes in cristobalite and only one (001) orientation in tridymite means that in a transformation from tridymite to cristobalite, the cristobalite will grow as parallel lamellae which may be twin-related. In the reverse transformation, four different lamellar orientations of tridymite are possible, each parallel to one of the {111} planes of cristobalite (Figure 12.6).

As is generally the case in reconstructive transitions, the relative rates of the different processes depend on the temperature conditions, as well as factors such as shear stresses on the crystal, grain size, defect density etc.

12.1.2 Transformation by partial dislocations – athermal and isothermal martensitic transformations

The fact that many mineral polymorphs can be described in terms of different stacking arrangements of planar modules, and the concept of polytypism, was introduced in Chapter 6 and discussed again in the context of planar defects in Chapter 7. Here we return briefly to the same topic in relation to the mechanisms of reconstructive phase transitions.

Example. Transitions between proto-, clino- and orthopyroxenes

In some mineral structures there is a clear polytypic relationship between the various structural types. Pyroxenes are a good example. The basic structure of proto-, clino- and orthopyroxene is described in terms of the stacking of 4.5Å thick slabs (Figure 6.15), which are related by a displacement vector **R** = 0.83[001] or 0.17[00$\bar{1}$] (Section 7.2.8). However, their structural differences require a reconstructive transition to transform one to another.

One mechanism of accomplishing transitions between such structures is to introduce shear displacements along the stacking planes. Such a displacement is equivalent to a stacking fault which can be generated by the movement of a partial dislocation. If the movement of such dislocations is coordinated it is possible to completely convert one structure to another. This has been demonstrated in the transformation of ortho to clinopyroxene (as well as the reverse) under shear stress. The presence of stacking faults in the transformed material confirms the role of partial dislocations in the reconstructive transition.

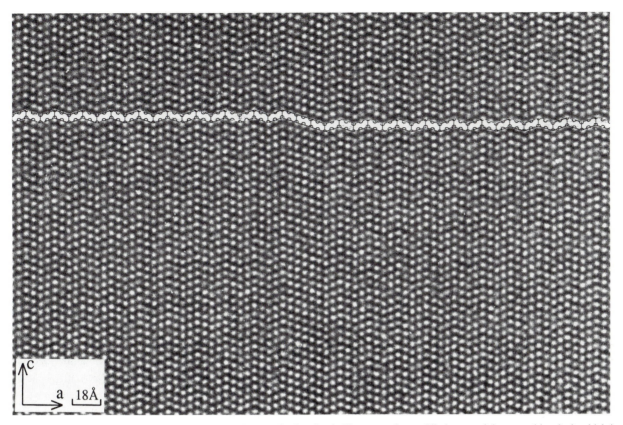

Figure 12.7. High resolution transmission electron micrograph of anthophyllite, an orthoamphibole, containing a stacking fault which is equivalent to a slab of clinoamphibole two unit cells thick. The arrangement of modules is outlined in the inset drawing and can be best seen when viewing the image at a low angle, parallel to the *a* axis.

A similar mechanism can be invoked to transform orthoamphibole to clinoamphibole. Figure 12.7 is a high resolution transmission electron micrograph of a (100) stacking fault in anthophyllite, which is equivalent to a two unit cell thick plate of clinoamphibole. Such defects may be formed during growth as well as by transformation.

A diffusionless transition in which there is a shear-induced change in shape of the unit cell is often termed a *martensitic transition* by analogy with similar transitions in metals. A partial dislocation can produce the required shear. Two types of martensitic transitions have been recognized:

(i) an *athermal martensitic transition* is unquenchable and the amount of transformation to the low temperature form depends on the temperature, and not on the time. At a given degree of undercooling a certain number of dislocations will form and move through the crystal. At a greater

undercooling more dislocations will form and the extent of transformation will be greater.

(ii) an *isothermal martensitic transition* is sluggish, the amount of transformation depending on time and not on the degree of undercooling. The transition can easily be suppressed by rapid cooling.

Examples of both types of martensitic transitions occur in pyroxenes. Protoenstatite is a stable phase above 1000°C at compositions around $MgSiO_3$ (Section 6.4.1). It is unquenchable, and rapid cooling results in the formation of highly faulted clinoenstatite. This proto-to-clino transformation has all the characteristics of an athermal martensitic transition. It is spontaneous and reversible. In the reverse reaction the amount of protoenstatite formed is independent of time and changes with temperature so that the transition appears to be smeared out over a temperature interval.

The stability field of clinoenstatite remains a problem. If the clinoenstatite is allowed to remain at high temperature (above about 650°C but less than 1000°C) it undergoes a slow, non-reversible, and time dependent transition to orthoenstatite. Clearly, orthopyroxene at compositions near $MgSiO_3$ must be more stable than clinopyroxene at these temperatures. When this orthoenstatite is then heated above 1000°C it undergoes a similar time dependent transition to protoenstatite. The mechanism of clino- to ortho-, and ortho- to proto- is consistent with an isothermal martensitic transition.

The relative stabilities of these three pyroxene structures as a function of temperature and composition is complicated by the role of shear stresses on the transition mechanism. As mentioned above, the ortho- to clino- transition can be accomplished by applying a shear stress, but it is not clear whether the stress merely provides a transition mechanism, or whether it plays a role in modifying the thermodynamics. The heating experiments in $MgSiO_3$ suggest that protoenstatite has a high temperature stability field above 1000°C below which orthoenstatite is stable. Clinoenstatite is most probably a metastable phase formed as a consequence of the rapid martensitic mechanism available as protoenstatite is cooled. Its metastability is consistent with the observation that when orthoenstatite is heated above 1000°C, its transition back to protoenstatite does not proceed via clinoenstatite as an intermediate.

In Ca,Fe bearing pyroxenes around the pigeonite composition the proto - structure does not appear at high temperature. Instead, clinopyroxene (high-pigeonite) has a high-temperature stability field and orthopyroxene is the stable low temperature phase (Figure 11.16). To be consistent with the previous paragraph, the clinopyroxene stability field must become narrower and finally disappear as the composition moves towards the $MgSiO_3$ corner of the pyroxene quadrilateral.

The problems of structurally similar polymorphs with very small free energy differences yet with large activation energies for transitions between them, discussed throughout Section 12.1.1, are all relevant to the discussion on clinopyroxene – orthopyroxene relations. The role of shear stress in promoting the movement of dislocations and the generation of stacking faults, clearly plays an important role. Note also that the stacking faults due to stresses associated with the high–low pigeonite transformation may be important in initiating the transformation of pigeonite to orthopyroxene (Section 11.1.6).

Other examples

We can make this discussion more general by including other minerals with polymorphs in which the different structures arise from the different modes of stacking of structurally compatible units (or structure modules). This subset of polymorphism is termed polytypism, and was introduced in Chapter 5 and amplified with examples in Chapter 6. Here we will restrict the discussion to those polytypes in which the structure modules are related by translations parallel to the stacking plane. The pyroxene structures are in this category, although there are only three different stacking modes possible. In other cases, such as the wollastonite polytypes (Section 6.4.2) different stacking sequences can lead to many different structures.

There are many minerals with polytypes. Often only two different structures have been observed, such as in clinozoisite and orthozoisite (ideal composition $Ca_2Al_3Si_3O_{12}OH$), in which modules are stacked parallel to (100) planes. Each successive module may be either undisplaced relative to the previous one or related by a displacement of ¼ [001] (Figure 12.8). In general, the two structures which result from such stacking arrangements have a clino–ortho or triclinic–monoclinic relationship with one structure doubling the lattice spacing of the planes parallel to the modules. Other examples include triclinic–monoclinic sapphirine and triclinic–monoclinic chloritoid.

In all such cases it is possible to devise a mechanism by which the passage of a partial dislocation carries out the required displacement. In some cases these shear displacements create an idealised configuration in which some of the atomic positions differ from those in the observed structure. For example, the shear may result in the correct anion distribution, but with an incorrect cation distribution. Some further local reorganisation of the cation positions may then be required following the shear. This process is termed *synchroshear* as it involves synchronised displacements of some cations as the partial dislocation moves along the plane. The transformation between clino and orthozoisite requires such a reorganisation of the Ca,Al positions although the partial dislocation with $\mathbf{b} = $ ¼ [001] produces the correct rearrangement of the Si and O positions.

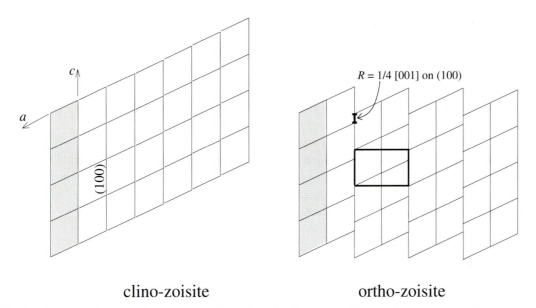

clino-zoisite   ortho-zoisite

Figure 12.8. The relationship between clinozoisite and orthozoisite unit cells in terms of the stacking arrangement of modules. The module (shaded) consists of a slab of structure one unit cell thick and parallel to the (100) plane. **(a)** In clinozoisite the modules are stacked along the *a* direction without any relative displacement along *b* or *c*. **(b)** In orthozoisite every alternate module is displaced along the *c* direction by the vector ¼[001] . The resultant unit cell (heavy line) is orthorhombic.

Dislocation glide mechanisms for reconstructive transitions have also been proposed for minerals which do not fall within the definition of polytypes. For example, the structural relationship between calcite and aragonite can be described in terms of a cubic close-packed array of Ca^{2+} in calcite and a hexagonal close-packed array in aragonite. The aragonite ⇔ calcite transition requires a change in the stacking sequence and this can be accomplished by the glide of partial dislocations. The CO_3 groups have to move independently of this shear by a 30° rotation and a small local displacement. Such a mechanism is able to explain the topotactic relationships between calcite and aragonite, as well as the well-known role of external stress discussed above (Section 12.1.1). Partial dislocations with the appropriate Burgers vector have been observed in calcite, but it is not known to what extent this mechanism operates in nature.

A dislocation glide mechanism has also been proposed for the transformation from sillimanite to kyanite. Stacking faults on (010) planes with displacements of ½ [100] have been observed in naturally and experimentally deformed sillimanite. Such faults, together with rotations of SiO_4 tetrahedra and a shift of the AlO_6 chains along the *c* axis produces a layer of the kyanite structure. To generate such a fault requires the dissociation of a disloca-

tion with a Burgers vector [100] on an (010) plane into two ½ [100] partials. The separation of the partials produces the stacking fault between them (Section 7.2.4). It has been experimentally observed that the separation of the partial dislocations is greatly increased as the sillimanite enters the kyanite stability field suggesting that the stacking fault energy must have decreased. This supports the argument that the structure around the stacking faults may be similar to that of kyanite. The stacking faults may then act as preferred nucleation sites for further kyanite growth. In this kind of mechanism for a direct transformation from sillimanite to kyanite, strain plays the important role of generating the dislocations which provide a local structural environment which promotes nucleation.

The combination of dislocation glide with either 'synchroshear' or some other coordinated atomic displacements (commonly called 'shuffles') to produce the correct new structure is clearly a very flexible mechanism and many polymorphs can be structurally related in such a way. While there is experimental evidence to support some of these mechanisms, it is not yet clear to what extent they operate in nature. A final example of such a shear-induced mechanism for a reconstructive transformation is the olivine – spinel transition discussed below.

12.1.3 High pressure transformations in the mantle

The earth's mantle has major discontinuities in density at 400 kms and 670 kms as observed by the discontinuities in seismic velocities. These discontinuities divide the mantle into the upper mantle, transition zone, and the lower mantle (Figure 12.9). Such discontinuities could be caused by abrupt changes in composition, i.e. a chemically stratified mantle, or may also be explained by a uniform upper and lower mantle composition with the discontinuities due to high-pressure phase transitions in the constituent minerals.

Petrological studies, based mainly on the study of mantle-derived rocks and experimental work on partial melting of mantle rocks to form basaltic lavas, suggests that the upper mantle is composed mostly of Mg-rich olivine, $(Mg_{1-x}Fe_x)_2SiO_4$ where $x \approx 0.1$, together with lesser amounts of pyroxene and garnet. With increasing depth, and hence pressure, these phases transform to higher density structures. It is possible to reproduce the high pressures and temperatures in the mantle experimentally, and Figure 12.10 outlines the general form of the phase relations in olivine at high pressure, determined

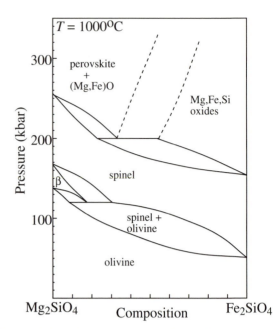

Figure 12.10. The high pressure phase relations in olivine $(Mg,Fe)_2SiO_4$ at 1000°C.

from a wide range of experimental and thermodynamic data. Such experiments have led to the general acceptance that the 400 km discontinuity is due to the transformation of the relatively open olivine structure to a denser polymorph with the spinel structure, without the need to invoke any bulk compositional changes. Figure 12.10 shows that the transition in Mg-rich compositions proceeds via the so-called β phase which has a modified spinel structure described below.

The 670 km discontinuity is more problematic, but is consistent with the disproportionation of the high pressure spinel phase to a mixture of magnesiowstite, $(Mg,Fe)O$, (which has the sodium chloride structure) and the perovskite phase $(Mg,Fe)SiO_3$. Thus:

at 400 km $(Mg,Fe)_2SiO_4$ (olivine) $\Rightarrow (Mg,Fe)_2SiO_4$ (spinel)

at 670 km $(Mg,Fe)_2SiO_4$ (spinel) $\Rightarrow (Mg,Fe)O + (Mg,Fe)SiO_3$ (perovskite)

Figure 12.11 shows the approximate volumetric constitution of the mantle. The other major phases stable in the upper mantle are garnet and pyroxene. As the pressure increases the clinopyroxene component in the mantle dissolves into the low-pressure garnet forming majorite which is a high-pressure Al-deficient garnet.

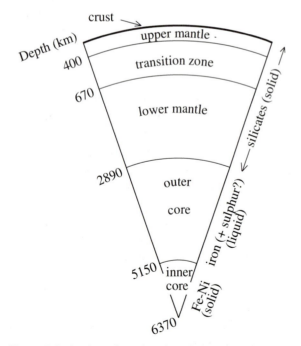

Figure 12.9. A schematic section through the earth. The seismic discontinuity at 400 km occurs at a temperature and pressure consistent with that for the transformation of olivine to spinel. The discontinuity at 670 km may be due to a change in chemical composition at this depth, with the perovskite phase of $MgSiO_3$ the primary constituent of the lower mantle.

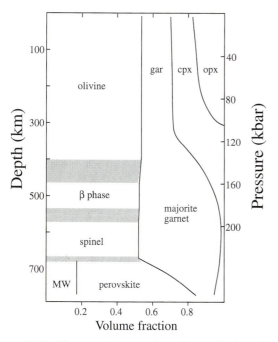

Figure 12.11. The approximate volumetric constitution of the mantle. The shaded areas are transition zones. MW is magnesiowstite, (Fe,Mg)O. (After Ito and Takahashi, 1987.)

As a final example of a reconstructive transition we discuss some aspects of the olivine to spinel transition.

Example. *The olivine – spinel transition*

The olivine structure can be described in terms of an hexagonal close-packed array of oxygen atoms, with Si occupying one-eighth of the tetrahedral sites and Mg,Fe in one half of the octahedral sites. The close-packed oxygen layers are parallel to (100) (Figure 6.3). In the spinel structure the oxygens are approximately cubic close-packed (Section 5.2.4) with the cations in the same coordination as in olivine. The transition from olivine to spinel thus does not change the coordination of the cations, an unusual situation in high-pressure phase transformations which involve a large increase in density. However, the density increase of around 8% is accomplished by changing the distribution of the cations in the available sites in such a way that the linkages between the cation polyhedra form a more compact structure.

The mechanism of the olivine – spinel transition has considerable importance to the mechanical properties of the mantle which in turn control the convective flow associated with plate tectonics, and

many experiments have been carried out on both silicates and their germanate analogues (used since they transform at lower pressures). Experiments are mainly of two types. The earliest experiments were on diamond-anvil cells in which a very small volume of material is compressed in a confined space between the faces of two opposing diamond crystals. Heating is generally by a laser beam. Later experiments on more recently developed large volume apparatus allow smaller shear stresses and more uniform heating rates using conventional resistance heaters. In both types of experiments X-ray diffraction can be performed on the samples in situ at high temperatures and pressures. The samples can also be quenched and hence examined at room conditions.

The experimental results suggest two different mechanisms for the transformation. One proposed mechanism is that of nucleation and growth in which the spinel crystals nucleate randomly on grain boundaries and grow into the olivine. The kinetics of growth is controlled by the diffusion of oxygen ions across the interface. No topotactic relationship between the olivine and spinel is required in such a mechanism. The other proposed mechanism is martensitic and hence diffusionless, involving the passage of partial dislocations on alternate close-packed oxygen layers to convert the hexagonal close-packing of olivine to cubic-close packing in spinel (see also Section 7.2.4). The cations are not however translated into correct positions by such a dislocation and additional cation displacements, i.e. synchroshear is required. In the martensitic mechanism the close-packed layers in the spinel and olivine are parallel, leading to a topotactic relationship between them with $(100)_{ol}$ parallel to $(111)_{sp}$. A variation on this mechanism is that the oxygen restacking precedes the much slower cation reordering, thus ruling out sychroshear in favour of a diffusion controlled cation migration. The volume reduction accompanies the oxygen restacking, indicating a reduction in the effective oxygen radius.

The present consensus is that the dislocation glide mechanism is most likely in experiments where there are large shear stresses, such as in laser-heated diamond anvil apparatus.

The role of the β phase in the transformation mechanism is not yet clear, although it is likely to be the prominent phase in the transition zone in the mantle between about 400 and 500 km depth (Figure

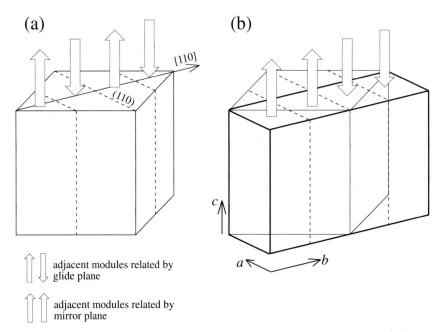

Figure 12.12. (a) A cubic unit cell of the spinel structure, showing the thickness and orientation of the (110) modules (one of which is shaded) which build up the structure. In spinel adjacent modules are related by a glide plane. **(b)** The same modular slabs are stacked differently in the β phase, being alternately related by glide and mirror planes. The new unit cell of the β phase is shown by the heavy line.

12.11). Its structure has been determined and can be described as a polytype of the spinel structure. As in spinel, the oxygen atoms are approximately cubic close-packed, but the cation distribution is different. The relationship between the two structures is best understood by first describing the spinel structure in a new way. The modular 'slab' which generates the spinel structure is an infinite sheet parallel to (110) as shown schematically in Figure 12.12. In spinel these modules are stacked such that adjacent sheets are related by a glide operation. This glide lies in the (110) plane and has a displacement of $\frac{1}{4}[1\bar{1}2]$ which is an anion– anion vector and so only affects the cation arrangement. In the β phase the same modular sheets are stacked in a different way, and are alternately related by glide and then mirror plane. The resulting structure has the same oxygen array as spinel but a new cation distribution which gives a unit cell which is orthorhombic and related to the cubic cell of spinel as shown in Figure 12.12(b).

A consequence of the structural relations between the β phase and the spinel is that one can be transformed to the other by the passage of partial dislocations with Burgers vector equal to this glide operation. Since an additional glide plane displacement on the plane converts an already existing glide to a mirror and vice versa, the resulting stacking fault has a local structure of the other phase. In other words, a stacking fault in the β phase has the spinel structure, and vice versa. As we have seen in other examples above, it is likely that stress plays an important role in generating such dislocations and providing nucleation sites for the transformation to proceed.

Microstructures associated with transformations between the $(Mg,Fe)_2SiO_4$ polymorphs can also be observed in meteorites which have been subjected to shock during some extra-terrestrial impact. Figure 12.13 is a low resolution electron micrograph of a grain of the spinel phase (termed ringwoodite) which contains complex stacking faults. The stacking faults in the spinel locally have the structure of the β phase and may have originated due to a partial transformation of the spinel back to the β phase on pressure release. Such stacking faults may provide suitable nucleation sites for further growth of the β phase as seen in Figure 12.14, where a fault-free grain of the β phase has nucleated within a highly faulted grain of ringwoodite.

The importance of determining the mechanism and hence kinetics of the olivine–β phase–spinel transition in the mantle lies in its effect on its mechanical properties and hence on the dynamics of mantle convection. This in turn has significant conse-

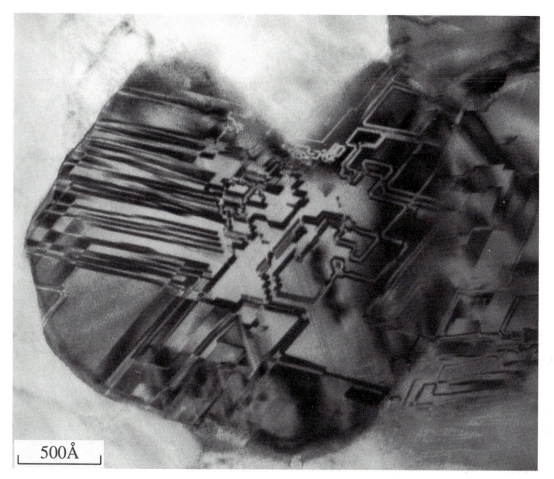

Figure 12.13. Transmission electron micrograph of a crystal of the spinel phase of Mg_2SiO_4 (ringwoodite) from a high pressure shock vein in the Tenham meteorite. The stacking faults crossing the crystal have a structure similar to that of the β phase. (See also Putnis and Price, 1979.)

quences for our understanding of the geochemical evolution of the mantle. The fact that olivine-bearing lithosphere is subducted at plate margins raises the question of the depth to which the slab will penetrate before the olivine transforms to spinel. A nucleation and growth mechanism would not be activated below some cut-off temperature, estimated at around 700°C. If the downgoing slab is cold, the activation energy for nucleation may be sufficiently large to inhibit the transition until higher pressures and hence greater depths are reached. If the plunging velocity of the downgoing slab of lithosphere is high, the temperature at its centre may remain below this kinetic cut-off even to depths in excess of 600 kms. Here the overpressure will be high and the olivine in a highly metastable state. As the slab heats up to above the cut-off temperature, the rate of the transition to spinel will be so rapid, and the change

in free energy so great, that the transition becomes implosive. The energy, released as seismic waves, is more than sufficient to generate deep-focus earthquakes in the downgoing slab.

The depth at which the transition occurs will decrease with decreasing plunging velocity of the slab. Figure 12.15 shows the temperature – depth profiles in downgoing slabs subducted at different rates, together with an approximate equilibrium P, T phase boundary for the olivine – spinel transition. If the kinetic cut-off for the transition is at 700°C, as shown by the dashed vertical line, a greater subduction rate means a greater depth and hence a greater overpressure before this kinetic temperature is exceeded.

If on the other hand, the transition to spinel takes place by a rapid martensitic transition involving dislocation glide, the activation energy may be too

Figure 12.14. The clear region in the centre of this transmission electron micrograph is the β phase which has nucleated from the heavily faulted spinel which surrounds it. (From Price *et al.*, 1982.)

small to allow significant metastability and the transition would take place under near-equilibrium conditions, with a small free energy change. In such a case the transformation to spinel in a rapidly subducted slab could take place at rather a low temperature (≈600°C). This has consequences regarding the mechanical properties of the slab which in turn are related to the grain-size of the spinel crystals. Such discussion is beyond the scope of this section, but the bibliography contains further reading on this topic.

12.2 Displacive polymorphic transitions in minerals

In a *displacive* transition the primary bonds in the structure are not broken but merely distorted. There is therefore a clear structural relationship between the high-temperature form which is usually more open and has higher symmetry, and the distorted low temperature form. This relationship allows a displacive transition to be continuously monitored in terms of the temperature (or pressure) variation of some structural parameter, such as a bond angle. The symmetry relations between high and low forms also enable a theoretical interpretation of the experimental observations to be made in a way which is not possible for reconstructive transitions.

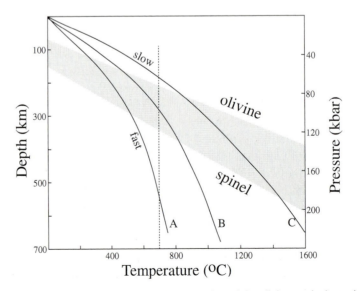

Figure 12.15. Temperature – depth diagram showing the approximate position of the olivine – spinel transition (shaded). If 700°C is the temperature required for the olivine – spinel transition to proceed at any significant rate, the pressure at which a downgoing slab passes through this kinetic barrier depends on the rate of subduction. Three different *P–T* profiles for downgoing slabs subducted at different rates are shown. (Modified from Rubie, 1984.)

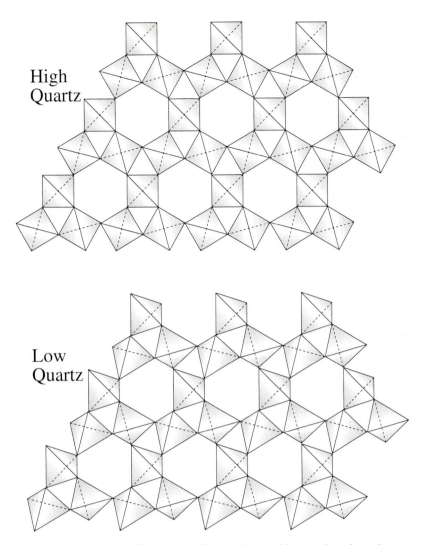

Figure 12.16. The structures of high and low quartz. The rotation of the relatively rigid tetrahedra reduces the symmetry from hexagonal to trigonal. Note how the distortion changes the shape of the six-membered ring of SiO_4 tetrahedra.

The fact that displacive transitions only involve small changes in energy and are usually fast and unquenchable may make them seem relatively insignificant compared with reconstructive transitions. However, displacive transitions provide a range of phenomena relevant to understanding the response of a mineral structure to changes in temperature and pressure, and to the development of domain structures in minerals. Most importantly, displacive transitions may interact with other aspects of the structure, such as the distribution of cations and their state of order. For example, the nature of the displacive transition in anorthite $CaAl_2Si_2O_8$ depends very sensitively on the degree of Al,Si order. Once this dependence is quantitatively understood it becomes possible to study the rapid and

hence experimentally accessible displacive transition to obtain information on the progress of the very sluggish Al,Si ordering transition. In the same context, the behaviour of albite, $NaAlSi_3O_8$, on cooling cannot be properly understood without considering the interaction of the displacive and the Al,Si order–disorder transitions. Such interactions will be discussed towards the end of this chapter, after the section on order–disorder transitions.

The most widely quoted mineralogical example of a displacive transition is the transition from high to low quartz at 573°C, and the technological importance of quartz as a stable-frequency oscillator in electronics has made this a very well-studied transition. Figure 12.16 compares the structures of high and low quartz, showing how the rotation of the relatively rigid SiO_4 tetrahedra results in a reduction in symmetry. Since there is no change in the translational symmetry i.e. the unit cell is essentially the same in both polymorphs (although with slightly changed dimensions), the transition is referred to as a zone centre transition (see Section 8.6). Another classical displacive transition is the cubic $\Rightarrow$ tetragonal transition in the perovskite structure, illustrated for the case of $SrTiO_3$ in Figures 5.23 and 8.18. In this case alternate octahedra rotate in opposite senses, resulting in a new unit cell which is a supercell of the cubic cell. This is therefore a zone boundary transition.

The common feature of both of these displacive transitions is that as the temperature falls, the high symmetry structure becomes unstable relative to some specific distortion. There are a number of different approaches used to describe how and why this distortion takes place. There have also been attempts, using computer simulations (Section 8.5.4), to predict how a given structure will distort on cooling. However, since there are so many different distortions possible in a complex structure, all with similar energies, it is not yet possible to discriminate between these many possibilities. We are left therefore to try to explain the observed experimental behaviour in terms of the structures of the polymorphs. In the rest of this section we will discuss some of these approaches, with examples from various mineral groups.

12.2.1 Displacive transitions and soft modes

In Section 4.4 we described the concept of normal vibrational modes of a crystal structure. Each mode is associated with a specific periodic distortion of the structure, in which the individual atoms undergo coordinated displacements. In a displacive transition one of these normal modes becomes 'frozen-in' as the temperature falls, superimposing this distortion on the high temperature form. In other words, when the high temperature symmetric phase is cooled, the frequency of this 'soft' mode decreases to zero, at which point the crystal is unable to withstand the corresponding distortion and undergoes a transition to a lower symmetry phase. The term 'soft' is used to suggest that the crystal is soft against the corresponding displacements of the atoms.

The distortion associated with the high–low transition in quartz and illustrated in Figure 12.16, is due to the softening of a normal lattice mode which occurs at 208 cm^{-1} in the high temperature form. Since this is a zone centre transition it can in principle be measured by Raman or IR spectroscopy, as pointed out in Section 4.4. However, not all modes can be detected in this way due to the symmetry-dependent selection rules which define which modes are active or inactive (Section 4.3). In the case of quartz, the soft mode in the high – low transition is Raman active.

In a zone boundary transition, such as that in $SrTiO_3$, the mode softening is associated with a wavelength which doubles the size of the high temperature unit cell. It becomes more convenient to discuss the situation in reciprocal space, in terms of the behaviour of the phonon dispersion curves as a function of temperature. The soft mode frequency goes to zero at the Brillouin zone boundary, at points corresponding to the appearance of the new reflections in the diffraction pattern of the low temperature form. To measure normal modes at wave vectors other than $\mathbf{k} = 0$, neutron scattering must be used.

Figure 12.17 illustrates soft mode behaviour for both zone centre and zone boundary transitions, as seen in schematic phonon dispersion curves. Soft modes may be optic modes or acoustic modes. In the case of acoustic modes, the frequency at $\mathbf{k} = 0$ is already zero, and the softening is due to the *gradient* of the dispersion curve falling to zero. This gradient defines the velocity of sound and the relevant elastic constant in the crystal, and a consequence of it becoming zero is a shear distortion of the unit cell.

The change in the crystal structure due to mode softening can be monitored by measuring the change in some order parameter as a function of temperature. The order parameter, Q, as defined in Section 8.6, may directly describe some structural parameter which changes during the transition or be related to it in some way. For example, in the high–low quartz transition, the order parameter could be related to the tilting angle η of the SiO_4 tetrahedra, while in $SrTiO_3$ it is the angle of rotation ϕ of the octahedra. The variation of Q with temperature allows the excess thermodynamic quantities due to the transition to be determined, using Landau theory as described in Section 8.6. Further examples will be discussed later in this section.

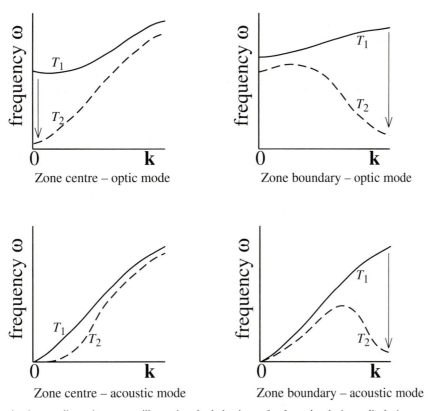

Figure 12.17. Schematic phonon dispersion curves illustrating the behaviour of soft modes during a displacive transition. Four different cases are illustrated depending on whether the transition is associated with a zone centre or zone boundary, and whether the soft mode is an optic or acoustic mode. In each case the full line shows the mode in the high temperature form at temperature T_1, while the dashed line shows the collapse of the mode in the low temperature form at T_2.

12.2.2 A geometric approach – polyhedral tilting

Many mineral structures can be described in terms of corner-linked, rigid polyhedra which enclose large cavities or cation sites which are weakly bonded compared with the rigid polyhedral framework. Examples include structures such as quartz, albite and perovskites. The effect of an increase in temperature on such structures is to preferentially expand the large sites, and this is achieved by rotation and tilting of the strongly bonded polyhedra. An increase in pressure tends to have the opposite effect and the large cavities are compressed. In this way of thinking about structures, it is assumed that there is some critical size below which the high temperature/low pressure structure is no longer stable. A displacive non-quenchable and reversible transition takes place when, on cooling or compression, these large sites reach some critical minimum size.

The low temperature, high pressure structure produced by polyhedral tilting is the low symmetry form, and in general there will be several symmetrically equivalent orientations in which the polyhedral tilting can occur. This results in the formation of transformation twins, a typical characteristic of this subset of displacive transitions.

In this polyhedral approach, since an increase in pressure has the opposite effect on the structure to an increase in temperature, the ratio of the thermal expansivity to the compressibility of the large sites determines the Clapeyron slope, dP/dT, of the phase transition. The presence of a cation in this site will also have an effect on its expansivity and compressibility. In general, the substitution of a larger cation expands the site and hence has a similar effect to an increase in temperature (or decrease in pressure). Assuming that these effects are linearly related we can define a transition surface in pressure–temperature–composition space at which the critical geometry for the transition is achieved.

Figure 12.18 shows the general form of such a transition surface for the monoclinic – triclinic dis-

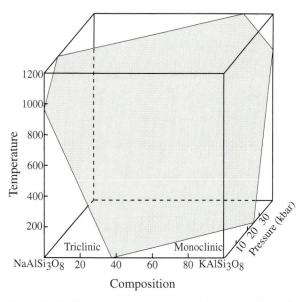

Figure 12.18. The general form of the transition surface due to the monoclinic to triclinic (C2/*m* to C$\bar{1}$) displacive transition in alkali feldspars as a function of temperature and pressure. High temperature and low pressure (the region in 'front' of the shaded surface) stabilizes the monoclinic form, while low temperature and high pressure (at the 'back' of the surface) stabilizes the triclinic collapsed structure. (After Hazen, 1976.)

placive transition in alkali feldspars with no Al,Si disorder (Section 6.8.3). In pure disordered albite NaAlSi$_3$O$_8$, the C2/*m* $\Rightarrow$ C$\bar{1}$ transition takes place at 980°C. With the substitution of K for Na the critical temperature drops sharply and falls to around 0°C at around 38 mole % K-feldspar. An increase in pressure promotes the transition to the triclinic state. K-rich feldspars, which in the Al,Si disordered state are monoclinic at room temperature, can be transformed to the triclinic form at higher pressure. At a composition 67 mole % K-feldspar the pressure required is about 12 kbar. At 82 mole % K-feldspar the transition to triclinic takes place at ~18 kbar, in agreement with the general prediction. Along the transition surface the structure has a critical geometry which is confirmed by the constant values of the lattice parameters.

The attraction of this description of a displacive transition is its conceptual simplicity, and given data on the temperature, pressure and composition dependence of the lattice parameters the slope of the transition surface can be calculated. In order to define its position in this *T–P–X* space, at least one experimental value of the transition temperature or pressure must be known, as in general, no specific value of the 'critical geometry' of the transition can

be independently determined. The relationship between the thermal expansivity and compressibility of the large weakly bonded site and the related changes in lattice parameters also depends on the structure type, and each case must be analysed separately.

The simplicity of this approach can be somewhat misleading if taken too literally. Detailed studies of the mechanisms of structural changes with changing temperature and pressure show that atomic movements which take place with a decrease in temperature are not *exactly* the same as those which take place with increasing pressure. Thus the transition surface and the concept of a 'critical geometry' may not survive rigorous crystallographic examination. Nevertheless, in many cases it can supply a qualitative prediction regarding the effect of temperature, pressure and compositional changes on the slope of phase boundaries.

12.2.3. Domain structures associated with displacive transitions

Transformation twins and antiphase domains were introduced in Sections 7.3.4 and 7.3.5 within the general context of planar defects. Both were related to phase transitions and the examples illustrated how both displacive and order–disorder transitions may give rise to domain structures. Here we describe some further aspects of domain structures with specific reference to displacive transitions.

Transformation twins may occur whenever a displacive transition involves a reduction in the point group symmetry of the crystal. A consequence of this symmetry loss is the possibility that a single crystal of the high temperature form will transform to different orientations of the low temperature unit cell. This results in a crystal made up of twin domains related to one another by the symmetry element lost during the transition. It is useful to make a distinction between two different types of transformation twins formed in displacive transitions, namely *ferroelectric* and *ferroelastic* twins.

The classic example of a *ferroelectric transition* is the transition in barium titanate, BaTiO$_3$, which above T_c = 120°C has the cubic perovskite structure. As outlined in Section 5.2.5, the displacive transition below this temperature involves the displacement of the Ti ions from the centre of the oxygen octahedra towards one of the apical oxygen ions (Figure

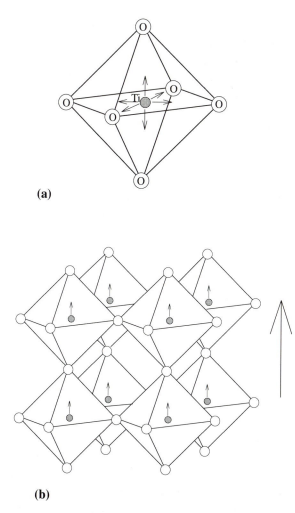

(a)

(b)

Figure 12.19. (a) In the $BaTiO_3$ structure, Ti ions at the centre of octahedra, move off-centre in one of six possible directions, towards one of the apical oxygen atoms. Each octahedron in the structure becomes an electric dipole. **(b)** A net spontaneous polarisation in the direction of the arrow results when the Ti displacements are all in the same direction.

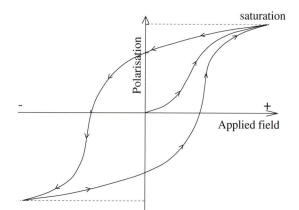

Figure 12.20. The hysteresis loop describing the relationship between an applied electric field and the polarisation of the $BaTiO_3$.

were all equally represented in the crystal the net polarization could be reduced to zero.

Applying an external electric field causes the individual TiO_6 dipoles to align themselves to the field, and this involves moving the twin-domain walls through the crystal, ultimately resulting in a single-domain crystal. At this point the spontaneous polarisation is a maximum i.e. the saturation value has been reached. Reversing the field direction creates domains with reverse polarisation, and again the twin-domain walls move in response to increasing the field. The relationship between the applied electric field and the spontaneous polarisation exhibits a hysteresis loop (Figure 12.20) whose shape depends on the ease with which domain walls can be moved.

The term *ferroic* is used to describe crystals which may contain two or more possible orientation states or domains which can be switched by the application of a suitable force or field. The term originates from the analogous phenomenon of ferromagnetism in which a spontaneous magnetisation arises below a certain critical temperature (the Curie temperature) due to the alignment of magnetic moments of ions containing unpaired electrons. Magnetite, Fe_3O_4, is a well-known mineral example. Ferromagnetic domains move under the influence of applied magnetic fields in a similar way to ferroelectric domains in an applied electric field, as described above. Ferromagnetic transitions are not displacive in the sense used here, and will be discussed in more detail in Section 12.6.

In a *ferroelastic transition* the structure distorts, changing the shape of the unit cell. A simple exam-

12.19(a)), reducing the symmetry to tetragonal. The cubic structure does not have a net electric dipole moment since the electrically charged ions are symmetrically positioned. The structure is termed *paraelectric* in this state. Below T_c the displacement of the Ti ions results in the formation of TiO_6 dipoles which tend to align themselves parallel to one another causing a spontaneous polarisation (Figure 12.19(b)). This state is termed *ferroelectric*. Ferroelectric twin domains arise when the direction of displacement and hence polarisation is not the same throughout the whole crystal, but occurs in two or more of the symmetrically equivalent directions in which the Ti atom could move from the centre of the octahedron. If the domains were small enough and

ple is the cubic–tetragonal transition in SrTiO$_3$ discussed above. Relative to the undistorted high temperature form, the low temperature structure has a spontaneous strain which defines the change in shape (see also Section 8.6). The twin domains which arise due to the different possible orientations of the tetragonal unit cell relative to the cubic parent are *ferroelastic domains*, whose walls may be moved by the application of an applied mechanical stress. This is clearly related to the phenomenon of mechanical (or deformation) twinning (Section 7.3.5), and it is important to note that the twin microstructure observed in minerals from deformed rocks will be influenced by both their thermal and deformational history. The relationship between the applied mechanical stress and the macroscopic strain of a ferroelastic crystal is described by a hysteresis loop directly analogous to that for the ferroelectric shown in Figure 12.20, and is due to the movement of ferroelastic domain walls.

In terms of the soft mode model of a displacive transition, a ferroelectric transition is due to the softening of a transverse optic mode at **k** = 0 (Figure 12.17(a)). If the size of the unit cell is doubled the softening takes place at a Brillouin zone boundary (Figure 12.17(b)). A ferroelastic transition is associated with the softening of an acoustic mode at **k** = 0.

The progress of a displacive transition as a function of temperature is monitored by measuring the value of the spontaneous polarisation (in the case of ferroelectrics), or spontaneous strain (in ferroelastics). These quantities define the order parameter as defined in Section 8.6, and using Landau theory, their variation can be used to determine the thermodynamic properties of the system.

Although the concept of ferroelasticity and spontaneous strain is very useful in describing displacive transitions in many minerals, there are important examples in which the strain arises as a secondary effect of a displacive transition. For example, in the $I\bar{1}$ to $P\bar{1}$ transition in anorthite at 240°C (discussed in more detail in next section) there is no change in the point group symmetry, but there is a spontaneous strain. Such transitions have been termed *co-elastic*.

As well as transformation twinning, displacive transitions may lead to the formation of *antiphase domains* if the transition involves a reduction in translational symmetry, such as increasing the size of the unit cell, or changing the lattice type i.e. is a zone boundary transition. As outlined in Section

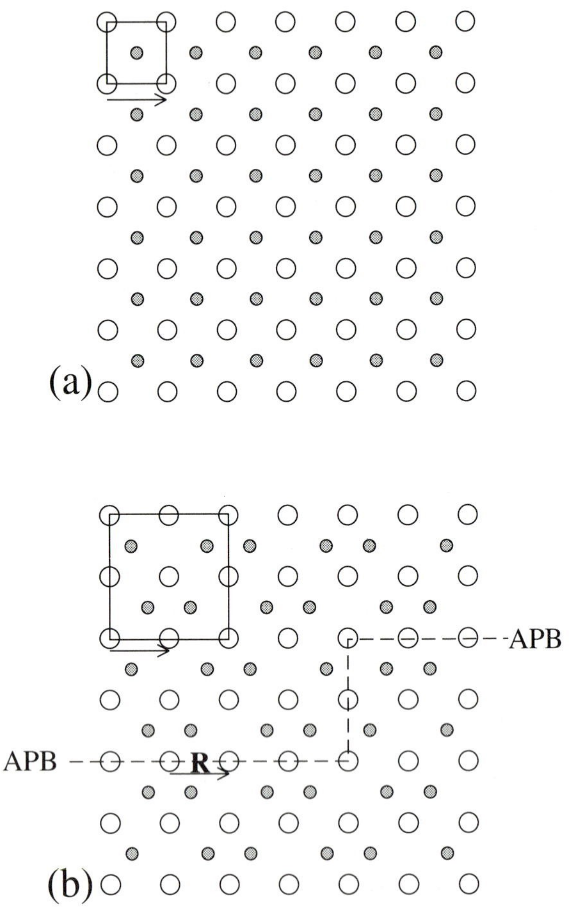

Figure 12.21. Schematic diagram illustrating the formation of antiphase domains in a displacive transition. **(a)** The high temperature form with the shaded atom in the centre of the outlined unit cell. The arrow indicates a lattice vector. **(b)** The transformation involves a displacement of the shaded atom in opposite directions in alternate cells. The supercell is outlined; the original lattice vector is no longer a lattice vector in the low temperature form. The antiphase domains are related by the vector **R** across the antiphase boundary (dashed).

7.3.4, an antiphase domain boundary is a type of stacking fault for which the translation vector **R** relating adjacent domains is a lattice vector for the high symmetry structure but not for the low symmetry structure. Figure 12.21 is a schematic diagram showing the formation of an antiphase boundary due to a displacive transition.

12.2.4 Examples of displacive transitions in minerals

Two examples of displacive transitions which have already been discussed from the point of view of the development of microstructure are the high – low pigeonite transition (Section 7.3.4) and the cubic –

tetragonal transition in leucite (Section 7.3.5). Here we describe three further examples, each illustrating different aspects of displacive transitions.

(i) The $I\bar{1} \Rightarrow P\bar{1}$ transition in anorthite $CaAl_2Si_2O_8$

When anorthite with complete Al,Si order is cooled below 237°C it undergoes a displacive transition which can be described as the softening of a vibrational mode mainly associated with the Ca atomic positions in the structure. The displacement of the Ca atoms, accompanied by a distortion of the unit cell, reduces the symmetry from $I\bar{1}$ to $P\bar{1}$. This causes the appearance of extra superlattice reflections whose intensity is a continuous and reversible function of temperature. The intensity I, of the superlattice reflections can be used to define the progress of a transition, and as was pointed out in Section 8.6 this intensity scales as Q^2, the square of the order parameter. Experimental measurements of I^2 are plotted against temperature in Figure 12.22(a). The plot is linear and hence $I^2 \propto (T - T_c)$. Since $Q \propto (T - T_c)^\beta$ the data indicate that $\beta = \frac{1}{4}$. This defines the transition as tricritical, i.e. thermodynamically intermediate between first- and second-order.

This result is confirmed by measurements of the changes in the lattice parameters as a function of temperature, from which the spontaneous strain tensor ε, can be calculated. For a zone boundary transition $\varepsilon \propto Q^2$ and since the plot of experimentally measured values of ε^2 against temperature is also linear (Figure 12.22(b)), this confirms that the critical exponent $\beta = \frac{1}{4}$, as expected for a tricritical transition. From this data the change in free energy due to the transition can be determined as a function of temperature, as discussed in Section 8.6.

This displacive transition is very strongly affected by the degree of Al,Si order in the anorthite, and hence by its thermal history. When the experiments are repeated on a volcanic anorthite containing around 3 mole% albite component and with less Al,Si order, both the critical temperature of the transition and its thermodynamic character change. As explained in Section 8.6, the effect of coupling between the order parameter Q for the pure displacive transition, and the order parameter Q_{od} which describes the degree of Al,Si order, results in a new critical temperature T_c^* for the coupled transition. In ordered anorthite $T_c = 510K$ and the transition is tricritical; in slightly less well-ordered anorthite T_c^*

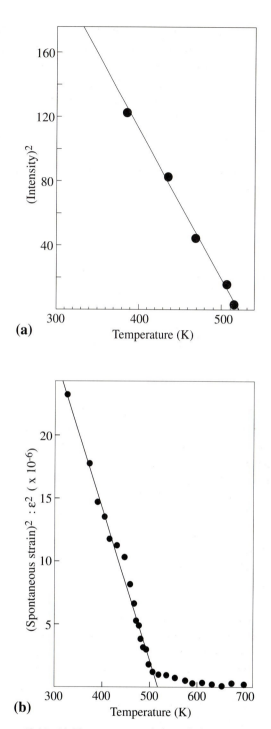

(a)

(b)

Figure 12.22. (a) Temperature variation of the square of the intensity of a superlattice reflection due to the $I\bar{1} \Rightarrow P\bar{1}$ displacive transition in pure ordered anorthite. (data from Adlhart *et al.*, 1980). (b) Variation with temperature of the square of the spontaneous strain due to the $I\bar{1} \Rightarrow P\bar{1}$ displacive transition in pure ordered anorthite. (From Redfern and Salje, 1987.)

$= 530K$ and the transition is second order. This is demonstrated by the difference in the way the order parameter Q changes as a function of temperature in

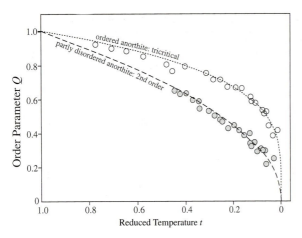

Figure 12.23. A comparison of the variation of the order parameter Q with temperature due to the $I\bar{1} \Rightarrow P\bar{1}$ displacive transition in pure ordered anorthite (open circles) and partly ordered anorthite (shaded circles). The dashed lines are the theoretically predicted curves for a tricritical transition and a second-order transition respectively, and illustrate the good correspondence between experiment and theory. (After Redfern *et al.*, 1988.)

the two anorthite samples. For such a comparison it is convenient to scale the data to a comparable temperature defined as the reduced temperature t where $t = (T_c^* - T)/T_c^*$, so that at the critical temperature for each sample $t = 0$ and at 0 K, $t = 1$. This is shown in Figure 12.23 together with the expected result for a tricritical model (dotted line) and second-order model (dashed line) (cf. Figure 8.46).

To justify how a small amount of Al,Si disorder can bring about such a pronounced change in the character of the displacive transition we note that both the displacive transition ($I\bar{1} \Rightarrow P\bar{1}$) and the Al,Si ordering transition ($C\bar{1} \Rightarrow I\bar{1}$) are zone boundary transitions and therefore the coupling of both order parameters via the strain ε must be quadratic. By adding the effect of this biquadratic coupling to the free energy expansion in Landau theory, the observed experimental data can be theoretically justified.

As expected for a zone boundary transition, an antiphase domain microstructure is formed as anorthite undergoes the $I\bar{1} \Rightarrow P\bar{1}$ transition. The antiphase domains and their imaging by transmission electron microscopy have been discussed in Section 7.3.4. The formation of antiphase domains is consistent with a soft-mode description of the transition with the frozen-in mode having a phase shift of π in different parts of the crystal.

(ii) The C2/m ⇒ C1̄ displacive transition in disordered albite

Above about 980°C, the equilibrium state of albite $NaAlSi_3O_8$ is monoclinic, space group C2/m. Below this temperature it transforms by a displacive transition to a triclinic (C$\bar{1}$) structure. The most important feature of this transition is that it is very strongly coupled to the degree of Al,Si order, and the overall behaviour must be considered in terms of this coupling (Section 8.6.4). Although both the displacive transition and the Al,Si ordering transition, taken separately, result in the same symmetry change, the Al,Si ordering transition is extremely slow compared with the unquenchable displacive transition. A more detailed description of the coupled displacive/order-disorder behaviour of albite will be given in Section 12.4.4, but here we merely point out some aspects of the displacive component.

The temperature of the displacive transition is strongly dependent on the substitution of K for Na (Figure 11.25) with the presence of the larger K ion stabilising the expanded monoclinic structure. This has led to the suggestion that the displacive transition is due to the reduction in the amplitude of the vibrations of the Na^+ in the large cation sites. In monoclinic albite, the large amplitude of the Na^+ vibration would support the feldspar framework. The larger K ion does not require such a large amplitude vibration to sustain the monoclinic framework which can then persist to lower temperatures. An alternative description of the driving mechanism is the inherent instability of the framework structure itself, with the cations playing a somewhat passive role.

A convenient order parameter Q to describe the distortion of the monoclinic unit cell is the change in one of the interaxial angles. From the symmetry rules of Landau theory, Q is directly proportional to $\cos \alpha^*$ where α^* is the interaxial angle between the reciprocal lattice axes z^* and y^*. High temperature lattice parameter measurements on albite with disordered Al,Si show a linear dependence of $\cos^2 \alpha^*$ with temperature (Figure 12.24), i.e. $Q^2 \propto (T - T_c)$ and hence the transition is thermodynamically second-order.

The domain structure associated with this transition arises due to the two different but equivalent distortions possible in a transition from monoclinic to triclinic symmetry, and the two different ways of relating these twin domains. Twin domains may be

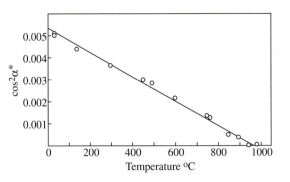

Figure 12.24. The change in the shape of the unit cell of albite due to the C2/m to C$\bar{1}$ displacive transition, plotted as the variation of $\cos^2\alpha^*$ as a function of temperature for pure albite with no Al,Si order. Since $\cos^2\alpha^*$ is proportional to Q^2, the linear plot indicates that the transition is thermodynamically second-order.

related either by the mirror plane lost in the transition (i.e. albite twins), or by the lost diad axis (pericline twins), and the geometry is the same as that described for the monoclinic to triclinic transition in K-feldspar, where the transition is driven by Al,Si ordering (see Section 7.3.5 and Figure 7.53). However, whereas in K-feldspar these two orthogonal twin planes lead to a fine 'tweed' texture due to the very slow growth of the domains, the rapid displacive transition in albite rich compositions results in a much coarser microstructure in which the energetically more favourable albite twins are dominant. Note however that the large strain associated with the transition means that applied stress is likely to have a large effect on the twin domain distribution.

(iii) High-low quartz

The mechanism of the transition from high to low quartz which occurs at 573°C has been observed *in situ* in a transmission electron microscope and illustrates some further aspects of the way in which the distortion in this 'classical' soft mode transition is

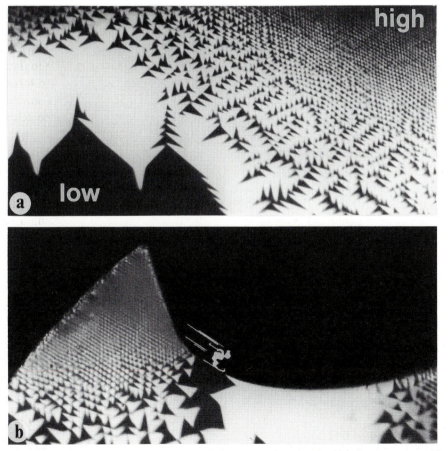

Figure 12.25. (a) Dark field electron micrograph across a crystal of quartz undergoing the high–low quartz transition. A temperature gradient has been set up across the crystal from lower left (<573°C) to the upper right (>573°C). In the low quartz, the black/white Dauphiné twin domains are large and become progressively finer as high quartz is approached. Within the high quartz no domain structure can be observed. Note the periodic distribution of domains in the transition zone. (b) Skiing from high to low quartz down the temperature gradient. The skier, 0.2 µm high, provides the scale. (Micrographs courtesy of G. Van Tendeloo. See Van Landuyt *et al.*, 1985; also Heaney and Veblen, 1991.)

distributed through the structure. Figure 12.16 shows the structures of high and low quartz and it is apparent that there are two possible alternatives for the sense of the distortion, resulting in two orientational variants related by a 180° rotation. The twinning which results from this transition is termed Dauphiné twinning after the province in France from which such twinned crystals were first observed.

At the transition temperature, these two twin alternatives ought to be equally represented. The development of the twin-domain structure is shown in the electron micrographs in Figure 12.25 in which a temperature gradient has been set up across a thin flake of quartz. In low quartz Dauphiné twins form a domain structure imaged here as black-white contrast. As the transition zone is approached from the low temperature structure, more twin boundaries forming triangular domains spontaneously appear. As the temperature increases the twin domains flip from one orientation to the other, and the scale of the twinning becomes progressively finer. As the high temperature form is approached the frequency of the structure reversal increases and the domain size decreases until the domains eventually blur into the homogeneous high temperature structure. This behaviour can be broadly correlated with the soft mode description of the transition, the distortion representing the frozen-in mode whose frequency has dropped to zero. Note also the periodicity of the twin domain structure, a feature to which we will refer in the next section.

The structure of a single twin boundary is shown

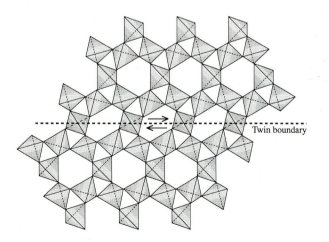

Figure 12.26. The structure of a Dauphiné twin boundary in low quartz. At the boundary the structure is locally sheared, and has a different symmetry to the low quartz structure.

in Figure 12.26. The feature to note here is that in order to maintain the integrity of the tetrahedra at the boundary, the two twin-related domains have to be shifted sideways relative to each other resulting in a local shear of the structure at the boundary. The structure along the boundary could be described as another mode of distortion of the high quartz structure which needs to be incorporated to stabilize the coexistence of the two energetically equivalent twin orientations along the twin plane. We develop this concept in the following section.

12.3 Incommensurate structures – quartz as an example

The high – low transition in quartz provides an opportunity to introduce a new structural principle which is an important aspect of the behaviour of many mineral groups. In a temperature interval of 1.3 K between the high and low quartz stability fields, a modulated structure with a well-defined periodicity develops. This periodicity is not an integral multiple of the translational periodicity of the underlying lattice and is therefore termed *incommensurate*. Although the existence of this structure over such a narrow temperature range may seem no more than a fine detail, the description of the structure and the reason for its formation has important implications.

Consider high quartz cooling towards the transition temperature to low quartz. As the soft mode reduces in frequency, two equivalent twin-related choices for the orientation of the low form exist. Since the probability of either orientation is the same, both possibilities coexist within the distortional fluctuations in the structure near T_c. As shown in Figure 12.26, the boundary between the two twin orientations must involve a local shear, which itself represents one of the vibrational modes of the high quartz structure. A shear mode is, by definition, an acoustic mode, and therefore as the transition temperature is approached the soft optic mode and the acoustic mode must intersect at some small value of the wave vector **k** (Figure 12.27). Since both of these vibrational modes have the same symmetry they can interact with one another producing a 'hybrid' structure which represents the coupling of the structure represented by the optic mode and that represented by the acoustic mode. This coupling results in a minimum in the dispersion curve for the acoustic

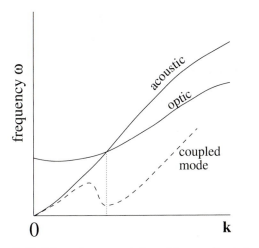

Figure 12.27. Schematic representation of the coupling between the soft optic mode and an acoustic mode in quartz near the high – low transition, resulting in a coupled mode with a minimum at wavevector **k′**. This wavevector corresponds to the wavelength of an incommensurate structure which is stabilized by the interaction.

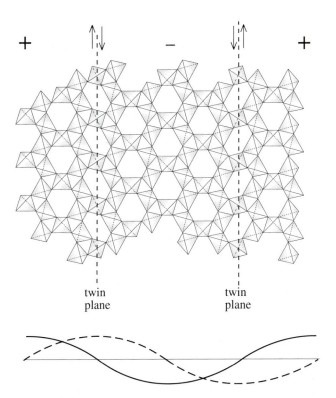

Figure 12.29. A simplified diagram of the incommensurate structure of quartz. Periodic twin planes result in a structure with alternating regions of distortion (+ and −) as well as alternating shears on the twin boundaries (↑ ↓ and ↓ ↑). The structure is represented as slabs of these four alternatives. A closer approximation to the real structure involves an interweaving of periodic distortions as indicated by the two waves, the dashed wave representing a periodic variation in the sense of shear, and the full line representing a periodic variation in the distortion from one twin alternative to the other. The wavelength of the periodicity is ~150Å.

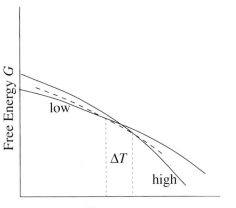

Figure 12.28. Schematic free energy – temperature curves for high quartz, low quartz, and the incommensurate phase (dashed) which is stable over a narrow temperature range (ΔT) around the cross-over between the high and low phases.

mode at a value of **k** which corresponds to the wavelength of the incommensurate modulation. The incommensurate structure exists because it represents a free energy minimum over the narrow temperature range between high and low quartz (Figure 12.28).

A simplified picture of the incommensurate structure formed by such coupling is that of a structure made up of periodic twin domains with the shear structure at the boundaries. However, narrow, closely-spaced twin boundaries such as shown in

Figure 12.29 would not result in a free energy reduction, rather the opposite. The two component structures do not exist as discrete slabs, but as two interwoven sinusoidal modulations whose coexistence results in the incommensurate structure. The two sine waves in Figure 12.29 represent the continuous structural variation of each component. The peaks and troughs represent the two twin-related distortions due to the optic mode (labelled + and −) and in the case of the shear structure (represented by the dashed sine curve) the two different senses of shear at adjacent boundaries. The position of the maximum of one component structure corresponds to zero in the other i.e.there is a phase shift of $\pi/2$ between the two waves. The actual coupled structure is the superposition of these two continuously varying components.

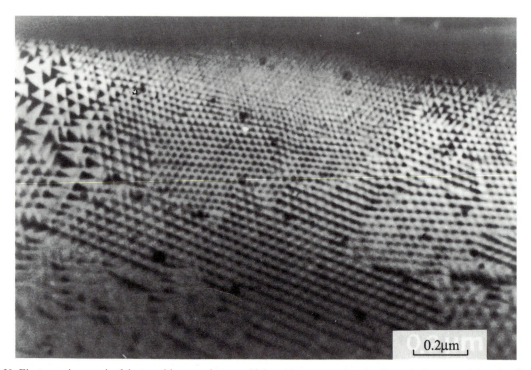

Figure 12.30. Electron micrograph of the transition zone between high and low quartz showing the periodic array of domains characteristic of the incommensurate structure. (Micrograph courtesy of G. Van Tendeloo.)

In quartz, the symmetry requires that the modulations exist in three directions giving the fine-scale triangular microstructure of the periodicity, as seen in the transmission electron micrograph in Figure 12.30. The incommensurate structure has a periodicity of about 150Å which increases as the temperature falls. This is consistent with the theoretical model in which the soft mode intersects the acoustic mode – as the soft mode frequency drops to zero the intersection point moves to smaller values of **k**.

The formation of the incommensurate structure is also explicable in terms of Landau theory in which the coupling is described between the order parameters of the two component structures. For an incommensurate structure however, this coupling is between one order parameter and the gradient of the other and stabilises the variations in the amplitudes of both as a function of distance in the crystal.

12.3.1 Some generalisations regarding the formation of incommensurate structures

In the quartz example above, the structural origin of the incommensurate phase is relatively easy to understand, as is the interaction model in which the dispersion curves of two vibrational modes intersect and define a point in reciprocal space at which they have the same energy. If the condition that the two modes also have the same symmetry is fulfilled, coupling between them can occur to produce an energy minimum for a 'hybrid' structure.

This generalisation is not restricted to soft mode transitions, and is equally applicable to coupling between order–disorder transitions, or between a distortion and an ordering transition. The quartz example was described in terms of a structural model involving periodic twin domains; other incommensurate structures can be understood in terms of periodic antiphase domains with the coupled structure forming at the boundaries. The latter case could arise when one, or both of the components involved an ordering transition lead to a loss of translational symmetry. Note again however that in the coupled structure the two components are described in terms of sinusoidal fluctuations in structure, or order parameter, rather than intergrowths with sharply defined boundaries. The maxima and minima of the fluctuations (labelled + and − in Figure 12.29) represent the two orientational or translational variants possible in the formation of the low temperature form.

There are many incommensurate structures

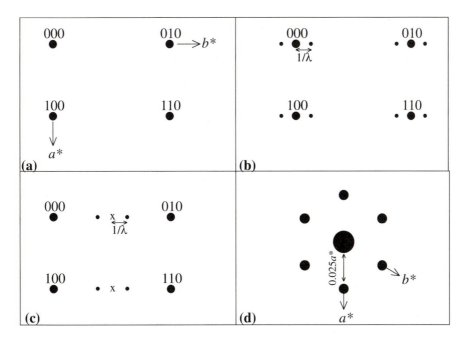

Figure 12.31. An illustration of the effect that an incommensurate structure has on the diffraction pattern using a small section of the reciprocal lattice. **(a)** A part of the a^*b^* section of the reciprocal lattice of an orthorhombic structure. **(b)** The formation of an incommensurate periodicity associated with a zone centre transition. If the incommensurate periodicity in the structure is in the [010] direction and has wavelength λ, pairs of weak satellite reflections form at a distance $1/\lambda$ around each Bragg reflection. **(c)** If the incommensurate periodicity is associated with a zone boundary transition, in this case doubling the cell along [010], the satellite reflections form about the absent superlattice reflection (marked x). **(d)** A single Bragg reflection in the incommensurate structure of quartz, with three pairs of satellites due to the three orientations of the modulations. The periodicity is normal to {100} planes with a wavelength in this case of 40 times the {100} lattice spacing.

known in minerals, and although their origin is not as well understood as in quartz, their temperature and composition range can be much more extensive. The behaviour of the plagioclase feldspar solid solution, the most abundant mineral group in the earth's crust, is dominated by the existence of incommensurate structures which exist in all natural samples. The stability of mullite is also due to the existence of an incommensurate structure. Both of these cases will be described in more detail below.

Incommensurate structures need not have a stability field, and their hybrid nature does suggest a kind of compromise structure which might result when, for kinetic reasons, the thermodynamically stable structure could not be attained. However, where detailed experiments are possible, as in the case of quartz, a true stability field has been defined. Furthermore, calorimetric measurements on incommensurate structures where the transitions cannot be directly observed also show that a considerable degree of energy reduction is associated with the coupling process. The configurational entropy reduction on forming an incommensurate structure

is also considerable, for although the structural variations suggest disorder by comparison with a normal structure, the coupling demands a well defined variation without the randomness associated with disorder. Our view of incommensurate structures is increasingly to regard them as a structurally and thermodynamically successful solution when several different transformation processes are possible.

12.3.2 Diffraction effects from incommensurate structures

The superposition of a long wavelength periodicity on the translational periodicity of the lattice results in the appearance of extra diffraction maxima as satellite reflections around the main Bragg maxima. This is most conveniently understood in reciprocal space and Figure 12.31(a),(b) illustrates the effect, on an a^*b^* reciprocal lattice section, of an incommensurate modulation along the [010] direction. The reciprocal vector from the Bragg reflection to the satellite defines the wave vector of the modulation and its distance from the Bragg position is the reciprocal of the modulation wavelength. In an incommensurate modulation this wavelength is not an integral multiple of the underlying lattice repeat.

When the satellite reflections are around a primary Bragg reflection the incommensurate structure is associated with a zone centre transition. However, incommensurate structures may also be associated with a zone boundary transition and the extra diffracted intensity appears as satellites around a potential superlattice reflection, such as could form in an ordering process (Figure 12.31(c)). Furthermore, the symmetry may require that the modulation exists in more than one direction, as in the case of quartz where the three pairs of satellite reflections around the Bragg peak (Figure 12.31(d)) are due to the three equivalent orientations for the incommensurate periodicity. This gives the domains a triangular shape, as seen in the electron micrographs in Figures 12.25 and 12.30.

12.4 Order–disorder transitions in minerals

The importance of cation ordering and the consequences to the structure of a mineral, its defects and its thermodynamics has been emphasised in various contexts throughout this book. In this section we review some of these ideas and add to them further by describing a number of examples or case studies of ordering transitions in minerals. Al,Si ordering in cordierite, $Mg_2Al_4Si_5O_{18}$, has been widely studied using a number of different techniques and illustrates the relationship between the degree of order, the microstructure and changes in the cell parameters during the process. There are many parallels between cordierite and K-feldspar, $KAlSi_3O_8$, in this respect. Al,Si ordering in albite, $NaAlSi_3O_8$, however, poses a very different problem. By discussing these examples in detail we can gain some insight into the issues which need to be considered when dealing with these processes.

12.4.1 Al,Si ordering and the hexagonal – orthorhombic transition in magnesium cordierite, $Mg_2Al_4Si_5O_{18}$: a case study

Cordierite is a widespread mineral in metamorphic rocks where it generally contains some Fe^{2+} substituted for Mg, although rarely with more than one Fe^{2+} atom per formula unit. The structure of cordierite has been described in Section 6.8.4 where it

was noted that it occurs in two polymorphic forms. The high temperature form is hexagonal and is sometimes referred to as indialite, and the low temperature form which is by far the most common in nature, is orthorhombic.

The symmetry of the high temperature form, P6/*mcc*, does not allow long-range Al,Si order, while complete Al,Si order can be attained in the orthorhombic structure, space group Cccm. Complete Al,Si order requires the elimination of all Al–O–Al bonds and the resulting structure is consistent with this orthorhombic space group. The relationship between the unit cells of the two structures, as shown in Figure 6.63 shows that the volume *per lattice point* has not changed, but that the distortion and loss of symmetry of the hexagonal structure requires a new description of the unit cell, which is C-face centred orthorhombic. Thus the transformation from hexagonal to orthorhombic does not involve the formation of a superlattice. However, sets of lattice planes which were equivalent under hexagonal symmetry, become non-equivalent in the orthorhombic form and will have slightly different d-spacings. Therefore, in a powder diffraction pattern, the transformation from hexagonal to orthorhombic symmetry will result in the formation of multiple peaks in the orthorhombic form in place of single peaks in the hexagonal form.

In the Mg end-member, the hexagonal form is stable only above about 1450°C, and so may not seem to be relevant to natural cordierite which forms at temperatures below about 600°C. However, as we shall see, there is evidence that many natural cordierites may have nucleated metastably in the hexagonal form before transforming to the orthorhombic structure.

Experimental work on Al,Si ordering in Mg-cordierite has been carried out within the stability field of orthorhombic cordierite. When Mg-cordierite is crystallised from a glass at temperatures between 1000°C and 1400°C, the first product to form is invariably hexagonal cordierite. Although it is not stable at this temperature, the nucleation of hexagonal cordierite is kinetically preferred, a point discussed in Section 12.4.2. When annealed for some time at these temperatures, the ordered orthorhombic structure eventually forms,

i.e.

Glass ⇒ Hexagonal cordierite ⇒
Orthorhombic cordierite

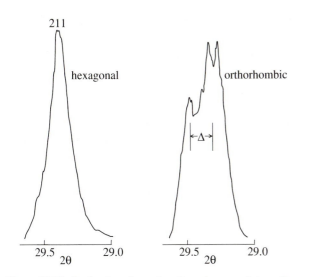

Figure 12.32. In the transformation from hexagonal to ortho-rhombic cordierite the 211 X-ray powder peak of the hexagonal form is split into three peaks (131, 511 and 421) in the orthorhombic form. A measure of the degree of distortion of the unit cell from hexagonal symmetry is provided by the distortion index, labelled Δ. (The 2θ angles refer to Cu K_α radiation.)

The kinetics of this process is sluggish, with a high activation energy consistent with Al,Si ordering. However, because the equilibrium transformation temperature is very high, it is possible to carry out experiments of this transformation at temperatures where Al,Si diffusion is fast enough for ordering to occur on a laboratory time scale. In the experiments, isothermally annealed samples are quenched to room temperature after progressively longer times and the products examined with a variety of techniques. The assumption is made that the Si,Al distribution is effectively frozen in during the quench. Such a study of the transformation from the hexagonal structure with no long range order, to the orthorhombic structure with complete long-range order, is the basis for the discussion which follows.

Al,Si ordering and the distortion of the unit cell

As mentioned above, the change from hexagonal to orthorhombic symmetry results in splitting of certain powder diffraction peaks due to the distortion of the unit cell. We would expect that the Al,Si ordering process would necessarily involve some distortion, because the AlO_4 tetrahedron is slightly larger that the SiO_4 tetrahedron. It is tempting therefore to equate the two phenomena and suggest that the degree of splitting of certain powder lines, in other words the strain associated with the symmetry

change, could be used as a measure of the degree of Al,Si order.

Early studies of Al,Si ordering in cordierite made this assumption, and the distortion index Δ, as it became known, was defined in terms of the splitting of the 211 peak of the hexagonal form into three peaks in the orthorhombic, namely 131, 511 and 421. Thus

$$\Delta = 2\theta_{131} - \frac{2\theta_{511} + 2\theta_{421}}{2}$$

as shown in Figure 12.32. The problem is that the Δ index never reaches more than about 0.25° 2θ and the three peaks overlap. It is even more difficult to measure the way in which the Δ index increases from 0 in the hexagonal form to its maximum value in the orthorhombic form. Overlap of the 211_{hex} peak with the three peaks in the orthorhombic form can make interpretation ambiguous and for many years it was believed that the Δ index increased continuously.

When X-ray powder diffraction patterns of the transformation sequence are made with synchrotron radiation the resolution is very much greater and it becomes apparent that the Δ index changes abruptly from $\Delta = 0$ to $\Delta \sim 0.2°$ at a particular point in the annealing sequence, as shown in Figure 12.33. This abrupt change marks the discontinuous structural change from hexagonal to orthorhombic symmetry and suggests the nucleation of orthorhombic cordierite within the hexagonal phase. This discontinuity is consistent with theoretical predictions from Landau theory that the symmetry change from P6/*mcc* to C*ccm* must be thermodynamically first-order.

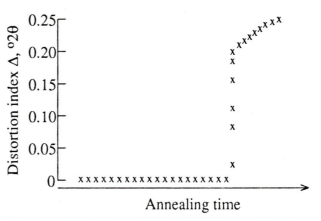

Figure 12.33. When hexagonal Mg-cordierite is annealed in the stability field of orthorhombic cordierite, the distortion index Δ changes discontinuously from 0 to ~0.2° 2θ as a function of annealing time, indicating an abrupt change in the shape of the unit cell.

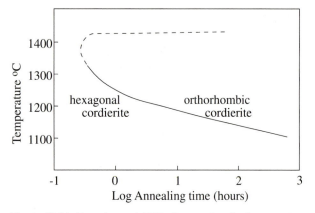

Figure 12.34. Experimental TTT diagram for the hexagonal – orthorhombic transformation in Mg-cordierite, measured from the splitting of X-ray diffraction peaks. (After Putnis *et al.*, 1987.)

The annealing time at which this transformation occurs depends on the temperature, as shown on an experimental TTT diagram (Figure 12.34) and indicates an activation energy of around 500 kJ mol^{-1} for the process. This is similar to the activation energy for Al,Si ordering measured in other framework minerals.

There would thus seem to be an *a priori* case for relating the Δ index to the Al,Si ordering. Does this suggest that the degree of Al,Si order changes abruptly as well? We will return to this question after seeing the evidence provided from transmission electron microscopy.

Transmission electron microscopy of the transformation sequence

Figure 12.35 shows a series of electron micrographs of the experimental samples of cordierite crystallised from a glass and annealed isothermally for increasing lengths of time. Figure 12.35(a) is hexagonal cordierite with no visible microstructure (apart from grain boundaries) and a generally homogeneous appearance. Samples annealed for a longer time have a very definite tweed microstructure. With continued annealing the microstructure coarsens. Throughout this tweed microstructure stage, the synchrotron X-ray powder peaks show no splitting, indicating that, on average, the cordierite is still hexagonal. Yet clearly there is a lot happening on a local scale.

The details of the later stages of the process depend on the annealing temperature: at high temperatures (~1400°C) orthorhombic cordierite nuc-

leates from this tweed microstructure, forming well defined domains (Figure 12.35(d)). At lower temperatures the tweed structure appears to coarsen progressively, the cross-hatched regions giving way to lamellar twins in two orthogonal orientations. Although in this latter case it is not obvious at which stage the cordierite becomes orthorhombic, X-ray powder measurements show that the Δ index in both cases shows the same abrupt change as described in Figure 12.33.

The interpretation of this microstructural evolution is similar to that described in K-feldspars (see Sections 7.3.5 and 7.3.6). The tweed microstructure represents modulations in the local distortion of the hexagonal unit cell due to short-range Al,Si order. X-ray diffraction experiments, which measure the long-range, correlated distortion of the structure only detect the transformation when clearly defined twin-related domains of the orthorhombic form exist. In other words, the Δ index only becomes apparent when the microstructure resembles Figure 12.35(f). The modulated structure is, on average, still hexagonal.

The microstructural evidence suggests the development of short-range Al,Si order in the early stages of the transformation. To measure the degree of Al,Si order, we need a short-range technique such as provided by spectroscopy, or from enthalpy of ordering measurements which depend on the energy of interatomic bonds. We next review the data provided by such techniques.

Spectroscopic measurements of the degree of Al,Si order

1. *^{29}Si magic angle spinning NMR spectroscopy* (Section 4.1) is capable of determining the relative proportions of Si in each different type of environment in a structure. The 'environment' which is sampled by NMR is restricted in distance to the nearest neighbouring polyhedra and so it particularly applicable to problems of short-range order. This is particularly well illustrated by a sequence of spectra from cordierite samples with increasing Si,Al order (Figure 12.36). Figure 12.36(a) is the earliest crystallisation product from the cordierite glass and is from the sample shown in the electron micrograph as Figure 12.35(a). Figure 12.36(f) is well ordered orthorhombic cordierite. The interpretation of the spectra begins with that of the fully ordered form.

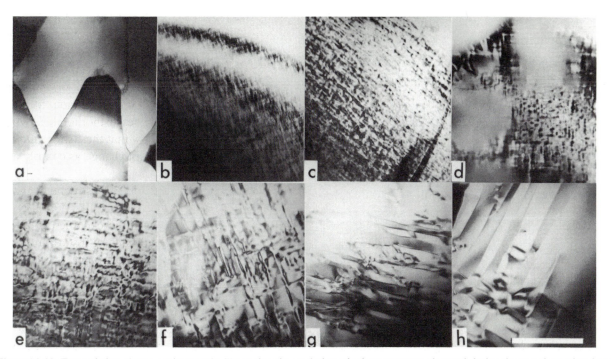

Figure 12.35. Transmission electron micrographs illustrating the evolution of microstructures observed during the transformation from hexagonal to orthorhombic cordierite. At 1400°C the transformation sequence is (a) ⇒ (b) ⇒ (c) ⇒ (d), where the orthorhombic cordierite nucleates within the modulated structure. At lower temperature the sequence is (a) ⇒ (b) ⇒ (c) ⇒ (e) ⇒ (f) ⇒ (g) ⇒ (h), the finer scale of nucleation or orthorhombic cordierite appearing to produce a continuously coarsening microstructure. The length of the scale bar is 0.2 μm. (From Putnis *et al.*, 1987.)

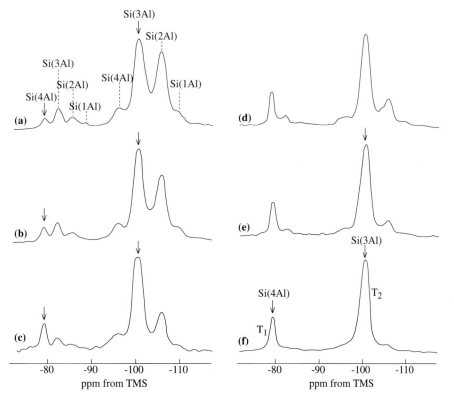

Figure 12.36. A sequence of ^{29}Si NMR spectra of synthetic cordierites isothermally annealed for increasing time to increase the degree of Al,Si order. In (a) the large number of Si(nAl) environments is due to the disorder in the first-formed hexagonal phase, while in (f) the orthorhombic cordierite is well ordered. (After Putnis *et al.*, 1987.)

In ordered orthorhombic cordierite Si occupies two types of tetrahedral sites. The ring sites, labelled T_2 in Figure 6.58 contain 4 Si atoms and each of these has 1 Si and 3 Al neighbours in adjacent tetrahedra, i.e. an environment denoted Si(3Al) in the terminology described in Section 4.1. Although these four sites are not crystallographically equivalent, the very small difference in their structural environment is not resolved and they are represented by a single peak in the NMR spectrum (Figure 12.36(f)) at a chemical shift of -100.2 ppm. The other Si atom in the formula unit occupies one of the T_1 sites which connect the six-fold rings. In ordered cordierite each of these Si atoms is surrounded by 4 Al tetrahedra i.e. Si(4Al) and gives a single NMR peak at -79.3 ppm in Figure 12.35(f). The ratio of the areas of these peaks is 4:1, the same as the ratio of Si atoms in T_2 sites to those in T_1 sites. Thus the assignment of these NMR peaks is made with reference to the known structure, determined by diffraction methods.

The ^{29}Si NMR spectra of progressively less well-ordered cordierites (Figure 12.36f $\Rightarrow$ Figure 12.36(a)) shows that while the $T_2 - T_1$ grouping of peaks is preserved, the number of peaks in each group increases. In Figure 12.36(a) four peaks in each group are discernible. These extra peaks can only arise due to different chemical environments and the assignment of these peaks can be made from the empirically established observation that the position of the Si peak shifts by about $+5$ppm for each Al atom added to the local environment (Figures 4.5 and 4.6). Thus the peaks can be labelled as shown in Figure 12.36(a).

Since the composition of the cordierite remains the same throughout the ordering process, and because it is a framework structure, the relative intensities of the peaks in these spectra can be used to calculate the number of Al–O–Al bonds which must be present. In completely ordered cordierite there are no Al–O–Al bonds, but any amount of disorder must introduce some Al–O–Al linkages. In completely disordered cordierite in which 4 Si atoms are randomly distributed over the six T_2 sites, and 1 Si atom is randomly distributed over the three T_1 sites, a statistical calculation shows that there would be on average 3.3 Al–O–Al bonds per formula unit (out of a total of 18 T–O–T bonds).

Figure 12.37 shows the number of Al–O–Al bonds per formula unit, N(Al–Al), in the experimental

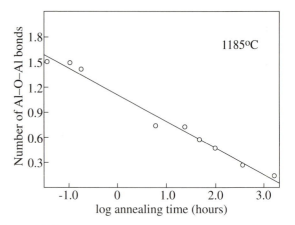

Figure 12.37. The numbers of Al–O–Al bonds per formula unit (9 tetrahedra) of cordierite as a function of the logarithm of the annealing time as it transforms isothermally from hexagonal (disordered) to orthorhombic (ordered). (After Putnis and Angel, 1985.)

cordierite samples as a function of annealing time i.e. increasing Al,Si order. There are two important observations to make here.

(i) At the shortest annealing times, the N(Al–Al) bonds in the hexagonal cordierite are around 1.5 compared to 3.3 for total disorder. Thus even before any tweed microstructure is apparent in the hexagonal cordierite, there is a considerable amount of short-range order.

(ii) The N(Al–Al) bonds decreases linearly with log time until eventually all are eliminated. There is no major discontinuity as there is in the distortion index Δ, nor is the change from hexagonal to orthorhombic symmetry apparent. The NMR spectra merely record a smoothly increasing degree of Al,Si order.

2. *Vibrational spectroscopies (IR and Raman)* (Section 4.3) operate on a longer length scale than NMR. The characteristic length scale of IR and Raman (i.e. the minimum length scale over which changes in structure affect the vibrational modes) is around 30Å, and therefore even if the structure is modulated, these techniques sample regions within the modulation wavelength which is usually around 200Å.

A sequence of IR spectra from progressively more ordered cordierites (Figure 12.38) are complex, but a number of conclusions can be drawn.

(i) The IR spectrum of modulated cordierite with the fine tweed microstructure is very similar to

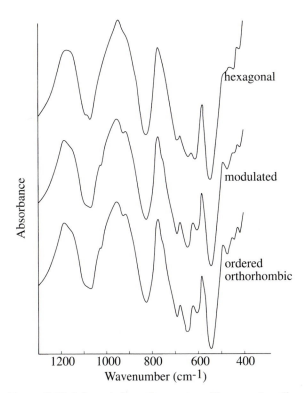

Figure 12.38. Infra-red absorption spectra of hexagonal cordierite, the modulated structure of cordierite and the ordered orthorhombic form. Although the average structure of modulated cordierite is hexagonal, the local structure from IR spectroscopy, is the same as that of orthorhombic cordierite. (After Putnis and Bish, 1983.)

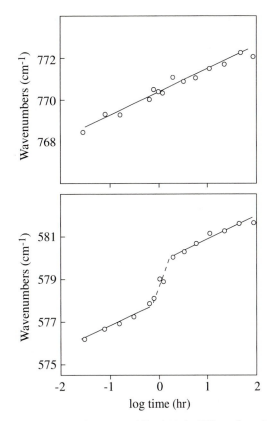

Figure 12.39. The frequency shift of **(a)** the 770 cm^{-1} mode and **(b)** the 580 cm^{-1} mode of cordierite as a function of the logarithm of the annealing time as it transforms isothermally from hexagonal (disordered) to orthorhombic (ordered). The frequency of the 770 cm^{-1} mode shifts continuously, just as the degree of Al,Si order, while the shift in the 580 cm^{-1} mode is an indicator of the local strain, being discontinuous at the point in time where the modulated structure forms. (After Güttler et al., 1989.)

that of orthorhombic cordierite. This suggests that although modulated cordierite is hexagonal on average, a local probe such as IR shows that small regions within the tweed structure are in fact already orthorhombic. Note, however, that all the IR measurements are carried on quenched samples. We will take this point up again below.

(ii) Different peaks in the IR spectra reflect different aspects of the structural changes taking place (Figure 12.39). For example, the peak at around 770 cm^{-1} shifts in frequency continuously throughout the whole annealing sequence, whereas the peak at around 580 cm^{-1} has a marked discontinuity at the same point in time as the tweed microstructure appears in electron microscopy. The 770 cm^{-1} peak can be correlated to progressive changes in local Si,Al environments, while the 580 cm^{-1} peak records the local distortion. The appearance of the modulations thus coincides

with a discontinuity in the local distortion of the structure.

The picture that emerges so far is that annealing hexagonal cordierite within the stability field of the orthorhombic form promotes Al,Si ordering, reducing the enthalpy by eliminating Al–O–Al bonds. This process is accompanied by distortions in the structure, and, while the Si,Al order is short range, the structure accommodates these distortions by modulations. These compensated distortions keep the macroscopic strain at zero. NMR spectroscopy can monitor the changes in local Al,Si order throughout the transformation. Selected peaks in the IR spectra can be used to observe the changes in Al,Si order on a somewhat longer length scale than NMR, as well as the local discontinuity in structural distortion that marks the hexagonal – modulated transition. TEM picks up the transformation

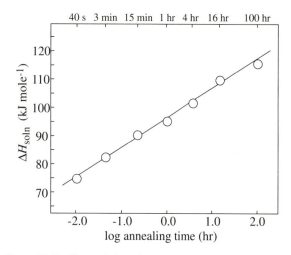

Figure 12.40. The enthalpy of solution ΔH_{soln} of cordierite as a function of the logarithm of the annealing time at 1200°C as it transforms from hexagonal (disordered) to orthorhombic (ordered). The linear plot can be directly related to the reduction in the number of Al–O–Al bonds (Figure 12.36). (After Carpenter *et al.*, 1983.)

sequence at this point, charting the microstructural changes which lead to the macroscopic transformation to orthorhombic symmetry as observed at a later stage in the synchrotron X-ray powder diffraction patterns.

Enthalpy of ordering measurements

The energy changes associated with this transformation can be measured from heat of solution measurements of the samples annealed for different times. Differences in the heat of solution of samples annealed for different times represent differences in their enthalpy (Section 8.5.1). If we assume that the strain energy due to the modulations is small (this is supported by the fact that the overall distortion of the structure is very small), we can equate the enthalpy difference to the enthalpy of Al,Si ordering.

As an example, Figure 12.40 shows a plot of $\Delta H_{soln.}$ against log annealing time for samples annealed at 1200°C, spanning the range from the first formed hexagonal structure to the stable orthorhombic state. The maximum difference in the $\Delta H_{soln.}$ is 41 ± 6 kJ mol^{-1}, which is the difference in enthalpy between the least ordered and most ordered sample.

It is a useful exercise to compare this enthalpy with the maximum which would be expected for a transition from complete disorder to complete

order. At the equilibrium temperature (1450°C) the ordered and disordered phases are in equilibrium, and hence the free energy difference is zero, i.e. $\Delta G_{ord} = \Delta H_{ord} - T\Delta S_{ord} = 0$ and $\Delta H_{ord} = T\Delta S_{ord}$

The configurational entropy of the disordered state is found from the random distribution of 4Si and 2Al atoms over six T_2 sites and 1Si and 2Al atoms over the three T_1 sites. This is calculated as an example in section 8.3.1 and gives a value of $\Delta S_{ord} = 47.61$ J mol^{-1}K^{-1}. Therefore $\Delta H_{ord} = 1723 \times 47.61 = 82$ kJ mol^{-1}.

If we compare this to the measured maximum enthalpy of ordering of 41 ± 6 kJ mol^{-1}, we can conclude that the first-formed hexagonal cordierite has around half of the configurational entropy of a completely random structure. This is consistent with the NMR measurements which also showed that the number of Al–O–Al bonds in the hexagonal cordierite was around half compared to a random structure.

If we assume that the measured enthalpy change is due to the reduction in the number of Al–O–Al bonds, which we can determine from NMR data, we can then estimate that the enthalpy change associated with a single Al–Si interchange i.e. for the reaction

$$(Al-O-Al) + (Si-O-Si) \Rightarrow 2(Al-O-Si)$$

is 34 kJ mol^{-1}.

Conclusions on the relationship between Al,Si ordering and lattice distortion

It is evident from the above discussion that cordierite manages to achieve a high degree of local Al,Si order without any change in the average symmetry or lattice parameters. This is possible due to the formation of the modulated structure which allows local Al,Si order within a periodically distorted structure, which on average retains the symmetry of the high temperature form. The symmetry of small-scale regions, as sampled by IR spectroscopy, does however appear to be that of the low temperature form. Eventually, as the degree of Al,Si order increases, local distortions can no longer accommodate the strain, and this triggers a macroscopic symmetry change.

Figure 12.41 summarizes this relationship by tracing a typical ordering path on a plot of distortion index Δ as measured by X-ray diffraction against the Al,Si order parameter Q_{od} based on the number of Al–O–Al bonds determined by NMR spectroscopy.

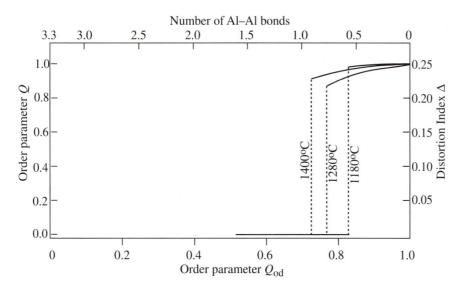

Figure 12.41. The experimentally determined relationship between the distortion index Δ, and the Al,Si order parameter Q_{od} during the hexagonal to orthorhombic transformation in Mg-cordierite at three different annealing temperatures. Q_{od} reaches about 0.8 before the sharp increase in the distortion of the average structure as measured by Q. (After Putnis *et al.*, 1987.)

The normalized order parameter (1 for complete order, 0 for complete disorder) is defined as

$$Q_{od} = \left\{ 1 - \frac{N(\text{Al–Al})}{3.3} \right\}^{1/2}$$

Only when the sample has transformed to orthorhombic symmetry is there a simple relationship between the degree of Al,Si order and the distortion index Δ.

The modulated structure is not however merely a microtwinned intergrowth of the low temperature structure. This point was made in Section 7.3.6 and will be discussed in a little more depth in Section 12.4.3 where we compare the ordering behaviour of cordierite and K-feldspar.

The importance of the length scale in the crystal over which changes in Al,Si order and structural distortions are correlated has been pointed out throughout this discussion, and emphasises the need for a multi-technique approach to such a study. Each experimental technique provides information on some aspect of the transformation and it is only from a combination of techniques that the full story emerges.

Relevance of experimental studies to natural cordierites
Is there any evidence that any of these experimental studies are relevant to natural cordierites? The cru-

cial question is whether cordierite always nucleates as ordered orthorhombic cordierite or whether there are any circumstances under which it might nucleate metastably as hexagonal cordierite, with no long-range order, and subsequently transform to the ordered orthorhombic form during annealing in metamorphic rocks. This is an important issue and has very significant repercussions on the estimates we can make of the temperatures and pressures at which metamorphic reactions have taken place.

The problem in trying to answer this question is that if the cordierite grew in the hexagonal structure and then transformed to the ordered structure we now find in rocks, what evidence of this transformation would be preserved? The one microstructural feature which is characteristic of cordierites formed in thermal metamorphism is sector twinning. In basal sections of (normal to the *c* axis) cordierite often has a hexagonal crystal shape with irregular and often complex twinning (Figure 12.42) made up of regions of three twin related orientations of the orthorhombic cell.

There has been considerable debate on the origin of this type of twinning, but the complexity of the twin intergrowths and their crystallographic relations suggest that they must have arisen as transformation twins associated with the hexagonal to orthorhombic transition. The cross-hatched twinning occasionally observed within the individual

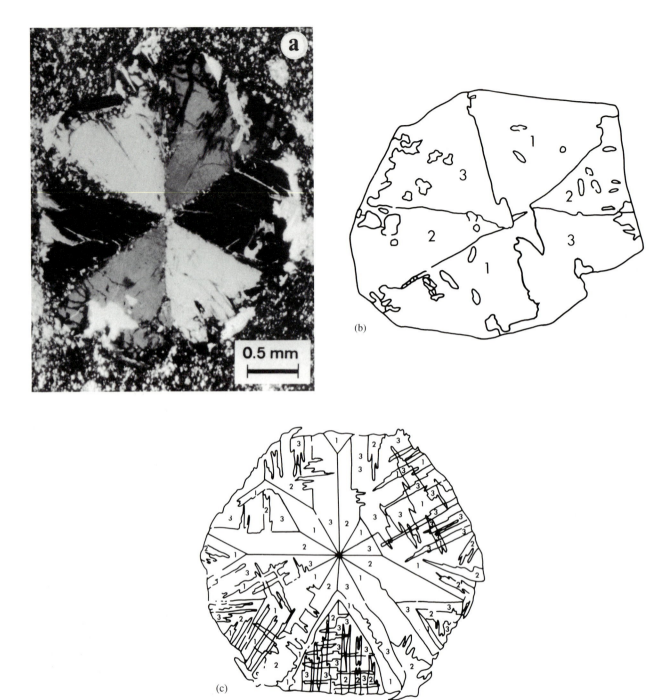

Figure 12.42. (a) Optical micrograph of sector trilling in cordierite from a contact metamorphic rock. (from Kitamura and Yamada,1987) **(b)** Sketch of sector trilling showing the irregularity of the twin interfaces. **(c)** Sketch of complex cross-hatched twinning within the sectors in a cordierite crystal. The three optical orientations related by 120° rotations are labelled 1,2 and 3. (After Putnis and Holland, 1986.)

sectors suggests a coarsened tweed structure and hence a microstructural sequence similar to that in the experiments above. Furthermore, sector twinning has been described as appearing during annealing of experimentally grown hexagonal prisms of cordierite. This also supports the argument that the sectors are associated with the symmetry change initiated by Al,Si ordering.

A mechanism by which sector twinning can arise via a tweed microstructure is described in Figure

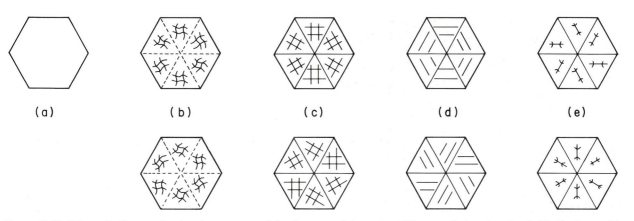

Figure 12.43. Schematic diagram showing the sequence of development of the sector trilling morphology in a simple idealized case. **(a)** Growth of a prism of hexagonal cordierite. **(b)** Development of modulated cordierite in the three possible orientations symmetrically distributed in sectors to reduce the overall strain in the crystal. Within any one sector there are two equivalent orientations of the tweed microstructure as shown by the upper and lower diagrams. **(c)** Coarsening of the tweed microstructure leads to cross-hatched twinning, as in Figure 12.42(c). **(d)** The loss of the energetically less favourable set of twins. **(e)** The final stages of coarsening in which only the sectors are preserved. The diagram shows the optic axial plane orientations resulting from the two possibilities in (b). (After Putnis and Holland, 1986.)

12.43. A coarsened tweed structure can only produce twins in two of the three possible orientations, and so three different pairs can exist. This is equivalent to saying that the orthogonal tweed structure can itself form in three orientations related by rotations of 60°. If these three orientations form as sectors in a hexagonal crystal, as the observations suggest, this would further reduce the macroscopic strain. The scenario depicted in Figure 12.43 is based on this assumption.

If in metamorphic rocks orthorhombic cordierite forms via a hexagonal precursor, we need to be able to estimate the enthalpy and entropy of the hexagonal polymorph before we can determine the effect this has on the temperature and pressure of cordierite-forming reactions. If we assume that the hexagonal cordierite has considerable short-range Al,Si order, as it does in laboratory experiments, we can take the values of enthalpy and entropy from the experimental data above. We assume therefore that hexagonal cordierite is enthalpically less stable than ordered orthorhombic cordierite by 41kJ mol^{-1} and has an additional configurational entropy contribution of 23 J mol^{-1}K^{-1}, i.e. 'half-ordered'.

The effect of these additional terms to the thermodynamics of metamorphic reactions involving cordierite is that hexagonal cordierite can only grow under conditions of considerable overstepping into the stability field of ordered orthorhombic cordierite. We illustrate this point with two examples.

1. Consider a typical metamorphic reaction responsible for introducing cordierite into magnesian gneisses of the granulite facies under decompression during uplift of parts of the deep crust of the Earth:

$$Mg_2Si_2O_6 + 2Al_2SiO_5 + SiO_2 \Rightarrow Mg_2Al_4Si_5O_{18}$$
enstatite   sillimanite   quartz   cordierite

The $P-T$ slope of this reaction is shown in Figure 12.44, for both ordered cordierite and hexagonal cordierite. (These reaction lines also take into account the fact that the enstatite becomes more aluminous as the temperature increases.) The lines intersect at 1450°C, the equilibrium order–disorder temperature, but diverge with reduced temperature, so that under granulite facies conditions a lowering of pressure of some 3 or more kilobars is required to nucleate hexagonal cordierite, relative to the stable ordered cordierite.

Tectonic interpretations, based on pressure estimates from cordierite bearing rocks, will be severely affected if the cordierite nucleated and grew in the hexagonal form before transformation. A 3 kbar pressure difference corresponds to about 10 km, or about one third of the crustal thickness.

This reaction is very sensitive to changes in the thermodynamics of one of the reactants, because the total change in the entropy (and enthalpy) of solid–solid reactions among anhydrous minerals is small.

2. In our second example we consider two dehyd-

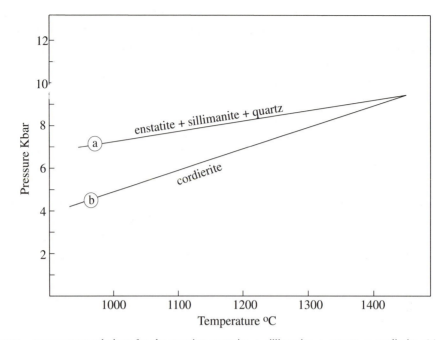

Figure 12.44. Pressure – temperature relations for the reaction enstatite + sillimanite + quartz ⇒ cordierite, **(a)** for the formation of ordered cordierite, and **(b)** for the formation of disordered cordierite. (After Putnis and Holland, 1986.)

ration reactions responsible for producing cordierites in thermally metamorphosed rocks:

muscovite + chlorite + quartz ⇒
$$\text{cordierite + biotite + water} \quad (1)$$

and at a somewhat higher metamorphic grade:

muscovite + chlorite + quartz ⇒
$$\text{cordierite + K-feldspar + water} \quad (2)$$

Figure 12.45 shows the position of these reaction lines for both ordered and hexagonal cordierite. These curves suggest that the nucleation of hexagonal cordierite during thermal metamorphism requires an overstepping of the equilibrium reaction boundaries (i.e. those for the stable ordered form) by about 80°–85°C. This is a very significant result, particularly as it is generally assumed that prograde metamorphic reactions take place at very near equilibrium conditions.

We should note that this is a controversial result, and there may be other factors which reduce the degree of overstepping required. Nevertheless, the principle is very important and must be kept in mind whenever we consider the thermodynamics of minerals which may have metastable cation order.

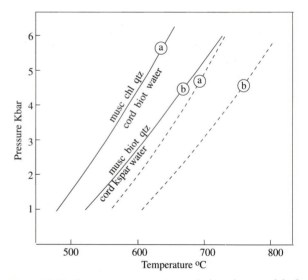

Figure 12.45. Pressure – temperature relations for two dehydration reactions (a) and (b) leading to the formation of cordierite. The full lines are the locations of the equilibria for the formation of ordered cordierite, and the dashed lines are the reaction lines for the formation of disordered cordierite. (After Putnis and Holland, 1986.)

12.4.2 Metastable nucleation of disordered phases

The issues raised in the previous section raise a general question: why do disordered phases nucleate and grow within the stability field of the ordered phase? There is no shortage of geological evidence that this is a fairly common phenomenon. In low temperature sedimentary environments authigenic K-feldspars can grow with almost any state of Al,Si order, depending on the nature of the sediment in which they grow. In ash-beds, authigenic feldspars tend to be highly disordered, while in sandstones and shales the wide variation in the degree of order presumably reflects variations in the fluid supersaturation during growth. Dolomite is an even more important example since it always grows with a disordered Ca,Mg distribution in sediments, although the ordered structure is much more stable. Ordered dolomite is only achieved after burial and diagenesis over geological time scales.

In hydrothermal environments adularia is the typical form of K-feldspar, distinguished externally by its crystal habit. Adularias found in low temperature hydrothermal veins have extremely variable degrees of Al,Si order and often show tweed microstructures indicating an intermediate stage in the transformation from monoclinic to triclinic structure. Again the degree of supersaturation in the fluid phase and hence overstepping of the equilibrium will determine the structural state.

As a final example, we note that omphacite, which forms in low temperature, high pressure blueschist metamorphic rocks, invariably contains antiphase domains indicating that an ordering transformation has occurred. The equilibrium order – disorder temperature is about 865°C, well above the crystallisation temperature of most blueschist pyroxenes.

The fact that observations of metastable disorder are not uncommon means that there must be some kinetic advantage in nucleating the disordered structure rather than the equilibrium ordered phase. Note however that nucleation of a metastable phase necessarily requires an overstepping into the stability field of the ordered phase.

The origin of the kinetic advantage can be seen by considering the expression for the rate of nucleation (eqn.(11.12)):

$$\text{Rate of nucleation } I \propto \exp\left(-\frac{\Delta G^*}{RT}\right) \exp\left(-\frac{G_A}{RT}\right)$$

where ΔG^* is the energy barrier for nucleation and G_A is the free energy of activation for the process. This leads to the C-shaped TTT curve for nucleation (Section 11.1.7). At high supersaturation ΔG^* becomes negligibly small for both ordered and disor-

dered crystals, and the variations in crystallisation kinetics depend on the magnitude of the G_A term. Since $G_A = E_A - T\Delta S_A$ we can write that

$$\text{Rate of nucleation } I = A \exp\left(-\frac{E_A}{RT}\right)$$

where the pre-exponential 'A' contains the terms associated with the entropy of activation. This is likely to be the most important term in discriminating between nucleation of ordered or disordered phases. As discussed in Section 10.1.1, the entropy of activation is related to the number of possible configurations of the activated state and hence describes the number of reaction pathways leading to the crystallisation. To produce an ordered nucleus a very specific configuration of atoms is required, but for a disordered nucleus the number of possible alternative structures will clearly be very much greater.

This phenomenon is merely a special case of the Ostwald step rule (Section 11.4) and illustrates the general principle that under non-equilibrium conditions, kinetics rather than thermodynamics controls mineral processes. Metastably disordered phases have the same status as the transition phases described in Section 11.4, and a slight modification of Figure 11.56 could equally well describe the relative kinetics of nucleation of ordered and disordered phases (Figure 12.46).

The significance of the presence of metastably disordered phases is the recognition that equilibrium phase boundaries have been overstepped, even in geological processes where the time scales are such

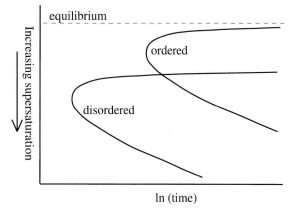

Figure 12.46. Schematic TTT diagram showing the relative kinetics of the onset of nucleation and growth of ordered and disordered polymorphs of a mineral. A family of such curves could be drawn for various degrees of order. (After Carpenter and Putnis, 1985.)

that it is often tacitly assumed that reactions take place very near equilibrium.

12.4.3 A comparison of the mechanism of Al,Si ordering in cordierite and K-feldspar

The tweed microstructures which form during the transformation from hexagonal to orthorhombic cordierite are virtually identical to those observed in some natural K-feldspars, and the whole microstructural sequence described in cordierite (Figure 12.35) could be matched with K-feldspars with a range of thermal histories.

In K-feldspars (see Sections 6.8.3, 7.3.5 and 7.3.6) Al,Si ordering results in a structural transformation from monoclinic to triclinic symmetry, producing the characteristic cross-hatched twinning observed in microcline, the ordered form. The equilibrium order-disorder transition in K-feldspar is about 450°C, compared with 1450°C in cordierite, and therefore we cannot do experiments on Al,Si ordering within the stability field of ordered microcline. The time-scale for ordering at temperatures below 450°C is very long, even geologically speaking, and ordered microcline only forms in deep plutonic rocks which have cooled very slowly. By choosing samples from a variety of geological environments, the full range of microstructures from disordered sanidine to fully ordered maximum microcline can be observed.

Many of the same analytical techniques used to study the experimental cordierite samples have been applied to K-feldspar, and a very similar picture emerges. The results are not as straightforward as in cordierite because, for example, ^{29}Si NMR spectroscopy cannot be interpreted to the same level. This is because the chemical shifts of the three Si sites namely T_{2m} with Si(2Al), and T_{2m} and T_{1m} with Si(1Al) are quite close together, and the increased number of peaks which exist in disordered material cannot be resolved. In terms of ^{29}Si NMR spectroscopy cordierite is rather exceptional in having well spaced chemical shifts, because it has an unusual structure for a framework aluminosilicate.

Conclusions from IR spectra, electron microscopy and X-ray diffraction of K-feldspars have many parallels with cordierite. The transformation sequence

$$\text{sanidine} \Rightarrow \text{orthoclase} \Rightarrow \text{microcline}$$

involves the development of a tweed structure in

orthoclase, which retains the high temperature monoclinic symmetry, although IR indicates that it is locally triclinic and has a high degree of local Al,Si order. The formation of microcline is marked by splitting of powder diffraction peaks, due to the monoclinic to triclinic transition on a macroscopic scale, although in this case the splitting is very much greater, indicating a greater strain and instability in the framework itself.

The usual way of measuring this degree of *triclinicity* in K-feldspar is to measure the splitting of the $131 - 1\bar{3}1$ reflections which have Δ values from 0° 2θ in monoclinic feldspar to about 2.0° 2θ in maximum microcline (compared with $\Delta = 0.25°$ 2θ in orthorhombic cordierite). There also appear to be a range of Δ values in microcline, defining a series of intermediate microclines in which the degree of splitting can be correlated to the degree of Al,Si order. In this respect there does not appear to be the same large discontinuity in lattice parameters in moving from orthoclase to microcline, although without experimental work this cannot be verified.

One significant difference between the transitions in cordierite and K-feldspar is that whereas in cordierite the symmetry change from P6/*mcc* to C*ccm* requires a first order transformation, there is no such constraint on the C2/*m* to C$\bar{1}$ transition in K-feldspar. The latter could therefore be second order, tricritical or first order.

Another related point which has general importance is the detailed description of the tweed structure in such cases. We first raised this issue in Section 7.3.6 in which we considered the effect of two orthogonal sinusoidal transverse modulations on the local symmetry. This can be analysed mathematically using group theory, or may be generated by computer, as done in Figure 7.65. The result is that a tweed modulation always contains two types of domains:

(i) domains sheared in opposite senses, connected by
(ii) regions with high temperature symmetry which are themselves rotated.

These regions are periodically arranged as on a checkerboard. In the case of K-feldspar, the sheared regions are triclinic and are separated by monoclinic domains. Therefore the modulation already contains 'nuclei' of the triclinic form. During coarsening these can grow continuously.

In cordierite the tweed structure also contains sheared regions separated by rotated hexagonal domains. However, the sheared regions do not have orthorhombic symmetry, but must be locally monoclinic. This is a simple geometric consequence which can also be verified theoretically. We have not as yet mentioned monoclinic cordierite, but there is no doubt that it must exist within the modulation as it develops. The fact that it does not have the symmetry of the eventual low temperature structure is not a particular problem, as evidently the modulated state represents an energetically and kinetically favourable compromise between increasing the degree of Al,Si order and reducing the macroscopic strain. The local Al,Si ordering scheme consistent with monoclinic symmetry will be different from that in the low temperature orthorhombic form, but the fact that it can exist within the modulation gives it an advantage.

Complete Al,Si order in cordierite cannot however be achieved within a monoclinic structure and eventually the orthorhombic structure forms by a process of nucleation and growth as observed by electron microscopy and X-ray diffraction, described above. This suggests that the modulated structure is a unique entity formed by a coupling between the ordering and the distortion and represents a free energy minimum in the overall transformation process. It is not merely a microtwinned version of the low temperature form.

We are left however with one problematic experimental observation which we must explain. IR spectroscopy of cordierite suggests that the local structure within the modulation is the same as that in the orthorhombic form. This seems to contradict the existence of monoclinic regions which would be obvious in the IR spectrum of the modulated structure. One possibility is that monoclinic regions do exist at high temperature but during the quench they distort to orthorhombic symmetry. If the sinusoidal modulation became a square wave during the quenching process, this would be equivalent to a microtwinned orthorhombic cordierite, and consistent with the IR observations. This question is not resolved and requires that the onset of the modulation be studied *in situ* by high temperature spectroscopy.

One final experimental observation which is relevant however, is that if the modulated state is quenched and then reheated to the original isother-mal run temperature, further coarsening of the modulation does not occur in the same way as it would without the quench. The kinetics of the transformation to the orthorhombic form is also very significantly retarded in the quenched material. These observations demonstrate that quenching the modulated structure alters it in some way and are consistent with the hypothesis that quenching the modulations produces interlocked twin domains.

12.4.4 Coupling between Al,Si ordering and structural instability – the case of albite, $NaAlSi_3O_8$

The key to understanding the Al,Si ordering process in albite is its relationship to the displacive transition (Section 12.2.4). Until the mid 1980's the displacive transition and the Al,Si ordering process were usually considered separately and the behaviour described as the following sequence: first the structural collapse of the high temperature monoclinic $C2/m$ structure to the triclinic $C\bar{1}$ structure (monalbite $\Rightarrow$ high albite), followed by the sluggish Al,Si ordering process within the triclinic structure (high albite $\Rightarrow$ low albite).

If the displacive transition could be suppressed, as in K-feldspar, the Al,Si ordering transition would still occur and result in the same symmetry change. The difference between Na-feldspar and K-feldspar is that in the former the displacive transition occurs at a higher temperature than the Al,Si ordering transition, while in K-feldspar the presence of the large cation effectively prevents the structural collapse.

In disordered high albite the four tetrahedral sites T_{1m}, T_{2m}, T_{1o} and T_{2o} are non-equivalent but have the same Al,Si occupancy i.e. 25% Al and 75% Si. Complete order can be attained with no further symmetry change and the process is essentially one of gradually increasing the amount of Al in the T_{1o} sites as the temperature is reduced. In low albite all of the Al is in T_{1o} sites. Given that there are no symmetry constraints on the Al,Si ordering process, we might expect that the degree of order would be simply dependent on temperature, as in non-convergent ordering.

This is not however the case. Experimental results for the equilibrium degree of order as a function of temperature (Figure 12.47(a)) show that the Al occupancy of the T_{1o} sites changes very markedly below about 700°C, and also that the kinetics of

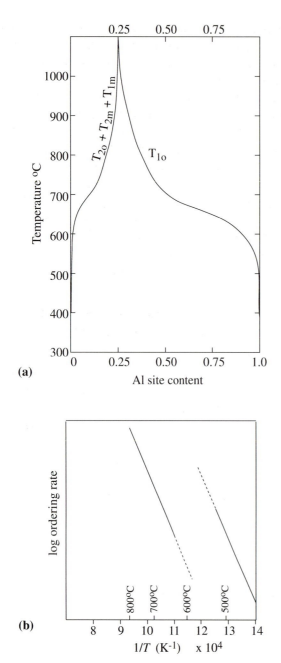

(a)

(b)

Figure 12.47. (a) The general form of the equilibrium Al distribution in albite as a function of temperature. In the completely disordered state the Al occupancy of each site is 0.25. On cooling, Al migrates into the T_{1o} site from the other three sites (in which the Al content is assumed to be equal). (After Smith and Brown, 1988). (b) A schematic Arrhenius plot for the rate of Al,Si ordering in albite, indicating a break in the plot between 700°C and 550°C. The low temperature line represents rates two orders of magnitude faster than would be expected from an extrapolation of the high temperature line. (After McConnell and McKie, 1960.)

ordering at low temperatures is much faster than would be expected by extrapolating the high temperature kinetic data (Figure 12.47(b)).

The solution to this problem, and an explanation of the behaviour of albite, was an early success for the application of Landau theory to minerals, particularly the way in which coupling between two processes is treated. The coupling between the degree of Al,Si order, described by an order parameter Q_{od}, and the degree of distortion in the structure due to the displacive transition, described by the parameter Q, has been referred to in earlier discussions on albite (see Sections 6.8.3, 8.6.4, and 12.2.4). To understand albite behaviour we must take into account that the displacive collapse of the structure is dependent on the degree of Al,Si order and conversely the degree of Al,Si order will influence the way in which the displacive transition occurs.

The method of handling this situation has been described in Section 8.6.4. The theoretical derivation for the excess free energy associated with the $C2/m \Rightarrow C\bar{1}$ transition is given in eqn. (8.46) as:

$$G = \frac{1}{2} a_1 (T - T_c^*) Q_2 + \frac{1}{4} B_1 Q^4 + \frac{1}{6} c_1 Q^6 + \dots$$
$$+ \frac{1}{2} a_2 (T - T_{c\,od}^*) Q_{od}^2 + \frac{1}{4} B_2 Q_{od}^4 +$$
$$+ \frac{1}{6} c_2 Q_{od}^6 + \dots - \lambda Q Q_{od}$$

from which we obtain equations (8.45) which define the equilibrium values of Q and Q_{od} as a function of temperature:

$$a_1(T - T_c^*) Q + B_1 Q^3 + c_1 Q^5 + \dots \lambda Q_{od} = 0$$
and
$$a_2(T - T_{c\,od}^*) Q_{od} + B_2 Q_{od}^3 + c_2 Q_{od}^5 + \dots \lambda Q = 0.$$

The temperature T_c^* is the equilibrium transition temperature for the displacive transition if the crystal is completely disordered i.e. $Q_{od} = 0$, and the temperature $T_{c\,od}^*$ would the equilibrium ordering transition temperature in a crystal if the displacive transition could be suppressed (i.e. $Q = 0$). T is the equilibrium transition temperature which is higher than either T_c^* or $T_{c\,od}^*$.

The unknown quantities in these expressions must be determined from experiments. Since the Al,Si process is very sluggish compared with the displacive transition, it is possible to measure values of Q as a function of temperature, for fixed values of the Al,Si order parameter Q_{od}.

Both Q_{od} and Q can be measured from the lattice parameters. The Al,Si ordering and the structural collapse have a different effect on the way in which the unit cell distorts from monoclinic symmetry. It has been found that Al,Si ordering particularly affects the interaxial angle γ which is 90° in the monoclinic form (i.e. complete disorder) and decreases to 87.7° in the fully ordered form. The order parameter Q_{od} can be equated to cos γ which varies from 0 to 0.04. The displacive transition has the most effect on the interaxial angle α, which is not very sensitive to the degree of Al,Si order. The order parameter Q is proportional to $-\cos \alpha^*$ where α^* is the interaxial angle between the reciprocal lattice axes z^* and y^*.

The results of such experiments provide sufficient data to be able to numerically fit values of T_c^*, $T_{c\ od}^*$ and the various coefficients in the above equations, and hence solve equations (8.45) simultaneously. This gives the temperature dependence of the Al,Si order parameter (Q_{od}) for different values of Q (Figure 12.48). Alternatively, this can be expressed as the temperature dependence of the displacive transition, described by the parameter Q, for different values of Q_{od} (Figure 12.49). These $Q - Q_{od} - T$ surfaces represent all possible stable and metastable

states of albite. The equilibrium states are defined by the intersection of the two surfaces in Figures 12.48 and 12.49, and this is shown as the heavy line across the surface of both diagrams.

These heavy lines are transferred to Figure 12.50 which shows the temperature dependence of the Al,Si order parameter Q_{od} and the displacive order parameter Q at equilibrium conditions. The fact that a relatively limited amount of experimental data can be combined with the theoretical description of the coupling process to produce a full description of the behaviour of albite, and hence of its thermodynamic properties, is a considerable achievement and emphasises the importance of this approach to studying mineral transformations.

12.5 Further examples of incommensurate structures in minerals: mullite, nepheline, and 'e' plagioclase

When a high temperature structure transforms on cooling, there may be a number of structural modifications which can reduce its free energy. Ultimately only one of these possible structures will be

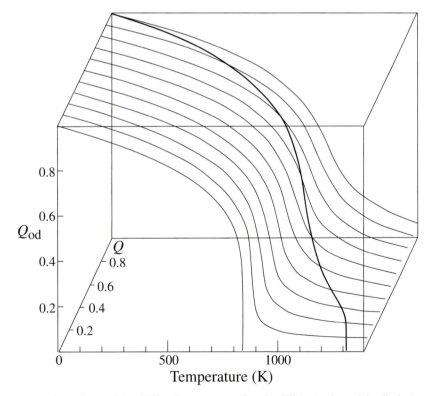

Figure 12.48. The temperature dependence of the Al,Si order parameter Q_{od}, for different values of the displacive order parameter Q, in pure albite. The curves define a surface which represents all possible stable and metastable states of albite. The heavy line on this surface represents the equilibrium states of albite. (After Salje *et al.*, 1985.)

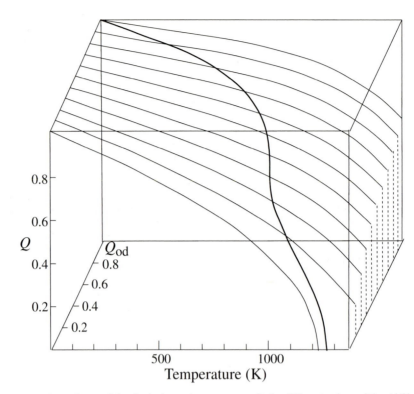

Figure 12.49. The temperature dependence of the displacive order parameter Q, for different values of the Al,Si order parameter Q_{od}, in pure albite. The curves define a surface which represents all possible stable and metastable states of albite. The heavy line on this surface represents the equilibrium states of albite. (After Salje *et al.*, 1985.)

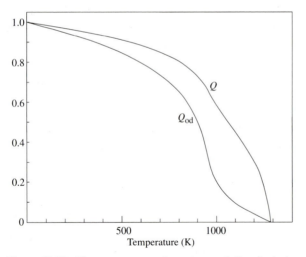

Figure 12.50. The temperature dependence of the displacive order parameter Q, and the Al,Si order parameter Q_{od} at equilibrium. The curves are derived from the intersection of the two surfaces in Figures 12.47 and 12.48. (After Salje *et al.*, 1985.)

the stable low temperature phase, but there may be a temperature range in which the minimum free energy is attained by a well-defined interaction between two of the structures. This can only happen if certain conditions are fulfilled, most importantly,

that the symmetry and energy of the two structures are compatible.

The result of such an interaction is a new structure with a periodicity superimposed on the high temperature form, rather like a superlattice, except that it is not commensurate with the periodicity of the lattice. In other words, the periodicity is a non-integral repeat of the underlying structure, and it may be in a direction which is not a lattice vector.

If we describe each of the two structures involved in the interaction in terms of order parameters Q_1 and Q_2 respectively, the formal description of the free energy reduction is that one order parameter couples with gradients in the other to stabilise the variations in the amplitudes of both as a function of distance in the crystal. Thus a Landau free energy expression for the free energy change in forming an incommensurate structure will contain coupling terms $Q_1(\nabla Q_2)$ and $Q_2(\nabla Q_1)$ which provide most of the stabilisation energy. (The gradient operator ∇ is defined in Section 10.5.2). Figure 12.51 is a schematic illustration of the variation in Q_1 and Q_2 with distance in the crystal.

We first encountered the concept of an incommen-

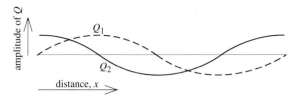

Figure 12.51. Schematic illustration of the variation in the amplitude of two interacting order parameters Q_1 and Q_2 with distance in an incommensurate structure. The periodicity varies from tens to hundreds of Ångstroms.

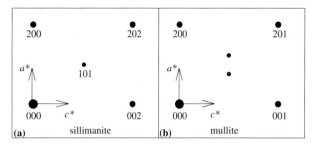

Figure 12.52. (a) A part of the a^*c^* reciprocal lattice of sillimanite which has a doubled c axis relative to the disordered unit structure. 101 is the superlattice reflection due to Al,Si ordering. Absences are due to the *Pbnm* space group of sillimanite. (b) The equivalent a^*c^* reciprocal lattice section of mullite. The main reflections are due to the disordered subcell, with the pair of satellites around the missing superlattice reflection.

surate structure in the case of quartz (Section 12.3), in which the interaction was between two different modes of distortion of the high quartz structure. Here we will describe how these same general ideas can be applied to incommensurate structures in other minerals.

12.5.1 Mullite

Mullite has a sillimanite-like structure with general composition $Al_{4+2x}Si_{2-2x}O_{10-x}$ (Section 6.3.3). Compared to the sillimanite composition ($Al_4Si_2O_{10}$), mullite contains excess Al atoms, electrostatically balanced by removing oxygen atoms. Although the composition of mullite does not allow it to achieve long-range Al,Si order, a comparison with sillimanite clearly shows that there will be a strong tendency for Al,Si ordering on a local scale. However, the distribution of Al,Si is not independent of the oxygen-vacancy distribution, and it is reasonable to assume that these will also tend to order.

An analysis of the symmetry changes involved in these ordering schemes shows that both would result in a doubling of the c axis of the cell, as is the case in sillimanite. However, neither of these ordering schemes can by themselves produce a long-range ordered structure, and the stabilisation achieved in mullite is by their interaction to form an incommensurate structure. Mullite is stable over a wide composition range ($0.17 < x < 0.6$), up to a temperature around 1850°C! This is a very impressive stability field for a 'defective' mineral structure, and emphasises the success of such interactions in stabilising the structure.

The incommensurate structure of mullite has a periodicity along the [100] direction with wavelengths ranging from ~ 10Å – 15Å depending on the composition. Since both the Al,Si ordering and the

oxygen-vacancy ordering schemes involved in the interaction tend to double the c axis the periodicity is associated with the superlattice reflection. This appears as a pair of satellite intensity maxima around the absent superlattice position, as shown in Figure 12.52.

12.5.2 Nepheline

Nepheline is a stuffed derivative of the tridymite structure with Al replacing Si in the tetrahedral sites, and Na and K ions added to balance the electrostatic charge (Section 6.8.2). If Al = Si the composition is $KNa_3Al_4Si_4O_{16}$, but generally Al < Si and the general formula may be written as $K_x [\ ]_{1-x} Na_3 Al_{4-x} Si_{4+x} O_{16}$ where [ ] is a vacancy on a K site, and $x \sim 1/3$.

Above 180°C, nepheline exists as a defect structure with hexagonal symmetry, $P6_3$. At lower temperatures this structure is no longer stable as two different factors become important. First, the Na atoms in the large cavities are not well coordinated with the oxygen atoms on all sides, and there is a tendency for some of the oxygen atoms which should lie on the triad axis in space group $P6_3$ to be slightly displaced from these symmetry positions in one of three different directions. These displacements are statistically distributed so that the triad axis is, on average, retained.

The second tendency in the nepheline structure is for the K ions and the vacancies to order, the large channels in the structure allowing considerable K migration at these low temperatures.

The incommensurate structure of nepheline below 180°C shows satellite intensity maxima paired about the points $\pm(\frac{1}{3},\frac{1}{3},0)$ in the hexagonal cell, at fractional coordinates of approximately ± 0.20 along c^*. Thus the incommensurate structure is associated with tripling the a axis of the hexagonal cell, and has a periodicity of around 5 times the c axis repeat along [001].

To determine whether the incommensurate structure is in fact due to the interaction of oxygen displacements with K-vacancy ordering, it is necessary to analyse the symmetry of each component structure to check whether the symmetry conditions for interaction are fulfilled. This has been discussed in a number of review papers by McConnell (see bibliography). Experiments on the kinetics of the high temperature to incommensurate phase transition give an activation energy which is consistent with that expected for K-vacancy ordering, lending further support to the model.

It is not always possible to determine the nature of the component structures which are interacting in an incommensurate structure. X-ray diffraction gives only the average structure, although it is possible in some simple cases to unravel the source of the extra intensity in the satellite reflections. This has been done in the case of mullite. The more common method is a combined approach in which the symmetry of likely structures is analysed, and combined with experimental and spectroscopic data on the phase transition to the incommensurate structure.

12.5.3 'e' plagioclase

By far the most important incommensurate structure in minerals is that of plagioclase, partly because it is such a common mineral, and partly because its existence dominates the behaviour of the plagioclase solid solution on cooling. As pointed out in Section 6.8.3, natural plagioclase is usually a very fine lamellar intergrowth producing irridescent optical effects seen in hand specimens and visible by electron microscopy. One or both of the phases in these intergrowths (Figure 6.59) are termed e plagioclase, and there are two similar, yet distinct, structural types: e_1 and e_2. (The term e plagioclase refers to the satellite reflections in the diffraction pattern, which are generally referred to as e reflections.)

To define the interactions which are responsible for the formation of the incommensurate structure of e plagioclase we need to look at the structural tendencies in the high temperature plagioclase solid

solution, examine their symmetries, and combine this with observations on natural plagioclases with various cooling histories, and experimental data from heat treated samples.

As we have discussed in previous sections (particularly 6.8.3, 7.3.4, 12.2.4 and 12.4.4.) the fully expanded $C2/m$ structure which exists in albite-rich plagioclase readily collapses to the triclinic $C\bar{1}$ structure. Within this structure there are four distinct tetrahedral sites T_{1o}, T_{1m}, T_{2o}, and T_{2m}, over which the Al,Si atoms are distributed. In albite with an Al:Si ratio of 1:3 complete order can be achieved within this structure, i.e. high $\Rightarrow$ low albite, but this ordering scheme becomes increasingly unfavourable as the amount of Al increases towards the anorthite composition. This is because any increase in the 1:3 ratio will necessarily produce Al–O–Al linkages. Thus although the $C\bar{1}$ structure exists over a wide range of plagioclase compositions at high temperature, no long range Al,Si order can be achieved within it, except near the pure albite composition.

In anorthite the problem of Al,Si ordering is solved by an alternation of Al and Si, not possible in the $C\bar{1}$ structure. This results in a new structure with a doubled c axis and space group $I\bar{1}$. Again, perfect long-range order within the $I\bar{1}$ structure is only achievable in pure anorthite, and as the composition moves towards albite, the degree of order necessarily decreases. Thus in the intermediate plagioclase composition range neither the albite nor the anorthite ordering scheme is a particularly satisfactory solution to the problem of achieving a maximum degree of Al,Si order and hence decreasing the free energy. The distribution of Na and Ca atoms required to achieve local charge balance must also follow the Al,Si distribution.

The satellite e reflections which define the incommensurate structure are paired around the position of the superlattice reflection associated with the anorthite ordering scheme, i.e. the periodicity is associated with a doubling of the c axis. This confirms that Al,Si ordering, according to the anorthite scheme, must be one of the component structures in the incommensurate structure. The most likely second component, which fulfills the symmetry constraints, is Al,Si ordering within the $C\bar{1}$ structure, i.e. according to the albite ordering scheme.

The interaction between the $C\bar{1}$ and $I\bar{1}$ ordering schemes appears to be a satisfactory solution to the problem, and produces an incommensurate struc-

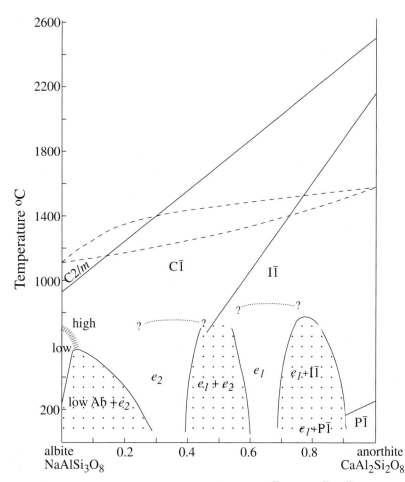

Figure 12.53. Schematic phase diagram for the plagioclase feldspars. The $C2/m \Rightarrow C\bar{1}$ and the $C\bar{1} \Rightarrow I\bar{1}$ transitions are full lines, truncated by the solidus and liquidus shown as dashed curves. The transitions to the incommensurate structures e_1 and e_2 are somewhat uncertain and shown as dotted lines. The shaded regions at lower temperatures are miscibility gaps. (After Carpenter, 1985.)

ture which can be regarded as a periodic array of antiphase domains associated with doubling the c axis (Section 7.3.4). The structure on the antiphase boundaries is necessarily $C\bar{1}$. Crudely, this could be described as slabs of anorthite-type structure alternating with slabs of albite-type structure, with a periodicity of $\sim$50Å. This however would only be approximately true as the stabilisation is achieved by interweaving the periodically fluctuating structures as implied by Figure 12.51. The Na and Ca ions must also be accommodated within this scheme, but it is thought that they play a supporting role by migrating to suitable sites, rather than determining the nature of the structure.

Another factor which is important in the stabilisation of the incommensurate structure is that the strain associated with ordering is minimised by the interaction. As pointed out in Section 12.4.4, Al,Si

ordering affects the shape of the unit cell, principally in the value of the interaxial angle γ. In low albite and anorthite the change in $\cos \gamma$ associated with Al,Si ordering is opposite in sign, and in the incommensurate structure $\cos \gamma$ is almost zero, suggesting that the interaction would be favourable in minimising the strain.

As the composition moves from anorthite-rich to albite rich, the wavelength and the orientation of the incommensurate periodicity change, presumably controlled by the need to keep the strain at a minimum while varying the relative influence of the two Al,Si ordering schemes in the interaction. Around the An_{50} composition there is a discontinuity in the orientation and spacing of the e reflections, as well as discontinuities in the lattice parameters. There are also breaks in the value of the heat of solution as a function of composition, incommensu-

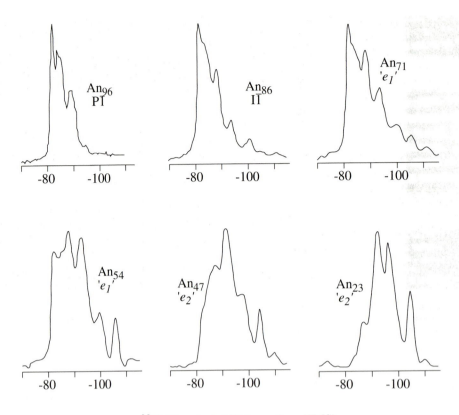

^{29}Si Chemical shift (ppm from TMS)

Figure 12.54. ^{29}Si NMR spectra of a suite of natural, relatively well ordered plagioclase feldspars, ranging in composition from An$_{96}$ to An$_{23}$. Although the equilibrium state of some of these compositions is likely to consist of two-phase intergrowths (see Figure 12.52), the samples used here were single phase, in the structural state indicated against each spectrum. The sharpness of the peaks in the incommensurate structures confirms a well-defined local ordering scheme. (After Kirkpatrick *et al.*, 1987.)

rate structures which are anorthite-rich having a larger excess enthalpy of ordering than those which are albite-rich. This has led to a distinction being made between e_1 and e_2 incommensurate structures which are anorthite-rich and albite-rich respectively.

At present the structural origin of this distinction is not clear, but reflects the increasing dominance of the C$\bar{1}$ structure, as suggested by the position of the C$\bar{1}$ ⇒ I$\bar{1}$ phase boundary (Figure 12.53). This is also supported by a sequence of ^{29}Si NMR spectra of compositions on either side of An$_{50}$, which show that on the nearest neighbour scale e_2 plagioclase shows spectra more similar to albite, while e_1 plagioclase has more anorthite-like NMR spectra (Figure 12.54). One manifestation of the difference in free energy between e_1 and e_2 plagioclase is the existence of a miscibility gap between them. Thus compositions around An$_{50}$ exsolve into two phases $e_1 + e_2$ which coexist as lamellar intergrowths (Figure 12.55).

Experiments and calorimetric measurements on intermediate plagioclases (see review by Carpenter), have confirmed that the stable transformation sequence in anorthite-rich regions is C$\bar{1}$ ⇒ I$\bar{1}$ ⇒ e_1. This suggests that the e_1 incommensurate structure is more stable at low temperatures than the imperfectly ordered I$\bar{1}$ structure. Similarly, in albite-rich compositions the stable transformation sequence is C$\bar{1}$ ⇒ e_2. This contrasts with the view that has been held for many years that the e plagioclase structure is a metastable compromise in compositions for which the two alternative ordering schemes are unsatisfactory.

The known phase relations in the plagioclase feldspars are summarised in the phase diagram (Figure 12.53). The shaded regions are the miscibility gaps:

(i) between albite and e_2 plagioclase (the peristerite gap),

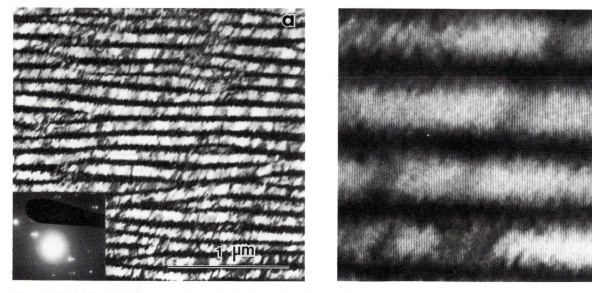

Figure 12.55. (a) Dark field electron micrograph showing the lamellar Bøggild intergrowth in a labradorite of composition An_{52}. (b) Higher magnification micrograph of the same specimen showing the 30Å incommensurate periodicity of the 'e' plagioclase structure, imaged in one of the components of the intergrowth. (From McLaren and Marshall, 1974.)

(ii) between e_2 and e_1 plagioclase (the Bøggild gap), and

(iii) between e_1 plagioclase and anorthite (the Huttenlocher gap).

Compositions between the miscibility gaps can exist as single phase e_1 or e_2 plagioclases at low temperatures.

12.6 Magnetic transitions and magnetic properties of minerals

We conclude this chapter with a brief discussion of the magnetic properties of some iron-bearing minerals. This is a major topic and here we will restrict ourselves to a few specific aspects of magnetic mineralogy, particularly the way in which the microstructure associated with phase transformations such as exsolution and cation ordering can influence the magnetic properties of a mineral. This will emphasise the general point that sub-solidus reactions in minerals play an important role in determining the magnetic properties of rocks, and hence the way in which the magnetic record is interpreted.

First we give an elementary account of the origin of magnetic properties in minerals, and draw parallels between magnetic ordering and the other types of structural phase transitions we have discussed in this chapter.

12.6.1 Origin of magnetic properties in minerals

An electron in an atom has a magnetic moment that results from its orbital motion, its spin, or a combination of both. In the structures we will be considering, the magnetic moment results mainly from electron spin, rather than orbital motion. Atoms and ions in which all electrons have paired spins have no net magnetic moment and are called *diamagnetic*. If there is a net magnetic moment due to one or more unpaired electrons, the atom is *paramagnetic*. The magnetic moment of a single unpaired electron has a value of one Bohr magneton (μ_B).

In the first transition metal series the $3d$ levels are each filled by a single electron, and only become doubly occupied after all five have been filled (see Section 4.5). For example, if we denote an electron spin by the arrow symbol ↑, a pair of electrons is shown as ↑↓, so that

Mn^{2+} and Fe^{3+} are ↑ ↑ ↑ ↑ ↑ with net moment $5\mu_B$

Fe^{2+} is ↑↓ ↑ ↑ ↑ ↑ with net moment $4\mu_B$

Ti^{3+} is ↑ with net moment $1\mu_B$

and Ti^{4+} with no unpaired electrons has zero moment.

The magnetic moment of other ions can be similarly determined with reference to Table 4.2.

These paramagnetic atoms and ions behave like permanent magnetic dipoles, but when they are incorporated in a crystal structure their magnetic

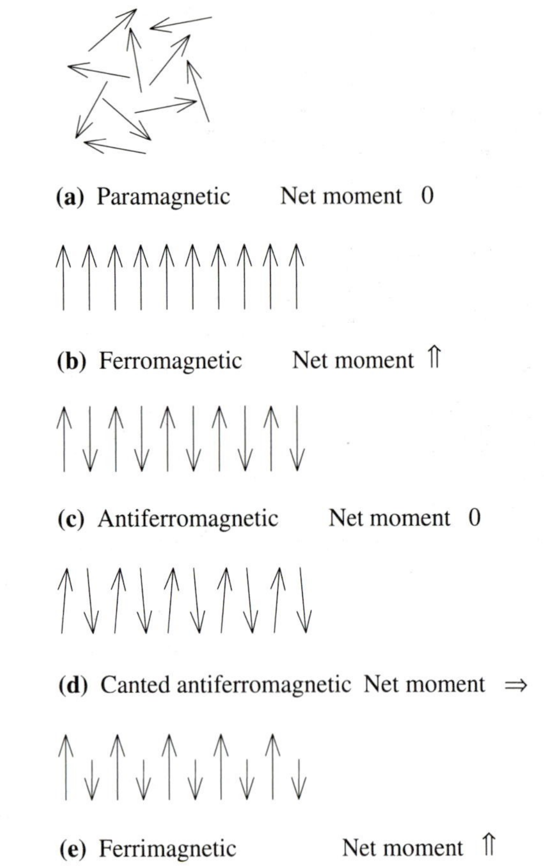

(a) Paramagnetic   Net moment 0

↑↑↑↑↑↑↑↑↑

(b) Ferromagnetic   Net moment ⇑

↑↓↑↓↑↓↑↓↑↓

(c) Antiferromagnetic   Net moment 0

↑↓↑↓↑↓↑↓↑↓

(d) Canted antiferromagnetic Net moment ⇒

↑↓↑↓↑↓↑↓↑↓

(e) Ferrimagnetic   Net moment ⇑

Figure 12.56. Different types of magnetic ordering in minerals.

moments may be randomly oriented (i.e. disordered) and the resulting solid has no net magnetic moment and is termed *paramagnetic* (Figure 12.56(a)). Below a certain critical temperature the interactions between the dipoles may cause the magnetic moments to become ordered in some way. In some metals, e.g. Fe, Ni and Co, there are strong exchange interactions between dipoles and they align parallel to, and in the same sense as each other (Figure 12.56(b)). These *ferromagnetic* materials have a permanent magnetic dipole. Above the critical temperature, termed the Curie temperature, this magnetic ordering is destroyed and the material becomes paramagnetic.

In oxides a different kind of exchange interaction occurs via the oxygen atoms, and the spins of neighbouring metal ions become aligned in an antiparallel way. This coupling is termed a superexchange mechanism, e.g. Fe ($\uparrow$) – O – Fe ($\downarrow$), and becomes most effective as the Fe–O–Fe bond angle approaches 180°. If these superexchange interactions cancel one another out, there is no net magnetic

moment and the material is termed an *antiferromagnetic* (Figure 12.56(c)). For an antiferromagnet, the critical temperature is termed the Néel temperature, above which the material becomes paramagnetic.

In some cases the spin directions are not precisely antiparallel and a small net magnetic moment results from this *spin canting* (Figure 12.56(d)).

Finally, a *ferrimagnetic* material has antiparallel interactions between the spins, but these are of unequal strength and there is a net magnetic moment (Figure 12.56(e)). Since antiferromagnetism is really a special case of ferrimagnetism, this is rather an unfortunate term and originates from the iron-bearing spinels (termed ferrites in materials science), in which the effect was first studied.

Ferrimagnets and canted antiferromagnets are the most important source of magnetism in rocks. Magnetite, Fe_3O_4 is a ferrimagnet with a Curie temperature of 578°C, above which it is paramagnetic. Hematite, Fe_2O_3 is a canted antiferromagnet with a Curie temperature of 680°C. Silicate minerals containing paramagnetic ions only become magnetically ordered at cryogenic temperatures and generally without any net magnetic moment (e.g. fayalite, Fe_2SiO_4, orders antiferromagnetically at 65K) and therefore do not play a role in the magnetic properties of rocks. We nevertheless do need to take this magnetic ordering into account when calculating the enthalpy and entropy of iron-bearing minerals. Here we will be concerned with iron–titanium oxide minerals, which are the major carriers of the magnetic signals in rocks.

The origin of the ferrimagnetism in magnetite arises due to the superexchange interaction between Fe ions on tetrahedral sites which couple antiferromagnetically with those on octahedral sites. Magnetite is an inverse spinel (Section 5.2.4) and the occupancy of tetrahedral and octahedral sites and the resulting magnetic moment are:

Tetrahedral	Octahedral
Fe^{3+}	Fe^{2+}, Fe^{3+}
5 ↑	4 ↓ 5 ↓

resulting in a ferrimagnet with a net magnetic moment of $4\mu_B$.

12.6.2 Magnetic structure and neutron diffraction

Magnetic ordering is studied by neutron diffraction. Since neutrons have a magnetic dipole moment they

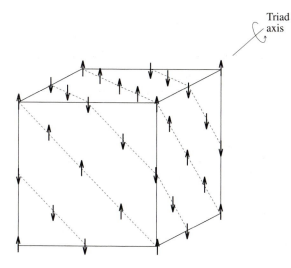

Triad axis

Figure 12.57. The antiferromagnetic structure of ordered MnO, FeO, CoO and NiO. The arrows indicate the electron spin directions of the cations which lie on layers parallel to (111) planes of the structure (shown as dashed lines). The alternation of spins in adjacent layers doubles the size of this unit cell relative to the disordered subcell.

interact with unpaired electrons, and this additional scattering effect, which is not present in X-ray diffraction, is superimposed on the atomic scattering. Thus when a material is cooled from the paramagnetic state to an ordered magnetic state, neutron diffraction is able to detect the distribution and the orientation of electron spins on different atoms in the crystal.

In some structures antiferromagnetic ordering gives rise to a magnetic superlattice. The simplest example is that of the transition metal oxides, MnO, FeO, CoO and NiO. These have the NaCl structure with the cations forming layers between oxygen atoms parallel to the {111} planes. Above the Néel temperature ($T_N = -153°C$ for MnO, $-75°C$ for FeO, $-2°C$ for CoO and $+250°C$ for NiO) the electron spins are disordered and the materials are paramagnetic. Below T_N, the spins order with spins in adjacent (111) layers aligned but antiparallel (Figure 12.57). From the point of view of the magnetic structure the symmetry is clearly lowered, with the loss of the tetrad axes and three of the triad axes. The structures also become slightly distorted on ordering.

The antiferromagnetic ordering in these oxides doubles the repeat of the unit cell along the remaining triad axis and this gives rise to extra reflections in the neutron diffraction pattern. Because X-rays cannot detect magnetic ordering, a distinction can be made between the magnetic symmetry and that determined by X-rays. In spinels, antiferromagnetic coupling between octahedral and tetrahedral cations does not change the symmetry because these sites are already non-equivalent.

12.6.3 Magnetic domains

In the absence of an applied magnetic field a ferro- or ferri- magnetic material is divided into macroscopic *magnetic domains*. There is a specific direction of magnetisation within each domain, but this changes from one domain to the next (Figure 12.58). The existence of these domains reduces the overall free energy associated with the magnetic ordering, and has the effect of minimising the magnetisation of the material, i.e. the external field associated with the alignment of magnetic dipoles. There are a number of different energy terms which arise when a mineral undergoes magnetic ordering, and we will briefly consider these in turn.

1. The direction of magnetisation within a single domain depends on the crystal structure (i.e. *magnetocrystalline anisotropy*). The energy is lower if the net magnetic moment is in a particular crystallographic direction, known as the *easy* direction of magnetisation. For example, the [111] direction is an easy direction in magnetite, compared to [100] in which magnetisation is energetically unfavourable. There are eight equivalent [111] directions in magnetite, and the magnetisation in each domain will be along one of these easy directions.

2. The *magnetostatic effect* is due to the existence of magnetic poles at a free surface or a domain boundary and gives rise to the magnetostatic energy from the Coulombic interaction between these poles. This energy depends on the shape of the crystal, and on the distribution of magnetic domains within it. To reduce this energy the external magnetisation is minimised by aligning the moments in each domain in opposite directions (Figure 12.58).

3. Finally, most magnetic materials change their shape slightly when they become magnetised. This elastic strain energy is termed *magnetostriction*, and in common with all the other transitions we have discussed, the macroscopic strain can be reduced by the formation of domains.

The net effect therefore, of the reduction in free energy on ordering and the associated increase in

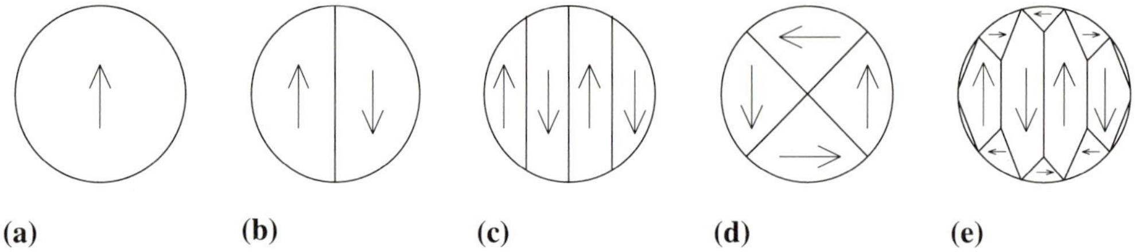

Figure 12.58. The formation of domains in a ferrimagnetic mineral such as magnetite. **(a)** Single domain with high coercivity, but with a large magnetostatic energy. **(b)** Two domains, with the coercivity and magnetostatic energy considerably reduced. **(c)** Further subdivision into domains continues until the reduction in magnetostatic energy is balanced by an increase in energy due to the domain boundaries. **(d)** Closure domains reduce magnetostatic energy but increase the magnetostriction and may be unfavourable due to magnetocrystalline anisotropy. **(e)** Too many small closure domains generate a large magnetostriction. (After McElhinny, 1973.)

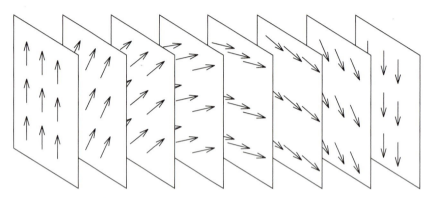

Figure 12.59. Schematic representation of the 180° rotation of the spin orientation in a magnetic domain boundary.

free energy due to the effects discussed above, is the formation of a magnetic domain structure which minimises the overall free energy.

The *magnetic domain boundaries*, across which there is a change in the direction of magnetisation, have a finite thickness because the exchange interactions between neighbouring atoms only allow a very small difference in the direction of spin from one atom to the next. The change in magnetisation must therefore be gradual (Figure 12.59), and the thickness of the domain boundary (also termed a *Bloch wall*) varies between about 100Å and 1000Å, depending on the magnitudes of the energy terms discussed above. Although little is known about the thickness of domain walls in minerals, it is an important concept because crystals smaller than the domain wall must be single domain grains, and these play a major role as recorders of palaeomagnetic information in rocks.

12.6.4 Magnetic hardness and hysteresis

When an external magnetic field is applied to a multi-domain crystal, the domain walls move in response to this applied field, increasing the net magnetisation of the crystal. As the applied field is increased, the magnetisation increases until saturation is reached when the crystal consists of a single magnetic domain with its magnetisation aligned to the applied field (Figure 12.60).

This *saturation magnetisation* M_s, is a property of the mineral and can be related to the cation distribution and the magnetic ordering (see Section 12.6.8). If the applied magnetic field is now reduced to zero, the magnetisation of the specimen will decrease to a value M_{rs}, the *remanent magnetisation* (Figure 12.61). In order to reduce the magnetisation to zero, a reverse field has to be applied. The value of this reverse field is termed the *coercivity* or *coercive force*, H_c. Increasing the reverse field saturates the sample in the reverse direction, and so on, giving a *hysteresis loop*, which describes the relationship between the applied field and the magnetisation (Figure 12.61).

For a multi-domain grain, the hysteresis loop describes the ease with which domain walls can move through the structure, and this defines the *magnetic hardness*. In a magnetically hard material

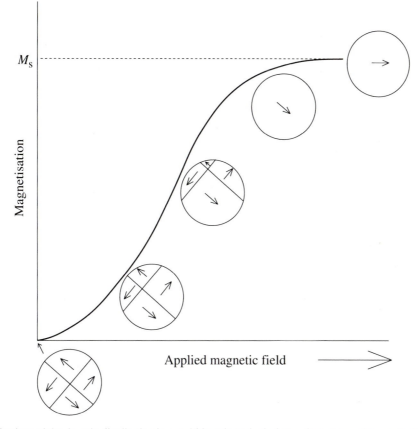

Figure 12.60. The behaviour of the domain distribution in a multidomain grain during various stages of magnetisation. M_s is the saturation magnetisation where the grain consists of a single domain magnetised in the direction of the applied magnetic field. (After Wyatt and Dew-Hughes, 1974.)

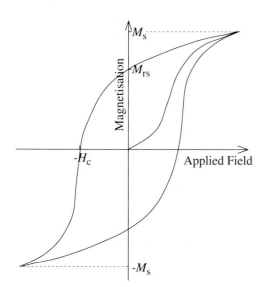

Figure 12.61. A typical hysteresis loop describing the relationship between the magnetisation and the applied field during a magnetisation – demagnetisation cycle. M_s is the saturation magnetisation, and M_{rs} is the remanent magnetisation on removal of the applied field. H_c is the coercivity, which is the reverse field applied to reduce the magnetisation to zero.

the domain walls are difficult to move and a high coercive force is required to remove the magnetisation. When the walls move easily the material is magnetically soft. This behaviour is analogous to that of ferroelectric and ferroelastic materials (Section 12.2.3) in which domain boundaries are moved by the application of an electric field or mechanical stress, respectively. Hysteresis loops for magnetically hard and soft materials are shown in Figure 12.62.

Magnetic hardness is a prerequisite for any material which is to act as a magnetic recording device. To preserve the magnetic record in rocks, the conditions under which the magnetisation was acquired must be 'frozen in' and not able to reequilibrate with any subsequent changes in the magnetic field, at least over the time-scale in which we are interested. In an igneous rock in which the magnetisation of the minerals is acquired by cooling in the earth's magnetic field, our interpretation of the magnetic

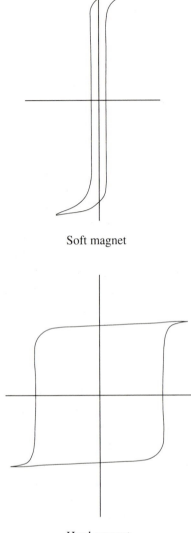

Soft magnet

Hard magnet

Figure 12.62. Typical hysteresis loops for magnetically soft and magnetically hard materials.

because if the magnetocrystalline anisotropy is high, the spins have to pass through high-energy directions. The resultant coercive force may be many orders of magnitude higher than for a multidomain particle, and single domain grains are therefore magnetically very hard.

Another source of magnetic anisotropy which will determine the coercivity of single domain grains, is *shape anisotropy* which arises from the magnetostatic effect. It is easier to magnetise a needle along its length than at right angles to it, because the area over which free poles exist must be minimised to reduce the magnetostatic energy. Thus to reverse the magnetisation of a needle requires it to pass through a high energy state i.e. there will be a large activation energy for the process, and the coercive force must increase. The coercivity of a uniaxial single-domain grain is simply related to the anisotropy and the value of the saturation magnetisation:

$$H_c = \frac{2K}{M_s}$$

where K is a constant which measures the degree of anisotropy.

Mechanical stress can also provide a source of anisotropy by increasing the effect of the magnetostriction when the magnetisation direction is changed.

It follows therefore, that in a rock, the magnetic memory will be preserved best in single domain, needle-shaped mineral grains. In magnetite, the crystalline anisotropy is not very strong and therefore high coercivities are mainly due to shape effects, while in hematite, the opposite is the case with the crystalline anisotropy being primarily responsible for its high coercivity.

Below a certain grain size the magnetic properties are lost due to thermal fluctuations overcoming the exchange forces. Thus very small grains cannot retain a net magnetic moment and are termed *superparamagnetic*. In magnetite, grains below about 0.05μm diameter are superparamagnetic, while above about a micron in diameter they will be multidomain. These values are temperature and shape dependent but give a general idea of the order of magnitude involved.

12.6.6 The relaxation time

If a random assemblage of single-domain grains is magnetised to saturation and the applied field

properties of the rock is based on the assumption that it has retained a memory of the direction and strength of the earth's field at the time, and that it has not been substantially erased by later events.

12.6.5 Single domain grains

If the size of a crystal falls below a certain critical value it becomes energetically unfavourable to have domain boundaries, and the particles exist as single magnetic domains. A single domain grain can only change its magnetisation by the simultaneous rotation of all the spin vectors. This is a difficult process

removed, the remanence will be M_{rs} as shown in Figure 12.61. However, this remanence is a non-equilibrium state and there is always a finite probability that the magnetisation within any domain will 'flip' into an opposite orientation, and hence the magnetisation will gradually decay to zero which is the equilibrium state in a zero field. The rate of this decay will depend on the size of the activation energy barrier between two different orientations of the magnetisation and the thermal energy within the sample.

The activation energy E_a to reverse the magnetisation of a uniaxial single domain grain is vK, where v is the volume of the grain, and K is the anisotropy constant, which takes into account the various forms of anisotropy mentioned in the previous section. The thermal energy is kT where k is the Boltzmann constant and T is the absolute temperature. The rate at which the magnetisation directions reverse, and hence the rate of decay of the magnetisation is then given by a rate equation of the usual form:

$$\text{Rate} = C . \exp\left(-E_a/kT\right)$$

where C is a frequency factor ($\sim 10^{10}\,\text{s}^{-1}$).

The *relaxation time* τ, is the reciprocal of the rate and is therefore

$$\tau = \frac{1}{C} \exp\left(E_a/kT\right)$$

$$= \frac{1}{C} \exp\left(\frac{vK}{kT}\right)$$

$$= \frac{1}{C} \exp\left(\frac{vH_cM_s}{2kT}\right)$$

The relaxation time is therefore related to the properties of the material (the coercivity and saturation magnetisation) as well as to the volume and temperature of the grain.

It is a measure of the time in which there is a significant probability that a domain will change its magnetisation. Expressed in another way, the time t for an initial magnetisation M_{rs} to be reduced to a value M_t is given by

$$M_t = M_{rs} \exp\left(-t/\tau\right),$$

so that the half-life for remanence (when $M_t = \frac{1}{2}M_{rs}$) is $t = 0.693\tau$.

If the relaxation time is small, the grain will very quickly re-equilibrate with any change in the applied magnetic field. A very small particle will have a such a short relaxation time ($\sim$seconds) that no remanence is possible and any acquired magnetisation is very quickly lost. This is the superparamagnetic state. At the other extreme, grains which retain their magnetisation over geological time-scales must have very long relaxation times ($\sim 10^{11}$ years).

The effect of the logarithmic dependence of the relaxation time on temperature and volume ($\log \tau \propto v/T$) is that the relaxation time for a given sized grain will very rapidly increase as the temperature falls, and alternatively, if a particle increases in size at a constant temperature the relaxation time will increase logarithmically with volume. If for example we increase the size of a magnetite grain from 0.05 μm to 0.1 μm, the relaxation time increases from 100 s to 10^{11} years. When the relaxation time is sufficiently long that re-equilibration does not take place on a geological time scale, the magnetisation is effectively 'frozen in' or 'blocked'. For each grain of volume v there is a *blocking temperature* below which the magnetic remanence is frozen in. Similarly, at any given temperature there is a *blocking volume*, above which the magnetisation is effectively permanent over the time scale being considered.

When the grain volume is sufficiently large for magnetic domains to form, the coercivity and the relaxation time become dependent on domain wall mobility and in general we can assume that multi-domain grains will not have sufficiently long relaxation times to be stable carriers of strong remanent magnetisation. The important grain size range in palaeomagnetic studies spans the single domain range.

12.6.7 Magnetic remanence

There are a number of different ways in which minerals in a rock can acquire a remanent magnetisation. In igneous rocks crystallisation takes place above the Curie temperature of the constituent magnetic minerals, and the rock, containing magnetic minerals with a range of grain sizes, becomes magnetised as it passes through the Curie temperatures during cooling. This is *thermoremanent magnetisation* (TRM). Just below the Curie temperature, even large single domain grains will have relatively short relaxation times, but as the temperature falls, progressively smaller grains will pass through their blocking temperatures. At room temperature, the larger single domain grains will have relaxation times of the order of millions of years, while the smallest grains may still have relaxation times of minutes, and so will not retain any record of their magnetic history.

In *chemical remanent magnetisation* (CRM) remanence is acquired during growth of the magnetic grain below the Curie temperature. Initially the growing crystal is superparamagnetic, then passes through the single domain stage with the relaxation time increasing until it passes through its blocking volume. With continued growth, the crystal becomes multidomain, at which point it is just visible by optical microscopy. Beyond this stage its relaxation time becomes somewhat unpredictable, and it no longer contributes significantly to the stable remanence of the rock. CRM can be acquired during many different processes including metamorphism, hydrothermal crystallisation, dehydration processes during diagenesis and late stage alteration.

In sedimentary rocks the magnetic grains themselves tend to be oriented by the earth's magnetic field as they are deposited in water. Subsequent lithification of the sediment locks in this palaeo field direction. This *detrital remanent magnetisation* (DRM) is also the cause of remanence in man-made mud-bricks which are the source of palaeomagnetic information about the more recent past.

12.6.8 Examples of the interaction of microstructure and magnetic properties

In the rest of this chapter we focus on the magnetic properties of the magnetite – ulvöspinel solid solution and the hematite – ilmenite solid solution, which are the major palaeomagnetic carriers in rocks. In both solid solutions (Figure 12.63), transformation processes take place in the minerals during cooling and these generate microstructures which are important when we come to consider the magnetic transitions.

(i) The magnetite – ulvöspinel solid solution

The behaviour of the magnetite (Fe_3O_4) – ulvöspinel (Fe_2TiO_4) solid solution has been described in Section 11.3.4 where two aspects were discussed. First, the distribution of Fe and Ti among the tetrahedral and octahedral sites in the solid solution, and second the microstructures associated with spinodal decomposition. Here we will comment on the implications to magnetic properties.

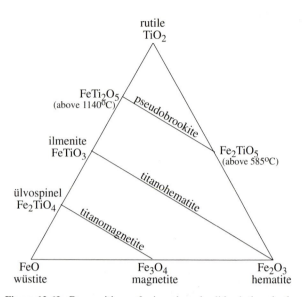

Figure 12.63. Compositions of minerals and solid solutions in the iron-titanium oxide system. The pseudobrookite solid solution is stable only at high temperatures. The most important solid solutions in mineral magnetism are the titanomagnetites and titanohematites.

The magnetic structure of magnetite is described above (Section 12.6.1). Magnetite is ferrimagnetic with a Curie temperature of 578°C and a net magnetic moment of $4\mu_B$. Ulvöspinel is a perfect antiferromagnetic with a zero net magnetic moment and a Néel temperature of −153°C i.e.

Tetrahedral	Octahedral
Fe^{2+}	Fe^{2+}, Ti^{4+}
4 ↑	4 ↓

Compositions between magnetite and ulvöspinel have increasing amounts of Fe^{3+} relative to Fe^{2+} and Ti^{4+} and three different models for the distribution of these ions have been proposed (Figure 11.51). For each of these models the total magnetic moment can be calculated and compared with measurements of the saturation magnetisation.

The magnetic moment can be converted to the macroscopic measure given by the saturation magnetisation (dipole moment per unit mass) by multiplying by the factor $N_A\beta/M$ where N_A is Avogadro's number, β is the value of 1 Bohr magneton (9.27×10^{-24} A m^2) and M is the molecular weight. It is therefore possible, in principle, to determine the cation distribution in a magnetic mineral from a measurement of the saturation magnetisation. This does require however, certain refinements to be

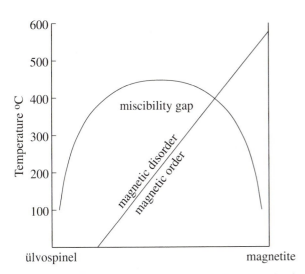

Figure 12.64. The approximate position of the solvus in the system ulvöspinel – magnetite, and the temperature dependence of the magnetic order – disorder phase transition.

made in the value of the magnetic moment assigned to each cation. The values quoted above are for spin only and do not take into account the orbital contribution. They also refer to free ions and the effect of covalent contributions to the bonding, which would reduce the magnetic moments, also needs to be considered in a rigorous treatment. Nevertheless, measurements of the saturation magnetisation of compositions across the solid solution support the model shown in Figure 11.51(a), commonly referred to as the Akimoto model.

The relationship between the magnetic order–disorder temperature and the solvus in the system is shown in Figure 12.64. The low solvus temperature means that even in plutonic rocks, the titanomagnetites have exsolution textures which are only barely visible optically. In dykes and sills which have intermediate cooling rates, exsolution in titanomagnetites has not progressed beyond the coarsened spinodal stage, which in cubic spinels produces a microstructure consisting of cubes of a magnetite-rich phase in an ulvöspinel-rich matrix (Figure 11.50).

The microstructural evolution of the solid solution on cooling controls the size and distribution of magnetic grains in a paramagnetic matrix. In rapidly cooled rocks, the titanomagnetites have not progressed beyond the earliest stages of spinodal decomposition and the magnetite-rich regions may behave superparamagnetically. In more slowly cooled rocks, coarsening proceeds during cooling and the cubes become richer in magnetite, passing through the Curie temperature during growth. Further coarsening produces plates of magnetite rather than cubes (Figure 11.48) increasing the shape anisotropy.

At some point in this sequence the magnetite grains reach their blocking volume and the the result is that the original large grain of titanomagnetite is now made up of an array of single domain magnetite particles in an ulvöspinel matrix. Only in the slowest cooled rocks will the magnetites coarsen enough to reach the multidomain stage. Finely exsolved titanomagnetites, with their high coercivity and long relaxation time, contribute significantly to the palaeomagnetic properties of the host rock.

Another consequence of the exsolution process is that the interface between the magnetite and ulvöspinel is likely to remain coherent throughout. This generates considerable local stresses and may be a major contributor to the high coercivities measured in titanomagnetites.

While the effect of the microstructure in titanomagnetites on magnetic properties is yet to be quantitatively evaluated, it is clear that it is likely to be significant. Commercial hard magnets based on Fe–Ni–Al alloys rely on the formation of precisely this same kind of microstructure for their high coercivities. The single-domain iron-rich precipitates are formed during spinodal decomposition.

(ii) Fe,Ti ordering in the hematite – ilmenite solid solution

The crystal structure of hematite (Fe_2O_3) and ilmenite ($FeTiO_3$) has been described in Section 5.2.4. Here we will be concerned with the Fe,Ti distribution in the ordered solid solution, the microstructure associated with Fe,Ti ordering, and the effect of this microstructure on the magnetic ordering transition.

The phase diagram for hematite – ilmenite (Figure 12.65) shows three principal features.

(i) The cation ordering transition which reduces the symmetry from $R\bar{3}c$ to $R\bar{3}$,
(ii) The miscibility gap (shaded) between disordered and ordered phases, and
(iii) The magnetic order-disorder transition.

The cation ordering and exsolution in this system is similar to that described in Section 9.7 and Figure 9.27, in which ordering is superimposed on a non-

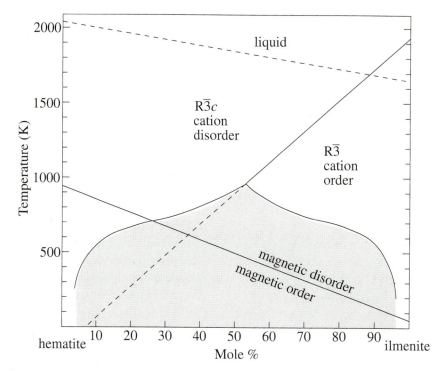

Figure 12.65. Phase diagram for the system hematite – ilmenite. The three principal features are (i) the R$\bar{3}c$ ⇒ R$\bar{3}$ Fe,Ti ordering transition, (ii) The miscibility gap in the system (shaded) and (iii) The magnetic order – disorder transition. (After Nord and Lawson, 1989.)

ideal solid solution and results in a conditional spinodal at low temperatures.

In the disordered R$\bar{3}c$ solid solution, which extends across the hematite side of the diagram, the cation layers between the close-packed oxygen atoms are equivalent (Figure 5.17). Therefore the substitution (Fe^{2+} + Ti^{4+}) for 2Fe^{3+} as the composition moves towards ilmenite takes place randomly. At ilmenite-rich compositions the cations become ordered so that alternate cation layers are no longer equivalent, the Ti^{4+} confined to alternate layers. In pure ilmenite alternate layers contain Fe^{2+} and Ti^{4+}.

One of the consequences of cation ordering is the likelihood of forming domains in which the cation layers are not matched across a boundary (Figure 12.66). These domains have the characteristics of antiphase domains, with a typically curved boundary between them. They are however also twin-related by a diad axis normal to the page in Figure 12.66, through oxygen atoms in the boundary. The domain boundaries are relatively strain-free and have only a slight tendency to adopt a preferred orientation, parallel to (001) planes, in the structure. The presence of these twin boundaries has important consequences to the magnetic ordering transition, as we will discuss below.

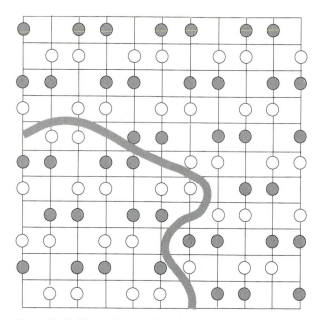

Figure 12.66. Schematic view of the structure of a twin boundary which arises due to Fe,Ti ordering in ilmenite. (After Nord and Lawson, 1989.)

First we need to consider the magnetic structure of hematite, ilmenite, and the solid solution. Above 685°C hematite is paramagnetic with disordered spins. Below this temperature the spins associated

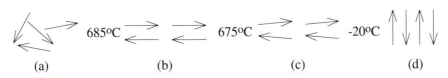

Figure 12.67. The sequence of magnetic transitions in hematite on cooling. **(a)** Above 685°C hematite is paramagnetic. **(b)** Between 685°C and 675°C it is antiferromagnetic with spins normal to the c axis. **(c)** Below 675°C the spins become canted, resulting in a small net magnetic moment parallel to the c axis. **(d)** At -20°C a 'spin-flop' transition results in an antiferromagnetic with spins parallel to the c axis.

with Fe^{3+} ions in alternate cation layers couple antiferromagnetically. The spins are oriented parallel to (001) planes. With perfect antiferromagnetic coupling there is no net magnetic moment, but below 675°C the spins become canted deviating about 10^{-4} radians from the (001) plane, resulting in a weak net magnetic moment parallel to the c axis (Figure 12.67). At -20°C, a 'spin-flop' transition changes the spin orientations to the c axis, but the antiferromagnetic coupling is retained.

Pure ilmenite on the other hand remains paramagnetic down to -218°C (55K) when the spins on the Fe^{2+} ions in alternate iron layers couple antiferromagnetically, with spins parallel to (001) planes. Since Fe^{2+} layers alternate with magnetically inert Ti^{4+} layers, the magnetic unit cell has a doubled c axis relative to the crystallographic unit cell. The temperature of the magnetic disorder – order transition decreases linearly from hematite to ilmenite and although the magnetic structure of ilmenite is of little interest in rock magnetism, it has an important effect on the properties of intermediate compositions.

Consider an intermediate composition $Hem_{50}Ilm_{50}$ which may be written $Fe^{3+}Fe^{2+}_{0.5}Ti^{4+}_{0.5}O_3$. In the ordered structure alternate layers contain Fe ions only. The Ti-bearing layers must also contain Fe, but the cation occupancies are not well known, although there appears to be no ordering of Ti and Fe within each layer. We can examine two possibilities. First we assume that the Ti^{4+}-bearing layers contain both Fe^{2+} and Fe^{3+} (Figure 12.68(a)). Thus a possible occupancy for alternate layers in the fully ordered structure is:

Layer A: $\frac{2}{3}(Fe^{3+} + Fe^{2+}_{0.5})$

Layer B: $Ti_{0.5} + \frac{1}{3}(Fe^{3+} + Fe^{2+}_{0.5})$

If the spins in alternate layers are antiferromagnetically coupled, the net magnetic moment will be due

to $\frac{1}{3}(Fe^{3+} + Fe^{2+}_{0.5})$ which is equal to $\frac{7}{3}\mu_B$. In other words the imbalance between the Fe occupancy in alternate layers leads to a relatively strong ferrimagnetism in intermediate compositions.

An alternative ordering model assumes that the Fe^{3+} occupies both A and B layers equally for any degree of order (Figure 12.68(b)). The occupancy in the fully ordered state would then be:

Layer A: $(Fe^{2+}_{0.5} + Fe^{3+}_{0.5})$
Layer B: $(Ti^{4+}_{0.5} + Fe^{3+}_{0.5})$

In this case the net magnetic moment will be due to $Fe^{2+}_{0.5} = 2\mu_B$. The value of the net magnetic moment clearly depends on the degree of cation order, as well as the ordering model proposed.

In the disordered $R\bar{3}c$ solid solution both A and B layers are equivalent and the hematite magnetic ordering scheme produces only a weak magnetic moment due to the spin canted antiferromagnetic coupling.

As the composition moves towards ilmenite, the amount of Fe^{3+} decreases, and in the ordered structure layer B eventually contains no Fe. In pure ilmenite, layer A has nothing to couple with in layer B, and the coupling is apparently between alternate iron-bearing layers, i.e. every second cation layer in the structure. As mentioned above, pure ilmenite is antiferromagnetic with a doubled c axis in the magnetic cell. The net magnetic moment would therefore be expected to fall as the ilmenite composition is approached. This is borne out by values of the saturation magnetisation as a function of composition (Figure 12.69).

We have so far ignored the presence of the miscibility gap in this discussion. In rapidly cooled volcanic rocks, no exsolution would occur, due to the relatively low temperature of the solvus. Ordering, on the other hand, requires only local diffusion and is kinetically a much faster process. We can assume therefore that rapidly cooled samples will achieve at least partial Fe,Ti order as they pass

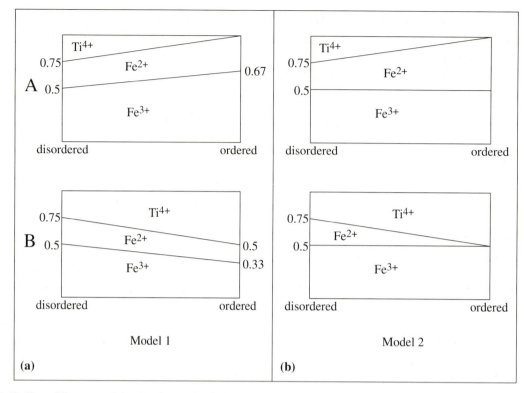

Figure 12.68. Two different models of cation order in both ordered and disordered phases with composition $Hem_{50}Ilm_{50}$ in the hematite-ilmenite solid solution. Alternate cation layers are labelled A and B. In both models the disordered structure contains Fe^{2+}, Fe^{3+} and Ti statistically distributed over both A and B layers, and in the ordered structure A layers are occupied by Fe only. **(a)** In model 1, Fe^{2+} and Fe^{3+} are present in both A and B layers **(b)** In model 2, Fe^{3+} occupies both A and B layers equally for any degree of order.

through the metastable extension of the cation ordering transition (shown dashed in Figure 12.65). Compositions more ilmenite-rich than Ilm_{50} pass through the ordering transition before the miscibility gap, whereas in more hematite-rich compositions ordering and local clustering compete as mechanisms of free energy reduction.

Self-reversed magnetisation in the ilmenohematites

We return now to the effect of microstructure on the magnetic properties of intermediate compositions which have cooled sufficiently fast to avoid any exsolution phenomena. A remarkable property of such compositions is that they exhibit *self-reversed magnetisation*. This means that when intermediate compositions are cooled through the Curie temperature in the presence of an applied magnetic field, the resultant remanent magnetisation is in the *opposite* direction to the applied field. For many years this was one of the major unsolved problems in geophysics and threatened the hypothesis that rocks with reversed magnetisation were evidence that the

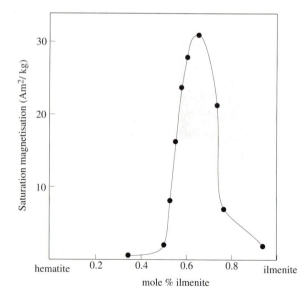

Figure 12.69. Experimental values of the saturation magnetisation as a function of composition across the hematite – ilmenite solid solution. (After O'Reilly, 1984.)

earth's magnetic field had reversed since their formation. Although there are still many unanswered questions regarding self-reversals, a satisfactory explanation has been found to explain this phenomenon in titanohematites.

First we briefly outline the general criteria which must be satisfied for self-reversal to occur. The basic requirement is the coexistence and magnetostatic interaction between two intergrown magnetic phases, which we can call X and Y, which have different Curie temperatures. In some circumstances the following sequence may take place.

(i) Phase X passes through its Curie temperature first and becomes magnetised in the same direction as the applied field.

(ii) As phase Y passes through its Curie temperature, magnetostatic forces at the interface between Y and X may cause a magnetic coupling, such that Y is magnetised in a reverse direction relative to X. This may reduce the overall magnetostatic energy of the intergrowth. In other words, the magnetisation of Y is less influenced by the applied field than it is by the presence of the magnetic phase X.

(iii) If the strength of the magnetisation of Y is greater than X, the net magnetic moment is reversed relative to the applied field.

For the magnetostatic interaction to be effective, the interface between the phases must be coherent or partially coherent, so that the strain associated with the magnetisation (i.e. the magnetostriction) can be transferred across the interface. Coherent intergrowths arise as a result of solid state phase transformations.

It has been known for several decades that the self-reversal in the ilmenohematites is connected with Fe,Ti ordering, although the identity of the X and Y phases has only recently been experimentally verified. Magnetic measurements and TEM observations on the same samples have demonstrated that the magnetisation is related to the surface area of the twin domain boundaries which arise during Fe,Ti ordering (Figure 12.66). The twin domain size and hence the area of domain wall can be controlled experimentally by annealing synthetic samples

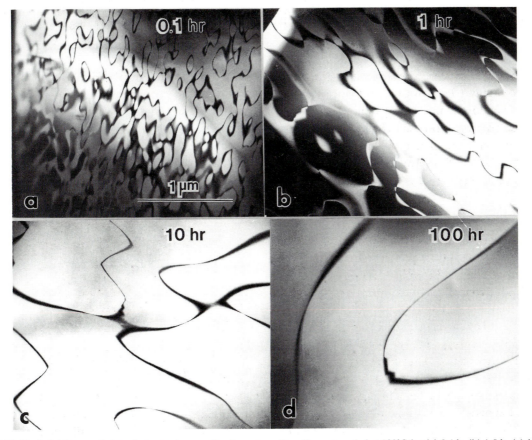

Figure 12.70. Dark field transmission electron micrographs of a composition Ilm_{70} annealed at 800°C for (a) 0.1 h. (b) 1.0 h. (c) 10.0 h. and (d) 100.0 h. (From Nord and Lawson, 1989.)

below the cation ordering transition temperature for increasing periods of time (Figure 12.70). When the surface area of the boundaries exceeds a critical value, samples cooled through the magnetic transition acquire a reversed thermoremanent magnetisation.

The relationship between twin domain boundary area and magnetisation suggests that the domain boundaries are the X phase and the domains themselves are the Y phase. To fulfill the criteria for self-reversal the domain boundaries must be coherent, and have a weaker magnetisation and higher Curie temperature than the domains themselves. The first two criteria are easily satisfied. The small strain associated with cation ordering results in curved, coherent domain boundaries. The structure within the boundaries is disordered $R\bar{3}c$, and hence is only weakly magnetised on the hematite magnetic ordering scheme.

To have a higher Curie temperature the domain boundaries must be enriched in iron relative to the domains. The evidence for this is less obvious. Depending on where the boundary is drawn in Figure 12.66, it could locally be enriched in Fe or Ti. Fe enrichment would suggest a preferred siting of the boundary in the structure. The small oxygen displacements associated with Fe,Ti ordering may control the boundary structure by reducing the surface strain energy. This might favour a boundary adjacent to Fe ions and hence control its local composition. Magnetic measurements of blocking temperatures in experimental cooling also suggest the existence of two magnetic phases, one enriched in Fe.

The evidence for the role of the domain boundaries in self-reversal is strong, and in natural samples the domain size is controlled both by composition, which defines the cation order–disorder temperature, and the cooling rate. In volcanic rocks only a limited range of ilmenohematite compositions between about Ilm_{51-73} appear to be capable of achieving the domain size necessary to produce self reversal. This composition range is found in acid volcanic rocks, such as dacites, and oxidised basalts.

In more slowly cooled rocks exsolution processes might be expected to play a role in determining the magnetic properties of ilmenohematites. Spinodal decomposition and coarsening will result in coherent intergrowths of Fe-rich and Fe-poor regions, which will have different Curie temperatures. Experimen-

tal study of this process is not possible due to kinetic problems, but self-reversals have been observed in natural samples in the compositional range Ilm_{15-25} which have cooled slowly in plutonic and metamorphic rocks. Although the origin of this self-reversal is not known, an exsolution intergrowth is a likely candidate.

12.6.9 A final comment on self-reversals and palaeomagnetism, and mineralogy

Although there was a time when self-reversals in minerals were thought to be responsible for reversed magnetisation in rocks, there is now overwhelming evidence from the correlation of reversal time-scales on land and sea-floor basalts, that the earth's magnetic field periodically reverses its polarity. The interest in self-reversals has subsequently declined and assumed to be a rather unusual phenomenon confined to a very limited range of compositions and geological environments. In other words 'it doesn't matter very much'. However, there are many likely situations in which self-reversals can occur. One problem is that they cannot always be reproduced in the laboratory because of the irreversibility of many solid state transformations on a laboratory time-scale.

For example, the oxidation of ilmenomagnetite during cooling of igneous rocks is a very common phenomenon. Under non-equilibrium conditions, magnetite–ulvospinel solid solutions are oxidised to defect spinels (titanomaghemites) which subsequently transform to the more stable hematite structure (Section 5.2.4). An intergrowth of titanohematite + titanomagnetite could satisfy the criteria for self-reversal, and there are documented geological examples of reversals which have been attributed to this transformation (see bibliography).

The magnetic properties of rocks are entirely dependent on the structural state and the microstructure of its constituent magnetic minerals. A scientific study of palaeomagnetism must address these underlying mineralogical problems. The same is true of many areas in Earth Sciences. High pressure transformations in the mantle control the seismic structure of the Earth; mantle convection proceeds by solid state transport and slip in mantle minerals; reactions between minerals in metamorphic events depend on the thermodynamics and hence structural state of the constituent minerals;

the rates of processes and their approach to equilibrium depend on the activation energies of available mechanisms... and so on. While the study of large-scale geological processes necessarily involves empirical and approximate methods, ultimately they all have a mineralogical basis. Only when we study Earth processes in an integrated way on a range of scales, from microscopic to macroscopic, will we move from speculation to scientific understanding.

Bibliography

CARPENTER, M.A. (1985). Order-disorder transformations in mineral solid solutions. In: *Macroscopic to microscopic*. S.W. Kieffer and A. Navrotsky (Eds.) Reviews in Mineralogy Vol. 14.

CARPENTER, M.A. (1988). Thermochemistry of Al/Si ordering in feldspar minerals. In: *Physical properties and thermodynamic behaviour of minerals*. E.K.H. Salje (Ed.) NATO ASI Series C. D.Reidel Publishing Co.

HEUER, A.H. AND NORD, G.L. (1976). Polymorphic phase transitions in minerals. In: *Electron microscopy in mineralogy*. H.-R.Wenk (Ed.) Springer-Verlag.

McCONNELL, J.D.C. (1985). Symmetry aspects of order-disorder and application of Landau theory. In: *Macroscopic to microscopic*. S.W. Kieffer and A. Navrotsky (Eds.) Reviews in Mineralogy Vol. 14.

McCONNELL, J.D.C. (1987). The thermodynamics of short range order. In: *Physical Properties and thermodynamic behaviour of minerals*. E.K.H. Salje (Ed.) NATO ASI Series C. D.Reidel Publishing Co.

McCONNELL, J.D.C. (1992). The stability of modulated structures. In: *Stability of Minerals*. G.D. Price and N.L. Ross (Eds) Chapman & Hall.

O'REILLY, W. (1984). *Rock and mineral magnetism*. Blackie.

POIRIER, J-P. (1991). *Introduction to the physics of the Earth's interior*. Cambridge University Press.

RAO, C.N.R. AND RAO, K.J. (1978). *Phase transitions in solids*. McGraw-Hill.

SALJE, E. (1990). *Phase transitions in ferroelastic and coelastic crystals*. Cambridge University Press.

TARLING, D.H. (1983). *Palaeomagnetism*. Chapman and Hall.

References

ADLHART, W., FREY, F. AND JAGODZINSKI, H. (1980). X-ray and neutron investigations of the $P\bar{1} \Rightarrow I\bar{1}$ transition in pure anorthite. *Acta Cryst.* **A36**, 450–60.

CARPENTER, M.A. (1988). Thermochemistry of Al/Si ordering in feldspar minerals. In: *Physical Properties and thermodynamic behaviour of minerals*. E.K.H. Salje (Ed.) NATO ASI Series C. D.Reidel Publishing Co.

CARPENTER, M.A. AND PUTNIS, A. (1985). Cation order and disorder during crystal growth: some implications for natural mineral assemblages. In: *Metamorphic Reactions. Advances in Physical Geochemistry Vol. 4*. A.B. Thompson and D.C. Rubie (Eds) Springer-Verlag.

CARPENTER, M.A., PUTNIS, A., NAVROTSKY, A. AND McCONNELL, J.D.C. (1983). Enthalpy effects associated with Al,Si ordering in anhydrous Mg-cordierite. *Geochim. Cosmochim. Acta* **47**, 899–906.

GÜTTLER, B., SALJE, E. AND PUTNIS, A. (1989). Structural states of cordierite III: Infrared spectroscopy and the nature of the hexagonal-modulated transition. *Phys. Chem. Minerals* **16**, 365–73.

HAZEN, R.M. (1976). Sanidine: predicted and observed monoclinic to triclinic reversible transformation at high pressure. *Science* **194**, 105–7.

HEANEY, P.J. AND VEBLEN, D.R. (1991). Observations of the α-β phase transition in quartz: A review of imaging and diffraction studies and some new results. *Amer. Mineral.* **76**, 1018–32.

HOLDAWAY, M.J. (1971). Stability of andalusite and the aluminium silicate phase diagram. *Am. J.Sci.* **271**, 97–131.

ITO, E. AND TAKAHASHI, E. (1987). Ultrahigh pressure phase transformations and the constitution of the deep mantle. In: *High Pressure Research in Mineral Physics*. M.H. Manghnani and Y.Syono (Eds) Geophysical Monograph 39. American Geophysical Union.

KIRKPATRICK, R.J., CARPENTER, M.A., YANG, WANG-HONG AND MONTEZ, B. (1987). ^{29}Si magic-angle NMR spectroscopy of low-temperature ordered plagioclase feldspars. *Nature*, **325**, 236–8.

KITAMURA, M. AND YAMADA, H. (1987). Origin

of sector trilling in cordierite in Daimonji hornfels, Kyoto, Japan. *Contrib. Mineral. Petrol.* **97**, 1–6.

McCONNELL, J.D.C. AND McKIE, D. (1960). The kinetics of the ordering process in triclinic NaAlSi$_3$O$_8$. *Mineral. Mag.* **32**, 436–46.

McELHINNY, M.W. (1973). *Palaeomagnetism and plate tectonics.* Cambridge University Press.

McLAREN, A.C. AND MARSHALL, D.B. (1974). Transmission electron microscope study of the domain structure associated with the b-,c-,d-,e- and f-reflections in plagioclase feldspars. *Contrib. Mineral. Petrol.* **44**, 237–49.

NORD, G.L. AND LAWSON, C.A. (1989). Order-disorder transition-induced twin domains and magnetic properties in ilmenite-hematite. *Amer. Mineral.* **74**, 160–76.

O'REILLY, W. (1984). *Rock and mineral magnetism.* Blackie.

PRICE, G.D., PUTNIS, A. AND SMITH, D.G.W. (1982). A spinel to β-phase transformation in (Mg,Fe)$_2$SiO$_4$. *Nature* **296**, 729–31.

PUTNIS, A. AND ANGEL, R.A. (1985). Al,Si ordering in cordierite using "magic angle spinning" NMR. II: Models of Al,Si order from NMR data. *Phys. Chem. Minerals* **12**, 217–22.

PUTNIS, A. AND BISH, D.L. (1983). The mechanism and kinetics of Al,Si ordering in Mg-cordierite. *Amer. Mineral.* **68**, 60–5.

PUTNIS, A. AND HOLLAND, T.J.B. (1986). Sector trilling in cordierite and equilibrium overstepping in metamorphism. *Contrib. Mineral. Petrol.* **9**, 265–72.

PUTNIS, A. AND PRICE, G.D. (1979). High pressure (Mg,Fe)$_2$SiO$_4$ phases in the Tenham chondritic meteorite. *Nature* **280**, 217–8.

PUTNIS, A., SALJE, E., REDFERN, S., FYFE, C.

AND STROBL, H. (1987). Structural states of Mg-cordierite I: Order parameters from synchrotron X-ray and NMR data. *Phys. Chem. Minerals* **14**, 446–54.

REDFERN, S.A.T., GRAEME-BARBER, A. AND SALJE E. (1988). Thermodynamics of plagioclase III: spontaneous strain at the I$\bar{1}$ ⇒ P$\bar{1}$ phase transition in Ca-rich plagioclase. *Phys. Chem. Minerals* **16**, 157–63.

REDFERN, S.A.T. AND SALJE, E. (1987). Thermodynamics of plagioclase II: Temperature evolution of the spontaneous strain at the I$\bar{1}$ ⇒ C$\bar{1}$ transition in anorthite. *Phys. Chem. Minerals* **14**, 189–95.

RICHARDSON, S.W., GILBERT, M.C. AND BELL, P.M. (1969). Experimental determination of kyanite-andalusite and andalusite-sillimanite equilibria; the aluminium silicate triple point. *Am. J.Sci.* **267**, 259–72.

RUBIE, D.C. (1984). The olivine – spinel transformation and the rheology of subducting lithosphere. *Nature* **308**, 505–8.

SALJE, E., KUSCHOLKE, B., WRUCK, B. AND KROLL, H. (1985). Thermodynamics of sodium feldspar II: experimental results and numerical calculations. *Phys. Chem. Minerals* **12**, 99–107.

SMITH, J.V. AND BROWN, W.L. (1988). *Feldspar minerals* Vol. 1 2nd Ed. Springer-Verlag.

VAN LANDUYT, J., VAN TENDELOO, G., AMELINCKX, S. AND WALKER, M.B. (1985). Interpretation of Dauphiné twin domain configurations resulting from the α-β phase transition in quartz and aluminium phosphate. *Physical Review B*, **31**, 2986–92.

WYATT, O.H. AND DEW-HUGHES, D. (1974). *Metals, ceramics and polymers.* Cambridge University Press.

Index